J.L. Wu, K. Ito, S. Tobimatsu, T. Nishida, H. Fukuyama (Eds.)

Complex Medical Engineering

J.L. Wu, K. Ito, S. Tobimatsu, T. Nishida, H. Fukuyama (Eds.)

Complex Medical Engineering

With 274 Figures, Including 7 in Color

Springer

Jing Long Wu, Ph.D., Professor
Department of Intelligent Mechanical Systems, Faculty of Engineering, Kagawa University
2217-20 Hayashi, Takamatsu 761-0396, Japan

Koji Ito, Dr.Eng., Professor
Complex Systems Analysis, Adaptive & Learning Systems, Tokyo Institute of Technology
4259 Nagatsuta, Midori-ku, Yokohama 226-8503, Japan

Shozo Tobimatsu, M.D., Professor and Chairman
Department of Clinical Neurophysiology, Neurological Institute, Graduate School of Medical Science, Kyushu University
3-1-1 Maidashi, Higashi-ku, Fukuoka 812-8582, Japan

Toyoaki Nishida, Dr.Eng., Professor
Department of Intelligence Science and Technology, Graduate School of Informatics Kyoto University
Yoshida Honmachi, Sakyo-ku, Kyoto 606-8501, Japan

Hidenao Fukuyama, M.D., Ph.D., Professor
Human Brain Research Center, Kyoto University Graduate School of Medicine
54 Shogoin Kawahara-cho, Sakyo-ku, Kyoto 606-8507, Japan

ISBN 4-431-30961-6 Springer Tokyo Berlin Heidelberg New York
ISBN 978-4-431-30961-1 Springer Tokyo Berlin Heidelberg New York
Library of Congress Control Number: 2006930401

Printed on acid-free paper

Springer is a part of Springer Science+Business Media
springer.com

Typesetting: Camera-ready by the editors and authors

Preface

In the twenty-first century, applications in medicine and engineering must acquire greater safety and flexibility if they are to yield better products at higher efficiency. To this end, complex science and technology must be integrated in medicine and engineering. Complex medical engineering (CME) is a new field comprising complex medical science and technology. Included are biomedical robotics and biomechatronics, complex virtual technology in medicine, information and communication technology in medicine, complex technology in rehabilitation, cognitive neuroscience and technology, and complex bioinformatics.

This book is a collection of chapters from experts in academia, industry, and government research laboratories who have pioneered the ideas and technologies associated with CME. Containing 54 research papers that were selected from 260 papers submitted to the First International Conference on Complex Medical Engineering (CME2005), the book offers a thorough introduction and a systematic overview of the new field. The papers are organized into six parts. Part 1 focuses on biomedical robotics and biomechatronics and discusses principles and applications associated with the micropump, tactile sensor, underwater robot, laser surgery, and noninvasive monitoring. Part 2 discusses complex virtual technology in medicine, which involves visualization, simulators, displays, robotic systems, and walking-training systems. In Part 3, the authors provide a comprehensive discussion of information and communication technology in medicine. In Part 4, complex technology in rehabilitation is discussed, with topics including rehabilitation robotics and neurorehabilitation. In Part 5, the authors discuss cognitive neuroscience and technology in five areas: complex medical imaging, including PET and MRI; human vision and technologies; brain science and cognitive technologies; transcranial magnetic stimulation (TMS); and electroencephalogram (EEG), neuron disease, and diagnostic technology. In Part 6, the authors discuss topics associated with complex bioinformatics.

We first proposed the new term "complex medical engineering" for the First International Conference on Complex Medical Engineering (CME2005), which was successfully held in Takamatsu, Japan, in 2005 (http://frontier.

eng.kagawa-u.ac.jp/CME2005/). When the conference was announced, we soon received a vast number of responses as well as support from the research community, industry, and many organizations. To meet the strong demands for participation and the growing interest in CME, the Institute of Complex Medical Engineering (ICME) was founded in 2005. The ICME is an international academic society (http://frontier.eng.kagawa-u.ac.jp/ICME/), the aim of which is to bring together researchers and practitioners from diverse fields related to complex medical science and technology. ICME conferences are expected to stimulate future research and development of new theories, new approaches, and new tools to expand the growing field of CME. The First Symposium on Complex Medical Engineering (http://frontier.eng.kagawa-u.ac.jp/SCME2006/) and The Second International Conference on Complex Medical Engineering (http://frontier.eng.kagawa-u.ac.jp/CME2007/) will be held in Kyoto, Japan, and Beijing, China, respectively.

This book is recommended by the ICME as the first book on CME research. It is a collaborative effort involving many leading researchers and practitioners who have contributed chapters on their areas of expertise. Here, we would like to thank all authors and reviewers for their contributions.

We are very grateful to people who joined or supported the CME-related research activities, and in particular, to the ICME council members: Y. Nishikawa, H. Shibasaki, H. Takeuchi, C.A. Hunt, J. Liu, J.C. Rothwell, M. Hashizume, M. Hallett, M.C. Lee, N. Franceschini, N. Zhong, P. Wen, R. Turner, S. Miyauchi, S. Tsumoto, W. Nowinski, Y. Deng, S. Doi, T. Touge, T. Kochiyama, M. Tanaka and R. Lastra. We thank them for their strong support.

Last but not least, we thank staffs at Springer Japan for their help in coordinating this monograph and for their editorial assistance.

Jing Long Wu
Koji Ito
Shozo Tobimatsu
Toyoaki Nishida
Hidenao Fukuyama

Contents

2. Complex Virtual Technology in Medicine

3. Information and Communication Technology in Medicine

4. Complex Technology in Rehabilitation

5. Cognitive Neuroscience and Technology

Improving the Performance of a Traveling Wave Micropump for Fluid Transport in Micro Total Analysis Systems

Takaaki Suzuki[1], Isaku Kanno[1], Hidetoshi Hata[1], Hirofumi Shintaku[1], Satoyuki Kawano[2], and Hidetoshi Kotera[1]

[1] Department of Microengineering, Kyoto University
[2] Department of Mechanical Science and Bioengineering, Osaka University

Chapter Overview. Micropumps are one of the most important microfluidic components in Micro Total Analysis System (μTAS). The authors have developed a traveling wave micropump that demonstrates high energy efficiency and does not require valves. A prototype valveless micropump was fabricated using microfabrication techniques to validate the pumping principle. The micropump uses piezoelectric bimorph cantilevers to deform a flexible microchannel wall. Traveling waves are induced on the surface of the microchannel by applying properly phased sinusoidal voltage to the piezoelectric cantilevers. The resulting peristaltic motion of the channel wall transports the fluid. The fluid flow in the micropump was numerically simulated with the computational fluid dynamics code, FLUENT. Comparing the experimental and numerical results confirmed that the proposed modeling method can accurately evaluate the performances of the traveling wave micropump. Based on the flow obtained in numerical analysis, an improvement in the pump efficiency is expected with optimization of the shape of the moving wall.

Key Words. Micropump, and Fluid Transport

1. Introduction

In recent years, the field of Micro Total Analysis Systems (μTAS) has become a highly active. The main advantages of μTASs compared to conventional macro-scale analysis systems are reduced sample and reagent volumes, high throughput screening capability (parallel analyses), lower cost, and portability.

A key component in μTASs is the micropump. A variety of micropumps have been proposed for fluid transportation and manipulation systems, but conventional diaphragm-type micropumps that use mechanical valves or diffuser / nozzle elements have a complicated structure and high fluidic impedance [1, 2]. To address these limitations, the authors have proposed and developed a novel valveless traveling wave micropump [3]. It uses piezoelectric bimorph beams to induce a traveling wave in a flexible microchannel made of silicon rubber. Despite the advantages of the micropump, its performance needs improvement to fully realize its potential for micro biological and chemical analysis in medical diagnosis.

In this paper, the performance of the valveless traveling wave micropump is improved by using PZT beams that generate larger deflection. In addition, the fluid flow in the micropump is numerically simulated to investigate the relationship between the fluid transportation characteristics, the deflection of the actuators, and the shape of the microchannel in the micropump. Based on these relationships, the performance of the traveling wave micropump is improved by optimizing the internal structure of the flexible microchannel.

2. Principle of Fluid Transportation

The traveling wave micropump is composed of the two components as shown in Fig. 1, i.e. a microchannel and an actuator array. The piezoelectric actuator array uses PZT beams to generate traveling waves on a flexible wall of the microchannel. The fluid flow was controlled by changing the sine wave forms of the applied voltage. By changing the phases of the voltage that is applied to the PZT beams, the micropump can induce bidirectional fluid flow without using mechanical valves.

The trajectory of a fluid particle near the moving wall of the microchannel follows an elliptic path. After a period of wave oscillation, a fluid particle travels along the channel a small amount from the initial position. The net fluid transport is achieved by repeating such motion in the viscous fluid. The proposed micropump can transport micro samples in carrier without collapsing sample surface, because the amplitude of the traveling wave on the microchannel wall is about 1/50 of the height of the microchannel and the micropump does not have mechanical valves.

Yin and Fung [4] theoretical analyzed the two-dimensional peristaltic pump velocity profile using the perturbation method. Following this analysis method, consider that the following traveling wave is applied to the fluid

$$\eta = a\cos\frac{2\pi}{\lambda}(x-ct), \tag{1}$$

where a is the amplitude of the traveling wave, λ the wavelength of the traveling wave, and c the velocity of the traveling wave. The time-averaged flow velocity of the x direction $\bar{u}$ is

$$\bar{u} = \frac{1}{2}a^2\Psi(z), \tag{2}$$

where $\Psi(z)$ is a function of the height of the microchannel. From Eq.(2), one can conclude that time-averaged flow velocity in the x-direction is proportional to the square of the amplitude of the microchannel traveling wave.

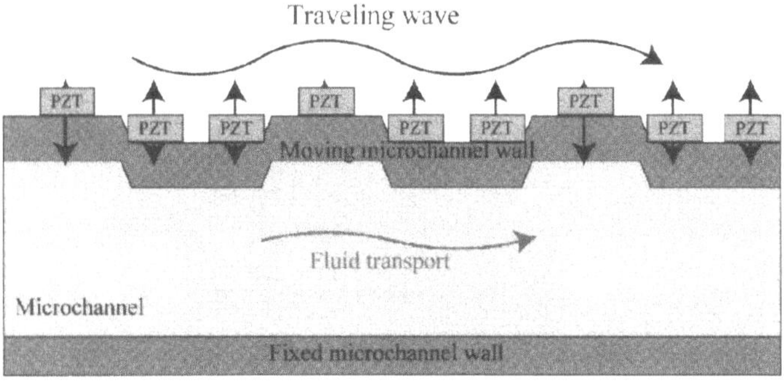

Fig. 1. Principle of fluid transport in a traveling wave micropump driven by piezoelectric actuators.

3. Experimental Development

3.1 Fabrication process

A valveless micropump driven by PZT beams was fabricated using soft-lithography [5]. The microchannel of the traveling wave micropump consists of two layers: a groove layer that constitutes the bottom and side walls of the microchannel, and a thin cover layer that has bumps. Each layer was fabricated by casting PDMS (Polydimethylsiloxane) onto master molds. Thick SU-8 photoresist was used to fabricate the master molds.

The fabrication process is shown in Fig. 2. The SU-8 was first spin coated with a thickness of 100μm on a glass substrate. The microchannel was fabricated on a glass substrate using photolithography. PDMS was poured on the master and degassed using vacuum-forming to mold the microchannel. Finally, the PDMS sheet was peeled off the master and bonded to the PDMS sheet with bumps. The cross-section of the microchannel was 200μm wide and 100μm high.

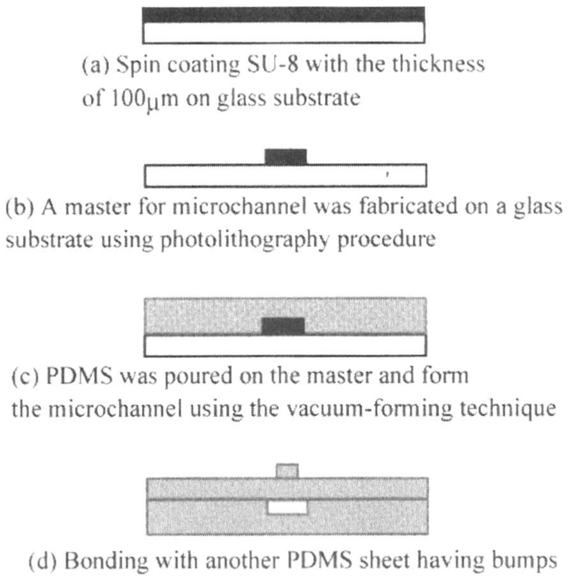

Fig. 2. Fabrication process for the flexible microchannel using soft-lithography.

3.2 Experimental observation of fluid flow generated by the traveling wave micropump

A picture of a prototype traveling wave micropump is shown in Fig.3. The top wall of the microchannel that is made from flexible material is deformed by the force generated by the PZT beams. Traveling waves were induced on the surface of the microchannel by applying sinusoidal signals with $2\pi/3$ phase differences to each PZT beam.

The 10mm long and 1.4mm wide PZT beams were subject to a 5V sinusoidal voltage and their resulting tip deflection was measured using a laser Doppler vibrometer. The tip deflection frequency response of the PZT beams is shown in Fig.4. The deflection of the current PZT beams is larger than that of the previously reported actuators [3]. From Eq. (2), the flow rate is expected to be proportional to the square of the amplitude of the traveling wave. Therefore, the current micropump is expected to have a significantly higher performance than the previous one.

For the flow rate measurement, 1μm in diameter fluorescent microbeads were added to the fluid. Using a fluorescent microscope, the movement of microbeads in accord with the applied traveling wave was confirmed. Based on initial flow observations, it was predicted that the continuous flow rate could be improved by optimization of the applied voltage and the system design.

Fig. 3. Photograph of the prototype micropump.

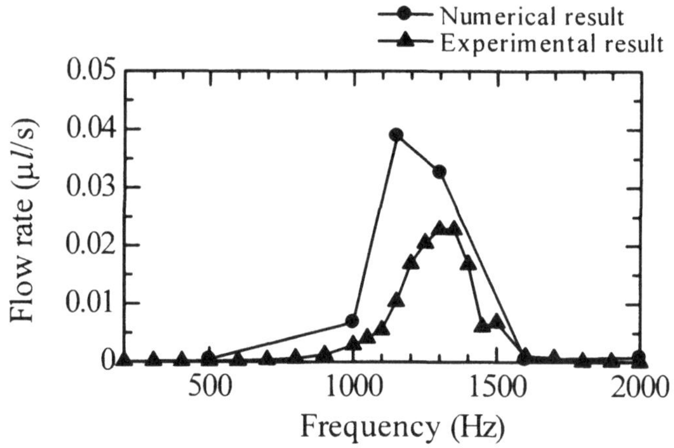

Fig. 7. Experimental and numerical flow rate as a function of frequency.

5. Geometry Optimization of the Microchannel

Based on the fluid flow obtained in the numerical analysis, it appears that the efficiency of the traveling wave micropump can be improved by modifying the surface of the moving wall as shown in Figs. 8 (b) and (c). For a microchannel excitation with a 1.4μm amplitude, 1.3kHz frequency, and 6000μm wavelength, the numerical flow rates for the wall designs (a), (b), and (c) are 0.040μl/s, 0.0510μl/s, and 0.0542μl/s, respectively. These results indicated that the flow rate of the micropump can be increased by 34% using the (c) design instead of the original (a) design.

The relationship between the "teeth" height and the time-averaged flow rate is shown in Fig. 9, where h is the teeth height, H the height of the microchannel, Q_0 the flow rate of the micropump (a), and Q_h the flow rate of the micropump with the "teeth" of height h, respectively. The shape of the microchannel is shown in Fig. 8 (b). The time-averaged flow rate is approximately proportional to the square of the "teeth" height. A significant improvement in pumping performance is expected by optimizing the microchannel structure using the proposed analytical procedure.

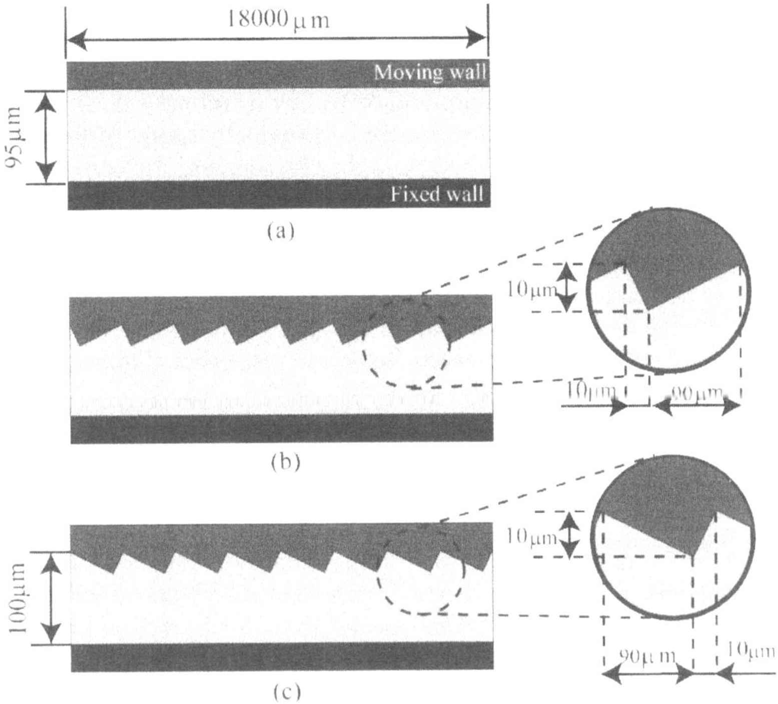

Fig. 8. Geometric optimization of traveling wave micropump.

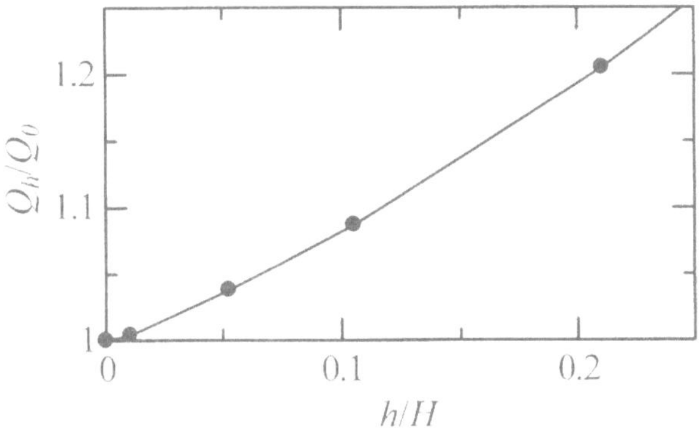

Fig. 9. Relationship between "teeth" height and the time-averaged flow rate.

6. Conclusion

In this paper, a prototype micropump driven by traveling waves was fabricated, experimentally characterized, and numerically modeled to improve the flow rate of the micropump. The higher flow rate of the current micropump compared to the previous one was achieved by using higher deflection PZT beams. Based on comparison of the experimental and numerical results, the proposed numerical modeling procedure was validated for simulating the pump performance. In order to improve the efficiency of the micropump, geometry optimization of the flexible microchannel wall was carried out. The flow rate was predicted to be proportional to the height of the "teeth" on the moving wall.

Acknowledgment

This study is a part of Kyoto City Collaboration of Regional Entities for the Advancement of Technological Excellence of JST on the basis of research results supported in part by grant-in-aids for Scientific Researches (A) (No.14205037 and No. 15201033) and Center of Excellence for Research and Education on Complex Functional Mechanical Systems (COE program) of MEXT, Japan. The authors would like to thank Jacob J. Loverich in Kyoto University for helpful discussions.

References

1. Linnemann R, Woias P, Senfft CD, Ditterich JA (1998) A self-priming and bubble-tolerant piezoelectric silicon micropump for liquids and gases. Proc. IEEE MEMS 1998:532-537
2. Cabuz C, Herb WR, Cabuz EI, Lu ST (2001) The dual diaphragm pump. Proc. IEEE MEMS 2001:519-522
3. Kanno I, Kawano S, Yakushiji S, Kotera H (2003) Characterization of piezoelectric micropump driven by traveling waves. Proc. μTAS 2003:997-1000
4. Yin FCP, Fung YC (1971) Comparison of theory and experiment in peristaltic transport. J. Fluid Mechanics 47:93-112
5. Sia SK, Whitesides GM (2003) Microfluidic devices fabricated in poly) dimethylsiloxane) for biological studies. Electrophoresis 24:3563-3576

Vision-based Tactile Sensor for Endoscopy

Kazuto Takashima[1], Kiyoshi Yoshinaka[2], and Ken Ikeuchi[1]

[1]Institute for Frontier Medical Sciences, Kyoto University
[2]National Institute of Advanced Industrial Science and Technology

Chapter Overview. Endoscopy would become more useful if the visual information obtained with it could be combined with tactile information. We therefore developed a new tactile sensor system with this feature using image processing of an infrared cut pattern. It is possible to install this sensor on the tip of an endoscope easily because wires for power delivery and transmission of signals are unnecessary. Doctors can use this sensor for visual diagnosis because it does not degrade diagnostic images. In this study, changes were made in the sensor to improve the accuracy of measurement and detect stiffness of living tissue. First, we corrected sensor output by considering sensor deformation characteristics, and were thus able to evaluate state of contact in inserting the endoscope into a vessel model. Second, we modified the sensor to detect not only three-axis force but also compressive modulus. The difference in compressive moduli between several industrial materials is discernible by measurement with the prototype of this sensor.

Key Words. Tactile Sensor, Endoscopy.

1. Introduction

Endoscopes are used for various medical treatments and are essential for low invasive surgery. However, it is very difficult to manipulate endoscopes, because the cavities in which endoscopes navigate are narrow and complex. In addition, the surgeon's sensory (visual and tactile) perception is severely reduced during manipulation in low invasive surgery, because these tools are long and flexible and have few degrees of freedom.

One method to improve manipulation of the endoscope is measurement of tactile force between the endoscope and the caliber wall, since surgery is essentially a visual and tactile experience [1]. However, no practical tactile sensor has been developed for intravascular treatment due to the small diameter of vessel. For example, the wires set in existing endoscopes to deliver power and transmit signals require additional endoscope volume. In addition, since correlation of information requires one coordinate per sensor, more space is needed to obtain more information.

A tactile sensor is also useful for detecting the stiffness of tissues, which can change due to disease. For example, softening of degenerated cartilage can be estimated in vivo by palpation of the articular surface with a blunt probe during arthroscopy. However, palpation is subjective and a low invasive method to measure quantitatively stiffness is desirable.

We therefore developed a new tactile sensor system that measures tactile force by image processing [2]. This sensor can detect three-axis force and stiffness. In this study, we tested the fundamental performance of two prototypes for two kinds of applications. We first manufactured one prototype to improve manipulation of intravascular endoscopes. We then manufactured another for arthroscopes to test sensing of living tissue in situ.

2. Measurements

2.1. Concept of tactile sensor

The tactile sensor shown in Fig. 1 is composed of a transparent window including infrared (IR) cut pattern, an elastic body, and an attachment between the sensor and endoscope. The transparent window is placed in the xy-plane, and the z-axis is aligned with the axis of the endoscope when no load is applied on the sensor tip. Using the relations between tactile force and displacement of the tip of the sensor, force along the z-axis (F_z) is expressed as a function of area of the IR cut pattern (S), and the forces along the x-axis and y-axis (F_x and F_y, respectively) are expressed as functions of displacement of the transparent window (x_i, y_i). If deformation characteristics of the elastic body are linear, these relations are expressed as:

$$F_x = k_x x_i, \quad (1)$$

$$F_y = k_y y_i, \quad (2)$$

$$F_z = k_z(1 - \sqrt{S_0 / S})(a - f), \tag{3}$$

where k_x, k_y and k_z are spring constants of the elastic body, a is the distance between the principal point of the objective lens of the endoscope and the transparent window, f is the front focal distance of the objective lens, and S_0 is the area when no load is applied on the sensor tip (see [2]). Displacement along the z-axis is:

$$z_i = (1 - \sqrt{S_0 / S})(a - f). \tag{4}$$

As visible light can pass through the IR cut pattern, this sensor does not degrade diagnostic images. In addition, optical absorption in the body has less effect on image processing of the pattern because IR light is hardly absorbed by the body [3].

For measurement of stiffness, we added a reference circle that is compressed on the tissue surface by a weaker force than the elastic body shown in Fig. 2. When a force (F) is applied to the sensor tip, the compressive modulus (E) can be determined as follows [4]:

$$E = \frac{9F}{16\sqrt{r}}\left(\frac{1 - e^{-0.42h/\sqrt{2rd-d^2}}}{d}\right)^{\frac{3}{2}}, \tag{5}$$

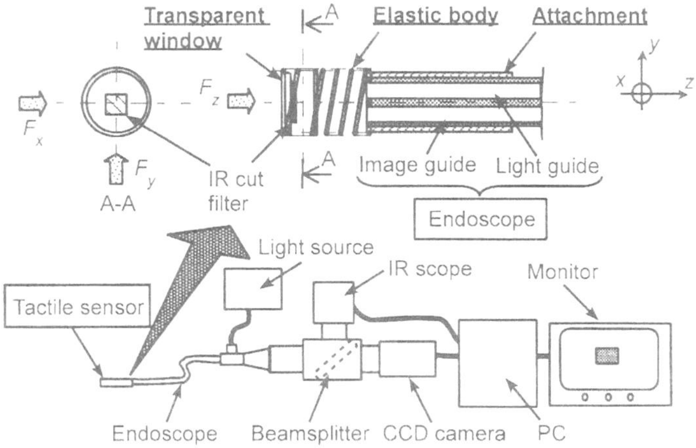

Fig. 1. Tactile sensor system

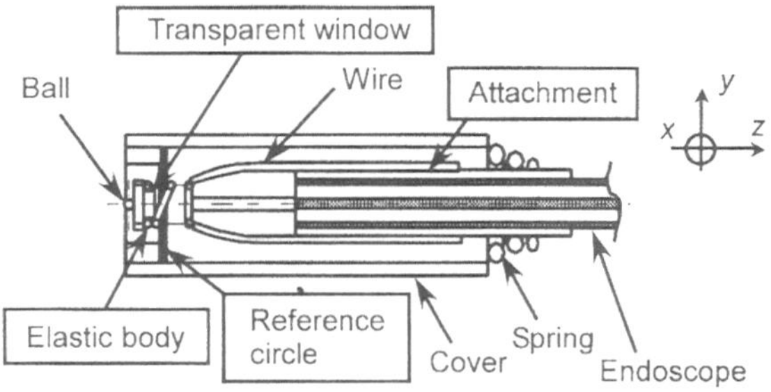

Fig. 2. Tactile sensor system modified to detect the stiffness

where r is the radius of the indenter, h is the thickness of the tissue before indentation, and d is the depth of indentation. In deriving Eq. (5), Poisson's ratio has been assumed to be 0.5, because instantaneously loaded cartilage behaves as an incompressible elastic solid [5]. In this study, r = 0.5 mm. When d = 0.2 mm and h = 2 mm, $(1 - e^{-0.42h/\sqrt{2rd-d^2}}) = 0.88$. Therefore, it is assumed that $(1 - e^{-0.42h/\sqrt{2rd-d^2}}) \fallingdotseq 1$ by considering the thickness of articular cartilage generally to be 1–3 mm.

Thus, when the endoscope moves only along the z-axis to apply only axial force on the sensor, the displacement of the reference circle along the z-axis (z') is also expressed by the area of reference circle (S') using Eq. (4). Therefore, d is expressed as follows:

$$d = (1-\sqrt{S_0'/S'})(a'-f)-(1-\sqrt{S_0/S})(a-f). \tag{6}$$

Substituting Eqs. (3) and (6) into Eq. (5), E is expressed as follows:

$$E = \frac{9k_z(1-\sqrt{S_0/S})(a-f)}{16\sqrt{r}\,\{(1-\sqrt{S_0'/S'})(a'-f)-(1-\sqrt{S_0/S})(a-f)\}^{3/2}}. \tag{7}$$

2.2. Methods of measurement to improve manipulation

First, we evaluated the performance of the sensor to improve manipulation of the endoscope. In the first prototype (Fig. 1, f 3.5 x 16 mm), an IR cut filter (1.5 x 1.5 mm, Kureha Chemical Co., Ltd., UCF 102) is attached to a transparent window. The elastic body is a compressive spring (0.18 mm of wire diameter, 3.32 mm of coil diameter, and 6.5 mm of length, effective winding number: 4, spring constant: 0.062 N/mm) and the attachment is a precision metal split sleeve. This tactile sensor is attached to the tip of a fiberscope (f 2.4 x 935 mm, 7000 fibers) and the image of the IR cut filter is captured through the fiberscope to an IR scope (H: 450 lines, V: 350 lines).

The light source of the fiberscope is a Tungsten Halogen fiber illuminator. The image was acquired from the IR scope to a computer in 8-bit format for image analysis illustrated in Fig. 3.

In a previous study [2], we noted that tactile force can be determined using the image of the IR cut pattern when axial or lateral force is applied separately. However, the sensor tip does not always contact the wall at the same angle and same point. Moreover, both axial and lateral forces are applied simultaneously. On the other hand, usual spring constants are calculated assuming that the force is applied at the center of the spring [6]. Consequently, the actual spring constants differ from the calculated values. It is thus necessary to correct the spring constant in our cases, as in other research [7]. For example, the axial and lateral spring constants used here are respectively calculated as 0.066 N/mm and 0.013 N/mm using the equation of Watari et al. [7] when each force is separately applied. In this study, we evaluated the deformation characteristics of the spring using finite element analysis (FEM). FEM was carried out using ANSYS version 8.1 (Ansys Corp.). In simulation models (Fig. 4), 116 3-D elastic beam elements were used and the contacts between nodes were considered. First, when the contact angle from the z-axis (a) was changed (Fig. 4a), we evaluated the axial and lateral rigidity. Second, when the angle from the end of the spring to the contact point (β) was changed (Fig. 4b), we evaluated lateral rigidity.

We then examined the correlation between the force and the image by considering the deformation characteristics of the spring. When a force was applied on this sensor by an arm connected to a load cell, we measured image change. The arm was manually displaced from 0 mm to 2 mm. Measurement was performed every 0.25 mm. The methods for application of force were as same as that of simulation model in Fig. 4a. White paper was attached on the transparent window to increase signal-to-noise ratio.

Lastly, We performed actual insertion of an endoscope with the sensor, using a vessel model (H+N-S-S-001, Elastrat Sarl). The values of sensor output were corrected by the results obtained in the previous paragraphs. The model is a carotid artery model for bifurcation stenosis.

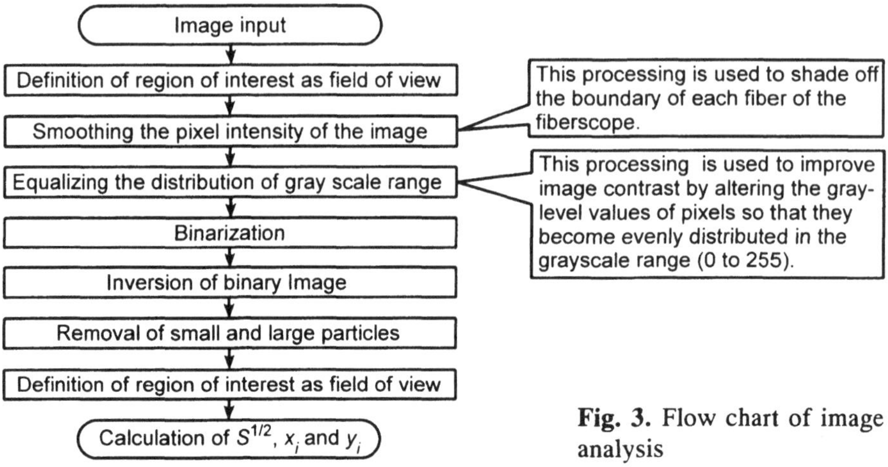

Fig. 3. Flow chart of image analysis

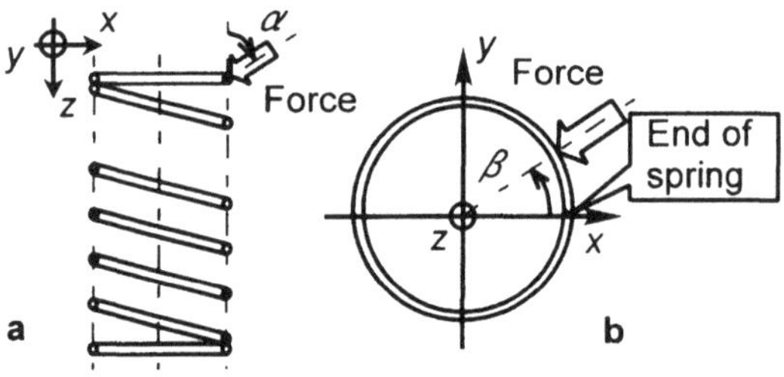

Fig. 4a, b. Simulation model to evaluate effect of: **a** contact angle from *z*-axis (α) and **b** angle from the end of the spring to the contact point (β)

2.3. Methods of measurement of tissue stiffness

Second, we investigated the stiffness of typical specimens. In the second prototype (Fig. 2, f 7 x 18 mm), two steel balls (f 1 mm) were attached to a transparent window from both sides. The ball contacting tissue was attached at the center of the transparent window to eliminate the effect of contact angle noted in the previous section. The elastic body is a compressive spring (spring constant: 4.95 N/mm). A reference circle made of acrylic cylinder (5mm of inner diameter) inserted in an aluminum pipe is connected to the attachment by a cone spring (spring constant: 0.32 N/mm). In this section, we replaced the IR scope with a high-speed camera (H: 504 lines, V: 243 lines, 200fps) to evaluate the compressive modulus of tissue in physiological conditions, whereas joint loading occurs within 10–150 ms [5]. When the prototype device attached to the fiberscope was automatically displaced from the contact point to 0.5 mm over 1 s at constant speed (0.5 mm/s), we measured image change. Measured values were averaged to minimize the effects of the measurement errors.

3. Results

3.1. Improvement of manipulation

Effects of contact angle are shown in Fig. 5. The force and displacement of the center of the end turn are divided into the x-axis and z-axis in Figs. 5a and 5b, respectively. The results show that the deformation characteristic along the x-axis is more affected than that along the z-axis. Thus, lateral force information from the sensor is more affected by a. The direction of displacement along the x-axis differs between $a < 30°$ and $a \geqq 30°$ though the directions of lateral force are the same for all angles of contact, since tilting of the end turn by axial force occurs when $a < 30°$ (Fig. 6). Fig. 5c shows effects on tilting of the end turn. Using Eq. (3), the axial force information of the sensor is calculated from not z but S, which is also affected by the tilt of the IR cut pattern. As the sensor contacts at the end turn, there is no tilt along the direction normal to the contact force. Therefore, we changed the IR cut pattern to a circle and calculated from the maximum diameter of the pattern (D). Then F_z is expressed as follows in replace of Eq. (3):

$$F_z = k_z(1 - D_0 / D)(a - f), \tag{8}$$

where D_0 is the diameter when no load is applied on the sensor tip. Fig. 5d shows the spring constant along the contact direction. In this figure, experimental values are also plotted. The spring constant becomes smaller as a becomes larger, since the lateral rigidity of the spring used in this study is smaller than its axial rigidity. The spring constants are respectively similar to the calculated values in Sect. 2.2 when $a = 0°$ or $a = 90°$.

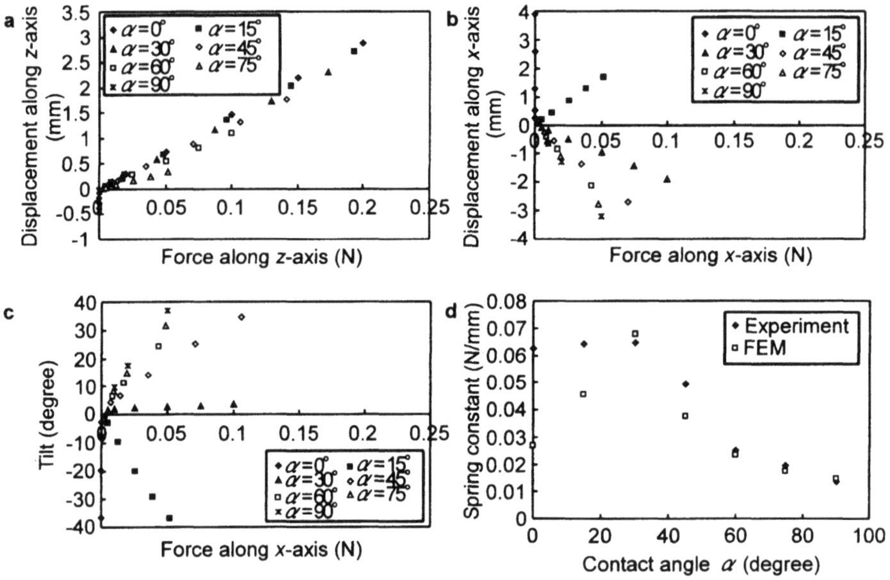

Fig. 5a, b, c, d. Effect of contact angle (*a*) on spring deformation. **a** Relation between axial force and displacement along z-axis. **b** Relation between lateral force and displacement along x-axis. **c** Relation between lateral force and tilt. **d** Relation between contact angle (*a*) and spring constant

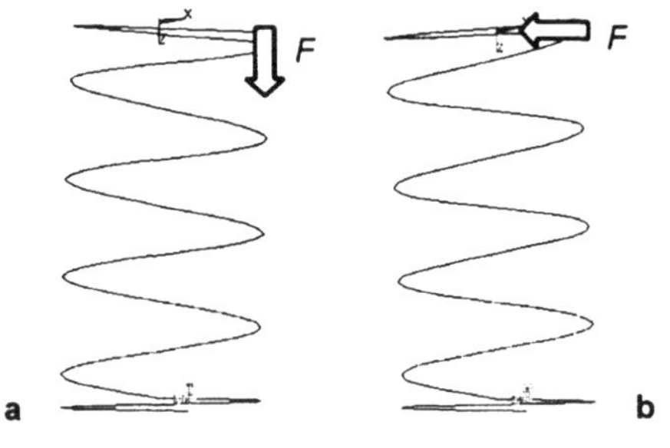

Fig. 6a, b. Deformation diagram of spring under different contact angles (*a*). **a** *a* = 0°. **b** *a* = 90°

Fig. 7 shows the lateral spring constants under different contact angles (β). The effects of β on lateral rigidity were not large.

Fig. 8 shows the relation between applied force and sensor output. These are the magnitudes of the resultant of the axial and lateral forces. In Figs. 8a and 8b, the axial force is calculated from the area ($S^{1/2}$) and the maximum diameter (D) of IR cut pattern, respectively. In these figures, it is appropriate for sensor output to be the same value (dash-dotted line). Before the experiment, we changed the IR cut pattern from a square to a circle (f 1.3 mm). It was proved that linear correlation becomes high using the diameter of IR cut pattern for calculation of tactile force. In these figures, linear correlation coefficients (R^2) between applied force and calculated values are 0.140 and 0.744, respectively. In addition, the effect of scattering of lateral force information shown in Fig. 5b is small, since the lateral force is smaller than the axial force under the angle when scattering occurs in the range of 0 to 30°. Thus, effects on lateral force appear unimportant.

The results of model experiments are shown in Fig. 9. The endoscope motions are shown in Fig. 9a. The reaction force measured by the sensor is shown in Fig. 9b. As shown in these figures, force is larger when the endoscope is inserted in the curved parts of the model. When the sensor contacted the wall laterally (Fig. 9a (ii)), the lateral force could be detected (Fig. 9b (ii)). When sensor tip contacted the wall (Fig. 9a (iii)), the axial force could be detected (Fig. 9b (iii)).

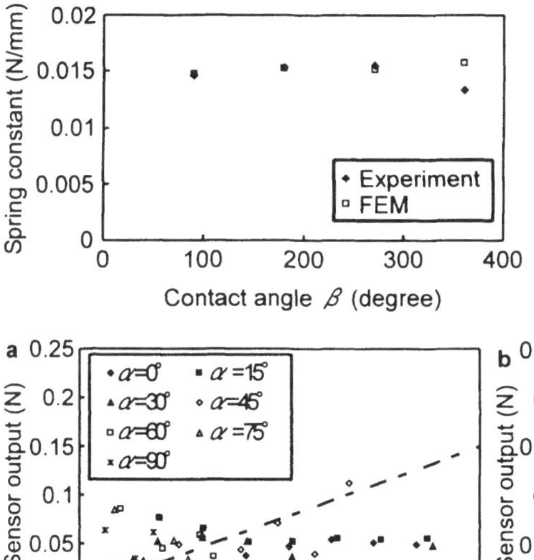

Fig. 7. Effect of contact angle (β) on deformation characteristics of spring

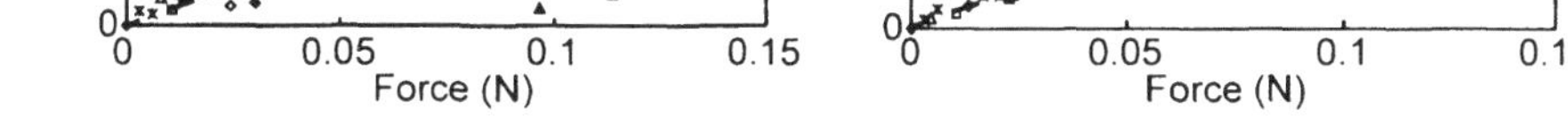

Fig. 8a, b. Effect of contact angle on deformation of spring. **a** Relation between applied force and sensor output. Axial force is calculated from the area of IR cut pattern ($S^{1/2}$). **b** Relation between applied force and sensor output. Axial force is calculated from the diameter of IR cut pattern (D)

3.2. Measurement of stiffness

Fig. 10 represents the transition of E for two specimens. The two transitions of E are similar, but the value for foam polystyrene is always lower than that for silicon rubber. These results show that this sensor can distinguish the stiffness of two different materials.

4. Discussion

As the simulation model used in this study is very simple, the FEM and

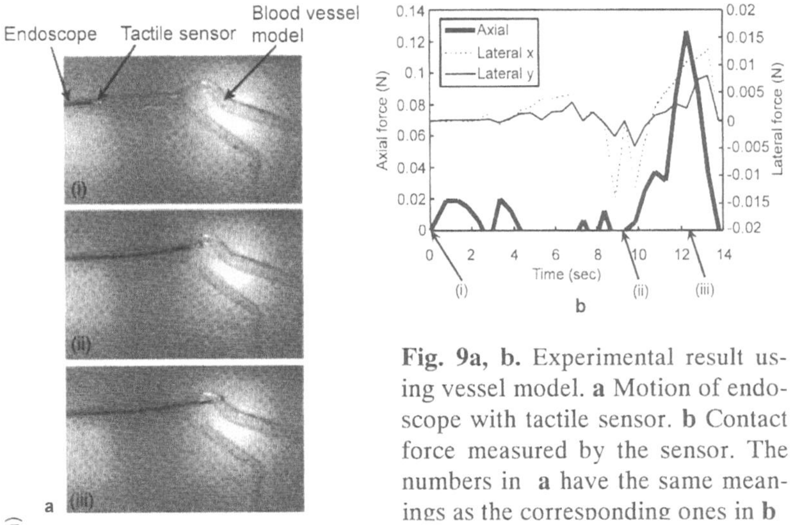

Fig. 9a, b. Experimental result using vessel model. **a** Motion of endoscope with tactile sensor. **b** Contact force measured by the sensor. The numbers in **a** have the same meanings as the corresponding ones in **b**

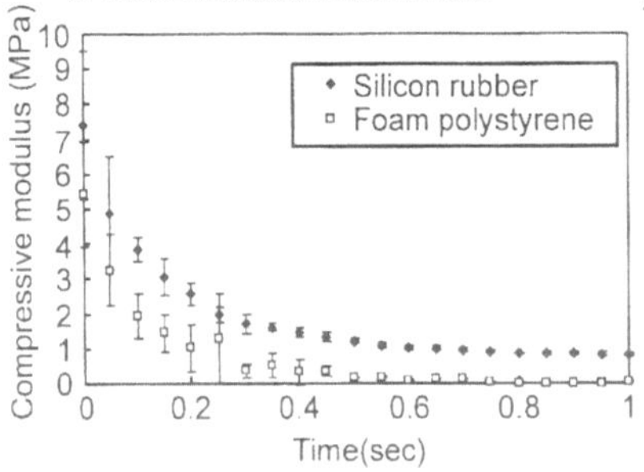

Fig. 10. Transition of compressive modulus of typical specimens

experimental results differ slightly, as shown in Fig. 5d and Fig. 7. Therefore, it may be necessary to determine the deformation characteristics carefully using a better simulation model. On the other hand, the light intensity at the transparent window changes because the light intensity is not evenly distributed and the angle between the axis of the sensor and the transparent window affects the intensity of reflection. Consequently, contact angle affects not only the deformation characteristics of the spring but also image processing. Evaluation of the effects of light is thus necessary.

The prototype device was manufactured based on the need for easy installation of the sensor in the tip of an existing endoscope. Consequently, the sensor looks like a probe or whisker attached to the original endoscope. If it is necessary to retain the original endoscope volume, one method would be to incorporate the IR cut pattern in endoscope optical system.

5. Conclusion

We tested the performance of our proposed tactile sensor system using image processing of the IR cut pattern. It was shown that tactile force can be determined using the IR cut pattern image. In addition, the sensor could detect not only three-axis force but also compressive modulus of the object with some modifications. The results of this study confirm that this system is potentially useful for endoscopic surgery.

References

1. Lee MH, Nicholls HR (1999) Tactile sensing for mechatronics—a state of the art survey. Mechatronics 9(1): 1–31
2. Takashima K, Yoshinaka K, Okazaki T, et al (2005) An endoscopic tactile sensor for low invasive surgery. Sens Actuators A 119(2): 372–383
3. Boas DA et al (2001) Imaging the body with diffuse optical tomography," Signal Process Mag IEEE 18(6): 57–75
4. Waters NE (1965) The indentation of thin rubber sheets by spherical indenters," Br J Physics 16: 557–563
5. Shepherd DE, Seedhom BB (1997) A technique for measuring the compressive modulus of articular cartilage under physiological loading rates with preliminary results. Proc Inst Mech Eng [H] 211(2): 155–165
6. Shimoseki M, Iwasaki S, Nakagiri S et al (1979) Analysis of a helical compression spring by matrix method (in Japanese). Trans JSME 45-C (396): 901–910
7. Watari A, Kobayashu S (1959) General equations in design of a helical compression spring (in Japanese). Trans JSSR 5: 89–93

Actually, with the increasing of the number of surgery and the advanced age population of society, there are numerous potential applications in which robot can work with human together.

In this paper, our purpose is to bring forward the concept of hybrid support system in response to the requirement of surgical operation, and to construct a prototype system based on 6-axis robot arm. We think that this support system in the face of decreasing economic support will likely be possible. The schematic diagram of the support system is shown in Figure 1.

2. Concept of Support System

The hybrid support system that we have constructed has two functions, including of learning from historical case data and carrying out the support manipulation. Firstly, the system should be able to predict the approximate inherent shakes period of manipulator's hand if it's given an initial situation. Secondly, system should be able to carry out the support with the manipulator together according to the results that system obtained from the initial situation by learning. Essentially, it's a process to relief the burden of manipulator. The safety of surgical operation can also be ensured because there is human attending the manipulation.

For this paper, it's just at the beginning. In order to discover the unaware shakes rule of hand, we firstly attempt to attain the data of the shakes of hand in the condition that the shake of hand is caused by the stimulation from outer body when a special mission is executed by the manipulator. Support system we developed can only be appropriate for the situation of aware shakes stimulated from outer body. What's the influence when there is the robot attending the manipulation is also discussed. The results are helpful to discover the unaware shakes rule of hand and find a suitable way to carry out the support by robot arm. Furthermore, it can become the important reference to what we shall do next.

2.1 Structure of the support system

The total system consists of six parts: 6-axis robot arm, sensor (force, torque, acceleration, laser and CCD), controlling device, computer, monitor, and interface mechanism. Figure 2 shows the structure of the support system. Force, torque and CCD sensors belong to parts of robot arm. Acceleration and laser displacement sensors are individual. The preliminary system prototype that has been developed is shown in Figure 3.

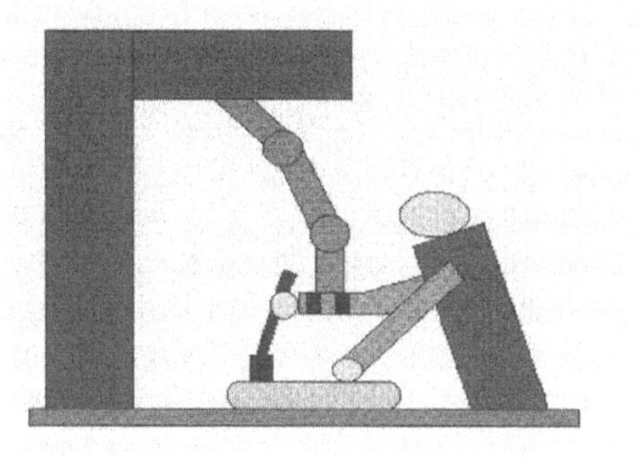

Fig.1. Schematic diagram of the support system

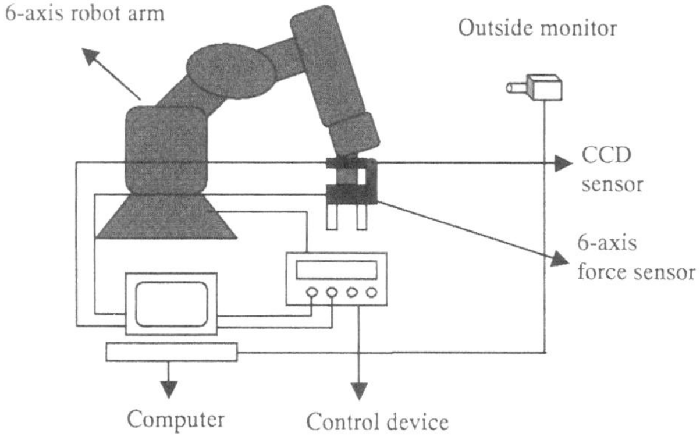

Fig.2. Structure of the support system

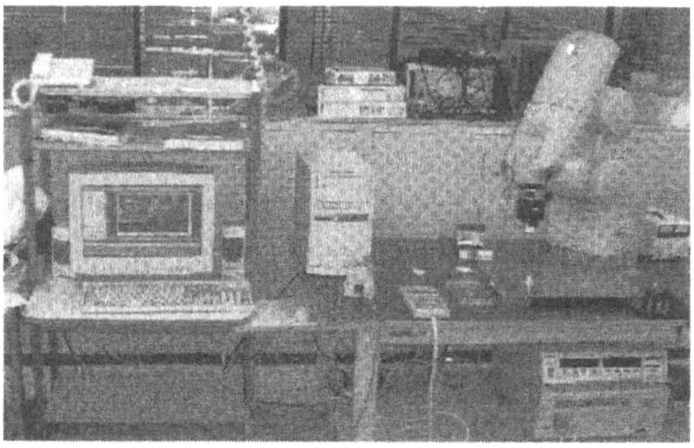

Fig.3. Developed prototype of the support system

2.2 Principle of support system

Figure 4 shows the architecture of the support system. There are two individual parts in the support system. The task of Part1 is to obtain the parameters about shakes of manipulator's hand in the condition that the hand is influenced by stimulation of outer body when the robot arm does not attend the support. The parameters of shakes of hand is primary important for robot arm to perform the support successfully. Depending on the obtained information, it makes decisions what kind of database about the manipulator can be established by probability analysis. The reason that causes shakes of hand dues to two aspects, including stimulation from outer body and inherent shakes of body. Our aim is to find out inherent shakes period of hand. For different manipulator, the different database needs to be built. It's a process to teacher robot how to accommodate to work with human together. Part2, as a real-time feedback control system, is designed to perform hybrid support based on 6-axis robot arm. The sensors for detecting changes of the real-time operating parameters are fixed on robot arm and hand of manipulator. In the case of improper manipulation caused by shakes of hand, robot arm can carry out the corresponding support to compensate it in terms of the obtained parameters and in comparison with the established database. Part1 also decides how to compensate part2.

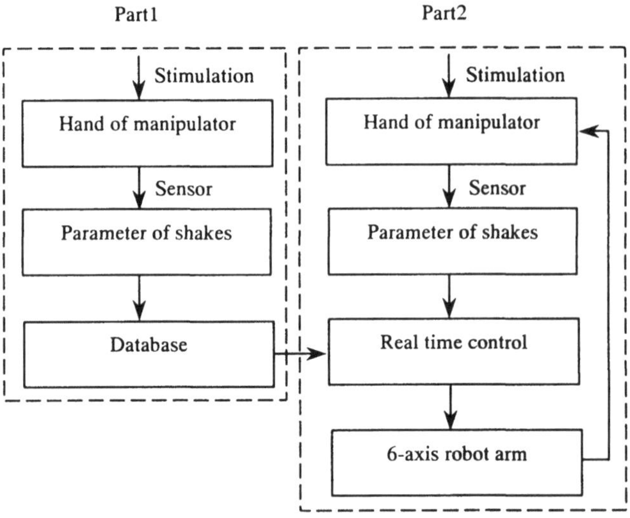

Fig.4. The architecture of the support system

3. Experimental Results of the System

3.1 Feasibility experiment

The 6-axis robot arm is the core of the system, and it directly makes a decision about the performance of the system. The first experiment is carried out to validate the feasibility if robot ram is appropriate to be used. In the experiment, a bean curd is adopted as the assumed skin. We insert the pipes as the assumed blood vessel into the side of bean curd, as shown in Figure 5.

The required task for the robot is to manipulate the scalpel to cut the bean curd. At the same time, the scalpel must evade the pipe obstacle to process ahead and avoid cutting the pipes. Figure 6 shows the reaction force model. As soon as the pipe obstacle is encountered against the scalpel, control system can get the reaction force signal via force and torque sensor. The accuracy of force and torque sensor can reach to 0.1N. Then the force value F can be calculated by equation (1).

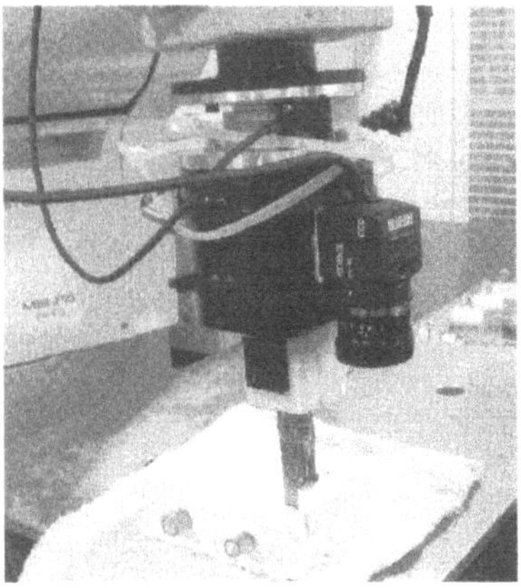

Fig.5. View of feasibility experiment

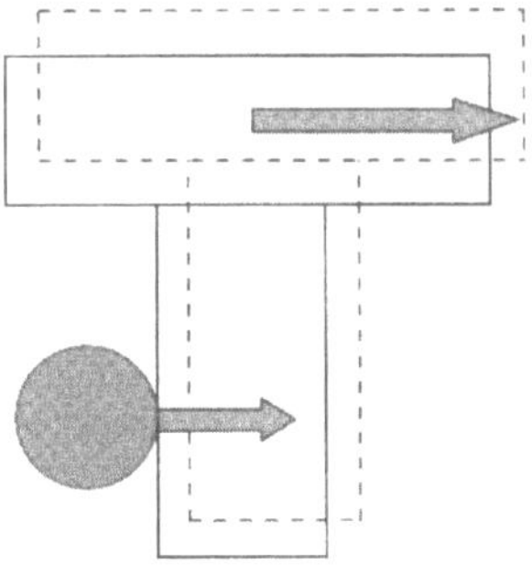

Fig.6. Reaction force model

$$F = CV \tag{1}$$

Where

$$F = \begin{vmatrix} F_x & F_y & F_z & T_x & T_y & T_z \end{vmatrix} \tag{2}$$

$$C = \begin{bmatrix} C_{11} & \bullet & \bullet & C_{66} \\ \bullet & \bullet & \bullet & \bullet \\ \bullet & \bullet & \bullet & \bullet \\ C_{61} & \bullet & \bullet & C_{66} \end{bmatrix} \tag{3}$$

$$V = \left| SG.0 \quad SG.1 \quad SG.2 \quad SG.3 \quad SG.4 \quad SG.5 \right| \tag{4}$$

C is the compensation parameter of sensor; V is 6- channel signal of sensors. When reaction force value is large than limit value, control commands will make the scalpel return and change the motion direction until to cross the obstacle step by step. The reaction torque value of the scalpel during the whole process is shown in Figure 7.

According to the preliminary experimental result, we can see robot arm is preliminary eligible to carry out the support task in unmanned condition.

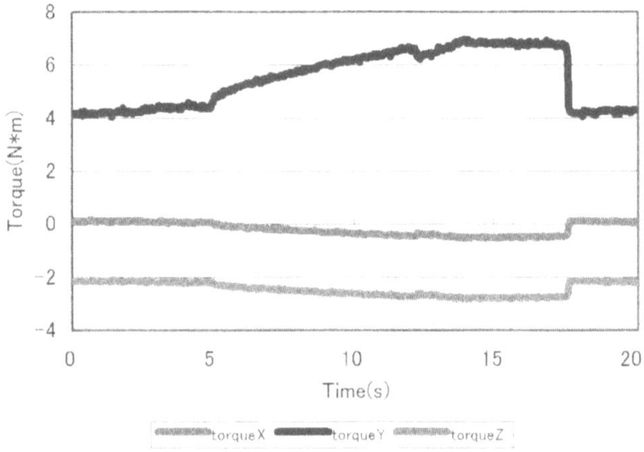

Fig.7. Reaction torque measurements

3.2 The Support experiment

As the beginning, a simple interface mechanism is adopted to combine the hand of manipulator with robot arm. Figure 8 shows the experiment process. An acceleration sensor is bound to the finger of manipulator, and the vibration generator used to stimulate the hand is placed on the surface of manipulator's arm. When the operating tool is operated, the person's hand will be compensated.

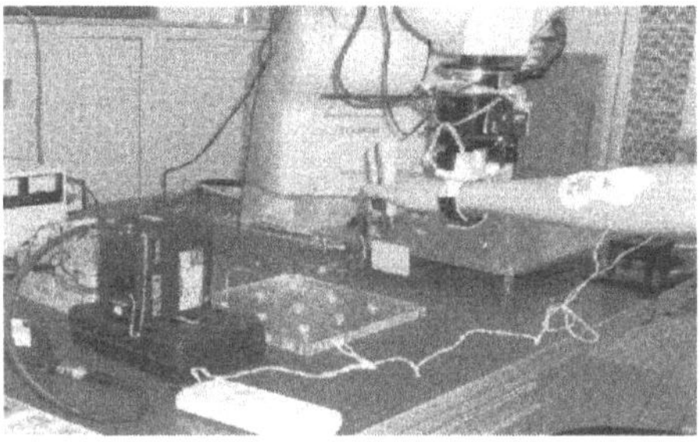

Fig.8. View of the experiment

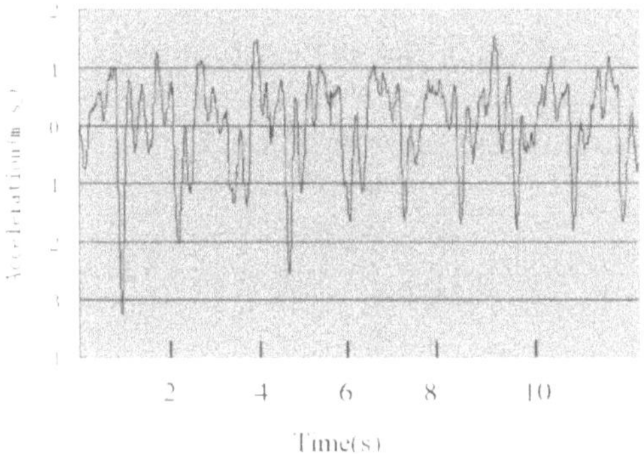

Fig.9. Acceleration of the hand of manipulator A

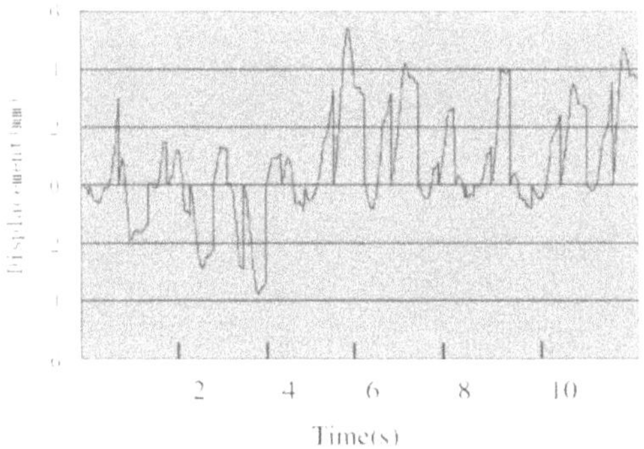

Fig.10. Displacement of the hand of manipulator A

The static location of hand, when there is no outer stimulation for manipulator, is set up as reference point. As the generator works, it makes hand generate the shakes resulting in the position shift of hand. The displacement of hand can be measured by laser displacement sensor. At the same time, the acceleration sensor can obtain the 3 directions (x, y, z) acceleration signal of hand. The accuracy of acceleration sensor can reach to $0.01m/sec^2$. The entire signal above is transmitted into computer. According to the obtained results, we construct the database by means of a statistical study of each variable and teach robot how to support the manipulation by itself.

Under the stimulation of vibration generator, we conducted the experiment for seven different manipulators to obtain the shakes parameters of hand in laboratory. In this paper, we give the measure results of manipulator A in Figure 9.

Then, we carried out the same experiments to the seven manipulators with real-time aid of 6-axis robot arm. The real-time information of manipulator can be obtained. It's the premise that how we analyze the attained information and infer the state of process. It contributes to the database constructed above. Figure 10 shows the displacement of manipulator A's hand with real-time support of 6-axis robot arm.

4. Analysis of Result

Through the experimental results, we can get the frequency of the shakes for different manipulators, and build a parameter database for them. From figure 9 and figure 10, we can see that the shake frequency is about 1Hz, and the robot arm will compensate the shake one time during every second.

Table 1. Displacement comparison before and after compensation

Subject	If there is support	Maximum (mm)	Minimum (mm)	Average (mm)
A	Y	7	1	4.2
	N	14.7	3.5	7.9
B	Y	16	9	10.7
	N	35.6	27.2	29.9
C	Y	4.4	1.5	2.3
	N	16	4.6	10.7
D	Y	8.1	2.1	4.8
	N	16.3	8.3	13.1
E	Y	7.5	2.4	5.6
	N	14.9	8.5	11.2
F	Y	5.6	2.4	4
	N	14.7	5.9	10.3
G	Y	5.6	2.1	3.5
	N	9.2	4	7

Table 1 lists the displacement (maximum, minimum and average) comparison of manipulator's hand (A, B, C, D, E, F) in two kinds of different instances, which there is no the robot attending the manipulation and there is the robot attending it.

5. Conclusion

In the medical field and in industry application, a novel type of biomedical support system has urgently been demanded. We attempt to discover the unaware shakes rule of hand, to build the shakes database about the manipulator, based on which we can carry out the 6-axis robot arm to compensate the operation error caused by the shakes of manipulator's hand. In this paper, it's just the beginning and the aware shakes stimulated from outer body are only discussed. By the comparison between two kinds of the different instance, in which there is no the robot attending the manipulation and there is the robot attending it, we attempt to find out the aware shakes rule of hand, to teach the whole support system have the ability of how to support the manipulation by itself. Also it's the important reference to discover the unaware shakes rule of hand. We have described the hybrid support system in response to surgical operation. In the primary experiment, a 6-DOF robot has been used to compensate the shake of hand, which is caused by the simulation from outer body. The experimental results indicate that the proposed biomedical support system can be used for medical application, and it can improve the operability of microsurgery.

The results are helpful to discover the unaware shakes rule of hand and find a suitable way to carry out the support by robot arm. Furthermore, it can be the important reference to what we shall do the next step.

References

1. Kiyoyuki Chinzei, Karol Miller (2001) MRI guided surgical robot.Proc.Australian Conference on Robotics and Automation 50-55
2. Jacob Rosen, Mitch Lum, Denny Trimble, et al (2005) Spherical Mechanism Analysis of a Surgical Robot for Minimally Invasive Surgery-Analytical and Experimental Approaches. Medicine Meets Virtual Reality 442-448
3. T. Fukuda, Hosokai (1991) Distributed Type of Actuator by Shape Memory Alloy and its Application to Underwater Mobile Robotic Mechanism. Proc. of the IEEE Conf. on Robotics and Automation 2:1316-1332
4. Fukuda, Kawamoto, Arai et al (1994) Mechanism and Swimming Experiment of Micro Mobile Robot in Water. Proc. Proc. of the IEEE Conf. on Robotics and Automation 1:814-819
5. Fukuda, Kawamoto, Arai et al (1995) Steering Mechanism of Underwater Micro Mobile Robot. Proc. of the IEEE Conf. on Robotics and Automation 1:363-368
6. M. Mojarrad and M. Shahinpoor (1997) Biomimetic Robot Propulsion Using Polymeric Artificial Muscles. Proc. of the IEEE International Conf. on Robotics and Automation 2152-2157
7. S.Guo, T. Fukuda, N. Kato et al (1998) Development of Underwater Microrobot Using ICPF Actuator. Proc. of the IEEE International Conf. on Robotics and Automation 1829-1834
8. Keisuke Oguro, Kinji Asaka and Hiroyasu Takenaka (1993) Polymer Film Actuator Driven by a Low Voltage. Proceedings of 4th International Symposium on Micro Machine and Human Science 39-40
9. S. Guo, T. Fukuda, K. Asaka (2003) A New Type of Fish-like Underwater Microrobot. IEEE/ASME Transactions on Mechatronics 8: 136-141
10. S. Guo, Y. Okuda, K. Asaka (2004) A Novel Type of Underwater Micro Biped Robot with Multi DOF. Proc. of the IEEE International Conference on Robotics and Automation 4881-4886
11. O. Sadao, M. Yoshinobu and E. Christos (2004), Real time robotic tactile sensor system for the determination of the physical properties of biomaterials. Sensors and Actuators 112:278–285
12. S. Tadokoro, S.Yamagami, T. Takamori (2000) An Actuator Model of ICPF for Robotic Applications on the Basis of Physicochemical Hypotheses. Proc.of the IEEE International Conference on Robotics & Automation 1340-1345

The Development of a New Kind of Underwater Walking Robot

Wei Zhang[1], Shuxiang Guo [2, 3]

[1] Graduate School of Engineering, Kagawa University, Japan
[2] Faculty of Engineering, Kagawa University, Japan
[3] Harbin Engineering University, China

Chapter Overview. In this paper, to approach a locomotion method for underwater rough surface, we designed a new kind of leg with two segments of ICPF (Ionic Conducting Polymer Film) actuators. And also, we carried out an underwater robot equipped a walk-style leg and two wheels. We have carried out the experiment of the robot. And the result shows that the robot has a good walking performance on rough surface. An improved structure of robot is designed and made out. Without a wheel, it is smaller and speed is higher. A speed model of the robot is carried out. On the other hand, the speed is also measured in experiments.

Key Words. Underwater Walking Robot, and Medical Robotics.

1. Introduction

Underwater vehicles almost exclusively use propellers or jets to move in water and are powered by electromechanical devices. At the same time, most of marine habitants use radically different means of locomotion. Smart materials, like ICPF, which can be used as artificial muscles, pave the way to a great variety of biomimetic approaches for underwater vehicle design. Theoretically, robots with ICPF actuators have several advantages [1-4]. They are lightweight, consume little energy and are environmental friendly. Until now, there are several kinds of underwater microrobots utilizing ICPF actuators in the world. ICPF actuators are used as artificial muscles to drive robots [7-8]. A kind of underwater microrobot with two tails has been worked out [9, 11-13]. A prototype of fish-like underwater robot with two pectoral fins is also reported [16]. ICPF actuators are also used for biped walking underwater robot and multi-DOF manipulator [15, 17].

Our purpose is to develop a structure for underwater rough surface. In this paper, we introduce a novel type of leg using ICPF actuators. And a robot utilizing the leg is worked out. To minimize the robot, its successor is worked out. The speed is also measured. The results show that it has a good speed characteristic, but the speed matrix is not similar to the theoretical model.

2. An Underwater Robot with a Walk-Style Leg

2.1 A Walk-style leg and the control method

Inspired by mantises' front legs, we designed a kind of walk-style leg with ICPF actuators, shown in Fig.1. A walk-style leg is made of two segments of ICPF actuators acted as the thigh A and the crus B. They are connected at point O_2. One end of the leg, O_1, is fixed on the robot.

The A and B, are controlled respectively. In our research, rectangle signals are used. The phase of signal for B is 90-degree delayed than for A, shown in Fig.2.

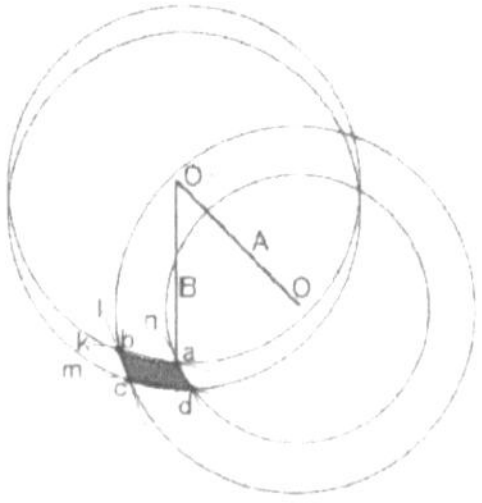

Fig. 1. Walking Mechanism

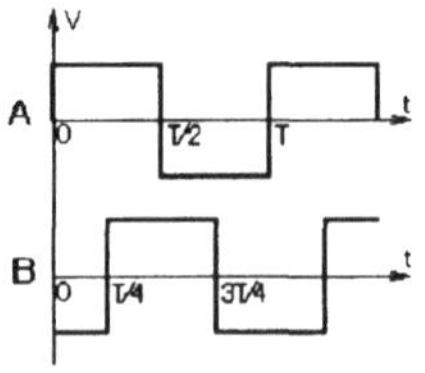

Fig. 2. Control Signals

2.2 Moving mechanism

The walk-style leg can stride over the rough surface, when the foot steps forward. And when the foot steps backward, the leg provides the driving force, and the roughness is favorable. One step circle can be described as the following:

- In the beginning of the first 1/4 cycle, segment A bends in clockwise, and the foot is moved from 'd' to 'a' along the arc 'da' (in Fig.1);
- In the beginning of the second 1/4 cycle, segment B bends in clockwise, and the foot is moved from 'a' to 'b' along the arc 'ab';
- In the beginning of the third 1/4 cycle, segment A bends in anticlockwise, and the foot is moved from 'b' to 'c' along the arc 'bc';
- In the beginning of the forth 1/4 cycle, segment B bends in anticlockwise, and the foot is moved from 'c' to 'd' along the arc 'cd'.

2.3 The structure of a robot equipped a walk-style leg

An underwater robot named Crawler-1 is made out with a walk-style leg and two wheels, shown in Fig.3 and in Fig.4.
Its dimension is 80*15*55mm^3. It is 6.03g (dried) in weight. Actuators A and B are 22*4*0.5mm^3, not including the joints. The angle between A and B is about 45 degree. The segment B is elongated by a plastic stick. A gum foot is equipped. The elongated segment is 27mm. The two wheels are ϕ 30*5mm^3. There are two aims of the wheels. The first is to reduce the resistance. And the second is to keep the body balance. The body is 50mm in length, 10mm in width, and 20mm in height.

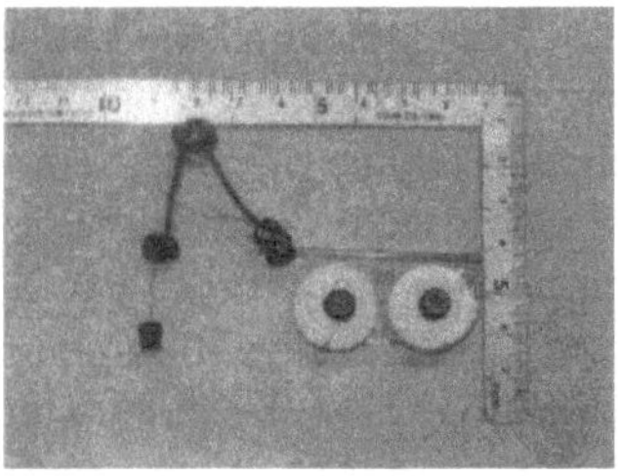

Fig. 3. A view of Crawler-1

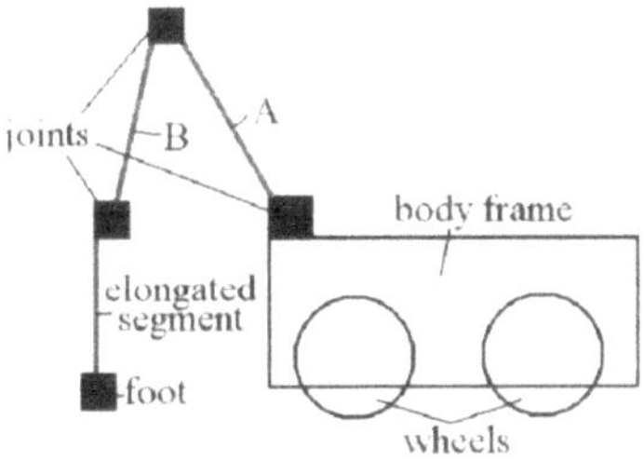

Fig. 4. Structure of Crawler-1

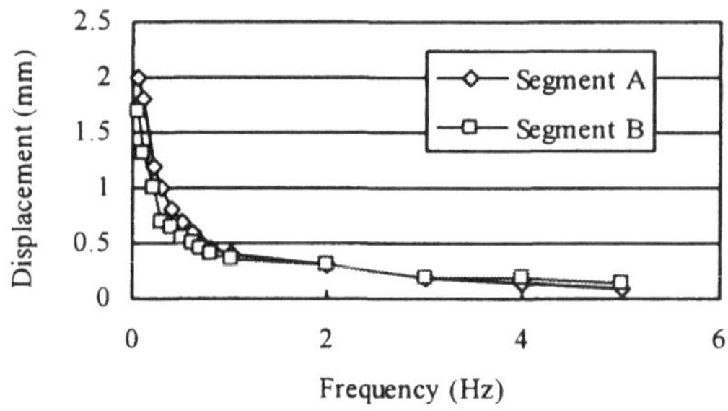

Fig. 5. The function between frequencies and displacements

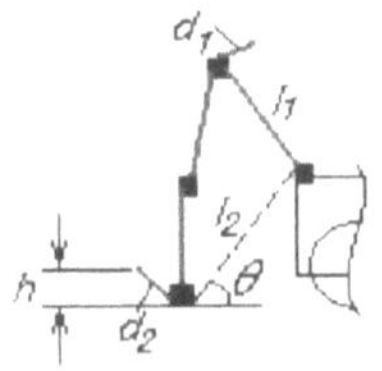

Fig. 6. Height that the foot can be lift

2.4 Characteristic measurement

The Fig.5 shows the curves between frequency and displacement of A and B. The signal is 10v rectangle. The theoretical displacement of one step is shown in Fig.6. l_1 is the length of segment A, l_2 is the distance from the foot to end of the leg, 38mm, d_1 is the displacement that segment A can move, d_2 is the displacement that the foot can move, θ is the angle between l_2 and horizontal line, about 45 degree, and h is the height that the foot can be lifted up. With the function between the frequency and displacement, the height of obstacles that the foot can stride over can be calculated with equation (1) and (2). And also, theoretical displacement of one step can be calculated with formula (3).

$$d_2 = \frac{l_2}{l_1} \times d_1 \tag{1}$$

$$h = d_2 \times \cos\theta \tag{2}$$

$$d_4 = \frac{l_4 + l_3}{l_3} \times d_3 \tag{3}$$

In formula (3), d_3 is the displacement of segment B, d_4 is the displacement of the end of elongated leg, l_3 is the length of segment B, and l_4 is the length of the elongated segment. With the displacement curve of one step, the theoretical speed curve can be calculated with formula (4).

$$s = d_4 \times f \tag{4}$$

In formula (4), d_4 is the displacement of one step, f is the frequency of signal and s is the theoretical speed of Crawler-1. The result is shown in Fig.8.

Experiment of Crawler-1 has been carried out. With a rectangle signal of 10-voltage and frequency of 0.2Hz, it walks at a speed of 2mm per step, 0.4mm per second, which is the top speed.

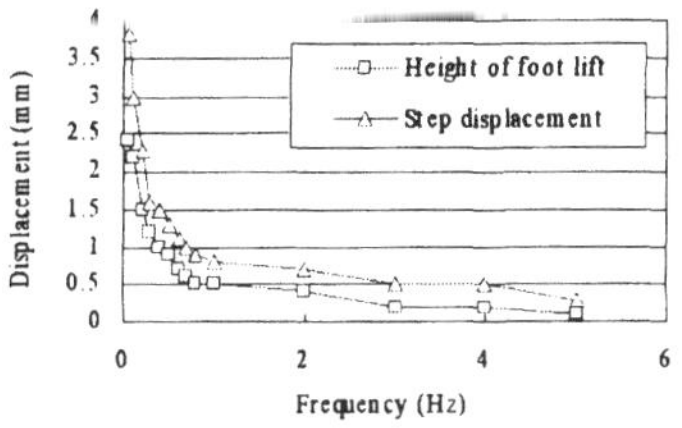

Fig. 7. Characteristics of the leg

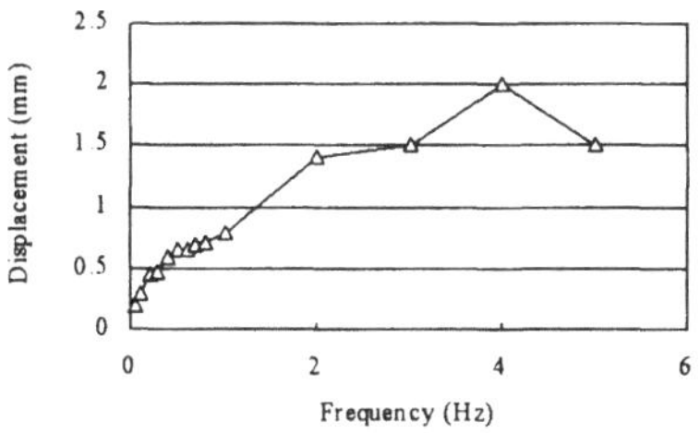

Fig. 8. Theoretical speed of Crawler-1

$$T = \begin{pmatrix} 0 & F \\ V^T & D \end{pmatrix} \tag{5}$$

$$F = \begin{pmatrix} 0.5 & 1 & 2 & 3 & 4 & 5 \end{pmatrix} \tag{6}$$

$$V = \begin{pmatrix} 1 & 2 & 3 & 4 & 5 & 6 & 7 & 8 & 9 & 10 \end{pmatrix} \tag{7}$$

$$D = \begin{pmatrix} 0.26 & 0.21 & 0.19 & 0.18 & 0.16 & 0.14 \\ 0.80 & 0.50 & 0.37 & 0.32 & 0.29 & 0.25 \\ 1.74 & 0.99 & 0.64 & 0.51 & 0.44 & 0.38 \\ 2.60 & 1.49 & 0.86 & 0.67 & 0.56 & 0.51 \\ 3.56 & 1.92 & 1.13 & 0.83 & 0.67 & 0.63 \\ 4.80 & 2.61 & 1.42 & 1.06 & 0.82 & 0.76 \\ 6.15 & 3.27 & 1.82 & 1.29 & 1.02 & 0.92 \\ 7.38 & 3.94 & 2.15 & 1.50 & 1.18 & 1.08 \\ 9.14 & 4.65 & 2.59 & 1.84 & 1.41 & 1.25 \\ 11.12 & 5.46 & 3.05 & 2.18 & 1.68 & 1.53 \end{pmatrix} \tag{8}$$

$$S = \begin{pmatrix} D_{n1} \times F_1 & D_{n2} \times F_2 & \cdots & D_{n6} \times F_6 \end{pmatrix} \quad (n = 1, \ 2, \ \cdots \ 10) \tag{9}$$

$$S = \begin{pmatrix} 0.13 & 0.21 & 0.37 & 0.55 & 0.65 & 0.68 \\ 0.40 & 0.5 & 0.73 & 0.97 & 1.15 & 1.23 \\ 0.87 & 0.99 & 1.28 & 1.53 & 1.74 & 1.88 \\ 1.30 & 1.49 & 1.71 & 2.01 & 2.26 & 2.53 \\ 1.78 & 1.92 & 2.26 & 2.48 & 2.68 & 3.13 \\ 2.40 & 2.61 & 2.83 & 3.17 & 3.29 & 3.81 \\ 3.08 & 3.27 & 3.65 & 3.86 & 4.10 & 4.62 \\ 3.69 & 3.94 & 4.29 & 4.51 & 4.74 & 5.38 \\ 4.57 & 4.65 & 5.18 & 5.52 & 5.63 & 6.24 \\ 5.56 & 5.46 & 6.11 & 6.53 & 6.71 & 7.66 \end{pmatrix} \tag{10}$$

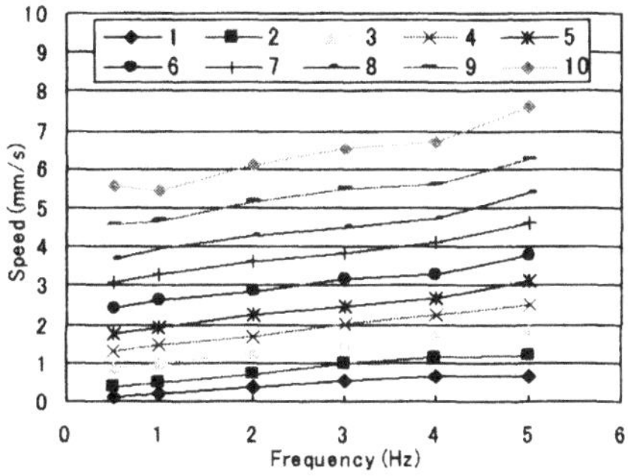

Fig. 12. Theoretical speed curves

Fig. 13. Scene of speed test experiments

3.4 Experiments result

To decrease the friction on the body and increase the friction on the driver, a special tramroad is used. The tramroad is a belt of sand paper, $61*12mm^2$, adhibited on an aluminum board. A view is shown in Fig.13. Changing the frequency and voltage of signals, the speeds of Crawler-3 on the tramroad have been tested, shown in Fig.14.

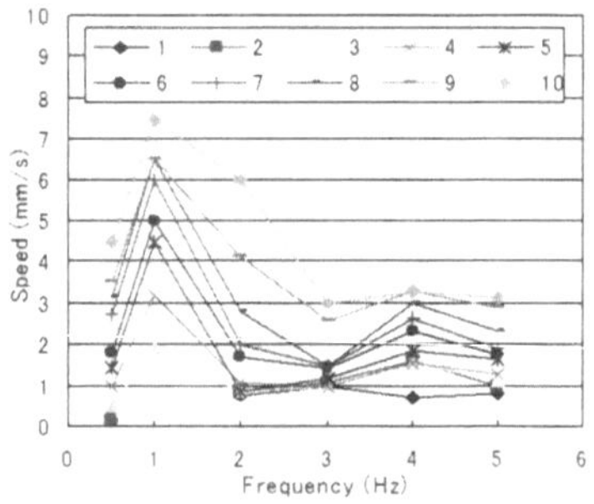

Fig. 14. Speed curves based on voltage from experiments

4. Conclusion

In this paper, inspired from rearhorse, we designed and carried out an underwater microrobot. It can move on underwater rough surface. And then a developed robot is made out. With a novel structure, the successor is faster in speed, smaller in dimension and higher in weight than its predecessor.
A speed model of the successor has been built up, but the model does not accord with the result of experiments. In our future research, we will improve the model of it.

References

1. Y. Hirose, T. Shiga, A.Okada, and T. Kurauchi, "Gel actuators driven by an electric field, " in Proc. 3rd Int. Symp. Micro Machine and Human Sciece, 1992, pp. 21-26.
2. Y. Osada, H. Okuzaki, and H. Hori, "A polymer gel of electrically driven moiety," Nature, vol.355, pp. 242-244,1992.
3. D. J. Segalman, W. R. Witkowski, D. B. Adolf, and M. Shahinpoor, "Theory and application of electrically controlled polymeric gels," print publication, Issue 1, March 1992.
4. K. Oguro, K. Asaka, and H. Takenaka, "Polymer film actuator driven by a low voltage," in Proc. 4th Int. Symp. Micro Machine and Human Science, Japan, 1993, pp.39-40.

5. T. Fukuda, A. Kawamoto, F. Arai, and H. Matsuura, "Mechanism and swimming experiment of micro mobile robot in water, " in Proc. IEEE Int. Conf. Robotics and Automation, vol.1, 1994, pp. 814-819.
6. T. Fukuda, A. Kawamoto, F. Arai, and H. Matsuura, "Steering mechanism of underwater micro mobile robot," in Proc. IEEE Int. Conf. Robotics and Automation, 1995, vol.1, pp.363-368.
7. M. Mojarrad and M. Shahinpoor, "Biomimetic robot propulsion using polymeric artificial muscles," in Proc. IEEE Int. Conf. Robotics and Automation, 1997, pp. 2152-2157.
8. M. Shahinpoor, Y. Bar-Cohen, J. O. Simpson, and J. Smith, "Ionic polymer-metal composites (IPMCs) as biomimetic sensors, actuators and artificial muscles- a review," print publication, Issue 6, December 1998.
9. S. Guo, T. Fukuda, N. Kato, and K. Oguro, "Development of underwater micro robot using ICPF actuator," in Proc. IEEE Int. Conf. Robotics and Automation, 1998, pp. 1829-1834.
10. S. Tadokoro, S. Yamagami, and T. Takamori, "An actuator model of ICPF for robotic applications on the basis of physicochemical hypotheses," in Proc. IEEE Int. Conf. Robotics and Automation, 2000, pp. 1340-1346
11. S. Guo, K. Sugimoto, S. Hata, J.Su, and K. Oguro, "A new type of underwater fish-like micro robot," in Proc. IEEE Int. Conf. Intelligent Robotics and Systems, 2000, pp. 862-867.
12. S. Guo, T. Fukuda, and K. Asaka, "Fish-like underwater micro robot with 3 DOF," in Proc. IEEE Int. Conf. Robotics and Automation, May 2002, pp. 738-743.
13. S. Guo, T. Fukuda, K. Asaka, "A New Type of Fish-like Underwater Microrobot," IEEE/ASME Transactions on Mechatronocs, Vol.8, pp.136-141, 2003.
14. S. Guo, Y. Okuda, "A Novel Type of Underwater Micro Biped Robot with Multi DOF," International Conference on Control Science and Engineering, December 2003.
15. N. Kamamichi, Y. Kaneda, M. Yamakita, K. Asaka, and ZW Luo, "Biped Walking of Passive Dynamic Walker with IPMC Linear Actuator," SICE Annual Conference in Fukui, August 2003.
16. M. Anton, A. Punning, A. Aabloo, and M. Kruusmaa, "Towards a Biomimetic EAP Robot," In Proc. of Towards the Autonomous Mobile Robots, TAROS 2004.
17. Yoshihiro Nakabo, Toshiharu Mukai, and Kinji Asaka, "A multi-DOF robot manipulator with a patterned artificial muscle," The Second Conference on Artificial Muscles, Proceedings, 2004.

Mid-Infrared Robotic Laser Surgery System in Neurosurgery

Shigeru Omori[1,3], Ryoichi Nakumura[1], Yoshihiro Muragaki[1], Ichiro Sakuma[2], Katsuhiro Miura[4], Masao Doi[4], and Hiroshi Iseki[1]

[1]Faculty of Advanced Techno-surgery, Institute of advanced Bio- medical engineering & Science, Graduate school of Medicine, Tokyo Women's Medical University
[2]Institute of Environmental Studies, Graduate School of Frontier Science, The University of Tokyo
[3]R&D Center, Terumo Corporation
[4]Mitaka Kohki Co., LTD.

Chapter Overview. There is certain limitation to do the perfect extraction of the tumors by conventional manual surgery, because the area close to the boundary between tumors and normal brain tissue is usually left in order to keep away from destruction of normal area. For the purpose to treat such the boundary area, a computer controlled robotic laser surgery system has been developed. This system is characterized by the mid-infrared laser device which can perform less invasive precise surgery with low output power (<1.0W), and the computer controlled system which can realize the ablation of designated area on brain surface within 0.5mm dislocation. For the further study, aiming to apply to use in vivo, we have developed an auto-focus system for laser irradiation. From the results of our animal study, the system enables to maintain the focal point of the laser head on brain surface to realize constant ablation in the designated area.

Key Words. Robotic surgery, Malignant brain tumors, Mid-Infrared, Computer controlled system

1. Introduction

Laser surgery in neurosurgical field has been performed for more than 20 years, and almost all of commercialized laser devices were used clinically in this field [1]. In neurosurgical field, however, the laser devices have been rarely used. Because the conventional laser irradiation is performed manually operated by surgeon's hand, the precise treatment can not be achieved. Besides such kind of devices are usually expensive.

We developed a robotic laser surgery system in order to remove tumor tissue which remains after manual resection. The remaining tissue is usually in a thin layer visualized by intraoperative MRI or navigation system [2,3,4]. In order to actualize this system it is necessary to fulfill following three conditions.

(1) The use of a laser device that does not cause collateral damages with non-contact irradiation.

(2) The development of a laser manipulation system that has a precise positioning function of less than 0.5 mm which is the resolution of MR image.

(3) The development of an auto-focus mechanism that can maintain a focal point on a given brain surface irrespective of its unevenness when laser beams are being irradiated.

With respect to (1) and (2) we developed a satisfied robotic surgery system and reported results of experiment in vitro [5].

This paper describes the concept of the newly developed robotic laser surgery system with confocal optical sensing system for laser auto-focussing in brain surface ablation.

2. Cntiguration of Laser-Irradiation System

2.1 Laser device

In the early history of laser application for neurosurgery, hemostasis was the major purpose in the resection of brain tumors. As bipolar electric cautery forceps were making progress, laser devices have lost their position for hemostasis treatment. In resent years, trials of precise ablation by using pulse laser have been reported [6,7,8]. But it has been pointed out that the pulse laser was difficult to treat eloquent area because of mechanical damage by shock wave rather than thermal damage [7]. We developed a laser device which was able to be used for tumor ablation adjacent to eloquent area.

Figure 1 shows the construction of our mid-infrared laser. This laser device is characterized by an Er-doped YSGG laser crystal chip on the distal end of a silica optical fiber. The laser crystal chip outputs 2.8μm wavelength laser beam pumped by the near infrared continuous wave (CW) laser diode (LD). Since an absorption peak of water exists at the wavelength of near 3μm [9], laser beams irradiated are absorbed by the water of the brain tissue before the beams reach deep into the brain so that heat generation occurs only in a local target area.

2.2 System configuration

Figure 2 shows a schematic diagram of the robotic laser surgery system. Once a surgeon determines the area of irradiation the system itself automatically carries out laser scanning inside the area. This system consists of the mid-infrared laser device, a 2-axis moving stage for laser scanning, a charge-coupled devise (CCD) camera, and a personal computer (PC). Image data of brain surface is acquired by the CCD camera, transferred to the PC, and displayed on a monitor screen. A surgeon designated a target area to be irradiated by tracing the outline of target area on the monitor screen. Positioning dislocation of the ablation is 0.5mm or less was achieved in in Vitro study with porcine brain.

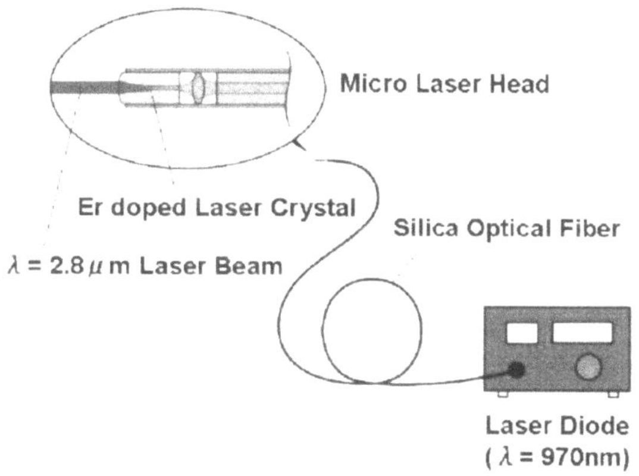

Fig. 1. Conceptual image of mid-infrared (λ=2.8μm) laser device

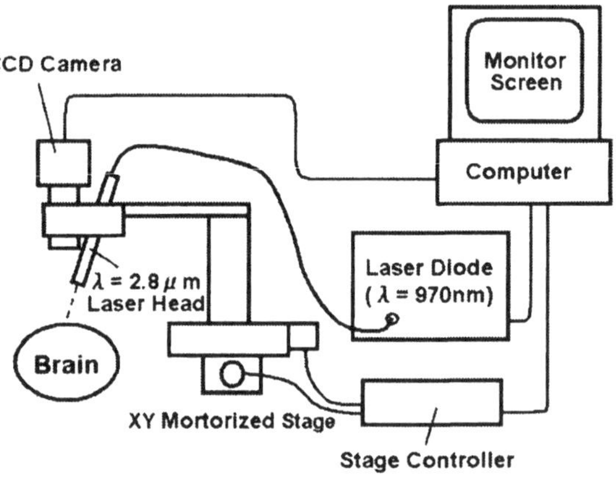

Fig. 2. Prototype of mid-infrared robotic laser surgery system

2.3 Effects of focusing status on ablation lesion

Since we used relatively flat brain surface in performing in vitro study, the laser beams were able to be focused on at any point in a target area. However, the brain surface to be treated in clinical cases is usually bumpy or the surface itself is slope away. Therefore it was necessary to evaluate ablation characteristics under defocused beam condition.

Figure 3 shows ablation of the porcine brain by several conditions in terms of defocused laser beam spots. The focused beam spot diameter was set to 0.16mm ($1/e^2$), the laser output power to 0.47W and the scanning speed of the laser beams to 2 mm/s. The sample is a cross section of the dissected porcine brain after laser irradiation with dislocation of 1mm to 4mm from just focussed position by 1mm intervals. In order to measure the ablation depth, the brain was fixed in formalin and cut it to expose all the incised lesions in one surface. Then, red ink was injected into each lesion.

Figure 4 shows the relation between the dislocation and the ablation depth. As the focal point was got away from the brain surface the depth of ablation lesion became shallower and shallower (Fig 3. right to left). Dislocation of 2mm brings more than 40% decrease of depth (third vertical line from the right) and this is not acceptable condition for clinical use. From the result of this experiment, it was considered that an auto-focussing system was necessary to maintain constant ablation depth, because the brain surface usually has over 5mm peaks and valleys in clinical cases.

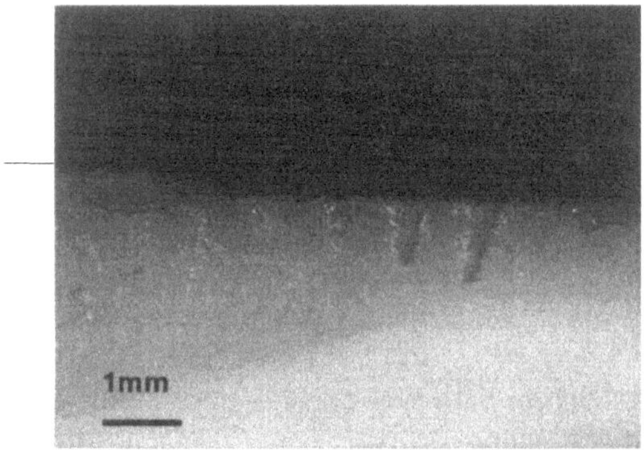

Fig. 3. Ablated lesion on a dissected porcine brain at various focus conditions

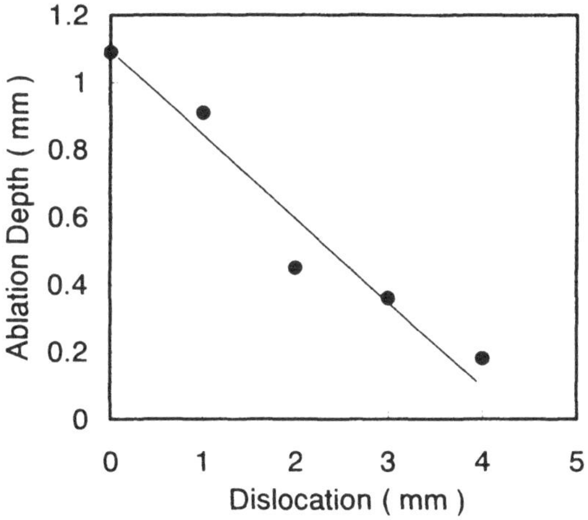

Fig. 4. Laser ablation depth with dislocation from just-focused position

3. Auto-focus System

In this study, we designed the auto-focus system to satisfy the following requirements:

a) To have a position sense function by non-contact fashion without being disturbed by the surface water.
b) To let the sensing point and the laser ablation point is at the same location.

c) to have a driving mechanism for a unit of the laser head and the CCD camera

In order to satisfy these requirements, we used an optical sensing system with another laser beam (λ=675nm) to measure and maintain a constant distance between the laser head and brain surface. Figure 5 shows the configuration of the sensing system. It has a confocal optical path. In this system, location of the sensing laser beam spot on the brain surface is projected on a silicon photo sensor. The beam spot on brain surface can be detected in spite of weak reflection because there is conjugate point on the sensor. The optical sensing path was built in an image capture optics of the CCD camera with the same optical axis. The computer controlled the movement of a linear motorized stage on that two laser heads were attached (Fig. 6). One of the laser heads was for backup.

4. Evaluation of Auto-focus System in Brain Ablation

An animal experiment was performed using porcine in order to demonstrate the function of the auto-focus system. Firstly, we drew an outline of about 5mm square on an irradiation area by computer "mouse" on the

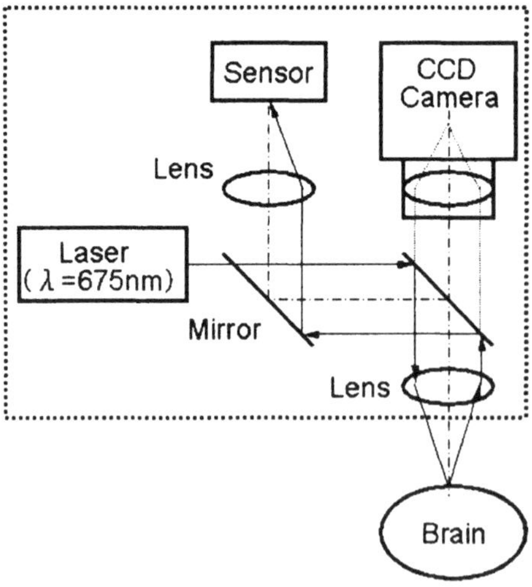

Fig. 5. Configuration of the optical sensing system for the auto-focus

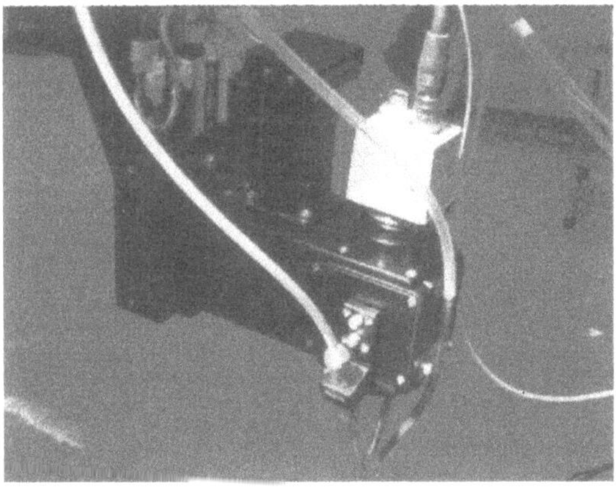

Fig. 6. Photograph of the auto-focus system with the mid-infrared laser heads

monitor screen. Secondary, ablation process was performed by laser irradiation with raster scanning. Physiological saline was poured before laser scanning in order to prevent carbonization. The focusing beam spot diameter was set to 0.14mm ($1/e^2$), the laser output power to 0.45W and the scanning speed of the laser beams to 4 mm/s. Figure 7 shows images displayed on the PC screen during laser ablation treatment. The auto-focus system functioned for the brain surface covered with water layer and surface ablation of the designated area was successfully performed.

5. Conclusions

We worked to remove residual tumor tissue in the resection of malignant brain tumors on a trial basis considering laser auto-focus system as improvement of our robotic laser surgery system. The detecting method using confocal optical design functioned to maintain the distance between the laser head and brain surface covered with slight water layer. And the precise focus of irradiation laser beam in ablation for designated area on porcine brain was achieved.

We are planning to evaluate the response of auto-focus function against laser scanning speed in order to apply for clinical study. Since it is considered that the countermeasure for unexpected bleeding in ablation process is most important, we are going to study surgical protocol to prevent such bleeding.

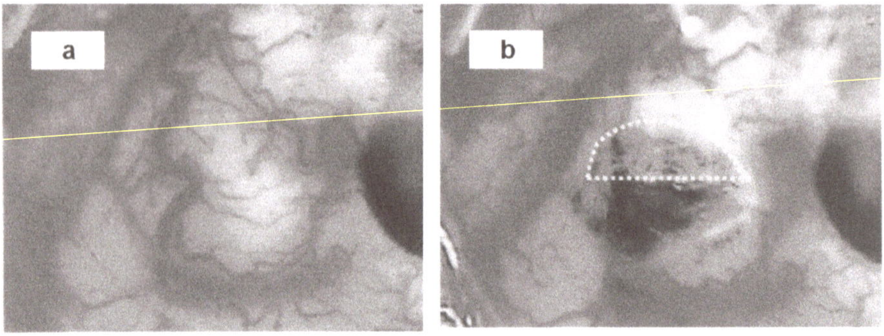

Fig. 7 a, b. Photographs of the porcine brain surface displayed on the monitor screen. **a** Before irradiation. **b** Laser ablation was performed inside the dashed line. Ablation depth is about 0.5mm.

Acknowledgements

This work was supported by Grants-in-Aid for Scientific Research from Ministry of Health, Labor and Welfare. And this study was also supported by Industrial Technology Research Grant Program in (00A45003a) from the New Energy and Industrial Technology Development Organization of Japan.

References

1. Satish K, Stephen K. P (1994) Lasers in Neurosurgery. Lasers in Surgery and Medicine 15:126
2. Iseki H. et al (2002) Intraoperative Examinations for Tumors required in the Neurosurgical Operating Theater of the 21st Century (in Japanese). Jpn J Neurosurgery 11 (8):508
3. Muragaki Y. et al (2001) Intraoperative brain mapping and intraoperative MRI for Glioma surgery. BRAIN MEDICAL 13 (3):255
4. Iseki H. et al (2001) New Possibilities for Stereotaxis Information-Guided Stereotaxis. Stereotactic and Functional Neurosurgery 76:159
5. Omori S. Muragaki Y, Iseki H, et al (2004) Robotic Laser Surgery with 2.8μm Microlaser in Neurosurgery. J Robotic and Mechatronics 16 (2):122

6. Andras C. et al (1997) Intracranial Pressure Waves Generated by High-Energy Short Laser Pulses Can Cause Morphological Damage in Remote Areas. Lasers in Surgery and Medicine 21:444
7. H.C. Ludwig et al (1998) Optimized Evaluation of a Pulsed 2.09μm Holmium: YAG Laser Impact on the Rat Brain and 3 D-Histomorphometry of the Collateral Damage. Minimally Invasive Neurosurgery 41:217
8. Janice O. et al (2003) Brain Ablation in the Rat Cerebral Cortex Using a Tunable-Free Electron Laser. Lasers in Surgery and Medicine 33:81
9. Bayly J.G., Kartha V.B., Stevence W.H. (1963) The absorption spectra of liquid phase H_2O , HDO and D_2O from 0.7μm to 10μm. Infrared Physics 3:211

Development of a Compact Automatic Focusing System for a Neurosurgical Laser Instrument

Masafumi Noguchi[1], Eisuke Aoki,[1], Etsuko Kobayashi[1], Shigeru Omori[2/3], Yoshihiro Muragaki[3], Hiroshi Iseki[3], and Ichiro Sakuma[1]

[1]Graduate School of Frontier Sciences, The University of Tokyo
[2]Terumo Corporation
[3]Faculty of Advanced Techno-surgery, Institute of Biomedical Engineering and Science, Graduate School of Medicine, Tokyo Women's Medical University

Chapter Overview. In neurosurgery such as the treatment of glioma, it is important to remove the tumor precisely and accurately, which can be achieved with a micro laser with a wavelength of 2.8 μm. It is necessary, however, to maintain a constant distance from the brain surface. We developed an automatic focusing system for the micro laser ablation system. This system is using a triangulation method with a guide laser and a small CCD camera. The guide laser spot was attenuated and penetrates the biomedical tissue according to its scattering and absorption features. Therefore, each measurement parameter such as the luminance threshold in the image processing was necessary to be appropriately controlled. In this research, we proposed the focusing method to appropriately control the luminance threshold. We confirmed that this method works effectively on biomedical tissue. In the future, we will carry out a combination test with the micro laser system and achieve a precise operation system for brain tumors.

Key Words. Neurosurgery, laser, focusing, threshold, and image processing.

1. Introduction

In current neurosurgical practice, surgeons can remove most of a tumor with an accuracy of a few millimeters by using a combination of conventional surgical instruments, such as an electric cautery, and a computer-aided navigation system. Nevertheless, residual tumor may induce recurrence and it is necessary to remove as much of the tumor as possible while keeping the normal tissue intact. However, it is difficult to know the exact boundary between tumor and normal tissue, and excessive ablation of the normal brain tissue will damage its function.

For treatment of residual tumor, or tumors that cannot be treated by surgical intervention, pharmacotherapy or radiotherapy is applied. These methods are effective; however, there are problems such as side effects.

To solve those problems, we have proposed a novel approach to therapy using 5-aminolevulinic acid (5-ALA) and a micro-laser ablation system, with the boundary between the tumor and the normal tissue distinguished by the 5-ALA fluorescence in the tumor [1, 2] and with accurate ablation of the tumor with the micro laser [3, 4]. The wavelength of the micro laser is 2.8 μm. Light with this wavelength is mostly absorbed by water, and therefore this laser is effective only on the surface of brain tissue, enabling precise ablation at the boundary between tumor and normal tissue. However, the focal depth of this laser is 1 mm, so the ablation depends on displacement from the focal point. Therefore, a robotic automatic focusing system is necessary for the micro-laser ablation system. By the combination of the robotic positioning system and precise laser ablation, more accurate and precise operation is realized. We developed an automatic focusing system for the micro laser ablation system, and the basic evaluations of its accuracy and following error were performed [5]. In this research, we have developed a new technique for controlling the threshold, which is appropriately designed for biomedical tissues.

2. Method

2.1 Desired specifications

We set the following requirements for the automatic focusing system.

1. Measurement without contact
2. Measurement under strong lighting
3. A compact mechanism that does not interfere with the surgical field
4. Intraoperative and immediate measurement

5. Following error within 0.5 , which is half of the focal depth of the micro laser, mm at inclines of up to 45 degrees
6. A velocity of the micro laser in operation of 2–4 mm/sec

2.2 System configuration and measurement method

Fig. 1 shows the system configuration of our automatic focusing system. In the system, position measurement was performed using a triangulation method, with a guide laser and a small CCD camera for measurement. The advantage of this method is that as light for treatment, with a wavelength of 2.8 μm, is not visible, the surgeon can identify the treated point by looking at the guide laser. In addition, as the mechanism is simple, it was easy to realize a compact device.

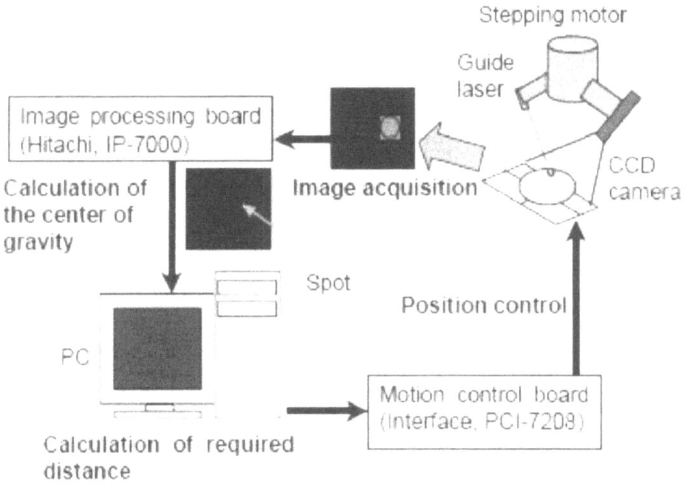

Fig. 1. Overview of the automatic focusing system. First, the system acquires an image using the CCD camera. The guide laser spot is extracted and the image processor calculates the coordinate of its center of gravity. The distance of the focal point from the target brain surface is calculated using a triangulation method. This error is sent to the motion controller and the stepping motor corrects the position.

The guide laser and the CCD camera were fixed so that the axes of each component crossed at the focal point of the ablation laser. If the focal point passes along the target surface, the spot of the guide laser is observed at the center of the screen of the CCD camera. The distance between the surface of the target and the focal point was calculated based on the displacement of the guide laser spot on the CCD image. Focusing was performed by continuously monitoring the displacement of the spot.

Fig. 2 shows a prototype of this system. The length of the device was 225 mm. The weight without the ablation laser module was 0.6 kg. A laser diode with a wavelength of 532 nm was used for the guide laser. Although a red laser diode is more practical, penetration into living tissue is stronger than with the green laser, and as this penetration causes positioning error, as described in the next chapter, we selected the green laser diode.

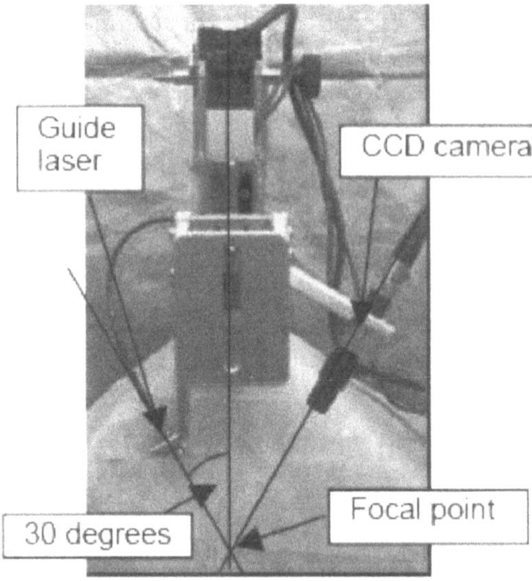

Fig. 2. Prototype of this system. The guide laser with wavelength of 532 nm and the CCD camera for measurement is set at the angle of 30 degrees

The operating frequency of the system was determined by the image acquisition cycle of 16.7 ms, the computing time for the image processing, and the positioning time of the stepping motor. The time for image processing was about 16 ms or more, and the time for positioning was often less than 50 ms. the drive of the motor was processed simultaneously with the image acquisition and processing. Therefore, the operating period of this system was approximately 50 ms.

The setting angle of the CCD camera and the guide laser was 30 degrees from the driving direction of the system. The resolution of the system, based on the resolution of the CCD camera, was 0.024 mm.

The electric shutter speed was normally set to 1/4000 s. The purpose of increasing the shutter speed is to cut noise from lighting and diffuse reflections. The image obtained was sent to the image processor, binarized by luminance level (eight bits), and spot extraction performed.

3. Image Processing Technique

The light radiated onto biomedical tissues is attenuated and penetrates the tissue according to its scattering and absorption features [6]. These optical properties vary with the type of target, for example brain cortex, white matter, or blood. We examined the relationship between target tissues and spot extraction. From this pilot study, we developed an appropriate spot extraction method for biomedical tissues.

3.1 Spot extraction issues

Blood has strong absorption feature around the wavelength of 532 nm [6]. Extraction of the spot was difficult in blood compared with brain parenchyma. To extract the spot successfully, it was necessary to lower the threshold in blood. On the other hand, if the threshold was too low, noise occurred, and extraction of the spot became impossible.

The coordinates of the spot are determined by measuring its center of gravity. These coordinates contain error caused by penetration of the biomedical tissue. This is because the camera catches the light from the inside of the tissue. If the area of the extracted spot was large, the error increased.

3.2 Threshold control

If a luminance threshold is fixed to a high level, for a target such as blood it was impossible to extract the spot in some conditions. However, if the luminance threshold is low, the area of the spot becomes large, and then the error becomes larger. It is desirable to control the luminance threshold to an appropriate value, so that the area of an extracted spot is constantly small. We developed a threshold controlling method based on the relationship between the luminance threshold and the area of an extracted spot.

We used a single mode fiber for the green guide laser with a wavelength of 532 nm. Its light intensity distribution depends on the normal distribution. If a target has only isotropic back scattering and no laser penetration efficiency, the relationship between light intensity and the distance from the center of a laser spot is showed in Eq. 3.1.

$$I \approx \exp\left(-d^2/2\sigma^2\right)/\sqrt{2\pi}\,\sigma \tag{3.1}$$

Where I is light intensity, d is distance from the center of a laser spot, and σ is standard deviation of the distribution.

Although the laser penetrates into brain tissues, the relationship between light intensity and the distance from the center of a laser spot is similar to Eq. 3.1. The relationship between the luminance threshold and the area of an extracted spot was derived from Eq. 3.1. The luminance threshold corresponds to light intensity and the area of an extracted spot corresponds to the square of the distance from the center of a laser spot (see Eq. 3.2).

$$Y = a\exp(-bA) + c \tag{3.2}$$

Where Y is luminance threshold, A is area of spot, and a, b, c are constants which are dependent on the target conditions.

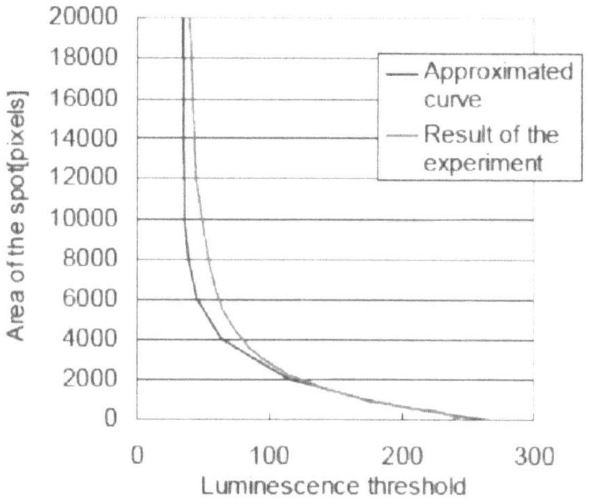

Fig. 3. Relation between the luminance threshold and the area of the extracted spot and an approximated curve in white matter

Fig. 3 shows the relationship between the luminance threshold and the area of an extracted spot, and a curve approximated from Eq. 3.2 for white matter. The values of a, b, c differ according to the target tissue conditions. These numbers were obtained with the luminance threshold Y, the area of an extracted spot A, and a histogram of luminance levels as follows:

1. A histogram of luminance levels was taken for a color image before binarization.
2. c is the asymptotic value of the luminance threshold. We determined it as the value 40% from the top of the histogram, where the area of the spot exponentially increased.
3. At the peak value of the histogram, Y_{max}, the area of the spot $A = 0$. a was determined as follows:

$$Y_{\max} - c = a\exp(-b \times 0) = a \tag{3.3}$$

4. Then b was determined from Y and A (Eq. 3.4).

$$b = -\ln((Y - c)/a)/A \tag{3.4}$$

When focusing, the approximated function was calculated using the above method for every process. We could assume the appropriate value of the luminance threshold used in the next process by this approximated function. The method of controlling the luminance threshold was as follows. Before focusing, A_0 was set as the desired value of the area of the spot: sufficiently small to reduce the error due to penetration.

1. Focusing started.
2. The area of a spot, A_n, was obtained with the luminance threshold Y_n.
3. The approximate function for the relationship between Y_n and A_n was calculated (Eq. 3.5).

$$Y_n = a_n \exp(-b_n A_n) + c_n \tag{3.5}$$

4. Y_{n+1} was computed (Eq. 3.6), and used for the next process.

$$Y_{n+1} = a_n \exp(-b_n A_0) + c_n \tag{3.6}$$

5. Return to 2).

As the device is moving at a speed of 2–4 mm/sec and the target tissue is being ablated while focusing, the condition of the target tissue is changing with time. Therefore, the assumed value of the threshold based on the current process is not perfectly adapted to the next process. However, the operating period of this system is 50 ms and the distance moved during this period is 0.1–0.2 mm, and thus the conditions of the target may not differ much in each consecutive process. Therefore, this method may be sufficient to allow stable extraction of the laser spot. In the following chapter, we compared this method with a method using a fixed threshold.

Acknowledgment

We wish to thank many people and groups who have contributed to this effort. One of them deserving special mention is Yasuo Oda of chief engineer of Biomedical Precision Engineering Laboratory, the University of Tokyo. The research funds were based in part on Grant-in-Aid for Scientific Research of JSPS (#15700349) and Terumo Lifescience Foundation.

References

1. T Maruyama, et al (2001) Intraoperative detection of malignant gliomas using 5-Aminolevulinic acid induced protoporphyrin fluorescence, openMRI and real-time navigation system. Computer Assisted Radiology and Surgery 15:270-275
2. K Shimizu, et al (2003) Application of blue semiconductor laser to measurement of 5-ALA induced fluorescence for intraoperative detection of brain tumor. Proceedings of 6th Japan-France Congress on Mechatronics and 4th Asia-Europe Congress on Mechatronics, pp 135-140
3. S Omori, Y Muragaki, I Sakuma, H Iseki (2004) Robotic laser surgery with $\lambda = 2.8$um microlaser in neurosurgery. Journal of Robotics and Mechatronics 16:122-128
4. S Omori, R Nakamura, Y Muragaki, H Iseki, K Takakura (2004) Improvement study of computer controlled mid-infrared laser system for neuro-surgery. Proceedings of 13th conference of Japan Society of Computer Aided Surgery, pp 37-38
5. M Noguchi, et. al (2004) Development of a Compact Automatic-Focusing System for a Neurosurgical Laser Instrument. Journal of Japan Society of Computer Aided Surgery 6:483-489
6. Sterenborg HJ, et al (1989) The spectral dependence of the optical properties of the human brain. Lasers Med Sci 4:221-227
7. J van Staveren, J M. Moes, J van Marle, A Prahl, J C van Germert (1991) Light scattering in Intralipid-10% in the wavelength range of 400-1100 nm. Appl Opt 30:4507-4514

Non-invasive Monitoring of Arterial Wall Impedance

Akira Sakane[1], Kenji Shiba[1], Toshio Tsuji[1], Noboru Saeki[2], and Masashi Kawamoto[2]

1 Department of Artificial Complex Systems Engineering, Hiroshima University, Hiroshima, Japan
2 Department of Anesthesiology and Critical Care, Hiroshima University, Hiroshima, Japan

1 . Introduction

The vascular system transports various substances necessary to maintain life, such as oxygen and nutrients, through arteries which exhibit funtional changes of constriction and dilation in response to various kinds of stimuli. In addition, the arteries react to organic changes, such as hardening and softening, according to aging and different physical conditions. Therefore, when dynamic characteristics of arteries can be measured quantitatively without unnatural stimulation, it is possible to estimate the internal condition of the body not only during surgical procedures, but also activities of daily living, such as physical training.

Many researchers have tried to reveal the dynamic characteristics of arterioles. For example, Sawada *et al.* analyzed the vascular tone and the effects of mental stress by estimating the vascular stiffness of the arterioles [1]. Also, Saeki *et al.* estimated the compliance of arterial wall with a plethysmogram and arterial pressure measurements, and other similar studies have followed [2],[3]. These studies, however, dealt only with either stiffness or compliance, and dynamic vascular characteristics were not sufficiently analyzed.

The arterial wall is composed of smooth muscles, in which dynamic characteristics have been analyzed in detail. Barra *et al.* measured time series of blood pressure and diameter of the brachial artery in conscious dogs, and calculated the stress-strain relations during vasoconstriction and vasodilation using a modified Maxwell model [4]. They described the similar changes in stiffness and viscosity of smooth muscles according to the degree of vascular contraction [5]. Meanwhile, the authors illustrated the dynamic characteristics of the human arterial wall by using mechanical impedance including stiffness, viscosity, and inertia, and developed a new technique to estimate the vascular impedance parameters in a *beat-to-beat* manner [6]. The method, however, was an invasive procedure because it was necessary to insert a catheter to measure the arterial pressure.

In this paper, for non-invasive estimation of the vascular dynamic characteristics, we propose a new method to estimate the vascular conditions *beat-to-beat* by measuring the arterial pressure non-invasively. In the proposed method, the impedance parameters can be assessed from electrocardiograms, arterial pressure, and plethysmograms with the linear least-squares method. Finally, we monitor vascular conditions using arterial wall impedance during surgical operations. The relevance of the proposed method is confirmed by comparison with the estimated results of the invasive method.

KeyWord Arterial Wall Impedance biosensor

2 . Impedance Model of the Arterial Wall

2.1 Arterial wall impedance

For extraction of vascular features including biological signals, we estimated the arterial wall impedance [6]. Fig.1 illustrates the proposed impedance model of the arterial wall. This model represents only the characteristics of the arterial wall in the arbitrary radial direction. The impedance characteristic can be described using stress and strain of the arterial wall as follows:

$$d\sigma(t) = Md\,\varepsilon(t) + Bd\,\varepsilon(t) + Kd\varepsilon(t) \tag{1}$$

where $d\sigma(t) = \sigma(t) - \sigma(t_0); d\varepsilon(t) = \varepsilon(t) - \varepsilon(t_0); \sigma(t)$ is the stress exerted on the arterial wall by arterial pressure; M . B , and K are the inertia, viscosity, and stiffness, respectively; $\varepsilon(t), \varepsilon(t)$, and $\varepsilon(t)$ are the strain, strain velocity, and strain acceleration of the arterial wall. t_0 denotes the time when the R wave appeared in each recognized ECG cycle.

In order to estimate the impedance parameters given in (1), it is necessary to measure $\sigma(t)$ and $\varepsilon(t)$.The stress exerted on the tunica intima of the arterial wall is equal to arterial pressure. Then, the stress can be given as follows:

$$\sigma(t) = P_b(t) \tag{2}$$

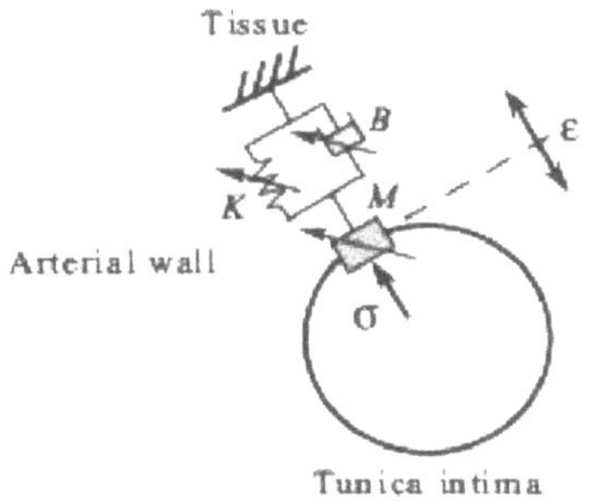

Fig. 1. Schematic description of the arterial wall impedance model

where $P_b(t)$ is the invasive arterial pressure. On the other hand, the strain of the arterial wall $\varepsilon(t)$ is difficult to measure directly and noninvasively. Therefore, a plethysmogram $P_l(t)$ is used instead of $\varepsilon(t)$ as follows:

$$\varepsilon(t) = \frac{P_l(t)}{A_0} \tag{3}$$

where A_0 is the mean value of absorbance [7].

2.2 Arterial wall impedance model

The stress exerted on the arterial wall, $\sigma(t)$, is expressed by the invasive arterial pressure, $P_b(t)$, given by (2), and the strain of the arterial wall, $\varepsilon(t)$, is represented by the plethysmogram, $P_l(t)$, in (3). Then, the arterial wall impedance is estimated by using $P_b(t)$ and $P_l(t)$ as follows:

$$dP_b(t) = \tilde{M}\, d\ddot{P}_l(t) + \tilde{B}\, d\dot{P}_l(t) + \tilde{K}\, dP_l(t) \tag{4}$$

where $dP_b(t) = P_b(t) - P_b(t_0); dP_l(t) = P_l(t) - P_l(t_0); d\dot{P}_l(t) = \dot{P}_l(t) - \dot{P}_l(t_0); d\ddot{P}_l(t) = \ddot{P}_l(t) - \ddot{P}_l(t_0)$. The impedance parameters included in

(4) are then given by:

$$\tilde{M} = \frac{M}{A_0},\ \tilde{B} = \frac{B}{A_0},\ \tilde{K} = \frac{K}{A_0} \tag{5}$$

where the parameter $\tilde{M}$ corresponds to the mass of the arterial wall existing in the measured part; $\tilde{B}$ and $\tilde{K}$ denote the viscoelastic properties, respectively. Noting that the inertia, M, is sufficiently small in the transversal arteries and can be ignored [8], the vascular dynamic characteristic at time t can be given as follows:

$$dP_b(t) = \tilde{B}\, dP_l(t) + \tilde{K}\, dP_l(t) \tag{6}$$

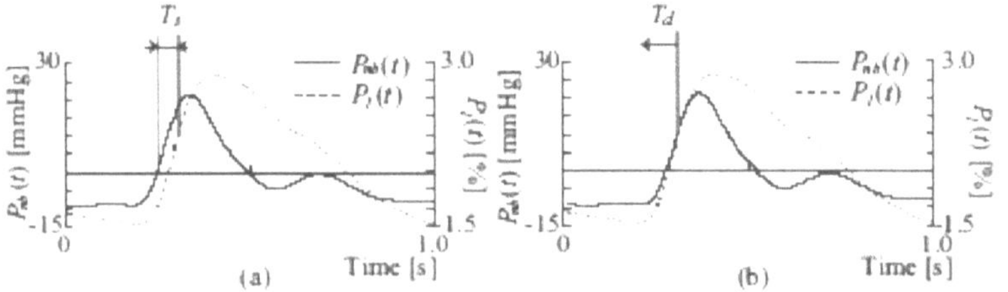

Fig. 2. The definition of shift time

To estimate arterial wall impedance using the previously proposed method of [6], it was necessary to insert a catheter to measure the arterial pressure. In this paper, the non-invasive arterial pressure measured by Finapres (Finapres 2300, Ohmeda Corp.) is utilized to evaluate the dynamic characteristics of the arterial wall. However, the arterial pressure measured non-invasively is delayed compared with the invasive method, and the delay time affects the estimated impedance parameters. Therefore, the delay time must be compensated to estimate the impedance parameters with a high degree of accuracy.

2.3 Compensation of the time delay

The method proposed in this paper is explained using Fig.2. The solid lines and dotted lines respectively indicate the measured non-invasive arterial pressure and plethysmogram. At first, the time difference of the inflection points between the measured non-invasive arterial pressure and plethysmogram is calculated, where the inflection points are determined from the leading curves to the peaks of the waves. The difference is defined as

T_s (Fig.2(a)) and is used to modify the time difference between the invasive arterial pressure and plethysmogram by leading the plethysmogram T_s (Fig.2(b)). Herewith, the system can compensate the variable time delay *beat-to-beat*.

Next, in order to compensate the constant delay included in the measurements, a shift time, T_d, is defined and is used to lead the noninvasive arterial pressure (Fig.2(b)). Consequently, the impedance parameters are estimated by using the shifted non-invasive arterial pressure and plethysmogram. The equation for estimating impedance parameters can be expressed as follows:

$$dP_{nb}(t+T_d) = \tilde{B}\, dP_l(t+T_s) + \tilde{K}\, dP_l(t+T_s) \tag{7}$$

where $dP_{nb}(t+T_d) = P_{nb}(t+T_d) - P_{nb}(t_0+T_d); dP_l(t+T_s) = P_l(t+T_s) - P_l(t_0+T_s)$; and $P_{nb}(t)$ is the non-invasive arterial pressure. The optimum shift time is determined by using the following cost function.

$$E_a(T_d) = \frac{1}{m}\sum_{p=1}^{m} E(T_d, p) \tag{8}$$

$$E(T_d, p) = \frac{\tilde{E}_B(T_d, p)}{\tilde{E}_{B\max}(p)} + \frac{\tilde{E}_K(T_d, p)}{\tilde{E}_{K\max}(p)} \tag{9}$$

$$\tilde{E}_B(T_d, p) = \frac{1}{n}\sum_{t=1}^{n}(\tilde{B}_i(t, T_d, p) - \tilde{B}_{ni}(t, T_d, p))^2 \tag{10}$$

$$\tilde{E}_K(T_d, p) = \frac{1}{n}\sum_{t=1}^{n}(\tilde{K}_i(t, T_d, p) - \tilde{K}_{ni}(t, T_d, p))^2 \tag{11}$$

where $\tilde{B}_i$ and $\tilde{K}_i$ are the impedance parameters estimated by using the invasive arterial pressure; $\tilde{B}_{ni}$ and $\tilde{K}_{ni}$ denote the estimated results of the proposed method; $\tilde{E}_B$ and $\tilde{E}_K$ are the mean square error of viscosity and stiffness; $\tilde{E}_{B\max}$ and $\tilde{E}_{K\max}$ signify the maximum of $\tilde{E}_B$ and $\tilde{E}_K$ when the shift time T_d is changed from 0 to 72ms; m is the number of subjects; and n is the number of estimated impedance parameters. When

Medical Physics and Biomedical Engineering, 531 (2003)
6. A.Sakane, T.Tsuji, Y.Tanaka, N.Saeki, and M.Kawamoto. Monitoring of Vascular Conditions Using Plethysmogram. The Society of Instrument and Control Engineers, 40, 12, 1236-1242 (2004)(in Japanease)
7. A.Sakane, K.Shiba, T.Tsuji, N.Saeki, and M.Kawamoto. Non-invasive monitoring of arterial wall impedance. in Proceedings of the First International Conference on Complex Medical Engineering, Takamatsu, 984-989 (2005)
8. W.W.Nichols and M.F.O'Rourke.: McDonald's Blood Flow in Arteries: Theoretical, experimental and clinical principles. Edward Arnold (1998)
9. Y.Koseki, T.Washio, K.Cchinzei, and H.Iseki.: Endoscope Manipulator for Trans-nasal Neurosurgery, Optimized for and Compatible to Vertical Field Open MRI. in Proc. of MICCAI 2002, Part I, 114-121 (2002)

Part 2
Complex Virtual Technology in Medicine

Advanced Volume Visualization for Interactive Extraction and Physics-based Modeling of Volume Data

Megumi Nakao[1], Tomohiro Kuroda[2], and Kotaro Minato[1]

[1]Graduate School of Information Science, Nara Institute of Science and Technology, Nara, Japan
[2]Department of Medical Informatics, Kyoto University Hospital, Kyoto, Japan

Chapter Overview. This paper proposes an advanced volume representation of 3D geometries and presents its application for volume visualization of medical images. 3D geometry is represented using cross sections of grids and visualized volumetrically using a slice-based volume rendering method. The developed framework achieves real-time visualization on general purpose PCs and graphic hardware. As practical application of the framework, on-demand extraction on 3D reconstructed images and volumetric representation in physics-based simulation like deformation and cutting are introduced. Performance and effectiveness of the proposed methods are confirmed in preliminary use.

Key Words. Volume representation, 3D geometry, Interactive extraction, physics-based modeling, Medical imaging

1. Introduction

Visualization software and systems with computer graphics or image processing technologies are widely utilized in medical fields. In case of computer aided diagnosis [1] and surgical planning [2], for example, medical doctors and researchers eager to visualize 3D region of interest flexibly and interactively on medical images. Because 2D images measured from CT or MRI are volumetric data composed by huge voxels, volumetric representation techniques [3, 4] are essential for enabling users to grasp complex 3D structures interactively and intuitively.

2D slice images or reconstructed volumetric images are commonly used in current preoperative planning or intraoperative navigation. Clipping representation cut by flat planes [5] are useful to observe internal structure of human bodies. However, to observe and to extract 3D region of interest on demand, current visualization environments are not enough sufficient. For example, surface and texture of inner organs on reconstructed images are difficult to be visualized on demand. Also, 3D shape is hard to be recognized through translucent volume rendering images. This situation consequently limits computer assisted diagnosis and geometrical modeling using volume data. Although other volumetric representation methods [6, 7] were proposed in the field of computer graphics, effective environment where users can define and change 3D geometry interactively on voxel data is not established.

Polygonal representation is a popular approach to define 3D geometry and mainly used for interactive systems like a VR-based surgery simulator [8]. Although surface rendering expresses shape of 3D geometry effectively, use of polygons has lots of problems in medical fields. Normally exhausting segmentation tasks are needed to define complex 3D shape of organs. More essential issue is large part of dataset including surface and internal textures are lost through surface reconstruction from CT/MRI images. Because any modification or loss of data is not allowed before diagnosis or preoperative planning, measured volume dataset should be directly visualized to physicians.

Considering these requirements and issues, volumetric representation of 3D geometries has possibility of improving current medical software used for computer aided diagnosis, intraoperative support and surgical simulation. However, there are few reports that aim to visualize 3D geometries volumetrically. Although some studies developed original geometry models and performed volume deformation using volume texture mapping. [9, 10], their models are not compatible with general geometrical representation (e.g. tetrahedral grid). To perform both flexibility of controlling geometry and adaptation to conventional physics-based modeling, compatibility with popular grid format should be taken into account.

This study aims to establish an advanced volume representation method that enables to visualize 3D geometries volumetrically. The proposed method supports on-demand extraction on 3D reconstructed image and enhances real-time graphic representation in physics-based simulation like deformation and cutting. In addition, real-time and fast volume visualization is achieved on general purpose PC and graphic hardware. This paper presents details of the methods and introduces some practical application in preoperative planning and surgical simulation.

2. Volumetric Representation of 3D Geometries on Medical Images

This section explains overview of our concept by introducing some application which utilizes both 3D geometry and medical images. Firstly, 3D geometry on voxel data can be used for setting 3D region of interest in preoperative planning or computer assisted diagnosis. On-demand extraction of 3D region on reconstructed volumetric image is performed.

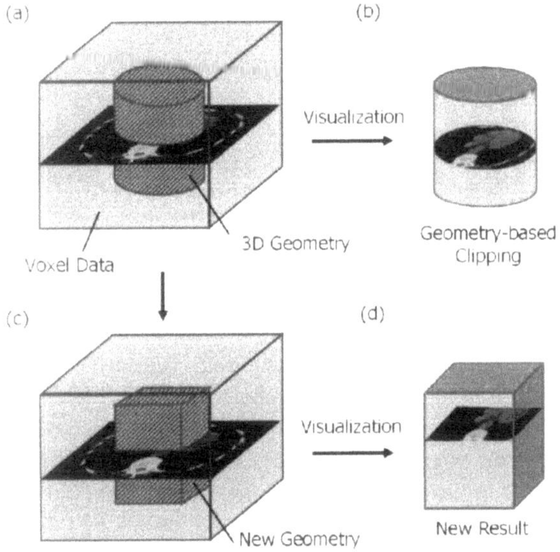

Fig. 1. Applying 3D geometry to on-demand extraction of medical images: (a) Cylinder-shaped geometry defined on voxel data, (b) Volume visualization inside the geometry performs geometry-based volume clipping. (c) The geometry is changed and (d) Next clipped result is interactively visualized.

Fig. 1 shows outline of on-demand extraction. Suppose voxel dataset of a human body measured from CT are prepared. In preoperative planning, inner structures like distribution of blood vessels and spatial relationship between some organs should be checked sometimes using reconstructed 3D images. However, blood vessels and inner organs like a heart cannot be observed unless users eliminate surrounded voxels of fat, skin, bone and other obstacle tissues. At this time, we prepare a 3D region mask. The mask is placed on voxel data (Fig. 1 (a)) and inside of the mask is visualized volumetrically in real time (Fig. 1 (b)). This task shows a basic idea on on-demand extraction by geometry-based volume clipping. If users can modify the 3D region mask by changing its 3D shape, size and position of the geometry, new clipped results are interactively displayed (Fig 1 (c), (d)). Such interactive clipping or modifications of 3D region of interest before or after algorithm-based segmentation, more flexible 3D extraction environment will be established.

Another use of volumetric representation of 3D geometries is for visualization of simulation results. Some general formats of volumetric geometries (e.g. tetrahedral grid) can be directly applied to conventional geometrical or physics-based models like finite element method (FEM). As shown in Fig. 2, we suppose users manipulate the geometry and geometrical transformation like deformation is simulated. Although currently popular visualization methods are based on polygonal rendering, the proposed framework visualizes simulation results volumetrically with deformation of surface texture and internal structures.

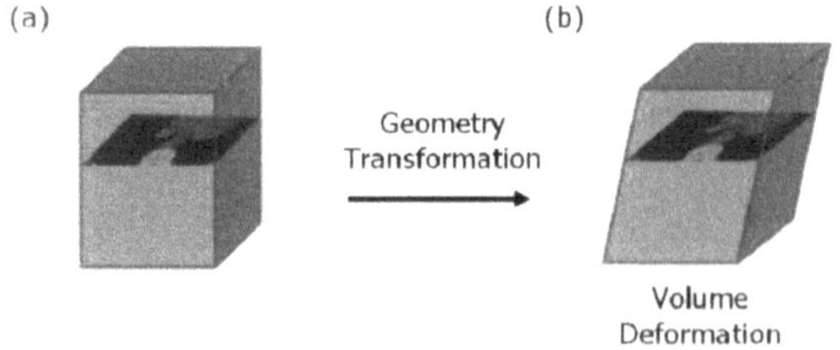

Fig. 2. Geometry transformation with the proposed framework enables volume visualization of simulation result. For example, volume deformation is performed using physics-based model like finite element method (FEM).

For integrating such geometrical representation into volume graphics, we use both 3D geometrical data and voxel data (i. e. medical images measured from CT or MRI). Because the geometry format must support volumetric representation, tetrahedral grid is employed for defining 3D geometry. Tetrahedral grid is suitable for 3D shape modeling and compatible with several grid control models as described above.

In order to realize this concept, fast volume visualization of 3D geometries is indispensable. As 3D geometry is composed by tetrahedral grids, volumetric representation of entire grid is required. We try to solve this issue using slice-based representation of grids and slice-based volume rendering approach. The details of representation method and volume visualization technique are described in the next section.

3. Fast Volume Visualization of Grids

3.1 Volumetric representation of a single grid

To explain the volume visualization technique of entire grids, volumetric representation of a single grid is first described. This paper deals with tetrahedral grid for an example of 3D geometries. A basic idea is that each element is represented using not polygonal surface surface but cross sections S1, S2 ... of the element (see Fig. 3) 3D shape is precisely reconstructed only if the distance between cross sections is sufficiently small. Physical distance is not recognized on the condition that the eye direction is vertical to the cross sections.

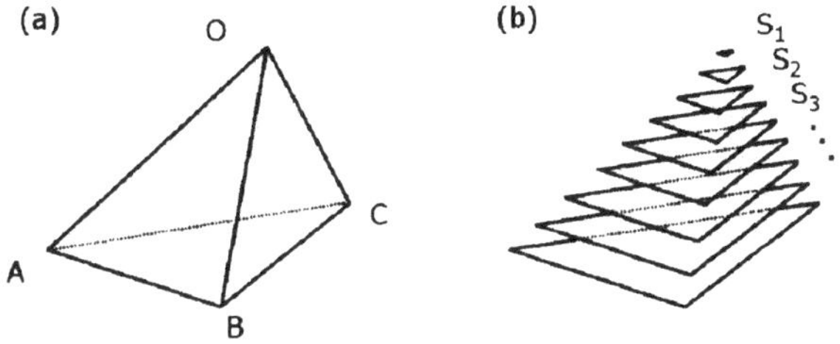

Fig. 3. Volumetric representation of a single grid element: (a) Polygonal representation and (b) Volumetric representation using cross sections.

Color information of the inside of the grid is visualized using a 3D texture mapping technique which is supported on standard graphic hardware. We suppose registration between grids and voxels is performed on world coordinate system in a virtual space. If current 3D positions of the vertices of cross sections are used as texture coordinates for referencing voxel data, volume clipping introduced in Fig. 1 is provided. In case of the situation when the grid is transformed and previous positions of vertices are used as texture coordinates, volume deformation illustrated in Fig. 2 is represented.

3.2 Slice-based volume rendering of entire grids

For applying this volume representation method to the entire 3D geometry which is composed by multiple grids, base planes are defined. (see Fig. 4) The base planes are placed where the grids are completely covered. The distance between neighboring base planes is constant D, and the normal vector parallels the eye vector in the virtual world.

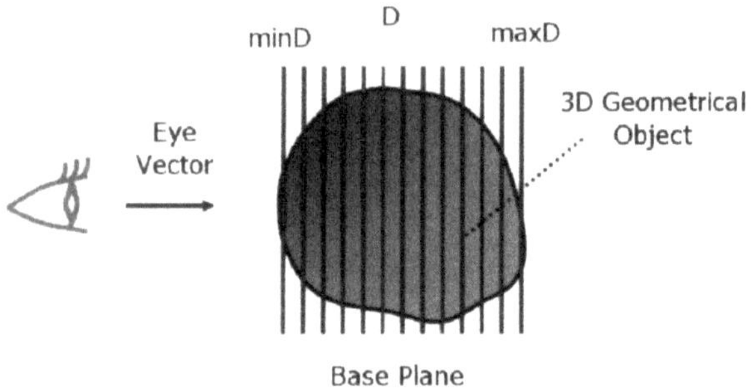

Fig. 4. Volume slicing of entire grids. The distance between neighboring base planes is constant D, and the normal vector parallels the eye vector in the virtual world.

Intersection between the base planes and the grids defines cross sections of all grid elements. Two patterns of cross sections (called as base polygons in this paper) are detected as illustrated in Fig. 5. The base polygons are registered in back to front order, and rendered using alpha blending while referring texture coordinates for all vertices on base polygons. Because base polygons of entire grids are created at each visualization step, slicing algorithms should be processed enough fast. Using distance-based arrays, all vertices detected by slicing algorithms are directly registered without sorting process.

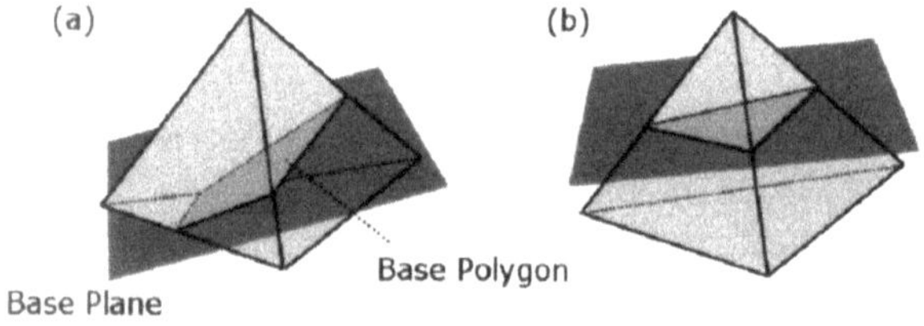

Fig. 5. Two patterns of intersection between a base plane and an element. Base polygons rendered with texture translucently in back to front order.

The overall approach is based on volume slicing and slice-based volume rendering technique [4]. So far, volume rendering of tetrahedral or hexahedral grids employ voxelization [11] or cell projection approach [12]. However, because more than 10 million voxels are to be updated in some case of 256^3 voxel data, fast visualization is difficult on current standard PCs. Since this approach renders base polygons of all base planes on normal graphic board, reconstruction of voxels is not required and fast visualization is performed. Also, definition of cross sections for each tetrahedral element is independently processed, which is suitable for parallel computing

4. Results and Application

Overall methods were implemented into a standard PC (CPU: Pentium4 2.4GHz, Memory: 1024MB, GPU: nVidia Quadro4 900XGL) and a prototype system supporting on-demand extraction and volume deformation was developed. As interface for inputting 3D coordinates in manipulating 3D geometry, the system prepares both window-based GUI with edit box and a PHANToM haptic device (SensAble Technologies Inc.).

Measured CT data (256^3 voxel) was applied to the developed system and several 3D extraction tasks were tried. Fig. 6 (a) shows volume rendering result of entire voxel dataset. For this data, volume clipping was performed using sphere-shaped tetrahedral grid illustrated in Fig. 6 (b). The extracted result is demonstrated in Fig. 6 (c). In this test, the number of the region mask is 216 and distance parameter between based planes is set by 2. Overall calculation time from creation of cross sections to visualization of clipped results was 23 msec. Thus, preparing grids whose shapes are close to region of interest effectively limits visible region and supports flexible observation of large volume data. The complexity of region mask is not needed only if visible region is sufficiently formed because level of details of the grid does not affect visualization results.

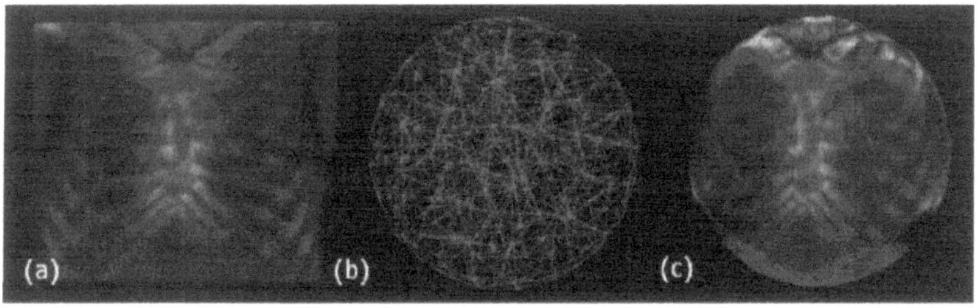

Fig. 6. Geometry-based volume clipping using a sphere-shaped grid (a) Full volume, (b) Tetrahedral grid of representing 3D sphere (c) extracted volume

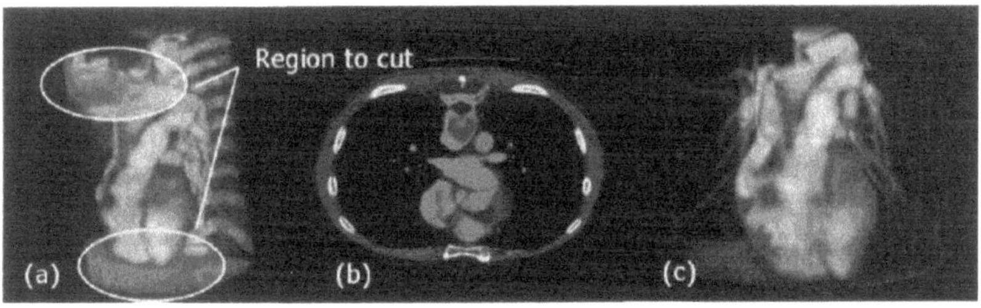

Fig. 7. On-demand modification of visible area by scaling and deforming cylinder-shaped boundary: (a) Extracted volume by cylinder-shaped boundary, (b) Selected region on 2D image and (c) Modified 3D visible region

Next, the authors tried to extract specific 3D region which includes only heart and blood vessels in lung area. As shown in Fig. 7 (a), cylinder shaped grid was placed on the 3D virtual body. The focus region is also confirmed through 2D sectional image. (see Fig. 7 (b)) The coloured area represents selected region. The large part of lung vessels is outside of the visualized area, and part of breast surface and other soft tissues should be eliminated. Changing scale of the mask contributed to covering entire lung region. Then, interactive deformation based on finite element based method eliminated surrounded soft tissue and enabled to fit visible region (Fig. 7 (c)). 84 sec was required throughout this extraction task. The trial confirms that volume clipping with deformable 3D grid efficiently supports interactive region extraction.

Fig. 8 shows another use of the methods for visualizing internal structure on visible human dataset [13]. If the shape of the grid is same as external 3D contour of visible human head, reducing size of the grid efficiently eliminates external frame (e. g. skin and bone) from the whole volume. Note that the external frame of human body structures is easily extracted using conventional region growing methods. Fig 8 (b-d) demonstrates extracted results by reducing the size of the boundary grid which has the same shape of visible human's head. Details of brain surface and texture are visualized interactively without any image filtering or segmentation algorithms.

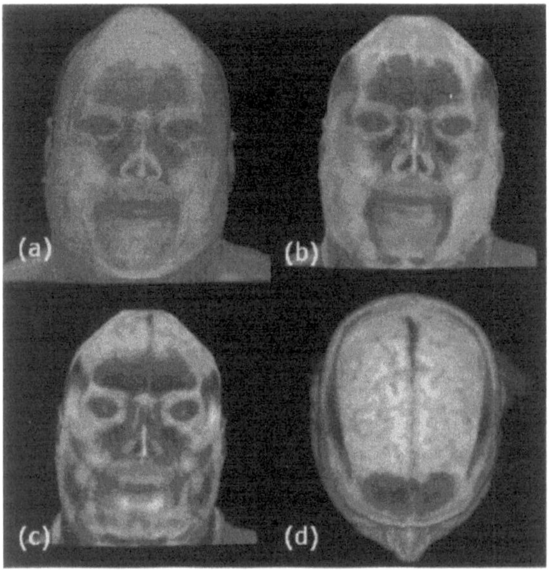

Fig. 8. Interactive extraction of inner structure of visible human head: (a) Full volume, (b-c) Reducing the size of boundary mesh. (d) Top view of extracted volume: brain surface and texture is interactively visualized.

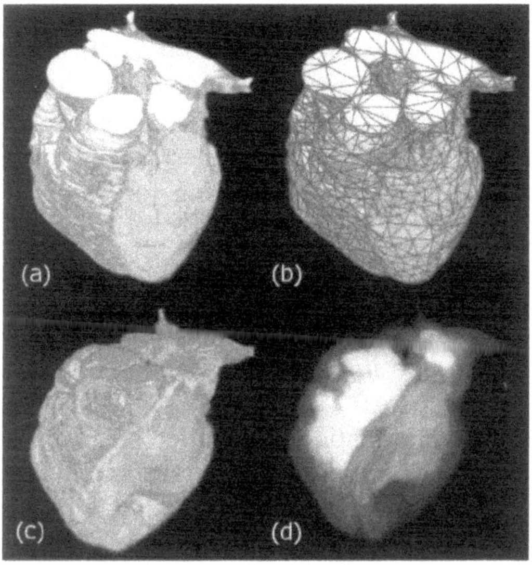

Fig. 9. Interactive volume deformation of finite element heart model: (a) Intial heart volume extracted from CT dataset, (b) Tetrahedral grid for representing 3D geometry of heart, (c) FEM-based deformation result. Coronary artery is effectively visualized as surface texture. (d) Translucent image of the deformed heart.

Finally, volume deformation example is demonstrated in Fig. 9. The 3D contour of the heart is roughly extracted using the developed volume clipping framework from CT dataset used in the previous test. After this interactive 3D extraction, color value filtering using specific thresholds and label smoothing is proceeded. Fig. 9 (b) shows tetrahedral grid with volume texture. The number of tetrahedra is 1134. Color look up table is also prepared. If the wall of left chamber is pushed through PHANToM, volumetric deformation is visualized as shown in Fig. 9 (c). The physical status is given by setting constant young modulus 5 MPa. Note that Fig. 9 demonstrates volume visualization result of the developed framework and does not consider deformation accuracy.

The developed framework performs more than 30 FPS and sufficient visual quality with not large number of tetrahedral elements. This achieved real-time performance is effective in the use of enhancing graphic representation of interactive systems like VR-based surgical simulation.

5. Discussion

This paper introduces 3D geometry and its volumetric representation on medical images, and demonstrated several applications for on-demand volume extraction and interactive volume deformation. Note that any segmentation or physics-based simulation algorithms are mentioned.

The results confirmed the effectiveness of on-demand extraction by manipulation of 3D geometry. Combining foregoing segmentation algorithms [14] with the developed framework will improve applicability of anatomical segmentation and reduce overall segmentation cost. Further verification of the proposed interactive extraction methods in actual use case is necessary. Clinical trial and evaluation is planned by introducing the developed system into image-based diagnosis and 3D shape modeling from clinical dataset.

Regarding volume deformation, simulation accuracy depends on the applied physics-based models. Because accuracy and interactive performance is incompatible, level of details of the grid should be chosen carefully by considering requirements of application. On the other hand, the proposed methods enable to enhance visualization quality of geometrical objects. Even a small number of elements are sufficient for visualizing surface and internal structure of organs.

6. Conclusion

This paper proposes an advanced volume representation of 3D geometries and presents its application for volume visualization of medical images. 3D geometry is represented using sectional slices of general tetrahedral grids and visualized using slice-based volume rendering method. Simulation results confirmed that the developed framework achieved fast volume visualization with surface and internal structure.

As practical application, on-demand extraction on 3D reconstructed image and volume deformation were introduced. Interactive manipulation of grid enables intuitive extraction of 3D region of interest. Even a small number of elements effectively visualize surface and internal structure of organs. Thus, the developed framework enables to advance graphical representation in geometry-based modeling and physics-based simulation. Application in image-based diagnosis and 3D shape modeling from clinical dataset is currently being developed. Clinical trial and evaluation of overall availability is also our future work.

Acknowledgment

This research is supported by Grant-in-Aid for Scientific Research (S) (16100001) and Young Scientists (A) (16680024) from The Ministry of Education, Culture, Sports, Science and Technology, Japan.

References

1. Pflesser B, Leuwer R, Tiede U, Hohne KH (2000) Planning and rehearsal of surgical interventions in the volume model. Medicine Meets Virtual Reality: 259-264
2. Toriwaki J, Mori K (1998) Visualization of the human body toward the navigation diagnosis with the virtualized human body. J Visualization 1(1): pp.111-124
3. Kaufman A, Cohen D, Yagel R (1993) Volume graphics. IEEE Computer 26(7): 51-64
4. Cabral B, Cam N, Foran J (1994) Accelerated volume rendering and tomographic reconstruction using texture mapping hardware, Proc. Symposium on Volume Visualization '94: 91-98
5. Chen W, Hua W, Bao H, QunSheng P (2003) Real-time ray casting rendering of volume clipping in medical visualization. J Computer Science and Technology 18(6): 804-814
6. Weiskopf D, Engel K and Ertl T (2003) Interactive clipping techniques for texture-based volume visualization and volume shading. IEEE Trans. on Visualization and Computer Graphics 9(3): 298-312
7. Owada S, Nielsen F, Okabe M Igarashi T (2004) Volumetric illustration: Designing 3D models with internal textures. ACM SIGGRAPH: 322-328
8. Kuhnapfel U, Cakmak HK, Mass H (2000) Endoscopic surgery training using virtual reality and deformable tissue simulation. Computers & Graphics 24(5): 671-682
9. Lin W, Robb RA (1999) Dynamic volume texture mapping and model deformation for visually realistic surgical simulation. Medicine Meets Virtual Reality: 198-204
10. Rezk-Salama C, Scheuering M, Soza G, Greiner G (2001) Fast volumetric deformation on general purpose hardware, The ACM SIGGRAPH/ EUROGRAPHICS workshop on graphics hardware: 17-24
11. Huang J, Yagel R, Filippov V, Kurzion Y (1998) An accurate method for voxelizing polygon meshes. ACM Symposium on Volume Visualization: 119-126
12. Weiler M. Kraus M. and Ertl T (2002) Hardware-based view-independent cell projection. Proc. the IEEE Symposium on Volume Visualization and Graphics: 13-22
13. Ackermann MJ (1998) The visible human project: a resource for anatomical visualization, 9th World Congress on Medical Informatics: 1030-1032

14 Suri JS, Setarehdan SK, Singh S (2002) Advanced algorithmic approaches to medical image segmentation, Springer

Development of a Prosthetic Walking Training System for Lower Extremity Amputees

Takahiro Wada[1], Satoshi Tanaka[1,2], Toyokazu Takeuchi[1,3], Kazuya Ikuta[1], and Kazuyoshi Tsukamoto[1]

[1]Faculty of Engineering, Kagawa Univ., Takamatsu, Kagawa, Japan
[2]Faculty of Health and Welfare, Prefectural Univ. of Hiroshima, Mihara, Hiroshima, Japan
[3]Ortho-Reha System Co.,Ltd. , Miki-cho, Kagawa, Japan

Chapter Overview. The number of the people affected with life-style related diseases such as diabetes and hypertension is increased recently. The number of the amputations of the lower limb cased by such life-style related diseases is increased as compared with the external wound such as the traffic and labor disasters. In this research, we develop a prosthetic walking training system for lower extremity amputees that can be utilized from just after the amputation surgery based on virtual reality technology. This system can realize the walking rehabilitation even before completion of the temporary prosthesis.

In this paper, structure of the developed system will be explained at first. Secondly, the experimental results of system evaluation by a subject with skillful prosthetic walking will be explained. Finally, application of the developed system to a patient just after a lower limb amputation will be described. The amputee can walk with prosthesis skillfully just after completion of the temporary prosthesis by training with the developed system.

Key Words. lower limb amputees, walking training system, lower extremity prosthesis, virtual reality

1. Introduction

In recent years, the number of the people affected with life-style related diseases such as diabetes and hypertension is increased. The number of the amputations of the lower limbs cased by such life-style related diseases is increased as compared with the external wound such as the traffic and labor disasters. Walking training with a lower extremity prosthesis is started after completion of the temporary prostheses and three weeks or more is needed for completion of the temporary prosthesis. Complications are often caused if the amputee spends this term in lying in bed especially for elderly people. Consequently, social reintegration tends to be later since acquisition of the walking ability with the prosthesis gets later. Therefore, it is important to realize walking training with the prosthesis as early as possible after the amputation surgery.

Many researches on walking training system for a person without amputation have been conducted. For example, Tani et al. developed the walking rehabilitation machine with virtual reality technique using a treadmill [1]. For prosthesis of lower limbs, there are many researches from many points of view. There are many researches on the mechanical aspects of the prosthesis. For examples, Goh et al. [2] evaluate the socket of the prosthesis by means of pressure exerted on stumps. Fridman et al. [3] studied on the influence of alignment of the prosthesis to the gait. Some researches on training method for prosthesis have been conducted recently. For examples, Schon et al. [4] proposed a walking training method with a temporary prosthesis. Chow et al. [5] proposed a prosthetic training method by using auditory biofeedback. Consequently, there is no research on a prosthetic training system that is utilized before having a temporary prosthesis.

In this research, we develop a prosthetic walking training system for lower extremity amputees that can be utilized from just after the amputation surgery. This system can realize the walking rehabilitation even before completion of the temporary prosthesis. Virtual reality technique is employed to realize virtual prosthetic walking environment. In this paper, structure of the developed system will be explained at first. Secondly, we propose a training procedure corresponding to individual's condition with the developed system. Then, the experimental results of system evaluation by a subject with skillful prosthetic walking will be explained. Finally, application of the developed system to a patient just after a lower limb amputation will be described. The amputee can walk with prosthesis skillfully just after completion of the temporary prosthesis by training with the developed system.

2. Prosthetic Walking Training System for Lower Limb Amputees

2.1 Structure of prosthetic walking training system

In this section, the structure of our developed prosthetic walking training system is described. Fig.1 shows the photo of the developed walking training system. Fig.2 illustrates the system diagram. As shown in Figs.1 and 2, the virtual prosthetic walking training system consists of a virtual prosthetic part, a visual display part, and a computer software part. In the prosthetic part, there exist an intact part and an amputation part. A user stands on the virtual prosthetic part. In the amputation part, there exists a socket for amputated end. The image on the display is changed by stepping on each part by turns.

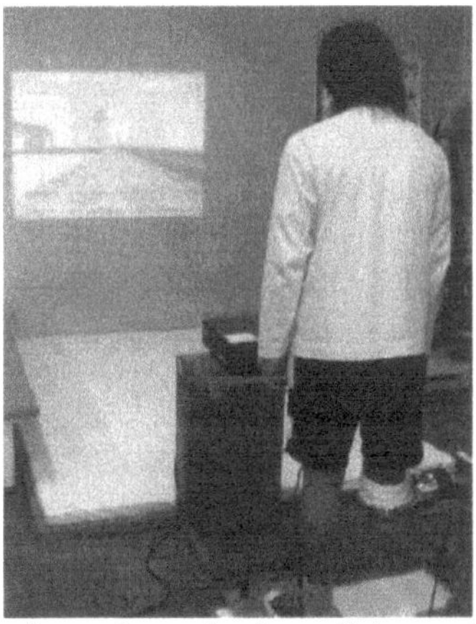

Fig. 1. Overview of prosthetic walking training system

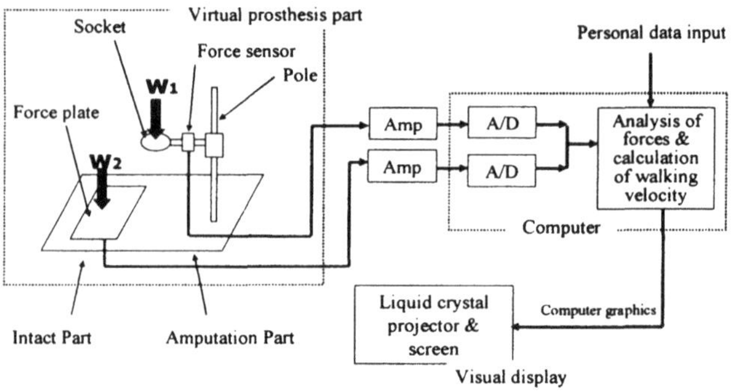

Fig. 2. System diagram

As shown in Fig.3, the amputation part of the virtual prosthetic part has the pole and the socket as an interface between the system and the amputated end in order to reduce pain by concentrated load to the end. A user inserts amputated end into the socket. The socket is made by a cast to adapt the differences among individuals and temporal conditions. The pole is fixed to the floor firmly and height of the socket can be changed to adaptive to users' height. There exist force sensors in the intact and amputation parts to measure stepping force. Output values of vertical forces are amplified and inputted to the computer through D/A converters. In the computer software part, inputted vertical forces of the intact and the amputated ends are analyzed and the walking velocity in the virtual environment is determined.

Fig.4 illustrates an example of the displayed graphic images of the walking road. A liquid crystal projector and a screen are utilized to display such images.

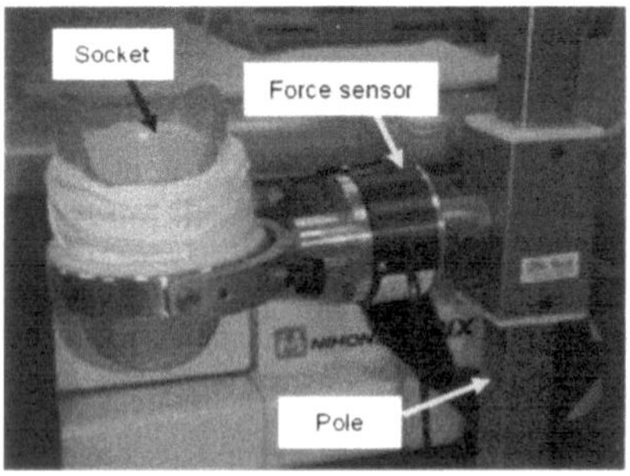

Fig. 3. Amputation part

Fig. 4. Computer graphics of virtual walking road

2.2 Walking rehabilitation using proposed system

In this section, we propose a method to realize prosthetic walking rehabilitation by using the developed system. Fig.5 illustrates the flow of the training. First, a medical doctor diagnoses the individual's condition of the user. Then, the medical doctor determines a threshold of stepping force of the amputated end to generate walking. Namely, computer graphics of the walking road are progressed only when the load to the socket exceeds the threshold that is determined corresponding to the individual's condition. At the early stage of the training, small threshold is employed. The threshold is increased gradually corresponding to the recovery of the condition. The detail is explained in the next section.

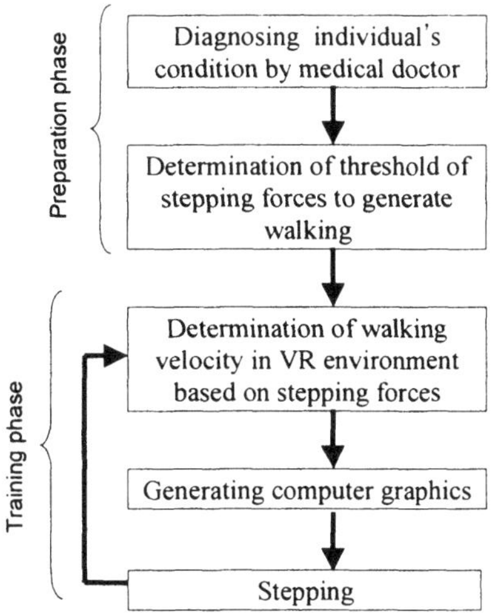

Fig. 5. Training procedure

2.3 Determination of walking velocity in virtual environment

In this system, walking velocity in the virtual environment is determined by analyzing the vertical forces exerted on the intact and amputated ends. In addition, the velocity can be adjusted to a training phase. Eq. (1) is utilized to determine the velocity.

$$v_i = L / T_{i-1} \tag{1}$$

where L denotes the step length of the user, and T_i denotes the time when the load exceeds the threshold determined by w/n as shown in Fig.6, where w denotes a user's weight and the training parameter n is arbitrary positive number and subscript i denotes the values of i-th step. Namely, quick load yields large velocity. Consequently, the user has to produce the force exceeding the threshold to generate walking velocity.

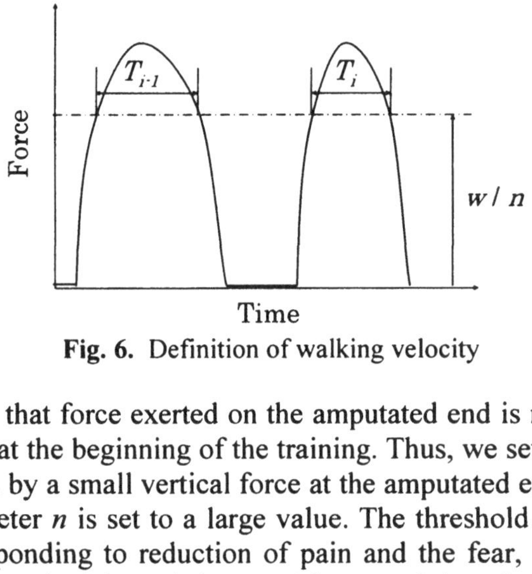

Fig. 6. Definition of walking velocity

It is expected that force exerted on the amputated end is not so large due to fear or pain at the beginning of the training. Thus, we set that walking is generated even by a small vertical force at the amputated end. Namely, the training parameter n is set to a large value. The threshold is gradually increased corresponding to reduction of pain and the fear, that is, n is decreased. At the end of the training, $n=1$ is set and the whole weight has to be loaded. This function allows us to realize the training adaptive to the individual conditions.

3. Evaluation of the system by skillful amputee of prosthetic walking

3.1 Experimental conditions

In this section, we examine whether the system can be applicable to the walking training. A male of 22 years old is employed as a subject. He had an above-knee amputation surgery of his right leg 4 years ago due to a traffic accident. He utilizes prosthesis usually and can run on the flat ground. The weight is 60[kgf] and the threshold of the vertical force is set to 20[kgf] by trial and error. Also, step length in eq. (1) is set to $L=0.6$[m]. The training parameter is set to $n=3$, namely, threshold of the vertical force is 20[kgf]. In order to investigate differences of force patterns by walking velocities, the subject walked in three velocities. The three velocities of slow, normal, and fast are determined by the subject's subjectivity.

First, the subject inserts the amputated end into the socket. After standing at rest, he starts the stepping foot and amputated end by turns for a minute to generate walking in the virtual environment. During the stepping, the subject gazes the screen.

3.2 Experimental results

The subject mentioned that stepping can be realized without pain on the amputation end. Figs.7-(a), (b), and (c) illustrate the time series patterns of forces of the intact side and the amputated side with slow, normal, and fast velocities, respectively. These figures show that this system can measure the forces smoothly. As shown in these figures, almost similar levels of forces are exerted on the amputated end and intact end.

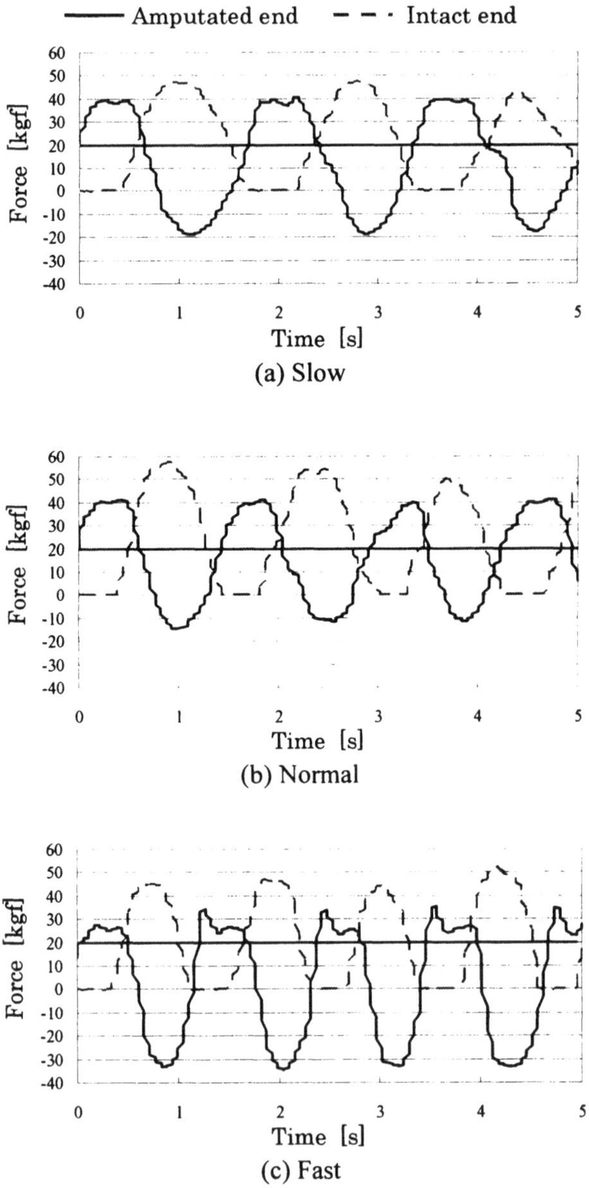

Fig. 7. Experimental results of skillful amputee

Fig.8 illustrates the calculated walking velocities by the three stepping velocities. As shown in this figure, walking velocity can be controlled by changing stepping speed.

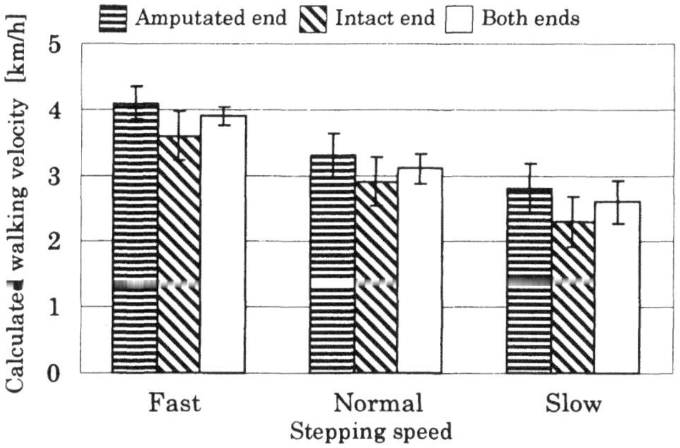

Fig. 8. Calculated walking velocities

4. Clinical Application

4.1 Contents of training

In this section, we apply the proposed system to an amputee just after an amputation surgery. The patient is 22 years old male. He had an above-knee amputation surgery of his right leg September, 2003. The weight is 60[kgf]. The application was done for a month from three weeks after the amputation. In the training, the step length in the virtual environment is set to $L=0.6$[m].

First, the patient inserts the amputation end into the socket. He starts the stepping foot and amputated end by turns for a minute to generate walking in the virtual environment. Handrails are equipped in order to reduce stump pain. The walking velocity is not designated in advance. Namely, the subject determined his walking velocity. The frequency of the training is 5 times per day and around 3 times per week. The training parameter n is changed by a therapist based on the patient's condition.

4.2 Training results

The subject mentioned that he had stump pain for three days from the start, but stepping can be realized gradually without pain on the amputated end.

Fig.9 illustrates the time series patterns of forces of the intact side and the amputated side. Fig.9-(a), (b) and (c) illustrate the results of the first day, two weeks and a month from the beginning, respectively.

As shown in these figures, he could not load force at the amputated end and loading speed was slow at the beginning of the training. Two week later, he could load much force at the end and loading speed was faster. Finally the force and loading speed reached the same level as that of intact side a month later and incline of the force was increased.

Fig.10 illustrates transition of the maximum forces at the intact and amputated ends. As shown in this figure, maximum forces at the amputated end are gradually increased day by day and finally the force reaches similar level of that at the intact end. Fig.11 illustrates the transition of the training parameter n. The parameter n can be decreased because the patient can generate larger vertical force at amputated end. Fig.12 shows transition of walking velocity in the virtual environment. The walking velocity is increased gradually by training with our system.

And the patient can easily walk with real temporary prosthesis just after its completion. Therefore, we can conclude that the system can apply the prosthetic walking rehabilitation with our proposed training procedure.

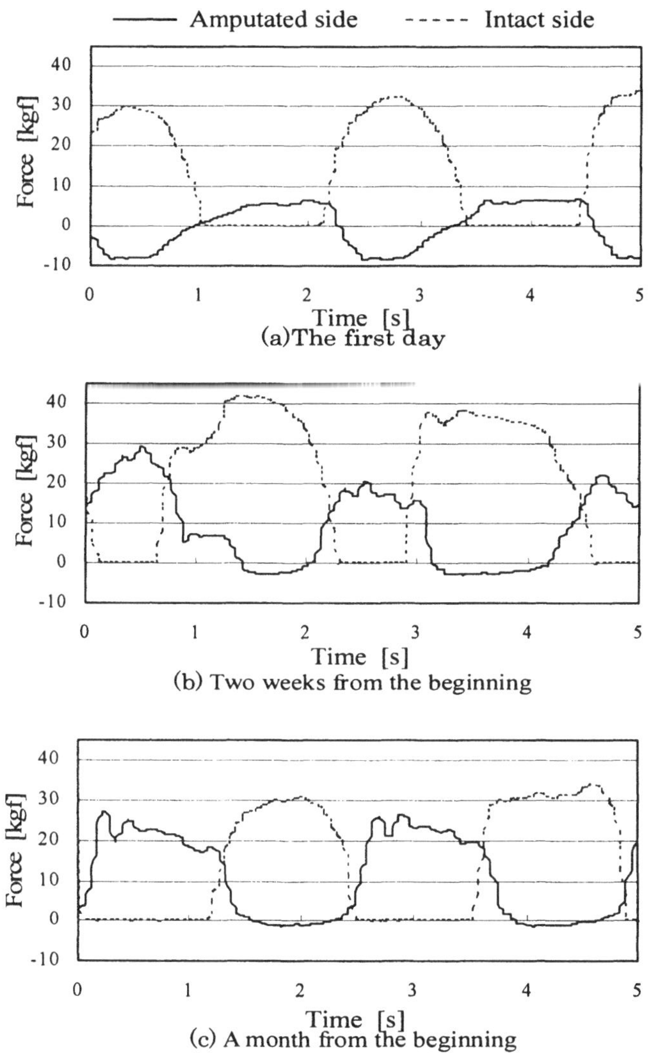

Fig. 9. Time series patterns of stepping forces during training

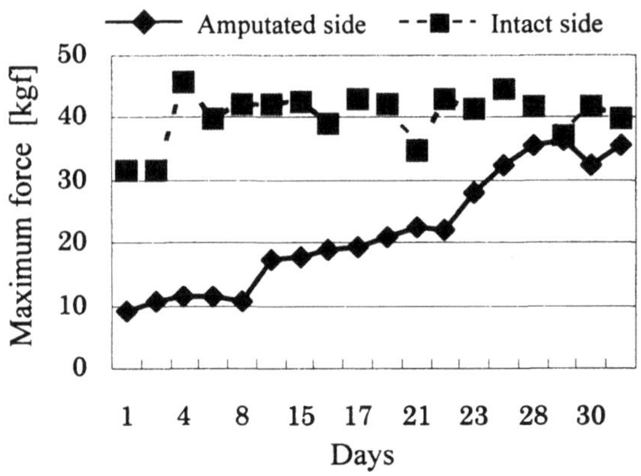

Fig. 10. Transition of maximum forces

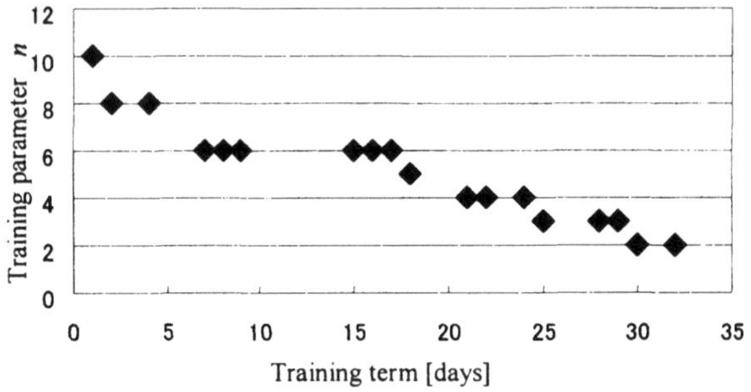

Fig. 11. Transition of training parameter *n*

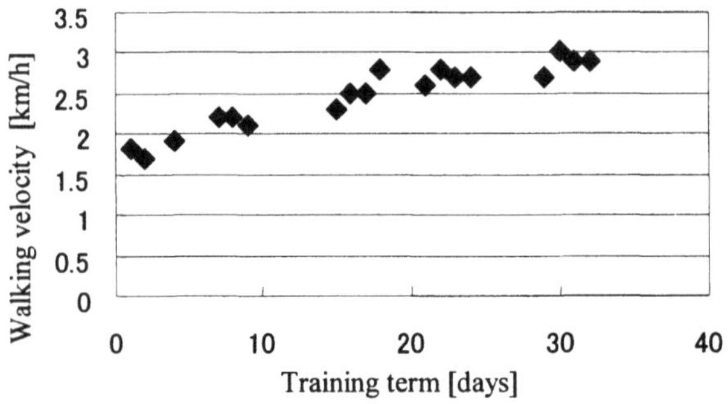

Fig. 12. Transition of walking velocity in VE

5. Conclusions

We proposed a new training system for prosthetic walking of lower limb amputee. Experimental results with a skillful amputee of prosthetic walking have shown that the developed system can measure forces exerted on the intact side and the amputated end and the smooth walking is realized. After that we have applied the system to clinical training of an amputee just after the amputation surgery. By using the system, the force exerted on the amputation end is increased gradually day by day and finally the forces reaches the same level as that at the intact end. Also, the training parameter *n* can be decreased because the patient can generate larger vertical force at amputated end. In addition, the walking velocity is increased gradually by training with our system. And the patient can walk with the real temporary prosthesis just after the training. Consequently, we can conclude that the system can apply clinical training.

As the future work, the system will be applied to many other patients in order to show the effectiveness of the system. Also, we will develop walking mechanism with free natural walking.

References

1. Tani, T., Sakai, A. Fujimoto, T. Fujie, M., "Walk training system: improvement of the ability of postural control", *IEEE/RSJ Int. Conf. on Intelligent Robots and Systems*, Vol.2, pp.627-631, 1997.
2. Goh, J.C., Lee, P.V., Chong, S.Y., "Static and dynamic pressure profiles of a patellar-tendon-bearing (PTB) socket", *Proceedings of the Institution of Mechanical Engineers:Journal of Engineering in Medicine Part H*, Vol.217, No.2, pp.121-126, 2003.
3. Fridman, A., Ona, I., Isakov, E., "The influence of prosthetic foot alignment on trans-tibial amputee gait", *Prosthetic Orthotic International*, Vol.27, No.1, pp.17-22, 2003.
4. Schon, L.C., Short, K.W., Soupiou, O., Noll,K.,Rheinstein, J., "Benefits of early prosthetic management of transtibial amputees: a prospective clinical study of a prefabricated prosthesis", *Foot & Ankle International*, Vol.23, No.6, pp.509-14, 2002.
5. Chow, D.H., Cheng, C.T., "Quantitative analysis of the effects of audio biofeedback on weight-bearing characteristics of persons with transtibial amputation during early prosthetic ambulation", *Journal of Rehabilitation Research and Development*, Vol.37, No.3, pp.255-260, 2000.

Electrophysiological Heart Simulator Equipped with Sketchy 3-D Modeling

Ryo Haraguchi[1,2], Takeo Igarashi[3,4], Shigeru Owada[3], Takenori Yao[2], Tsunetoyo Namba[2], Takashi Ashihara[2], Takanori Ikeda[2], and Kazuo Nakazawa[1,2]

[1] National Cardiovascular Center Research Institute, Japan.
[2] Working Group on Cardiac Simulation and Mapping, Japan
[3] Graduate School of Information Science and Technology, The University of Tokyo, Japan
[4] PRESTO, Japan Science and Technology Agency, Tokyo, Japan

Chapter Overview. We present an electrophysiological heart simulator equipped with sketchy 3-D modeling interface. It has been tedious and time-consuming to create the shape of heart for use it in the simulator. In this study, we developed a new simulator that is combined with a sketch-based 3-D modeling interface for the shape transformation. We also developed a semiautomatic method in order to save labor for pre-process of the simulation. The sketchy 3-D modeling interface increases the facility of computer simulation.

Key Words. Electrophysiology, Simulation, and Modeling

1. Introduction

To better understand the mechanisms of fatal arrhythmias, we had previously developed a sophisticated human ventricular-shaped model with the use of computational techniques and visualization technology [1]. The 3-dimensional model contains about 5.64 million volumetric myocardial units, of which membrane kinetics was represented by the Luo-Rudy-1 equations (LR1) [2]. The LR1 consists of the nonlinear ordinary differential equations with 8 variables, based the ionic channel activities of ventricular myocardium. We had developed an electrophysiological heart simulator for solving the differential equations holding about 45 million variables in total, and had implemented the actual calculation on a high-performance supercomputer. Using this simulator, we had simulated physiological electrocardiogram, and also had visualized the various electrophysiological phenomena of arrhythmias in our ventricular-shaped model.

The simulation technology should be applied to the clinical use in the future, and therefore it is imperative that we can use an individual ventricular-shaped model based on the patient's data, and can execute the simulation as quickly as the need arises. As the computer hardware and software technology are progressing rapidly, it is possible to execute the electrophysiological heart simulation more quickly and precisely. However, the preliminary setup is required before calculating the simulation. We have to preliminarily construct the individual shape of patient's ventricles. Not only the acquisition of patient's 3-D data but also the transformation of the volumetric data is important in order to simulate heart disease such as a morphological abnormality like hypertrophic cardiomyopathy. However, skilled operations are needed to transform the volumetric data. We consider that the shape-transformation could be an obstacle to executing the simulation using the individual data. Furthermore, we have to preliminarily determine the parameter setting of ionic channel activity, the excitation sequences of the ventricular subendocardial layer, the position of extrasystole in the ventricular-shaped model, and so on. The subendocardial excitation sequences are key in substitution for incorporating an actual Purkinje fiber network in the model. We had manually built the subendocardial excitation sequences based on the references of real human ventricles [3]. When we use the individual heart shape in the simulation, we need to set up some parameters and embed the subendocardial excitation sequences in the ventricular-shaped model. However, such pre-processes require a lot of tedious work and time. We consider those pre-processes to be potential obstacles to the repetitive trial simulation.

In this study, we developed a new electrophysiological heart simulator that has been combined with a sketch-based interactive 3-D modeling interface for the shape transformation, and with a semiautomatic method to save labor for pre-process of the simulation. This study's objective is different from the former large-scale simulation performed on the high-performance supercomputer. Instead, it is a simple, real-time simulation combined with an interactive modeling interface. The aim of this study is not to perform an accurate and detailed simulation, but to enhance the capability of the computer simulation by using a 3-D modeling interface.

2. Method

2.1 Sketchy 3-D modeling

Our new simulator equips a sketchy 3-D modeling interface for the transformation of ventricular volumetric data. The 3-D modeling interface is based on the "Teddy" [4] software. Teddy is a sketch-based interactive, quick and intuitive 3-D modeling software. When the user draws 2-D freeform strokes interactively on the display like sketching, the system infers the user's intent and executes the appropriate editing operations (Fig.1). Teddy is suitable for the rapid construction of simple and approximate models.

By using the Teddy software, we can construct a simple ventricular model very quickly (Fig.2). First, the user draws a closed stroke on a blank canvas (Fig.2(1)), and then the system automatically generates plausible 3-D polygonal surfaces (Fig.2(2)). Next, the user draws a stroke that runs across the object, and then the system cuts the object on a plane that de-fined by the viewpoint and the stroke (Fig.2(3)). Fig.2 (5) and Fig.2 (6) show the digging operation by the two strokes. In this way, we can obtain a rough shape of ventricles within minutes.

We can also edit the object after loading pre-existing shape data. Figure 3 shows the transforming operations of the ventricular model that obtained from a real human volumetric data. In this study, we added the "expand" operation to original Teddy software as a suitable operation for this study. Using the expanding operation, we can expand the ventricular thickness, assuming the hypertrophic cardiomyopathy.

The sketchy 3-D modeling interface was written in Java™ and we can create and modify the model interactively on an ordinary personal computer (PC).

2.2 Semiautomatic pre-process method for the simulation

Next, we obtained a ventricular model as voxels converted from the polygonal data. And we extracted the subendocardial layer automatically from the ventricular volumetric data by using image-processing techniques e.g. region growing method. Then we built the excitation sequences onto the extracted subendocardial layer by manually specifying the earliest excitation sites, which correspond to the distal ends of the left and right Purkinje bundles. Finally, we defined the extrasystole position for the arrhythmic simulation.

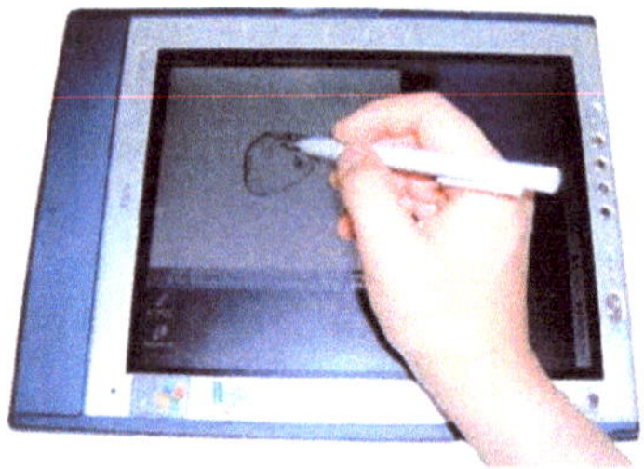

Fig. 1. Sketchy 3-D modeling interface on a Tablet PC.

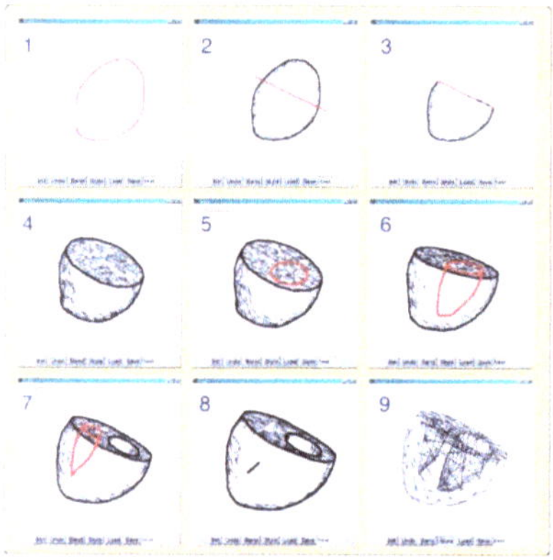

Fig. 2. Creating a simple ventricular model by the sketchy 3-D modeling, (1-8) Create, cut, and digging operations by input strokes (red), (9) A generated object with mesh display.

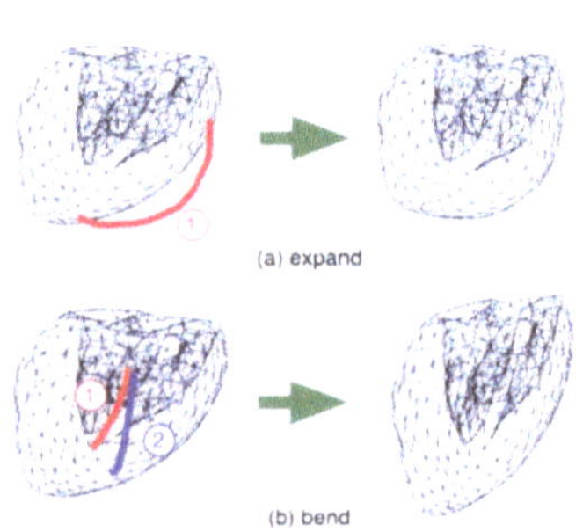

Fig. 3. Modifying the actual human's ventricular model, (a) expanding the thickness by one input stroke (red), (b) bending the model by two strokes (red and blue).

2.3 Electrophysiological heart simulator on a standard PC

We have developed new application software that implements the simple electrophysiological heart simulator on a standard PC. By using this software, we can execute the simulation immediately after modeling the shape of the heart and we can visualize the results in real-time.

The simulation software handles ventricular volumetric data as 64x64x64 voxels. The membrane kinetics of those myocardial units is represented by the FitzHugh-Nagumo equations (FHN) [5, 6]. The FHN consists of the nonlinear ordinary differential equations with 2 variables and needs lower computational resource than LR1. The FHN is not directly based on the ionic channel activities of ventricular myocardium, but it gives an approximate representation properly for our simple electrophysiological simulation on a PC.

Our program's screenshot is shown in Fig.4. In our new application, we can execute the simulation using two models simultaneously for easy comparison. The Marching Cubes [7] method is used for visualizing ventricular volumetric data. The simulator was created using C++, OpenMP™, OpenGL® and so on.

3. Results

We run the sketchy 3-D modeling interface and the simple electrophysiological simulator on the same PC. The original shape of ventricles was constructed based on geometrical data of the actual human ventricle (Viewpoint DIGITAL™). By using the sketchy 3-D modeling interface, we were able to create some shape-transformed ventricular models easily and quickly. Using our new simulator, we built the excitation sequences semiautomatically onto the subendocardial layer of the ventricular models. Then we were able to observe the visualized results of the electrophysiological simulations immediately. We observed various electrophysiological phenomena (dissipation, breakup, sustainment, meandering, and so on) in the difference of the heart shapes (Fig.5). Table 1 shows the time required for numerical calculation and visualization per 1 frame on the PCs. The frame rate is from 0.5 to 2 frame/sec.

Some time-consuming procedures are necessary for obtaining a 3-D model. In this study, because we have integrated the modeling interface with the simulator, we can handle the shape transformation of the ventricular model easily in the simulation research. For example, we can "expand" the shape of the model with only one stroke. Then we can execute the simulation easily with estimating morphological abnormalities, such as hypertrophic cardiomyopathy. We observed that the ventricular shapes do influence the electrophysiological phenomena.

The sketchy 3-D modeling interface is suitable for not only transforming the existing model, but also creating a new model. In the future, the sketchy interface can be applied to assist the extraction procedure from medical image sequences obtained from MRI, CT and echocardiography.
Usually, a large-scale electrophysiological simulation requires powerful computational resource and much time. Our simple system that can be created on an ordinary PC has some limitations in accuracy and scalability because of limited computational resources for numerical equations representing myocardial membrane kinetics. However, our system has unique merits that can be observed in real-time as soon as modeling the shape of heart has been completed. Our system enables us to execute repetitive trial simulation and makes it easy to observe the significant data. It also enables us to increase the potential for a large-scale precise simulation (Fig.6). Additionally, we ran a large-scale electrophysiological simulation that was previously developed. We used an NEC SX-6/8A supercomputer for the computation. The polygonal data of the transformed ventricular model were converted to 300x300x300 voxels and each unit was represented by the LR1. LR1 is more complex than FHN and needs remarkably higher computational resources. Required time to perform a 1,000msec simulation was about 3 hours. We used volume-rendering techniques [8] for visualizing the results. In the large-scale simulation, we observed various electrophysiological phenomena in the difference of the heart shapes. In Fig.7, the original shape (left) became sustainment caused by extrasystole, and the transformed shape (right) changed over from sustainment to breakup.

5. Conclusion

Integrating the sketchy 3-D modeling interface with the electrophysiological heart simulator enables us to save the labor required for repetitive trial simulations using the individual and transformed ventricular models. Our system equipped with a sketchy 3-D modeling interface enables us to increase the facilities of the computer simulation. We expect that, in the future, our simulator would be used for the research and clinical purposes, and also as an educational tool for medical staffs.

Acknowledgments

This research was partially supported by the Ministry of Education, Culture, Sports, Science and Technology, Grant-in-Aid for Scientific Research (C)(2)16500311 and Leading Project for Biosimulation. This work was partially supported by the Research Grant for Cardiovascular Diseases (15-6) from the ministry of Health, Labour and Welfare. This work was partially supported by grants from PRESTO/JST (Japan Science and Technology Agency).

References

1. Nakazawa K, Suzuki T, Ashihara T, et al. (2000) Computational analysis and visualization of spiral wave reentry in a virtual heart model, Clinical Application of Computational Mechanics to the Cardiovascular System, Yamaguchi T (ed), Springer: 217–241
2. Luo CH, Rudy Y (1991) A model of the ventricular cardiac action potential: Depolarization, repolarization, and their interaction, Circ Res 68:1501–1526
3. Cassidy DM, Vassalo JA, Marchlinski FE, et al. (1984) Endocardial mapping in humans in sinus rhythm with normal left ventricles: activation patterns and characteristics of electrograms, Circulation 70:37-42
4. Igarashi T, Matuoka S, Tanaka H (1999) Teddy: A sketching interface for 3d freeform design, ACM SIGGRAPH'99, Los Angles, ACM, pp.409–416
5. FitzHugh R (1961) Impulses and physiological states in theoretical models of nerve membrane, Biophysical J 1:445-446
6. Nagumo J, Arimoto S, Yoshizawa S (1962) An active pulse transmission line simulating nerve axon, Institute of Radio Engineers 50:2061-2070
7. Lorensen WE, Cline HE (1987) Marching cubes: A high resolution 3d surface construction algorithm, ACM SIGGRAPH'87, Los Angles, ACM, pp.163-169

8. Levoy M (1988) Display of surfaces from volume data, IEEE Computer Graphics and Applications 8(3):29-37

Three-dimensional Display System of Individual Mandibular Movement

Michihiko Koseki[1], Akira Niitsuma[2], Norio Inou[1], and Koutarou Maki[3]

[1] Department of Mechanical and Control Engineering, Tokyo Institute of Technology
[2] Department of Mechanical and Environmental Informatics, Tokyo Institute of Technology
[3] Department of Orthodontics, Showa University

Chapter Overview. It is expected to develop an intelligible diagnostic system of temporomandibular disorders (TMD) for both medical doctors and patients. This study proposes a display system that visualizes motion of the human mandible. The system integrates two engineering methods. One is an optical motion capture technique for measuring the mandibular movements. The other is an individual modeling method based on the X-ray CT data. It is important to know exact mandibular movements for the proper diagnosis. This paper discusses experimental verification of the total performance of the system using a device of hinge movement. The verification clearly shows that precision of the model has a great effect on accuracy of the movements. The total performance of the system is achieved within an accuracy of 0.2mm at the hinge of the device. The system provides not only three-dimensional visual information of the mandibular movements as animations but also quantitative information of position, velocity and acceleration at an arbitral point of the model. The system will be useful for informed consent in medical treatments of TMD.

Key Words. Medical imaging, Temporomandibular disorders, Patient-specific modeling, X-ray CT

Introduction

It is very important to understand a motion of a mandibular condyle individually for proper diagnosis and treatment of temporomandibular disorders (TMD). Recently, some optical devices to record mandibular movements have been proposed. The non-contacting measuring devices have an advantage of permitting a masticatory movement of a patient under almost natural conditions.

There are some reports to evaluate mandibular movements using Gnatho-hexagraph (JM-1000, Ono-sokki Co.) [1][2]. Gnatho-hexagraph is an opto-electronic system with six degrees of freedom. This device consists of a facebow and a headframe with three LEDs on each frame. Two CCD cameras take time-series pictures of the positions of the LEDs, and the motions of the facebow and the headframe are calculated. Latest version of the system enables to display the movement with several three-dimensional mandibular models prepared in advance [3]. However the system has several problems although the visualization is useful for clinic. One of the problems is that the shape of the prepared mandibular model is different from that of the subject in the motion capturing. The geometrical difference of the mandibular shape causes an incorrect visualization such that a condyle and a mandibular fossa overlap even if motion-capturing data are fully accurate. The other problem is that the headframe possibly causes inaccuracy of motion capture because they are not directly fixed the skull.

In order to compute exact mandibular motions, it is essential to obtain geometrical relationship between motion-capturing devices and anatomical structures. For this purpose, the X-ray CT image of a patient with the devices is most appropriate solution. The original research of the authors aimed to indicate the mandibular movement using a cephalometric radiograph or laminated layer image obtained from X-ray CT data [4]. In the successive studies, we developed a display system of individual mandibular movement using patient-specific finite element models [5][6]. In the recent past, another research group reported a similar system with the same concept with us [7]. However, the quantitative accuracy of the system was not described in the paper as well as our previous studies. This study focuses on the improvement of the accuracy of the system.

Display System of Mandibular Movement

Overview of the System

The procedure of our display system of individual mandibular movement is shown in Fig. 1. There are several tasks to display mandibular movements. Firstly, optical markers are provided for motion capture of the human mandible. As the markers, we employ acrylic fluorescent balls that glow under UV-A lights often called "black lights." A facebow equipped with the markers is designed to fit the human face as shown in Fig. 2. The facebows are attached to the labial surfaces of upper and lower incisors with cyanoacrylate adhesive. Three-dimensional mandibular movements are measured by two cameras.

Secondly, the patient with facebows is subjected to X-ray CT. Based on the CT images, individual models of a skull and a mandibular bone are generated.

Thirdly, coordinate transformation of the mandibular model is performed according to the three-dimensional motion data, and an animation of the mandibular movement is displayed. The following subsections describe details of these tasks.

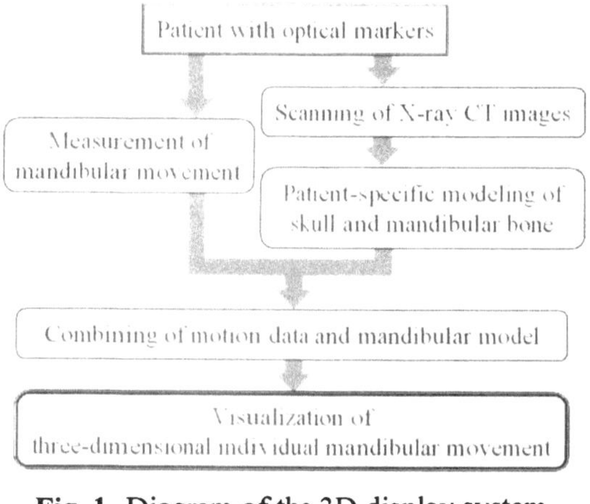

Fig. 1. Diagram of the 3D display system.

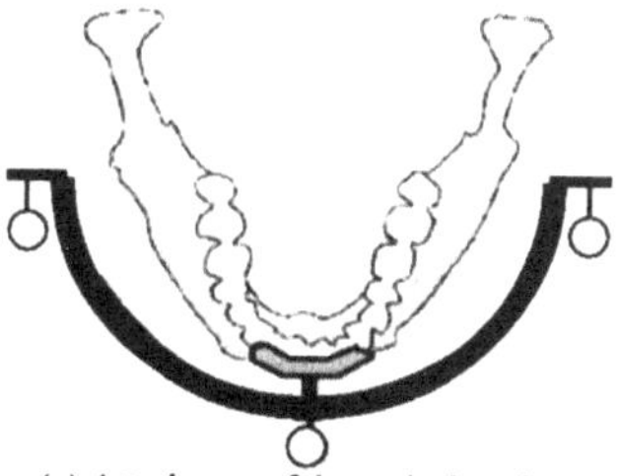

(a) Attachment of the optical markers.

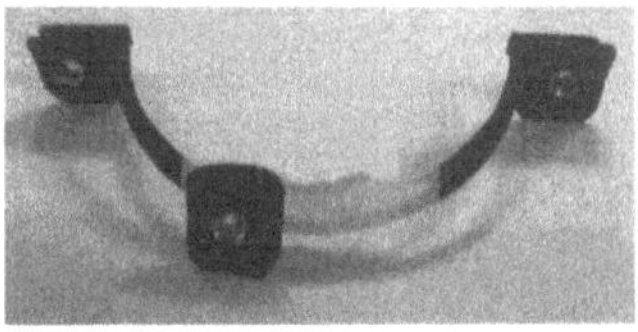

(b) The facebow with the optical markers.

Fig. 2. Schematic diagram and fabrication of the facebow.

1. Introduction

Laparoscopic surgery is widely performed as a means of minimally invasive surgery. In this method, surgeons make 3-4 holes on the abdominal wall, and entire operations are conducted inside the cavity through the holes using thin scope and long-handled surgical tools such as forceps and scalpel. Compared with the conventional laparotomy requiring large incision on the abdomen, laparoscopic surgery has benefits for patients because of its small invasion; reduction of postoperative pain, discomfort, medication, and the time needed for recovery [1]. It has, however, some difficulties for surgeons. Forceps have the limited Degree of Freedom (DOFs) (two DOFs for orientation of forceps, two DOFs for insertion and rotation of forceps) and limited DOFs reduce the dexterity of the surgeon. Procedure is operated symmetrically around the incision hole, so that surgeon gets confused. Thus, this procedure requires great skill and stress for surgeons. To solve these problems, surgery- assisting robots with master-slave system, such as da Vinci® [2], have been clinically applied to thoracic and abdominal surgery. These robots enhanced the dexterity and ability of surgeons, which contributed to the higher quality and more precise operation that could not be realized using conventional forceps. They, however, have some problems, such as;

- Large size for conventional operating theatres.
- Occupancy of operating space above the abdomen.
- Collision with manipulators or surgeons.
- Lack of end effectors with advanced surgical instruments such as laser surgical instruments and ultrasonic scalpel.

The purpose of this study was to develop a compact surgical-assisting robot system with enough working space, which functions as a master-slave robotic system. We developed a new master-slave robotic system with three forceps operating cooperatively that corresponded to both hands of surgeon and assistant. We also developed functional forceps with as semi-conductor laser coagulator and bipolar electric cautery. We evaluated the feasibility of the system.

The robot consisted of three modules; manipulator-positioning arm, forceps manipulator, bending forceps with 2 DOFs including integrated semi-conductor laser coagulator and bipolar electric cautery. Manipulator-positioning arm had 6 DOF. Each DOF was passive and had only pneumatic-drive braking mechanism. The system was designed so that three manipulators are operated cooperatively. Thus, the three positioning arms were mounted at the manipulator positioning arm.

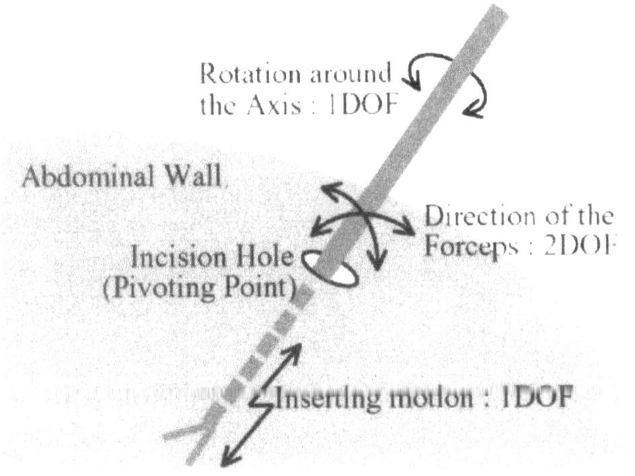

Fig. 1. Limited degrees of freedom in laparoscopic surgery: Mechanical constraint is set at the insertion hole. Thus surgeons must manipulate forceps opposite to the motion of the forceps's tip.

2. Configuration of Robotic System for Less Invasive Abdominal Surgery

2.1 Manipulator-positioning arm

We used a surgical microscope arm (CYGNUS®, Mitaka Kohki, Japan) for manipulator positioning arm. The arm was approved as a surgical device and had shown actual performance in the operating theater. It had six DOFs using parallel linkage mechanism for the position and spherical bearing for attitude. Each DOF was passive and had a disk brake with pneumatic-drive releasing mechanism. It had advantages in the points that it would keep braking and never release in case of electric power down and that we could move the arm with larger torque than the rated torque. We used this arm for the rough positioning of the whole manipulator. In an abovementioned emergency, we will only move it so that surgeons can resume the operation immediately [3].

This system aimed to operate three forceps manipulators equivalent to the surgeon's both hands and an assistant's hand cooperatively. Thus, at the edge of microscope arm, three 6- DOF arms were mounted for each forceps manipulator; passive 2-DOF positioning in the horizontal plane using Selective Compliance Assembly Robot Arm (SCARA), 1-DOF positioning for vertical motion using linear actuator, passive 3-DOF (roll, pitch, yaw) positioning for the direction of the manipulator using spherical bearing with pneumatic-drive breaks. These arms were used for precise positioning of each forceps manipulator respectively. The working space of each arm was R300 [mm] x H200 [mm] and wide enough to cover the whole abdomen. 6-DOF arms enabled the intuitive arbitrary positioning.

The pneumatic-drive braking had two switches and could be released by pushing both buttons at the same time. Multiple buttons using an AND logical function enhanced the safety and the brake would never released by accident. The position and orientation of each arm was measured using optical tracking system (Polaris®, Northern Digital Inc. Canada). Optical marker was mounted at the end of the arm and measured from the sensing unit that emitted the infrared light and received reflected light. No rotation angle sensors such as rotary encoder or potentiometer were mounted.

2.2 Forceps manipulator

Forceps manipulator had 4 DOFs (2 DOFs for orientation and 2 DOFs for insertion and rotation of forceps). It adopted Remote Center of Motion (RCM) mechanism so that forceps could rotate around the pivoting point (the incision hole). We proposed new two-linear-actuation mechanism for RCM mechanism [4]. In this mechanism, rotational motion around the pivoting point was realized using two linear actuators (ball screw and motor). When the feeding ratio of two actuators was constant, the lines running through tips of both actuators crossed at a certain point (pivoting point) as shown in Fig. 2.

The insertion of the forceps along the shaft was realized by the belt-driven "double-stage mechanism". It consisted of a couple of linear guide. One was mounted on the other, and it expanded and contracted like a ladder truck. This mechanism realized twice as long traveling distance as the length of the linear guide, so that the size of manipulator was miniaturized. In this manipulator, the inserting distance of 300 [mm] was realized using the 150 [mm]-long linear guide. Rotational motion of the forceps was realized by transmitting the rotation of the motor with gear.

The size of RCM part was W250 x H110 x D60 [mm^3], and forceps driving unit was W140 x H410 x D70 [mm^3]. The whole weight was 4.5 [kg]. The actuators used in this study were pulse motors. In case of malfunction of software system, pulse motors had higher possibility of stopping rather than getting out of control seen in the case of DC motors. Stop is the best action to avoid enlarging the risk of damage on the patient, thus we adopted pulse motors for this robotic system.

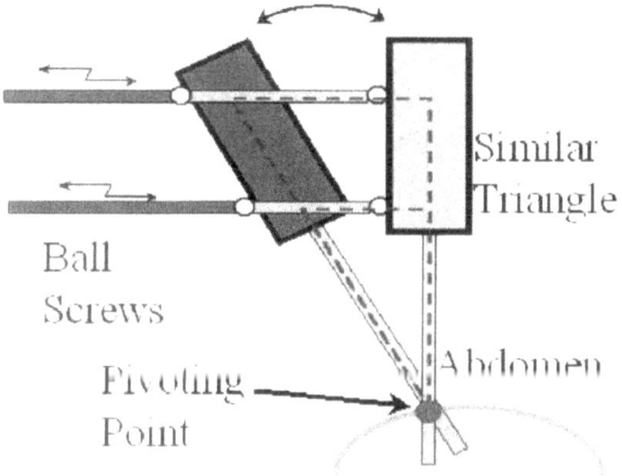

Fig. 2. New remote center of motion (RCM): The feeding ratio of the two linear actuators (ball screw and motor). Was set constant to make the lines running through tips of both actuators crossed at a certain point (pivoting point).

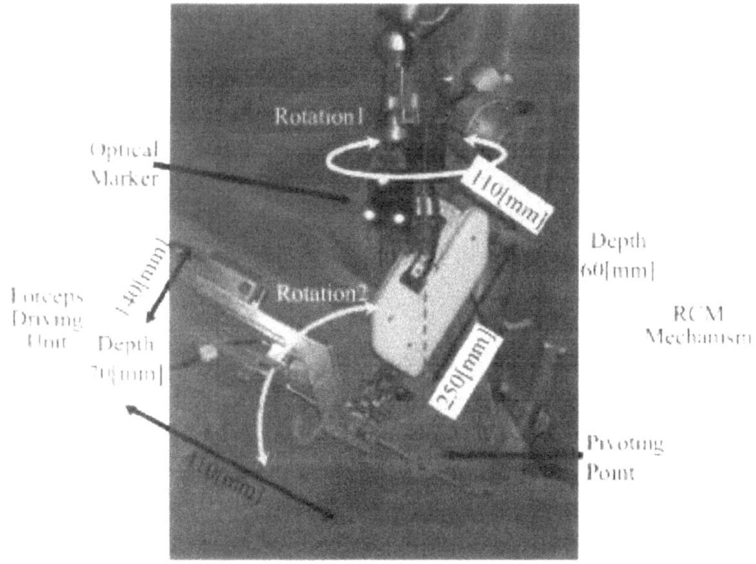

Fig. 3. Prototype of forceps manipulator

2.3 Bending forceps and functional forceps

We have developed two types of bending forceps with 2 DOF, which were actuated by a tendon mechanism [5, 6] and linkage mechanism [7]. Furthermore, functional forceps such as semi-conductor laser forceps [8] and bipolar forceps [9] were developed. We have developed a novel robotic laser coagulator with a CCD (Charge coupled device) video endoscope and a bending joint. Endoscope presents the detail of target. Bending joint realized the irradiation in any direction. We adopted two laser diodes; infrared light for coagulation and red color light to point out the target.

Bending forceps consisted of "forceps part" and "motor driving unit". The forceps was 10 [mm] in diameter. Coupling mechanism that connects bending forceps and forceps manipulator is standardized so that the various types of forceps are interchangeable.

2.4 Variable view angle laparoscope

Many robotic endoscope systems have been developed to provide surgeon with desired laparoscopic view of surgical field [10-12]. Most of them manipulate existing endoscopes. There are two types: rigid and flexible. Rigid endoscopes need to rotate the insertion point to observe a wide range of views, whereas flexible types rotate the tip of the endoscope to achieve the same end. Thus, to obtain a wide field of view a large operating area is required. We proposed a new type of robotic wide-angle view endoscope that does not require rotation or bending of the endoscope to move the field of view. To achieve the above requirements, we used two wedge prisms and a laparoscope with zoom capacity. Details of the mechanism are found in the literatures [13, 14].

2.5 Control system

The master-slave manipulator system consisted of a pair of master arms (MASTER) developed by Mitsuishi et al [15] and a slave manipulator system (SLAVE). In the MASTER system, an operator sits in front of a monitor showing a laparoscopic view and controls the master manipulators, using footswitches to command the SLAVE. The SLAVE system uses three slave robotic manipulators with seven degrees of freedom that include a grasping function, and an endoscope with a variable viewing angle [13]. The slave robotic manipulators are located on passive positioning arms for easier presetting of the manipulators before surgical operation. The position and orientation of the manipulators and the endoscope were measured using an optical positioning sensor (Polaris, Northern Digital Inc).

The motion of the master arms was recognized as the motion commands of the endoscope's coordinates, and this motion was transformed to the manipulator's coordinates. The required computation for the SLAVE

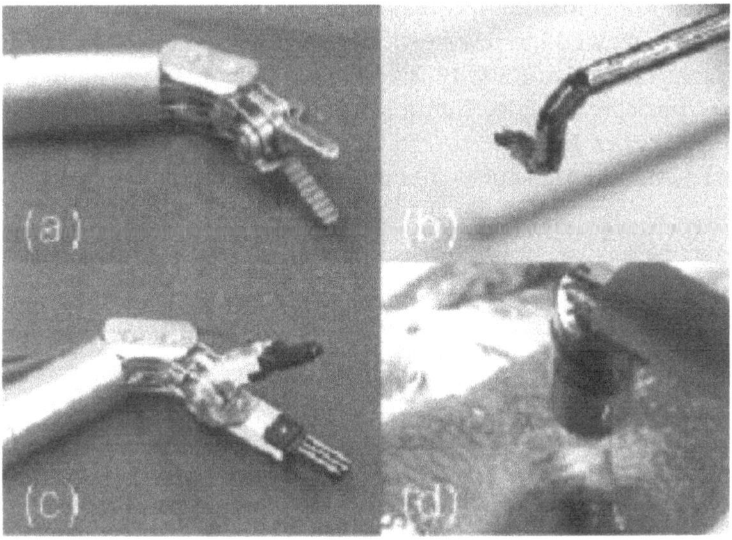

Fig. 4. Bending forceps (a) Wire mechanism (b) linkage mechanism, (c) bipolar electro cautery forceps (d) semi-conductor laser forceps

control was conducted using a real time base. The MASTER and SLAVE were connected using a dedicated communication channel employing TCP/IP protocol. These two systems were tightly connected as a single client device and a server robot. This system was also connected to the other distributed objects using CORBA. Using this pseudo-MASTER, the SLAVE could accept motion commands from the other objects using CORBA [16].

3. Experimental Results and Discussion

3.1 Evaluation of the slave robot system

Fig.5 (a) shows the Slave robot system for laparoscopic surgery. Evaluation of the basic performance of the manipulator, working range, force, torque, and speed was conducted. We measured the working range, maximum speed, and torque or force of forceps manipulator. Each motion except insertion had enough power to manipulate the 0.6 [kg] load. The working space of forceps was sector shape whose radius was 340[mm] and whose vertex angel was 180[deg] in horizontal plane, and vertical depth was 360[mm]. Bending forceps driven by a tendon mechanism had two independent joints that realized easy control and stable motion. The results of evaluation showed the low positioning accuracy and large backlash that was thought to arise from wire-driven mechanism. We need modification to control the wire tension. Output torque did not fill the required specification (0.6 [kg] or 6 [N]). Friction force between wire and pulley or wire path obstructed the transmission, and led to low output torque. The bending forceps utilizing linkage mechanism showed excellent performance with good reproducibility of motion and sufficient holding power. Histological study indicated the necrosis of the liver tissue and in-vivo experiment showed the feasibility of the robotic laser coagulator with CCD video endoscope [8].

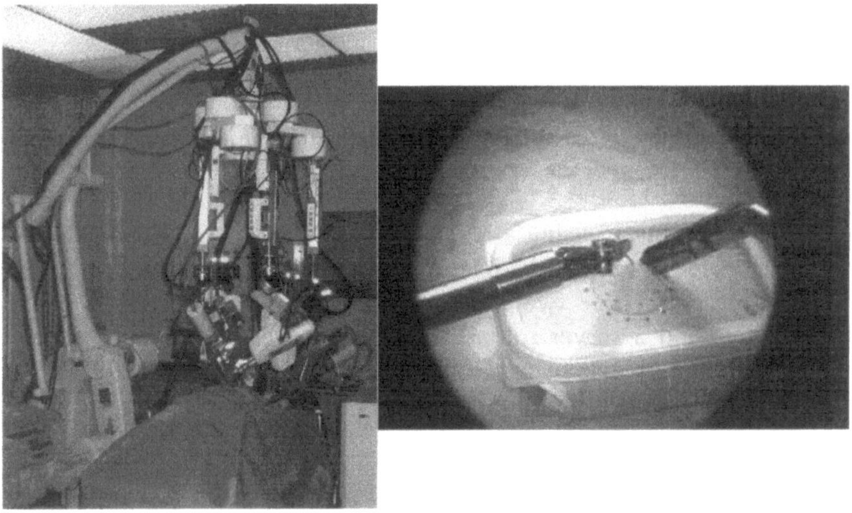

(a) (b)

Fig. 5. (a) Slave robot system for laparoscopic surgery (b) Suturing procedure conducted by two different types of bending forceps

Forceps manipulator had pivoting point using a new compact RCM mechanism realized by the two sets of linear actuators. Its pivoting point was bound mechanically at the trocar port, so that manipulator never injures the abdominal wall even if unexpected motion occurred because of electronic control errors. The danger of collision between manipulators found in conventional surgical robot would be reduced in our new robotic system. This was because moving swinging parts of manipulator outside the abdominal cavity needed relatively less space, and the space was located outside the surgical table.

3.2 Master-Slave operation of the system

We could successfully operate the master-slave manipulator and were able to record the control data without any problems. The manipulator was only activated when the foot switch was activated. The system demonstrated the expected behavior of the slave manipulator. We also could record the motion data of the system by the data logging system connected with CORBA system. Suturing procedure could be conducted with two different types of forceps can be used at the same time as shown in Fig.5 (b). This shows the flexibility of our master-slave surgical robot system.

4. Conclusion

We developed a new compact robotic system as a slave robot in a master-slave system. It consisted of three modules; manipulator positioning arm, forceps manipulator, and bending forceps with two DOFs. Manipulator positioning arm realized easy setting up by the combination of rough and precise positioning, and had wide working space. Forceps manipulator realized 4-DOF motion of the forceps around the incision hole and realized wide working space. As a RCM mechanism, we proposed a new "two liner actuator mechanism". Its miniaturized mechanism realized mechanical fixing of the pivoting point at the incision hole on the abdominal wall. Newly developed variable view angle laparoscope also contributed to reduce the volume occupied by the surgical robots above the patient. We have developed two types of bending forceps with 2 DOF, which were actuated by a tendon mechanism and linkage mechanism. Furthermore, functional forceps such as semi-conductor laser forceps and bipolar forceps were developed. We constructed a master-slave surgical robot system and demonstrated the feasibility of the system.

Acknowledgment

This study is supported by Research for the Future Program 99I00904 by Japan Society for the Promotion of Science and Grant in Aid for Scientific Research by Ministry of Health, Labor and Welfare (H15-Physi-002).

References

1. HashimotoD (1995) Advanced Technique in Gasless Laparoscopic Surgery. World Scientific
2. http://intuitivesurgical.com/
3. Kobayashi Y, et al. (2002) Small Occupancy Robotic Mechanisms for Endoscopic Surgery. Proc. of 5th International Conference on Medical Image Computing and Computer-Assisted Intervention - MICCAI2002 (I), pp75-82
4. Kim D, et al (2002): A new, compact MR-compatible Surgical Manipulator for minimally invasive liver surgery, Proc. of MICCAI2002 (I), pp99-106
5. Suzuki T, et al (2004) Development of forceps manipulator for assisting laparoscopic surgery, CARS2004 - Computer assisted radiology and surgery, Proc. of the 18th International congress and exhibition, pp.1338,
6. K.Nishizawa, et al.(2004) Development of Interference -Free Wire-Driven Joint Mechanism for Surgical Manipulator Systems, Journal of Robotics and Mechatronics, Vol.16, No.2, pp116-121
7. Yamashita Y, et al (2003) Multi-Slider Linkage Mechanism for Endoscopic Forceps Manipulator , in Proceedings of the 2003 IEEE/RSJ International Conference on Intelligent Robots and Systems (IROS 2003), Vol.3, pp2577-2582
8. Suzuki T, et al (2004): Development of a Robotic Laser Surgical Tool with an Integrated Video Endoscope, Proc. of 7th MICCAI, Lecture Note in Computer Science 3217, Springer, pp25-32
9. Suzuki T, et al (2003) Basic Study for automatic ligation and resection of bile duct with electric cautery, Journal of Japan Society of Computer Aided Surgery Vol.5, No.3, pp197-198
10. Taylor RH et al (1995) Telerobotic Assistant for Laparoscopic Surgery. Computer Integrated Surgery: The MIT Press, pp581-592
11. Sackier JM, Wang Y (1995): Robotically Assisted Laparoscopic Surgery: From Concept to Development. Computer Integrated Surgery: The MIT Press, pp577-580
12. Finlay PA, Ornstein MH (1995): Controlling the Movement of a Surgical Laparoscope. IEEE Engineering in Medicine and Biology Vol.14, .pp289-299
13. Hashimoto T et al (2004) Development of a Wide-angle View Laparoscope using Wedge Prisms, Journal of Robotics and Mechatronics 16(2): pp129-137

14. Kobayashi E et al (1999) A New Safe Laparoscopic Manipulator System with a Five-Bar Linkage Mechanism and an Optical Zoom Journal of Computer Aided Surgery, 4(4), pp182-192
15. Mitsuishi M, et al (2003) Development of a Remote Minimally -Invasive Surgical System with Operational Environment Transmission Capability, Proc. of the 2003 IEEE International Conference on Robotics and Automation, pp2663-2670
16. E.Aoki et al (2004) System Design for Implementing Distributed Modular Architecture to Relaible Surgical Robotic System , Proceedings of 7th International conference, Medical Image Computing and Computer-Assisted Intervention - MICCAI 2004, Part II,Lecture Note in Computer Science 3217, pp184-191

Part 3
Information and Communication Technology in Medicine

Attribute Selection Measures with Possibility and Their Application to Classifying MRSA from MSSA

Koulchi Hirata[1], Masateru Harao[1], Minoru Wada[2], Shogo Ozaki[1], Shigeki Yokoyama[3], and Kimiko Matsuoka[4]

[1] Kyushu Institute of Technology, Iizuka, Japan
[2] infoSense Co., Fukuoka, Japan
[3] Koden Industrial Co., Ltd, Tokyo, Japan
[4] Osaka Prefectural General Medical Center, Osaka, Japan

Chapter Overview. In this paper, we investigate the construction of decision trees with possibility. First we introduce a threshold d of possibility ($0 \leq \delta \leq 1$). Then, we formulate new measures for an attribute selection called a gain with possibility, a gain ratio with possibility and a GINI index with possibility. An intuitive idea for the gain and the gain ratio with possibility is that, for a probability p of the class +, the probability 1-p of the class - is replaced with max(d-p,0) and one for the GINI index with possibility that the value 1is replaced with d. Under the above new measures, we design a new algorithm to construct decision trees with possibility of which leaf is labeled by either d+ or -. Finally, we construct decision trees separating MRSA data from MSSA data in bacterial culture data by the new algorithm.

Key Words. decision tree, machine learning, data mining

Introduction

A decision tree, which is one of the most famous hypothesis in Machine Learning, is a directed tree consisting of nodes labeled by an attribute, edges labeled by an attribute value, and leaves labeled by a class value. In Machine Learning, it is the main purpose how to construct small decision trees. From the theoretical viewpoint, however, it is well known that constructing the smallest decision tree from a given database is intractable

[HJLT96, HR76]. Then, the introduction of several heuristics to construct small decision trees is one of the important research areas in Machine Learning [BFOS84, Min89, Qui86, Qui93, Utg89].

The heuristics is based on an *attribute selection* from a database. Quinlan has introduced a *gain* in his system ID3 [Qui86] and a *gain ratio* in his system C4.5 [Qui93] based on information theory. On the other hand, Breiman et al. have introduced a GINI *index* in their system CART [BFOS84] based on *impurity*. There are other heuristics as $\chi 2$, G value and Marshall value summarized in [Min89].

Consider a database D with two classes + and - .Let A be an attribute and p a probability of data with + in D, that is, $|D \ |/|D|$. Then, an *information* $I(D)$and an *impurity* $im(D)$of D are formulated as $- p \log p - (1 - p)\log(1 - p)$and $1 - p - (1 - p)$, respectively. Also *information* $I(A)$(*resp.*, *impurity* $im(A)$) of A is formulated as a summation of product of information (*resp.*, impurity) and frequency of the database separated by every attribute value of A. Then, a *gain* of A is formulated as $I(D) - I(A)$and a *GINI index* of A is formulated as $im(D) - im(A)$. Also a *gain ratio* of A is the normalization of the gain of A by *split information*. In constructing decision trees, we select the attribute of which gain, gain ratio or GINI index is minimum.

Note that all of them are based on *symmetric* functions. This symmetry is possible to be disadvantage if we know admissible and inadmissible to misclassify to classes. In order to construct decision trees under this setting, in this paper, we introduce a threshold d of possibility $(0 = d = 1)$ and deal with a label d+ instead of + as a leaf of decision trees, which means that $|D \ |= d|D|$. Here, - is admissible to misclassify to +. It is a kind of *cost-sensitive learning* (*cf.*, [Tur95]).

Then, we formulate a new information $I(A, d)$ and a new impurity $im(A, d)$, instead of $I(A)$and $im(A)$, respectively. Intuitively, $I(A, d)$is formulated by replacing the probability $1 - p$ in $I(A)$with $\max(d - p, 0)$ ($\log 0$ is assumed to be 0) and $im(A, d)$ is formulated by replacing 1 in $im(A)$with d. By using $I(A, d)$ and $im(A, d)$, we also introduce a *gain with possibility*, a *gain ratio with possibility* and a *GINI index with possibility*. Under these new measures, we design the algorithm $DT(D, d)$ to construct decision trees with possibility of which leaf is labeled by either d+ or - .

Finally, we apply the algorithm $DT(D, d)$ to bacterial culture data, in particular, MRSA (methicillin-resistant Staphylococcus aureus) data of which number is 118 and MSSA (methicillin-susceptible Staphylococcus aureus) data of which number is 4886, by expanding 93 attributes to 109. When we construct decision trees separating MRSA data from MSSA data, MSSA is admissible to misclassify to MRSA, while MRSA is inadmissible to misclassify to MSSA from the viewpoint of emerging infection in a clinical environment.

Note here that the data contain mutually the same patient, and the number of data only depends on the doctor's determination and is independent from the patients' situation. Then, in this paper, we also use *initial data* from a patient consisting of 35 MRSA data and 1613 MSSA data. Hence, we apply *DT(D, d)* to the data sampling all MRSA data and the MSSA data with the same number of the MRSA data randomly from both data and evaluate the constructed decision trees by *DT(D, d)*.

Attribute Selection Measures

In this section, we introduce the measures of an *attribute selection* according to [Min89].

Let D be a database and suppose that the classes of D is just two, + and - . The number of data in D is denoted by $|D|$. For an attribute A of D and an attribute value $a_i(1 = i = n)$of A, we denote the restriction of D that $A = a_i$ by $D_{A=a_i}$. We denote the restriction of D of which class is + (*resp.*, -) by D_+ (*resp.*, D). Note that each $D_{A=a_i}$ is mutually disjoint for i, and D and D are disjoint. Then, the following statement holds.

$$D = \bigcup_{i=1}^{n} D_{A=a_i}, D^+ = \bigcup_{i=1}^{n} D^+_{A=a_i}, D^- = \bigcup_{i=1}^{n} D^-_{A=a_i}.$$

Furthermore, we describe the separation of D for an attribute A and its value a_i as Figure 1.

A	+	-	$\cup$
a_1	$D^+_{A=a_1}$	$D^-_{A=a_1}$	$D_{A=a_1}$
...	...	...	...
a_i	$D^+_{A=a_i}$	$D^-_{A=a_i}$	$D_{A=a_i}$
...	...	...	...
a_n	$D^+_{A=a_n}$	$D^-_{A=a_n}$	$D_{A=a_n}$
$\cup$	D^+	D^-	D

Fig.1. The separation of a database D for an attribute A and its value a_i.

Let p, q_i and r_i be the following probabilities:

$$p = \frac{|D^+|}{|D|}, q_i = \frac{|D^+_{A=a_i}|}{|D_{A=a_i}|}, r_i = \frac{|D_{A=a_i}|}{|D|}$$

Hence, the probability of data with - in D and the probability of data with - in $D_{A=a_i}$ are $1 - p$ and $1 - q_i$, respectively. Furthermore, it holds that $q_1 + \cdots + q_n = r_1 + \cdots + r_n = 1$. Quinlan has introduced two measures, a *gain* [Qui86] and a *gain ratio* [Qui93] based on *information theory*. An *information* $I(D)$ of a database D is defined as follows.

$$I(D) = -p\log p - (1-p)\log(1-p).$$

An *information* $I(A)$ of an attribute A is also defined as follows.

$$I(A) = \sum_{i=1}^{n} r_i(D_{A=a_i}) = \sum_{i=1}^{n} r_i(-q_i\log q_i - (1-q_i)\log q_i).$$

Then, a *gain* $gain(A)$ and a *gain ratio* $gain\_ratio(A)$ of A are defined as follows.

$$gain(A) = I(D) - I(A),$$

$$gain\_ratio(A) = \frac{gain(A)}{split(A)}.$$

Here, *split*(A) are:

$$split(A) = -\sum_{i=1}^{n} r_i \log r_i.$$

Note that an attribute selection based on the gain tends to select an attribute of which value varies widely [Qui93]. In order to avoid such variance, an attribute selection based on the gain ratio adopts the normalization by split information.

On the other hand, Breiman *et al.* [BFOS84] have introduced a *GINI index* based on *impurity*. An *impurity* $im(D)$ of a database D is defined as follows.

$$im(D) = 1 - p^2 - (1-p)^2.$$

An *impurity* $im(A, a_i)$ of an attribute A and its value a_i and an *impurity* $im(A)$ of an attribute A are defined as follows.

$$im(A, a_i) = 1 - q_i^2 - (1-q_i)^2 = 2q_i(1-q_i),$$

$$im(A) = \sum_{i=1}^{n} r_i im(A, a_i) = 2\sum_{i=1}^{n} r_i q_i (1-q_i).$$

Then, a *GINI index* $gini(A)$ of A is defined as follows.

$$gini(A) = im(D) - im(A).$$

For an attribute selection in constructing decision trees, we select the attribute of which value of a gain, a gain ratio or a GINI index is maximum.

Attribute Selection Measures with Possibility

In this section, we introduce a new attribute selection based on possibility.

For a database *D* with just two class + and - ,let *p* be a probability of data with + in *D*. Then, an information *I*(*D*) and an impurity *im*(*D*) of *D* are represented in the following simple equations.

$$I(D) = -p\log p - (1-p)\log(1-p),$$
$$im(D) = 1 - p^2 - (1-p)^2.$$

Then, *I*(*D*)and *im*(*D*) satisfy the following properties.

1. They increase monotonically for 0 = *p* = 1/2 and decrease monotonically for 1/2 = *p* = 1.
2. They have the maximum value for *p* =1/2 and the minimum value 0for *p* =0 and *p* =1.
3. They are symmetric with respect to *p* =1/2.

For an attribute selection in constructing decision trees, we select the attribute *A* that *gain*(*A*), *gain_ratio*(*A*)or *gini*(*A*)is maximum. By the definitions, we select the attribute A that *I*(A), *I*(A)/*split*(A) or *im*(*A*) is minimum.

In this paper, we introduce a threshold *d* of possibility (0 = *d* = 1) and a leaf labeled by *d*+ instead of + in the decision trees. Then, we formulate the following new information *I*(*A*,*d*) and impurity *im*(*A*,*d*) of *A*. Note that, in order to set the value of measures to be 0 for q_i=0 and *d* = q_i= 1, it is necessary to add - *d*log *d* to *I*(*A*,*d*)and to maximize with 0

$$I(A,\delta) = \sum_{i=1}^{n} r_i\{\delta\log\delta + \max(0, -q_i\log q_i - \max(\delta - q_i, 0)\log\max(\delta - q_i, 0))\},$$

$$im(A,\delta) = \sum_{i=1}^{n} r_i\left(\max\left(0, \delta^2 - q_i^2 - (\delta - q_i)^2\right)\right)$$

Hence, we introduce the following *gain with possibility gain*(A,d), *gain ratio with possibility gain_ratio*(A,d) and GINI *index with possibility gini*(*A*,*d*) , *gain ratio with possibility gain_ratio*(A,d) and GINI *index with possibility gini*(*A*,*d*).

$$gain(A,\delta) = I(D) - I(A,\delta),$$
$$gain\_ratio(A,\delta) = \frac{gain(A,\delta)}{split(A)},$$
$$gini(A,\delta) = im(D) - im(A,\delta).$$

Furthermore, suppose that *G*(*A*, *d*) denotes one of the above new measures. In this paper, by using the above new measures, we design the algorithm *DT(A,d)* to construct decision trees with possibility of which leaf is labeled by either *d*+ or - , described as Figure 2. Here, *DT(A,* 1) means the standard algorithms to construct decision trees such as ID3 (if *G(A,*1) = *gain*(*A*)) [Qui86], C4.5 (if *G*(*A*,1) = *gain_ratio*(A)) [Qui93] or CART (if G(*A*,1)=*gini*(A)) [BFO84].

```
procedure DT(D, d)
if D = D+ then label to - ; /* leaf */
elseif |D⁺|=d|D| then label to d+; /* leaf */
else /* internal node */
  select an attribute A of D such that G(A, d) is minimum;
  label to A;
  forall an attribute value a_i of the attribute A do begin
    branch an edge labeled by a_i;
    DT(D_{A=a_i}, d);
   end /* forall */
```

Fig.2. Construction of decision trees with possibility

Empirical Results

In this section, we use two kinds of MRSA and MSSA data from Osaka Prefectural General Medical Center. One is the all data consisting of 118 MRSA data and 4886 MSSA data which we refer to a_MRSA and a_MSSA, and another is the initial data from a patient consisting of 35 MRSA data and 1613 MSSA data which we refer to i_MRSA and a_MSSA.

Figure 3 and 4 describe the evaluations of constructed decision trees from a_MRSA and i_MRSA, respectively, in 1000 times. Here, `nodes' and `depth' denote the average of the number of nodes and the average of depth in 1000 times, and `root' and `occ.' denote the most frequent attribute of the root and the number of its occurrence.

Furthermore, the attributes Dis26 and Dis32 denote the diseases of patients that is gynecology and poisoning, respectively. The attribute Drn4 denotes that the drainage is from a belly, and the attribute Dept denotes the department. The attributes Cep1, Cep4 and ML denote the sensitivity of antibiotics of 1st generation cephems, 4th generation cephems and macrolides, respectively.

The constructed decision trees by the gain and the GINI index are similar and become small by decreasing the value of δ. On the other hand, ones by the gain ratio grow large by decreasing the value of δ ; When δ =0.8 and 0.7, our algorithm constructs the largest trees for all data and initial data, respectively. The size of decision trees by the gain ratio is larger than one by the gain and the GINI index. Furthermore, the frequent attributes of the root for the gain ratio is different between δ =1 and δ =0.9.

gain	d=1.0	0.9	0.8	0.7	0.6
nodes	120.1	84.7	60.6	46.3	40.4
depth	3.83	3.63	3.46	2.76	2.43
root	ML	→	→	→	Cep4
occ.	628	758	658	922	686

gain ratio	d=1.0	0.9	0.8	0.7	0.6
nodes	91.1	167.3	183.9	175.1	132.6
depth	10.89	43.53	47.57	47.65	42.59
root	ML	Dis32	→	→	→
occ.	240	232	→	→	→

GINI	d=1.0	0.9	0.8	0.7	0.6
nodes	115.3	81.7	58.2	45.3	37.5
depth	4.37	3.87	3.55	2.78	2.46
root	ML	→	→	Cep1	Cep4
occ.	761	758	841	762	686

Fig.3. The evaluations of constructed decision trees from a_MRSA.

gain	d=1.0	0.9	0.8	0.7	0.6
nodes	41.4	28	18.6	17.1	12.7
depth	2.2	2.25	1.78	1.55	1.13
root	Dept	→	→	→	→
occ.	886	924	922	945	744

gain ratio	d=1.0	0.9	0.8	0.7	0.6
nodes	31.7	74.5	88.3	100.4	92.9
depth	6.21	30.39	35.5	39.04	35.98
root	Drn4	Dis26	→	→	→
occ.	237	218	→	→	→

GINI index	d=1.0	0.9	0.8	0.7	0.6
nodes	40.1	27.6	19	17.1	11.9
depth	2.36	2.41	1.78	1.53	1.17
root	Dept	→	→	→	→
occ.	842	904	920	940	660

Fig.4. The evaluations of constructed decision trees from i_MRSA.

A Virtual Schooling System for Hospitalized Children

Eisuke Hanada[1], Mamoru Miyamoto[2], and Kenji Moriyama[2]

[1]Department of Medical Informatics, Shimane University Hospital
[2]San-in Denko Co., ltd., Izumo, Japan

Chapter Overview. Children hospitalized for long periods of time often study in a room designated as the hospital classroom. Unfortunately, they are isolated from their classmates, which can lead to stress and dissatisfaction with hospital life. Therefore, a system was developed to help these children participate, as much as possible, in normal school life.

The system consists of a computer terminal in the hospital, a terminal in the school classroom, and a communication system connecting the terminals through the Internet. This system can send video/audio data and can operate the zoom function of a camera. The display and camera direction can be adjusted by the patient. The school-side terminal has a lamp that is lit when the patient terminal is in operation. With this system, the children can feel as if they are participating in the lessons and events of their school. Also, free conversation with classmates is possible.

Key Words. Hospitalized Children, Schooling, Remote control, and Multimedia communication.

1. Introduction

Patients hospitalized for long periods of time at the age of compulsory education (6-15 years old in Japan) can compensate for missed lessons by studying in "hospital classrooms", which can commonly be found in Japan in large hospitals with pediatric wards and in children's hospitals. Nearby elementary and junior high schools dispatch teachers who hold lessons at the hospital classroom. However, the management of these hospital classes has many problems. For example, the patient admission period is not fixed, some patients have limited mobility, some cannot be moved from their rooms, and some are isolated because of the possibility of the spread of infection.

For these and other reasons specific to each hospital, the lessons taught per day in a hospital are often fewer and shorter than those given in the usual elementary or junior high school setting. Also, many science experiments and physical education lessons cannot be done. Hospitalized students also cannot participate in various educational events done outside of the classroom [1].

Teleconferencing systems have been used to address these problems [2]. Although high effectiveness was shown, few have the flexibility to respond to individual interests and needs, so they tend to be limited to single events, possibly for the following reasons.

1. Many of the trial systems used a fixed screen with a single background. For example, a scene of a hospital is shown on a monitor placed at the front of a classroom. Thereby, the patient feels as if he/she is being watched at all times during the session. In this situation the mental stress of the patient might increase.
2. In many cases the position and direction of the camera at the school is fixed and can be operated only from the school side. Therefore, inpatients cannot change the direction of their camera so that they can see freely.
3. Even if a lesson uses a special room, neither the camera nor the screen supports movement.
4. The purpose of the test systems was in many cases only participation in lessons, and the patient could not converse with friends easily, such as during breaks, at meal times, or after school.

In recent years, concern has arisen that children often only talk within limited circles, leading to further concern that long-term patients may not be able to return to normal interaction after leaving the hospital.

Therefore, "a virtual schooling system" was constructed using digital teleconference techniques to solve the above problems. The aim of this system is to allow a hospitalized child to participate in school life as normally as possible. This system also aims to enable patients to participate in lessons and conversations, and to participate in special school events.

2. Construction and Functions of the System

This system consists of three parts, a hospital side terminal, a school side terminal, and communication functions. The communication function uses TCP/IP.

The system supports a single hospitalized child who uses a patient controlled terminal for communication with a school side terminal, which allows the student to feel as if he is in the regular classroom.

3. The Functions of Each Terminal

Both terminals have following functions.

- An image display of the counter side.
- Sound from the counter side.

Fig.1. First version terminals
Left: A client terminal for school side
Right: A control terminal for a patient

Only the school side terminal has the following functions, which can be operated by the patient from the hospital side terminal.

- Horizontal direction control of the movable cart swivel base.
- On/off control of the declaration lamp
- Vertical direction control of the camera
- Zoom function of the camera

For compression and extension of sound/audio data, MPEG4 real-time network codec (DM-NC40, Victor Company of Japan Ltd.) was used. The compression rate of this equipment is variable by prior setup: In this case 1Mbps for one direction.

(1) Movable cart swivel base (horizontal direction control facility)
The school side terminal (Fig.1, Left) has a movable cart with a swivel base. On the swivel base is a microphone for sending sound from the school to the patient, a camera for sending video of the school side, a dis-

play that shows the picture of the patient, and a speaker for the sound of the hospital side.

The swivel base rotates as directed by the signal from the patient terminal. This enables the patient to freely turn the camera and display in the desired direction and allows the school side to know in which direction the patient is looking, which indicates the student's interest to the teacher. Also, conversation with friends is enabled by this rotation, without changing the direction of the whole terminal.

For apparatus such as codec and displays that have not been put on a swivel slide or on a swivel slide connected using a cable, the degree of rotation is limited into 90 degrees left and right. A signal control encoding and transmission device for the swivel base rotation was developed. Signal transmission for this manipulation uses TCP/IP.

The hospital side terminal (Fig. 1, Right) does not have a swivel slide for protection of the privacy of patients other than the system user: Patients seldom have single rooms in Japanese hospitals. For patients who do not wish to have picture transmission, it is possible to set up only a sound transmission at the time of connection.

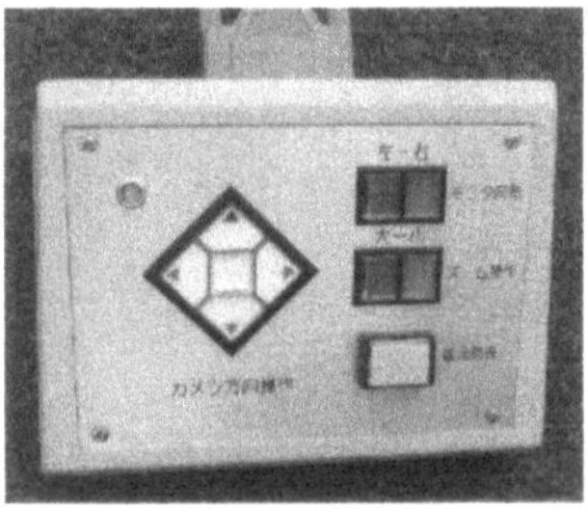

Fig.2. Patient control unit (Second version)

(2) Vertical direction and the zoom control of the school side camera
For the vertical direction control of the camera on the school terminal, we used remote control software that attached to the same camera as used in the first version of the terminal. Control software was installed in a notebook type personal computer connected to the hospital terminal. The zoom control of the camera was also enabled by this software.

For the second edition of the school terminal, the camera was replaced and control of the vertical direction and zoom was enabled by use of the signal control device shown in Fig. 2.

(3) A lamp for calling attention to the patient. (Declaration lamp)
A lamp that can be turned on or off by re mote control was installed on the school side terminal. This lamp simulates the raising of a hand when the

patient wishes to speak during a lesson. It not only enables the possibility of spontaneous response to a question, but also can display of the intention of the patient in other situations, such as participation to voting. Signal transmission to manipulate this lamp is the same as the above-mentioned vertical direction control.

4. Communication Function

Communication through the system was designed in consideration of the following points.

(1) Data communication in hospitals

The hospital terminal needs to be mobile so that it can be moved through the hospital building. For example, when an inpatient can sit up in bed but cannot move from the room (i.e., because of communicable infection), a terminal needs be moved to the patient's room.

Although the safety of wireless LAN use in hospitals has been reported [3], a consensus concerning such use has not yet been obtained [4]. Therefore, the probability that a bedside or hospital classroom information socket needs to be installed is high. In our test, the hospital terminal was connected directly to a LAN.

(2) Data communication in a schoolroom

In a schoolroom, in order to make it possible to experience the whole of school life, which is the purpose of this system, it is necessary to enable use of the school side terminal not only in the patient's classroom, but in special classrooms, the gymnasium, and the schoolyard.

Unfortunately, not only is it difficult for financial reasons to put a terminal on all classrooms, cases may occur in which two or more hospitalized children may need to participate in lesson or event in the same time. , It is desirable for the terminal to be made as mobile as possible.

It was thought that there would be little necessity for the school terminal to move during a session. Thus, in the prototype, an infrared wireless LAN system (OA-C301, OA-M301, 100Mbps (half duplex), Victor Company of Japan Ltd.) with an automatic-tracking function was used.

In addition, the use of a wireless LAN using an electromagnetic wave is possible as long as there are no special restrictions on the school side. However, since the data transfer rate is slower compared with LAN using infrared radiation, it is necessary to investigate the present wireless LAN system products in terms of compression and the display quality of the picture.

(3) Communication between the hospital and the school

For experimental use, LAN cables were wired to enable the school side terminal to be put in one of two places, a classroom on the third floor or in the school library. The hospital side terminal was placed in a laboratory on the third floor of the university cooperative research center (see Fig. 3).

The distance between the two buildings in this communication experiment is about 270m, and no buildings are between them. For financial reasons, outdoors type infrared modem systems (SIL155 M/L300, 155Mbps and 300m in maximum, Showa Electric Wire & Cable Co., Ltd.) were used, one installed in the laboratory at the university and the other on the third floor balcony of the school building.

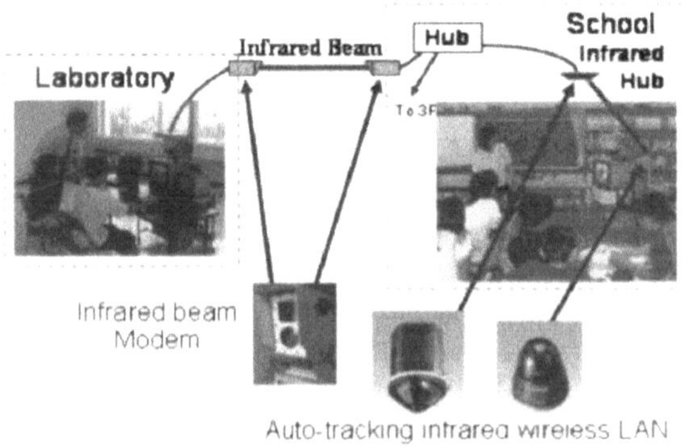

Fig.3. Experimental LAN

In the laboratory, the modem and the terminal were directly connected using a twisted-pair cable. In the elementary school, after connecting the switching hub to the infrared modem, twisted-pair cables were wired behind the ceilings of the two rooms. Indoor infrared hubs were installed in the ceilings of both rooms. This network was isolated from the public Internet.

5. Evaluation

5.1 Method

In order to obtain a performance evaluation of the first version of the system and to determine which functions should be added, a trial lesson was performed, after which an evaluation meeting with the teachers of the elementary school was held. The target elementary school is a large school with 931 children (as of May, 2003) and has routinely dispatched a teacher to the hospital classroom.

5.2 Evaluation by teachers

After an explanation of the function of the system, opinions were elicited from six teachers in a free discussion forum concerning the practical use of the system. Three of the participants had experience teaching in the hospital classroom.

The opinions concerning the basic functioning were very positive. Also, the following opinions were stated about possible additional functions.

・A system for sending supplementary materials (mainly print).
Supplementary materials are used for many lessons and examinations are regularly given for all subjects at Japanese elementary schools. Sending these supplementary materials and examination papers to the hospital in advance requires excessive time and effort for preparation. A function for sending the materials necessary for each lesson and tests would be easy to attach to this system.

・A function for responding to students needing individual assistance
Classrooms involve not only note-taking, but many times the students need teacher attention on an individual basis when doing lessons in an elementary school setting.

Furthermore, the following opinion was stated.

・Since children's minds tend to wander if they are not immediately engaged in what is happening in a lesson, smooth operation of the system is needed.

・In a hospital classroom, flexibility is important because changes in the planned syllabus often occur because of unforeseeable events.

5.3 Evaluation by children

With the cooperation of a sixth grade class (18 boys and 15 girls), one special lesson was performed. In the lesson, the function of the system was explained, and then two children were moved to the laboratory of our university. Questions and answers between the student and classroom teacher and questions and answers between children were performed.

The expectations for and an evaluation of the system were elicited at the start and end of the lesson. The questions and answers were as follows.

(1) Questionnaire and answers for/by children, before the lesson (33 answers)
Q1: Have you experienced hospitalization? ("Yes" or "No")
A1: "Yes" 6 boys, 5 girls "No" 12 boys, 10 girls

Q2: Supposing you were hospitalized, what would make you anxious? (Choose from below, two or more answers are allowed)

A2:1. Cannot study
10 boys, 9 girls, total 57.6%
2. Cannot meet friends (Lose contact with friends)
11 boys, 12 girls, total 69.7%
3. Cannot participate in school events
7 boys, 8 girls, total 45.5%
4. Other than above
Cannot participate in club activities (2 boys, 2girls),

Q3: Did you know about the hospital classroom? ("Yes" or "No")

A3: "Yes" 4 boys, 6 girls, total 30.3%

Q4: Do you think that you would be able to study effectively in a hospital classroom? ("Yes" or "No")

A4: "Yes" 2 boys, 5 girls, total 21.2%

Q5: What events in school other than lessons would you like to participate in when you are in the hospital? (Free answer)
Athletic meets: 7 boys. School trips: 5 boys, 4 girls

For Q1, girls rated "cannot meet friends" higher than boys. (61.1% of boys, 80% of girls)

The percentage of children who had experienced hospitalization and the percentage of children who knew of the hospital classroom were almost equivalent.

Most children thought that study in a hospital would not be as good as an everyday lesson. (79%).

(2) Questionnaire and answers for/by children, after the lesson (32 Answers)

Q1: Do you think a quality lesson is possible with this system? : (Choose one answer)

A1: "Yes": 24 children (75%) "No": 1 child (3.1%) "Don't know" 7 children (21.9%)

Q2: Please point out the bad points of this system.

A2: "It is hard to hear the voice": 14 children (43.8%)
"Breaking off of communication": 4 children (12.5%)

Q3: Please point out any functions you feel should be added to this system.

A3: "Move by itself by remote control": (2 children), "A support for books being read in the hospital", "A fax. type function", "Larger display"

Q4: Please point out uses for the system other than for lessons.

A4: "Chat" (16 children), "Visiting the hospital", "Playing", etc.

As mentioned above, conversation from a position distant from the terminal seemed difficult. Also, it was observed that adjustment or improvement of the speaker and microphone was necessary.

The reason for the breaking off of communication was the automatic-tracking function of the indoor type infrared equipment, which occurred when the position of the school side terminal moved during a lesson. It took almost three seconds to adjust.

6. Improvements and the Addition of Functions Based on the Evaluation Results

Improvements and additional functions as pointed out in the above evaluation of the first trial product are as follows.

(1) Changing the system casings and smoothing out the operation
In the first trial product, as a result of thinking that the function of the camera and the display were most important, the gross weight of the items carried on the swivel base was about 25kg, resulting in trouble turning. It was also dangerous for children when awkward revolutions occurred. The cart was, therefore, changed to a wooden box instead of a rack, and the revolving superstructure was stored in the cart. The size of cart was also reconsidered. The sizes of the new carts are 70cm x 65cm x 130cm for the school side robot and 50cm x 55cm x 120cm for the hospital side robot. (See Fig. 4)

(2) Simplification of the zoom and focus functions of the camera
In the first product, in order to perform vertical movements of the camera and control of the zoom and focus functions, software on the PC connected to the hospital terminal was used. However, the software was complicated and increased the operating procedures at the beginning and end of system, use, cause causing mistakes and difficulties for the student operators.

Therefore, the camera model was changed to one for which the manipulation control signal is public. A new remote-operation interface was developed that is easy for children to use. (See Fig. 2)

(3) Adding a print function
As mentioned in the evaluation, since supplementary material (printed copies) is often distributed during lessons at elementary and junior high schools, a function for the transmission of printed copies is indispensable.

To accomplish this, an easy to use scanner and printer were connected to the system, and a transceiver interface was developed. In addition, this function was made so that it could be added and removed easily.

(4) Techniques for avoiding communication blackouts
As some children pointed out in the evaluation, the automatic tracking infrared transmission equipment requires about three seconds for reconnection when communication is stopped for some reason, such as an

obstacle moving between the equipment.

Compared with hospitals, anxiety over EMI by electromagnetic waves is small in schools. Therefore, replacement of the high-speed wireless LAN with electric wave types, such as IEEE 802.11g may be preferable.

7. Discussion

As shown by the results of the questionnaires, children regard being "unable to meet friends" to be equally or more stressful than "study continuity is broken" if they were to be hospitalized for a medium-to-long period of time. It also became clear that they want to attend lessons that are held in other places than in the classroom and to participate in school events. It was shown that when considering distance education, participating in lessons in the regular school classroom is important and that conversation with friends should also be made possible.

Fig.4. Second version terminals Left: School, Right: Hospital

When using a fixed camera, smooth conversation cannot be done if the partner is not directly in front of the camera. However, our system with a swivel base that can be operated by remote control results in almost no stress in holding conversations between two or more friends, giving this system advantages in comparison with previous systems.

8. Conclusion

This system has possibilities for other than school education. We intend to explore the system for use as a visiting service for aged persons living alone.

In order to produce this system commercially, we plan to improve its communication ability. To realize our goals, verifying transmission speed is needed and encryption of communication data should be considered for communication using public networks.

Acknowledgment

The authors wish to express our thankfulness to ACK, Inc., Victor Company of Japan Ltd. Showa Electric Wire & Cable Co., Ltd., and LTEL Co., all of which assisted in the development of this system and to the teachers and children of Enya elementary school who cooperated in the experiment.

References

1. M. Takano (2003) "Parental Evaluations of Education Services on the Hospital-classroom," (in Japanese) Japan Journal of Child Health 62, (1):43-49
2. H. Yamakami, Y. HIgashibara, Y. Harada, et al (1999) "Construction of an inter-hospital School Network in Nagano," Proc. 19th Joint Conference of Medical Informatics, P-31
3. E. Hanada, Y. Hoshino, K. Takano, and T. Kudou (2004) "A Pilot Study on Electromagnetic Interference between Radio Waves Used in Wireless LAN Communication and Medical Electronic Equipment," Journal of Information Technology in Healthcare, 2(4): 281-291
4. E. Hanada, S. Hirano, Y. Watanabe, and S. Tsumoto (2002) "Wireless LAN Systems in Medical Environment," (in Japanese) Japan Journal of Medical Informatics, 22(3):287-294

Using Computational Intelligence Methods in a Web-Based Drug Safety Information Community

A. Ammar Ghaibeh[1], Mikio Sasaki[1], Eiko Nakata Doolin[2], Kumiko Sakamoto[3], Hiroshi Chuman[3], and Aiko Yamauchi[3]

[1]Saila Systems Inc. Tokyo.

[2]NTT-East Kanto Medical Center.

[3]Institute of Health Biosciences, Tokushima University, Japan.

Chapter Overview. A web-based drug safety information community is introduced. The developed community enables the circulation of drug teratogenicity information among health care consumers and health care professionals. In addition, the system provides data mining and decision support tools for use in epidemiologic research and drug discovery.

Key Words. Drug safety, Computational Intelligence, Pharmacoepidemiology, and Drug Discovery

1. Introduction

The wide and fast spread of the Internet has very much affected the society in all aspects because it provides an easy to access source of information and offers a virtual space where people from different locations, backgrounds, ages, and professions can meet, and exchange opinions, knowledge, or experience.

By its nature activities related to the public health involve a large number of participants such as doctors, patients, pharmacists, and of course researchers. It is obvious that applications in this area can benefit from the Internet to ease communication and improve research and service quality. For example the Internet can be used to remotely monitor patients, collect measurements data, and send it to hospitals or health centers for observation. It can be also used for remote consultation via ordinary emails or form oriented sites. In addition it provides a powerful tool for researchers to share ideas and enhance research results. For example in epidemiologic studies, web based technologies are more advantageous than classical approaches because they allow for the simultaneous access to information to a wider number of investigators regardless of their location. On the other

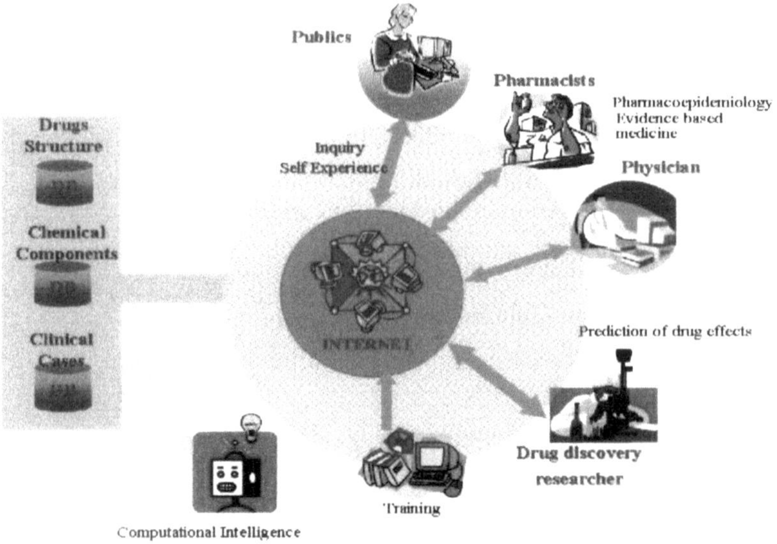

Fig. 1. The system provides useful tools for health care consumers, health care professionals, trainees, and epidemiological and drug discovery researchers.

hand, the Internet creates new challenges such as data security and individuals' privacy protection that should be addressed.

This chapter describes a web-based drug safety information community developed as a joint project of Tokushima University, NTT-East Kanto Medical Center, and Saila Systems Inc. The system aims to provide a common drug teratogenicity database available to professionals and the public. The public are provided with search and self experience sharing forms, while researchers are provided with computational intelligence and decision support tools for data mining, data analysis, and chemical structure similarity search.

2. System Description

Usually information about drug effects, chemical structures, and clinical cases, are collected in separate closed databases. The objective of this project is to join these databases into one common database accessed via the Internet by health care consumers, health care professionals, trainees, and epidemiological and drug discovery researchers.

The importance of the project emerges from its multifaceted benefits to different users (Fig. 1):

- Hospitals, Pharmacies, and Public users (particularly pregnant women) can use the system to collect information about a certain drug and its side effects and teratogenic risk. In addition the system enables the public to add their experience to the database. As opposed to other systems reports about the safe usage cases of medicines are also collected. However, these reports are not intended to be an evidence of the absence of bad effects, but simply an indication which should be further investigated.
- Researchers from different fields such as pharmacoepidemiology, evidence based medicine, and drug discovery can benefit from the collected data and analysis tools provided in their researches.
- Drug industries can also benefit from the data and computational intelligence tools provided to investigate the possible degree of risk of an underdevelopment drug and avoid components with high teratogenic risks.
- In addition, the system can be used for training students and medical staff on chemoinformatics, risk identification, risk management, and drug information service.

The project is faced with many ethical and technical challenges. Ethically, the validation of the provided information and the determination of what portion of the data can be revealed to a certain user group without any violation of members' privacy are critical issues that need to be addressed. Technically, designing the system to meet the requirements of different users with different objectives and interests, providing high level of security and privacy protection, and the implementation of data mining and decision support algorithms are very challenging problems.

Ethical problems are solved by an Information Evaluation Committee with professional members who will be responsible of validating the entered data and making the decisions addressing privacy issues.

The next section briefly addresses the technical issues and describes the tools provided to the system users.

5. System Clients

The system is intended to be used by a wide range of users with different interests. To meet users' requirements different access methods have been developed. Basic database search functionality is provided to pharmacists, hospitals, health care consumers and researchers via a simple web site (Fig.3). A web server works as a broker between the Internet browser and the system's web service. The web server requests the proper data manipulation functions from the web service providing it with the correct parameters transformed into XML format and converts the output retrieved also in XML format into HTML format understandable by the commonly used Internet browsers.

The system's more complex functions or sensitive information can be accessed via stand-alone programs needed to be installed on the client machine. The use of stand-alone programs has the following benefits:

1. Stand-alone programs provide more flexibility for graphical manipulation.
2. The use of local resources will reduce the computation server load especially for time and computational consuming processes.
3. Limiting the number of users that can access the data to users who have the program installed on their machines will provide a better overall system security.

As mentioned earlier other systems can also act as clients and integrate the web service into their systems.

6. Computational Intelligence and Drug Design

It is very useful for drug discovery researchers to inspect the presence or absence of a chemical structure or substructure in the molecules database. This is because it is expected that molecules with similar structure exhibit similar physical properties or similar biological activities. Searching for similar chemical structures [1] is a very computationally demanding problem. The problem is much more demanding when the search for substructures is required.

A tool for conducting a 2 D similarity search in the molecules database has been developed. The researcher has the option to load the required structure from a file of a CT (Chemical Table) or MOL formats [2, 3] or to draw the structure directly using the developed program as shown in Fig.4.

When the user selects the required similarity algorithm the program converts the chemical structure into XML format and sends the data to the

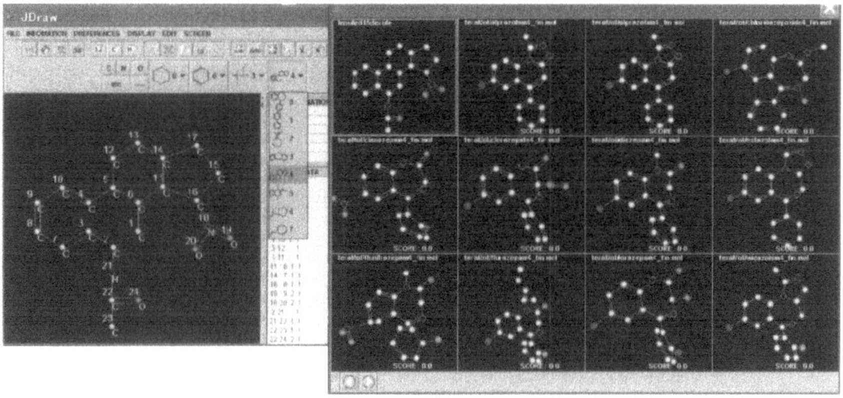

Fig. 4. The chemical substructure similarity search results as provided by the developed program.

web service asking for the execution of the selected similarity search function. The web service in its turn requests the execution of the selected search algorithm from the computation server, receives the results, and sends it back to the program to be converted from XML to its graphical presentation. The use of SOA makes it possible to easily add new similarity search algorithms to the system.

6.1 Data mining

The system provides data mining methods to be used for analyzing the collected clinical cases. Data mining (also referred to as knowledge discovery) uses machine learning and statistics to extract useful information implicitly embedded in the data. Those methods are particularly useful for researchers in pharmacoepidemiology and evidence-base medicine.

7. Summary

The Internet provides a new opportunity for multidisciplinary researches, and facilitates the cooperation of large number of participants regardless of their geographical location. The public health sector is one of those areas where large number of participants with different background is involved.

In this paper a web-base drug safety information community has been presented. The community provides an opportunity to share data among drug industries, the publics, and interested researchers from different fields. The system implements computational intelligence methods to facilitate data analysis and provide tools for data mining and chemical structure and substructure similarity search for use in epidemiological research and drug discovery process. A web service oriented structure was used to provide the system with good reliability, usability, extendibility, and interoperability with other systems.

Acknowledgement

This project is supported by the Research Institute of Science and Technology for Society (Ristex) of Japan Science and Technology Agency (JST).

References

1. P. Willett, J. M. Barnard, and G. M. Downs, "Chemical similarity searching," Journal of Chemical Information and Computer Science, vol. 38, pp. 983-997, 1998.
2. CT files Formats MDL information System, Solutions White Papers http://mdl.com/solutions/white papers/ctfile_formats.jsp.
3. A. Dalby, J.G. Nourse, W.D. Hounshell, Aki Gushurst, D.L. Grier, B.A. Leland and J. Laufer, "Description of Several Chemical-Structure File Formats Used by Computer-Programs developed at Molecular Design," Journal of Chemical Information and Computer Science, vol. 32, No.3 pp. 244-255, 1992.

Analysis of Hospital Management Data using Generalized Linear Model

Yuko Tsumoto[1] and Shusaku Tsumoto[2]

[1] Department of Fundamental Nursing, Shimane University, School of Nursing, Shimane, Japan
[2] Department of Medical Informatics, Shimane University, School of Medicine, Shimane, Japan

Chapter Overview. Rapid progress in information technology has come to enable us to store all the information in a hospital information system, including management data, patient records, discharge summary and laboratory data. Although the reuse of those data has not started, it has been expected that the stored data will contribute to analysis of hospital management. In this paper, the discharge summary of Chiba University Hospital, which has been stored since 1980's were analyzed to characterize the university hospital. The results show several interesting results, which suggests that the reuse of stored data will give a powerful tool to support a long-period management of a university hospital.

Key Words. Medical Data Mining, Multivariate analysis, Hospital Management

Introduction

It has passed about twenty years since clinical information are stored electronically as a hospital information system since 1980's. Stored data includes from accounting information to laboratory data and even patient records are now stared to be accumulated: in other words, a hospital cannot function without the information system, where almost all the pieces of medical information are stored as multimedia databases[1]. Especially, if the implementation of electronic patient records is progressedinto the improvement on the efficiency of information retrieval, it may not be a dream for each patient to benefit from the personal database with all the healthcare information, "from cradle to tomb". However, although the studies on electronic patient record has been progressed rapidly, reuse of the stored data has not yet been discussed in details, except for laboratory data and accounting information to which OLAP methodologies are applied. Even in these databases, more intelligent techniques for reuse of the data, such as data mining and classical statistical methods has just started to be applied from 1990's[2,3]. Human data analysis is characterized by a deep and short-range investigation based on their experienced "cases", whereas one of the most distinguished features of computer-based data analysis is to enable us to understand from the different viewpoints by using "cross-sectional" search. It is expected that the intelligent reuse of data in the hospital information system provides us to grasp the all the characteristics of university hospital and to acquire objective knowledge about how the hospital management should be and what kind of medical care should be served in the university hospital. This paper focuses on the following two points for analysis. One is what kind of knowledge can be extracted by statistical methods from the datasets stored for about twenty years in Chiba University Hospital. The other is how these pieces of knowledge areuseful for the future hospital management and decision support. The analysis gives interesting and results. For example, the discharge summaries show that most of the patients are admitted to the university hospital because of the diseases which requiresthe advanced treatment, such as malignant neoplasm. Combination of the discharge summaries and data in the hospital accounting systems shows that the profitability significantly differs in each disease, but that within each disease, the number of days inthe hospital is a principal factor for the profitability. Also, most of the distributions of the number of days in the hospital for the diseases do not follow the normal distributions, but log-normal distributions, which influence the profitability of the university hospital. The reason why the distributions follow the log-normal distribution should be investigated in the near future.

Objective

The objectives of this research is to investigate what kind of knowledge can be extracted by statistical methods from the datasets stored in the hospital information system of Chiba University Hospital, especially useful for future hospital management anddecision support. Especially, since the revenue of Japanese hospital is based on NHI points of Japanese medical care, it is important to investigate the factor which determines the amount of NHI points.

Methods

Representation of Discharge Summaries

When the hospital information system for discharge summaries is introduced in Chiba University Hospital in 1978, a discharge summary is distributed to doctors as a paper sheet for each patient admitted to the hospital. Doctors fill in each sheet just after the patient leaves the hospital, the parts of this sheet which can be coded are stored electronically. A sheet for discharge summary is composed of the items common to all the departments and the items specific to each department. For example, the itemsspecific to neurology consists of the results of neurological examinations and the severity of diseases. The common items consist of those in which codes or numerical values should be filled in and those in which texts should be input. After the doctor in charge fill in those items and submit to the division of medical records, the staff input codes and numerical values into a database system. These processes are continued until a new hospital information system was introduced in 2000, which means that the non-text items common to all the departments has been stored for about 20 years. There are 16 items for codes or numerical values: patient ID, the department in charge, occupation, height and weight on admission, height and weight just before hospital discharge, a motivation for visit, outcome, autopsy or not, cause of death, the date of first visit, the date of admission, the date of discharge, the name of disease (ICD-9 code [4]), treatment method. However, height and weight just before hospital discharge are not input in the database. Concerning the items specific to each department, only those of surgery and ophthalmology are stored electronically.

Extraction of Datasets from Hospital Information System

The databases in the hospital information system of Chiba University Hospital are described by MUMPS[5]. MUMPS is a programming language which can construct a database with a hierarchical structure based on a binary tree. By using the characteristics of a binary tree, each item can store several data as a tree, which makes the data management and retrieval more efficient than relational databases. Datasets for analysis are extracted from the database on discharge summaries and the database on patient basic information by using patient ID and the date on admission as keys. The program for extraction is developed by the first author due to the following reasons. Since NHI points, which stands for National Healthcare Insurance points, are stored for each patient ID and each month, the total points for each admission for each patient are calculated from NHI points for each month. The total points are combined with the dataset extracted from the discharge summaries by using patient ID and the date on admission as keys. The number of the records of the dataset extracted from the global: MRMG, which is a database on discharge summaries, is 157,636 for 21 years from 1978.4 to 2000.3. The time needed for computation is about one hour by SUN Workstation (Enterprise 450). Concerning the dataset combined with NHI points, the number of the records is 20,146 for three years from 1997.4 to 2000.3.

Methods for Statistical Analysis

Descriptive statistics, exploratory data analysis and statistical tests were applied to the dataset extracted only from the discharge summaries for the analysis of patient basic information (gender, age and occupation), outcome, the number of the days in hospitals and diseases, including their chronological trends. Concerning the datasets combined with accounting information for three years (1997.4 to 2000.3), the relations among NHI points and items in the discharge summaries were analyzed by descriptivestatistics, exploratory data analysis, statistical tests, regression analysis and generalized linear model. SPSS 11.0J for windows was used for these analyses.

Results

Due to the limitation of the spaces, the most interesting results are shown in this section. In the subsequent subsections, the results of the whole cases, and two levels of ICD-9 code, called major and minor divisions, are compared. Especially, concerning the results for the major and minor divisions, malignant neoplasm and the following three largest minor divisions of the malignant neoplasm are focused on: neoplasm of trachea, bronchus, and lung, neoplasm of stomach, and neoplasm of liver and intrahepatic bile ducts. In the subsequent sessions, neoplasm of lung, stomach and liver denotes the above three divisions for short.

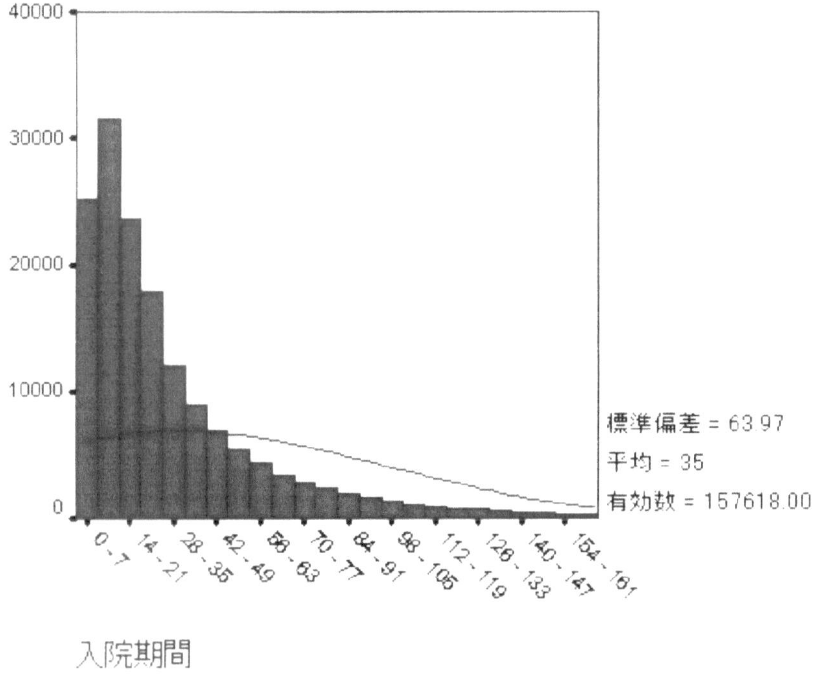

Fig. 1. Distribution of Length of Stay (Raw Data, Whole Cases)

Distribution of Length of Stay

Figure 1 shows the distribution of the length of stay of the whole cases, which skewed with the long tail to the right. Thus, the logarithm of the length of stay is taken to tame this skewness. Figure 2 gives the result of this transformation, whose distribution is very close to normal distribution: this means that the distribution of the whole cases is log-normal. This observation holds even when the major divisions are taken as sample. It is notable that the nature of this distribution holds even in the minor divisions, three largest three diseases.

Table 1 summarizes the descriptive statistics of length of stay with respect to the whole cases, major and minor divisions. The natures of these distributions are not significantly changed ifthese cases are stratified by the condition whethera surgical operation is applied to a case or not.

0.1 Distribution of NHI Points

Since the variance of raw data of NHI points are very large and the distribution is skewed, the raw data are transformed into the "median index", which is defined as the ratio of the total points to the median of the whole cases. Figure 3 and 4 show the distribution of the raw data and that of the logarithm of the raw data of median index. Those figures suggests that the NHI points of the whole cases follow log-normal distribution. On the other hand, the distributions for minor divisions are different. The same observations are obtained from the distribution of NHI Points of Neoplasm.

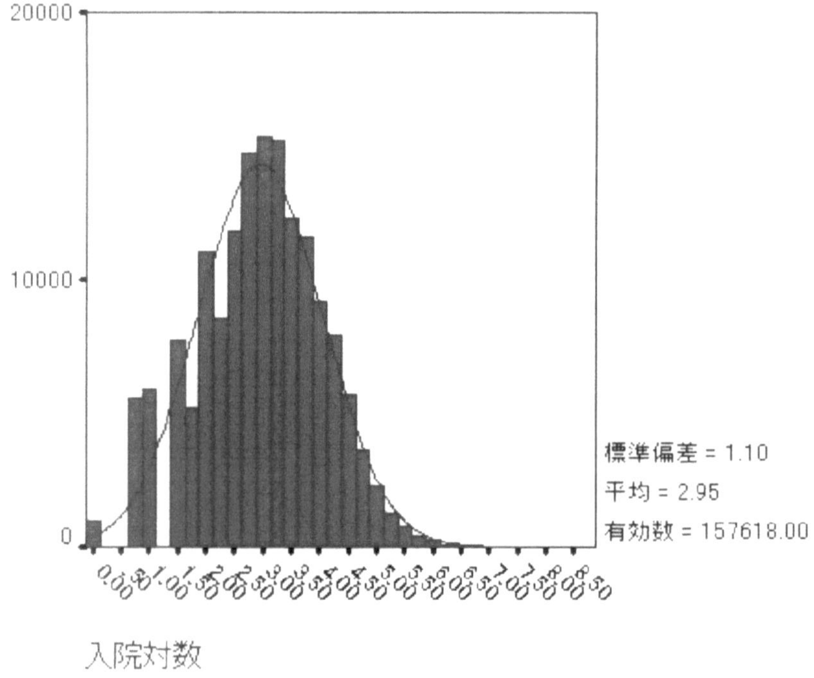

Fig. 2. Distribution of Length of Stay (Logarithm Transformation, Whole Cases)

Table 1. Descriptive Statistics of Length of Stay

	Average	Median	SD	Skewness	Kurtosis
Whole Cases					
Raw Data	26.46	16.00	33.67	4.34	34.15
Logarithmic Transformation	2.74	2.77	1.06	-0.06	-0.28
Neoplasm					
Raw Data	37.54	25.00	38.72	2.90	13.21
Logarithmic Transformation	3.19	3.22	0.98	-0.32	0.08
Malignant Neoplasm of Lung					
Raw Data	49.65	39.00	43.42	2.57	10.82
Logarithmic Transformation	3.57	3.66	0.88	-0.79	2.00
Malignant Neoplasm of Stomach					
Raw Data	36.44	36.00	19.18	0.46	0.37
Logarithmic Transformation	3.40	3.58	0.72	-1.42	2.55
Malignant Neoplasm of Liver					
Raw Data	35.93	33.00	21.40	1.19	2.70
Logarithmic Transformation	3.38	3.50	0.71	-1.18	3.03

Correlation between Length of Stay and NHI Points

Figure 5 depicts the scattergram between the length of stay and NHI points of the whole cases, which suggests a high correlation between two variables. For simplicity, the vertical and horizontal axes show the logarithm of raw values. Actually, The coefficient of correlation is calculated as 0.837, which means that the correlation is very strong.

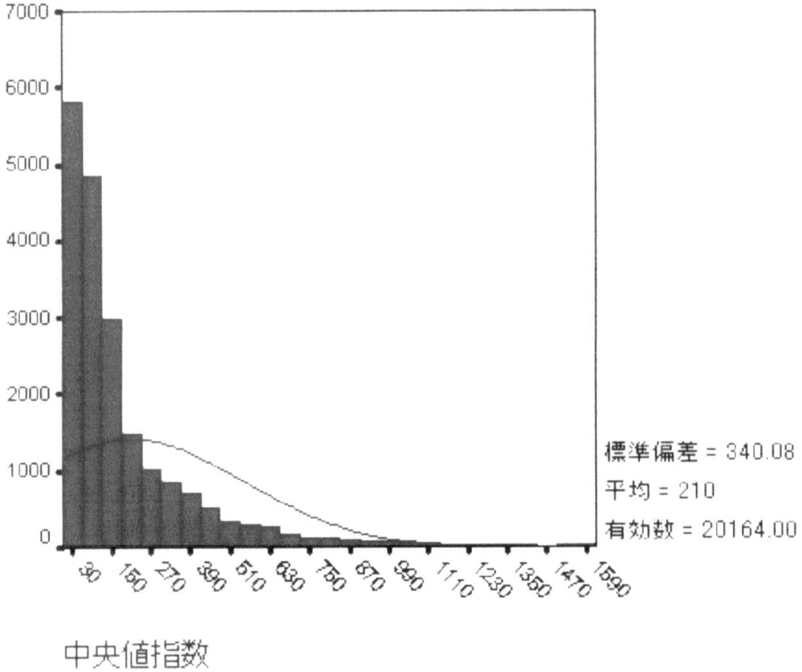

Fig. 3. Distribution of NHI Points (Raw Data, Whole Cases)

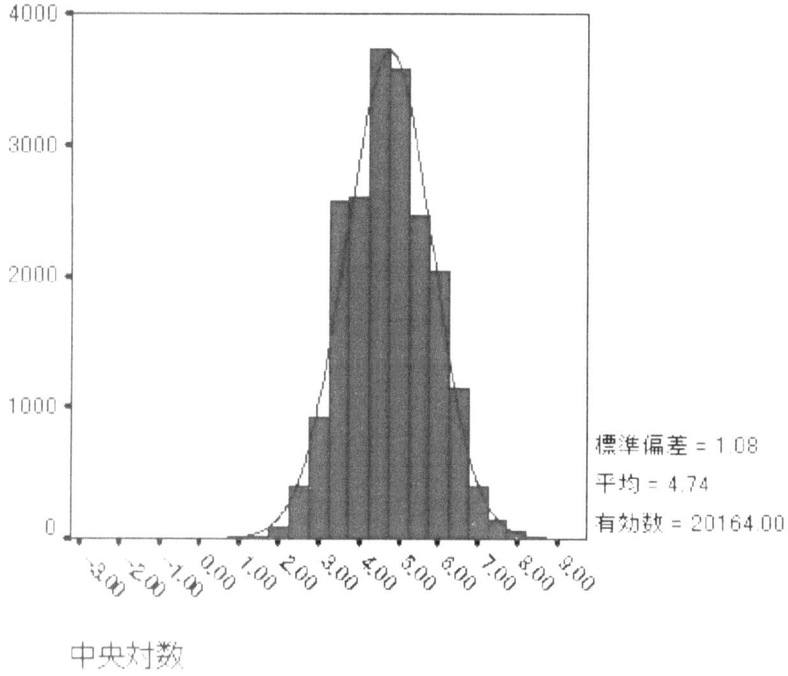

Fig. 4. Distribution of NHI Points (Logarithm Transformation, Whole Cases)

Table 2. Correlation between Length of Stay and NHI Points (After Logarithm Transformation)

	Total	With Operation	Without Operation
Total Cases	0.837	0.829	0.779
Neoplasm	0.867	0.844	0.826
Lung Cancer	0.838	0.648	0.903
Stomach Cancer	0.827	0.738	0.801
Liver Cancer	0.711	0.577	0.755

Table 2 summarized the correlation coffecients between NHI points and Length of Stay with respect to the whole cases, neoplasm and three major types of malignant neoplasm: lung, stomach and liver.

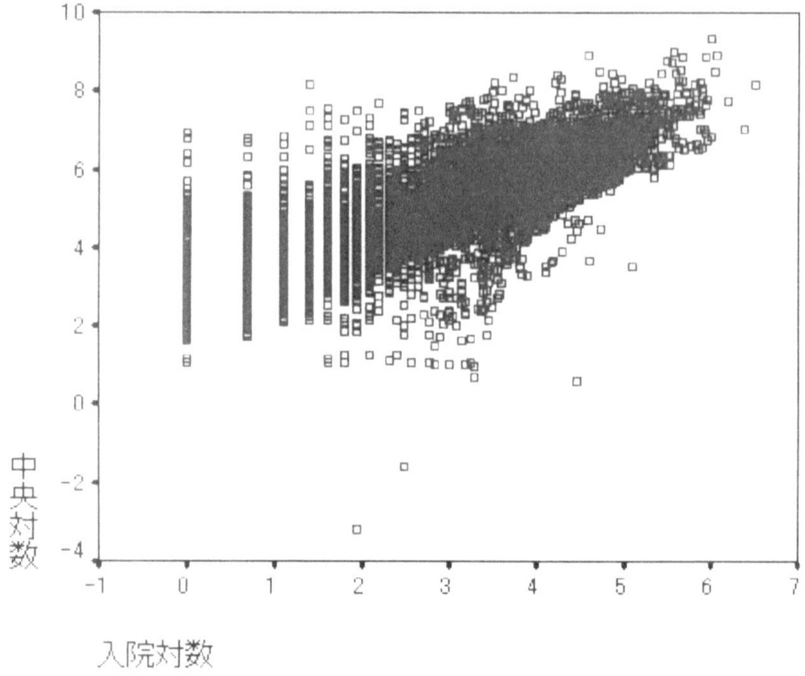

Fig. 5. Scattergram of Length of Stay and NHI Points (Logarithm Transformation, Whole Data)

Comparison of the coefficient of correlation between the group with and without a surgical operation shows that the group without an operation has higher correlations than that with an operation, which suggests that NHI points of the treatment methods other than surgical operations should be-strongly dependent on the lengths of stay.

Generalized Linear Model

Since all the items except for the length of stay are categorical variables, conventional regression models cannot be applied to the study on relations between NHI points and other items. For this purpose, generalized linear model [6] was applied to the dataset on combination of accounting data and discharge summaries. NHI point was selected as a target variable and the following four variables were selected as explanatory variables: outcome, treatment method, major division of ICD-9 codes and thecategorized length of stay. The length of stay is categorized so that the distribution of the transformed variable is close to normal distribution, where the width of windows is set to 0.5 for the logarithmic value of the length of stay. Treatment, outcome and major divisions of ICD codes are transformed into dummy variables to clarify the contributions of these values to a target variable. For example, the outcomes of discharge are split into the following six dummy variable: D1: recovered, D2: improved, D3: unchanged, D4: worsened, D5: dead and D6: others. Table 3 shows the results of GLM on the total cases, whose target variable is NHI points. All the variables are sorted by the F value. The most contributing factor is the length of stay, whereas the

Table 3. GLM Analysis on NHI Points (Total Cases, Logarithmic Transformed Data)

	F - value
Length of Stay	1590.3
Death	347.7
Psychiatric Disorders	264.4
Pregnancy Complications	241.7
Circulatory Diseases	228.2
Operations	119.4
Perinatal Disorders	98.2
Precipitating	56.2
Treatment	53.9
Neurological Diseases	52.1
$R^2 = 0.741$	

Table 4. GLM Analysis on NHI Points (Neoplasm, Logarithmic Transformed Data)

	F - value
Length of Stay	625.8
Death	36.0
Operation	23.2
Changeless	16.4
Other Therapies	9.8
Radiation	8.5
Exacerbation	8.2
$R^2 = 0.753$	

contributions of the other factors are small. Table 4 gives the results of GLM on major division (malignant neoplasm), and three minor divisions, whose target variable is NHI points. Compared with Table 3, the number of the factors which gives the significant contributionsto NHI points are very small, which suggest that the variabilities of NHI points in major and minor divisions are very low, compared with that of total cases.

Conclusion

This paper analyzes the following two datasets extracted from the hospital information system in Chiba University Hospital. One is the dataset extracted from the database on discharge summaries stored for about twenty years from 1978 to 2000. The other iscombination of data from discharge summaries and accounting information system. The analysis gave the following results: (1) malignant neoplasm is the first major category which determines the profitability of Chiba University Hospital, which is stable for twenty years. (2) In a global view, the length of stay is the principle factor for the revenue of the hospital, whose distribution follows the log-normal distribution. (3) Treatment method may be a secondary factor to determine the distribution of the length of stay for each disease, which may be correlated with the property that the length of stay follows log-normal distribution for each minor division in total. (4) Treatment without a surgical operation should be more examined by additional information, which is also important to evaluate the profitability of the university hospital.

Reference

1. Institute of Medicine Committee on Improving the Patient Record (1997) The Computer-based Patient record: An Essential Technology for Health Care. National Academy Press, Washington DC.
2. Tsumoto S (2000) Knowledge discovery in clinical databases and evaluation of discovered knowledge in outpatient clinic. Inf. Sci, 124(1-4): 125-137.
3. Tsumoto S (2001) Chapter G5: Data mining in medicine, In: Kloesgen W, Zytkow J, editors. Handbook of Data Mining and Knowledge Discovery, pp.798–807, Oxford University Press, Oxford.
4. Online ICD9/ICD9CM codes and Medical Dictionary. URL. http://icd9cm.chrisendres.com/
5. Walters R (1997) M Programming: A Comprehensive Guide. Butterworth-Heinemann, Woburn.
6. McCullagh P and Nelder JA (1990) Generalized Linear Models. 2nd ed. ,CRC Press, Boca Raton.

Data Mining Approach on Clinical/Pharmaceutical Information accumulated in the Drug Safety Information Community

Aiko Yamauchi, Kumiko Sakamoto, and Hiroshi Chuman

Graduate School of Pharmaceutical Sciences, The University of Tokushima, Tokushima Japan

Chapter Overview. Congenital malformations or birth defects are a major public concern. We have undertaken a new research/development project to create an information-community for teratogenic agents, as a social system, consisting of pregnant women, pharmacists, physicians and research scientists working on medicinal products. We have designed a web-based database system which performs not only retrieval of teratogenic information and chemical structures of drugs/medicinal products, but also registration of clinical case reports and teratogenic information on various chemicals. Using the database, we predicted the drug excretion into human breast milk by means of QSAR, as a data mining approach on clinical/pharmaceutical information.

Key Words. Information-community system,
Pregnancy and teratogenic agents, Medicinal products, Human breast milk, and Clinical QSAR.

1. Introduction

The progress of the latest science and technology, so-called "IT, medical and genomic revolution", is innovative. The knowledge from science and technology should be used to support people's safe social lives, and the information from them needs to be supplied for the development of novel science and technology. Thereby it is important to make a social system in which the information from citizen's life can be fed back effectively to the progress of society (Fig. 1).

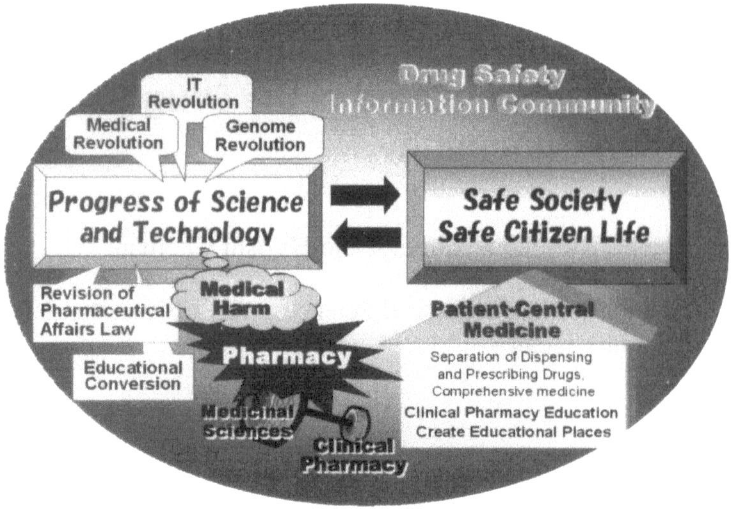

Fig. 1. A trial to create a new social system.

In the medical field, an epoch-making new medicine has been developed one after another, and thus it is difficult for the doctor to maintain the drug safety to patients, independently from other medical staffs. Actually, sorrowful medical harms happened until now. Therefore, the separation of dispensary from medical practice has been increasing up to more than 50% in Japan. The Japanese pharmacy schools have started reforming the educational system in order to balance the pharmaceutical sciences and the clinical pharmacy.

From such background, we have undertaken a new research/development project, "Construction of a Drug Safety Information Community", supported by Research Institute of Science and Technology for Society in Japan Science and Technology Agency [1]. The objective of this research is to develop a drug safety information community as a social system. This information community system allows an effective circulation of drug safety information among health care consumers, health care professionals, and drug discovery researchers through sharing the knowledge each other using the web-based system (Fig. 2) [2].

In a future study of pharmaceutical information sciences, the technique of computational chemistry should be positively introduced in order to predict unknown adverse effects of a drug using the precious clinical data. Thus, we have carried out a data mining approach on clinical/pharmaceutical information accumulated in the drug safety information community.

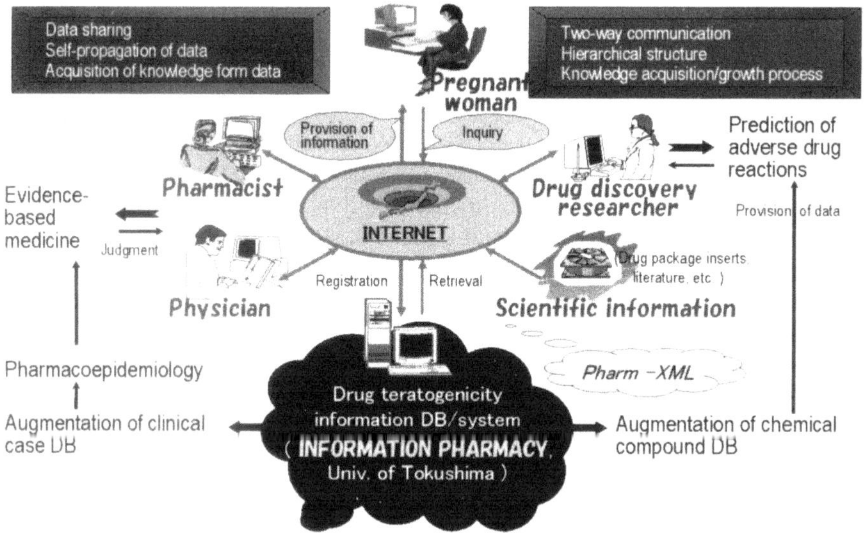

Fig. 2. Circulation of drug teratogenicity information within the drug safety information community [2].

2. Construction of a Drug Safety Information Community

Recently declining birthrate and a growing proportion of elderly people have been rapidly advanced in Japan. Especially, the decrease in the birthrate is one of the serious issues for the future of society. In such a society, the medication is one of the most important therapeutic methods for the patient. On the other hand, congenital malformations or birth defects are major public concern, since the thalidomide catastrophe (limb defects) had occurred in the beginning of the 1960's when this drug was administered to pregnant women as an anti–anxiety agent during first trimester [3,4]. However, lack of proper information may panic a pregnant woman who had taken medicines without noticing the pregnancy, and thus may result in an unnecessary abortion.

In 1963, the Japanese Ministry of Health and Welfare imposed the teratogenicity test using experimental animals in manufacturing approval applications of medicinal products. The guideline regarding detection of toxicity to reproduction was also determined in 1989. Therefore, during the development of new drugs, the reproductive risk is necessarily examined using experimental animals. After the cessation of development, however, the reproductive data of medicinal products generally have been held privileged by pharmaceutical companies without dissemination efficiently and widely.

The clinical trial program almost never includes pregnant women. Therefore little data are available to evaluate reproductive risk to women in their pregnant or childbearing period just after a new medicinal product is approved for marketing. In our country, the special surveillance regarding pregnant women has been carried out since 1997 to assure efficacy and safety of new drugs according to the good post-marketing surveillance practice (GPMSP). However, many more case reports are required to assess the risk of drug-exposure to pregnant women. Therefore, we focused on the information of drug teratogenicity and started making a database system regarding drugs and pregnancy in this project.

The integrated bi-directional database system has three classes of information; 1) documentary information of the medicinal drugs on teratogenicity, 2) prediction of teratogenicity for a given compound, 3) information of the clinical case reports (Fig. 3). The information regarding teratogenicity, a drug adverse effect which must never happen in medications although the frequency is low, is featured. The data can be easily retrieved and registered by pregnant women and clinical researchers on the web site. The information provided by this net community system contributes to the practice of evidence-based medicine, an efficient drug research and devel-

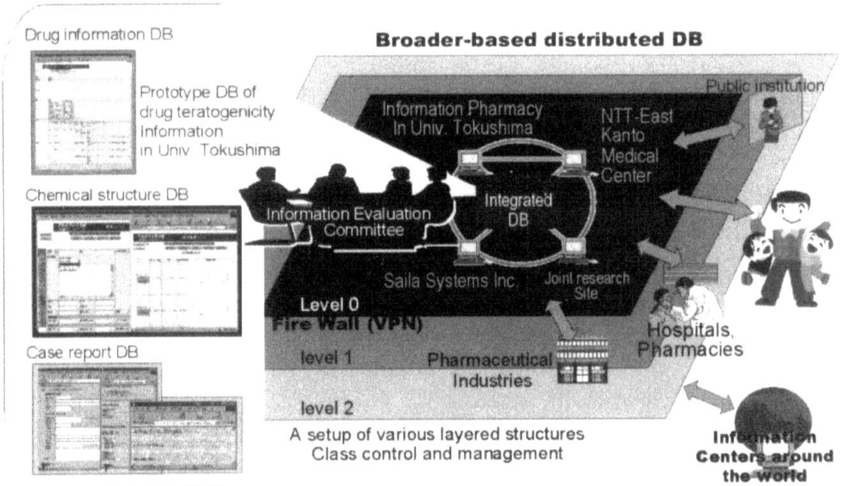

Fig. 3. Human-based infrastructure of teratogenic information network. Figure shows the structure of bi-directional / growing information community.

opment, and training for "the informational pharmacist" who will play important roles in risk communication.

3. Data Mining Approach on Clinical/Pharmaceutical Information

In order to use our database for the risk-assessment of drug, we tried to predict the drug excretion into human breast milk using a method of quantitative structure activity relationship (QSAR), as a data mining approach on clinical/pharmaceutical information.

When lactating mothers need to take medicines, it is very important to evaluate risk potential of drugs to their babies. However, there is little information about the drug excretion into human breast-milk. Meskin and Fleishaker have reported that the transfer of drug into breast milk depends on a concentration-gradient that allows passive diffusion of unionized, protein-unbound drugs [5, 6]. In that case, the drug permeability is affected by the condition of molecule, such as the molecular size, the lipophilicity, degree of ionization and others.

In this study, the human M/P (milk to plasma drug concentration) ratios were collected from literatures as the index of the drug permeability through the mammary membrane. The correlation of M/P values with the various physicochemical characteristics of drugs were analyzed using the QSAR method.

The M/P values for 45 drugs were collected from the clinical case reports in literatures. The information regarding the compound name, CAS number, chemical structure, log P (measured value or calculated CLOGP), pKa were obtained from SciFinder Scholar (ver. 2004, ACS) etc. The molecular volume was computed using Sparan'02 for Window (Wavefunction).

An equation 1 was led by the multiple linear regression analysis, in which log M/P was correlated with the logarithm of molecular weight (log MW), the partition coefficient (log P) and ratio of the fraction unionized (log $[f^{un}(\mathrm{P}) / f^{un}(\mathrm{M})]$). The equation was significant at a 95% confidence level as indicated by the F-test, with a highly correlation coefficient. A negative-dependence on log MW, and a positive-dependence on both log P and log $[f^{un}(\mathrm{P}) / f^{un}(\mathrm{M})]$ seemed to be agreeable with the predicted mechanism of membrane transport.

$$\log M/P = 1.71_{(\pm 0.56)} \log MW + 0.07_{(\pm 0.06)} \log P + 1.00_{(\pm 0.39)} \log[f^{un}(\mathrm{P}) / f^{un}(\mathrm{M})] + 3.64_{(\pm 1.34)} \quad (1)$$

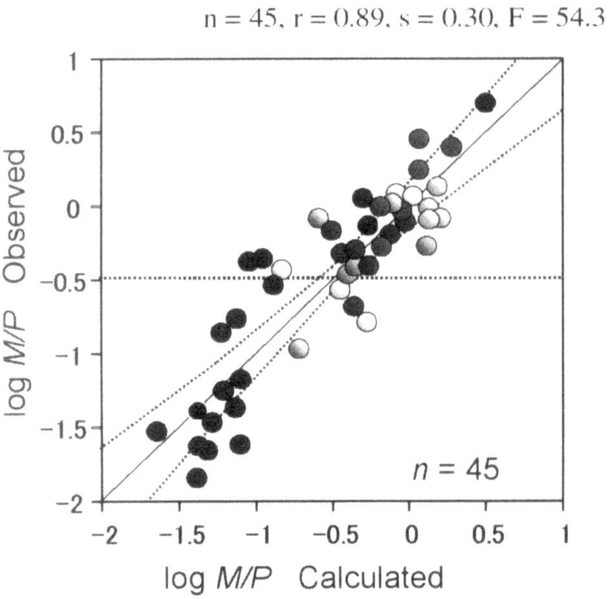

Fig. 4. Comparison between Observed and Calculated log *M*/*P*.

The results from present study show that M/P calculated form Equation 1 utilizing physicochemical parameters predicted the M/P observed in the human (Fig. 4).

Present results indicate that the QSAR methodology allows us to predict the drug excretion into breast milk using the clinical data of human, suggesting a future development as the "Clinical QSAR".

4. Spin-off Benefit of Present Project

There already exist a number of databases available regarding drug toxicities. The construction of a bi-directional growing information-community for teratogenic agents may be one of the most unique and relevant plans in the world. Our database can be a common platform for many people, including laypersons to expand not only the database but also new knowledge/wisdom.

Our project will bring about the following spin-off benefits,

1. Our database will provide high-grade information for physicians and pharmacists to employ tested the evidence-based medicine with the pregnant women.

2. Increased clinical case reports in the database will be used to develop the pharmacoepidemiologic study for drug-induced birth defects, resulting in the feedback of information to bedside.
3. Relationship between the structural formula of medicinal product and its risk potential such as teratogenic activity or excretion into breast-milk in human will be analyzed as information of QSAR. The data will be useful to research/develop new chemicals in the stage of drug design in silico.
4. Our activities have also an educational meaning for the pharmacy students or pharmacists to learn informatics, risk identification/risk management/risk communication skills, and to receive training of the drug information service linked bedside.

Our system will be useful for medical workers who perform the clinical risk-assessment of teratogenic exposures during pregnancy. Thus the pregnant woman and her family will be able to carry out family planning without worry. In addition, it will be beneficial for the pharmaceutical scientists in the development of new drugs based on chemical information and computer science.

Acknowledgement

This research is supported by Research Institute of Science and Technology for Society in Japan Science and Technology Agency, and Grant-in-Aid for Scientific Research from Ministry of Education, Culture, Sports, Science and Technology, Japan.

References

1. http://www.ristex.jp/, accessed on March 18, 2005
2. Yamauchi A, Nakata E, Chuman H (2003) Construction of a Growing Information-Community for Teratogenic Agents. J Comput Chem Jpn 2:71-78
3. Mcbride WG (1961) Thalidomide and congenital abnormalities. Lancet 278: 1358
4. Burley DM and Lenz W (1962) Thalidomide and congenital abnormalities. Lancet 279:271-272
5. Meskin MS and Lien EJ (1985) QSAR analysis of drug excretion into human breast milk. J Clin Hosp Pharm 10:269-278
6. Fleishaker JC, Desai N, McNamara PJ (1987) Factors affecting the milk-to-plasma drug concentration ratio in lactating women: physical interactions with protein and fat. J Pharm Sci 76:189-193

Clinical Decision Support based on Mobile Telecommunication Systems

Shusaku Tsumoto [1], Shoji Hirano[1] and Eiisuke Hanada [2]

[1] Department of Medical Informatics, Shimane University, School of Medicine, Shimane, Japan
[2] Division of Medical Informatics, Shimane University Hospital, Shimane, Japan

Chapter Overview. In early 1980s, many medical expert systems were developed with knowledgebase which are acquired from medical experts, and their performance was almost as good as domain experts. However, they were not frequently used mainly due to the poor user interfaceand the lack in learning new knowledge. However, the solutions of these two problems have been introduced since 1990s. For the latter problem, machine learning methods have provided several solutions, and for the former problem, the rapid progress of webtechnologies enables us to implement a good user interface. Furthermore, recent advances in computer resources strengthen these two solutions. In this paper, we focus on the application of knowledge engineering techniques to medical mobile communication network, where the web intelligence technologies are used for an efficient interface of medical expert system. Then, the system was put on the internet to provide an intelligent decision support in telemedicine and is now being evaluated by region medicalhome doctors. The results show that such an internet-based medical decision support enables home doctors to take a quick action to the applied domain.

Key Words. E-health, medical expert system, clinical decision Support, telemedicine

Introduction

Medical knowledge grows very rapidly and is complicated: medical expertise usually needs the number of cases in which a medical doctor has met before. For example, in general, more than thousand patients on one symptom or disease requires that a medical doctor is called as an "expert", which implicitly means that a population of the region where a medical doctor works is an important factor for acquiring a sufficient degree of expertise. However, if the population is very low, it is very difficult for a doctor to have such knowledge. On the other hand, the doctors in the medical center of large hospitals have enough experiences. Thus, medical decision support for people very far from the medical center of large hospitals is very important to achieve the high quality of healthcare. However, until 1980s it is very difficult to provide neither such medical decision support nor data transportation to medical experts in the center. Now, technically, broad-band network services enable us to send large amount of data, without any concern about the distance. On the other hand, in early 1980s, many medical expert system were developed with knowledgebase which are acquired from medical experts, and their performance was almost as good as domain experts. However, they were not frequently used mainly due to the poor user interface and the lack in learning new knowledge[1]. However, the solutions of these two problems have been introduced since 1990s. For the latter problem, machine learning methods have provided several solutions, and for the former problem, the rapid progress of web technologies enables us to implement a good user interface. Furthermore, recent advances in computer resources strengthen these two solutions. In this paper, the internet-based decision support system is introduced with intelligent web interface. The recent advances in web technologies were used for an efficient interface of medical expert system. Then, the system was put on the internet to provide an intelligent decision support in telemedicine and is now being evaluated by region medical home doctors. The results show that such an internet-based medical decision support enables home doctors to take a quick action to the applied domain.

The paper is organized as follows: Section 2 discusses the motivation of this study from the medical side: telemedicine. Section 3 introduces a diagnostic model for medical expert system, the architecture of the expert systems and shows how clinical decision support system works. Section 4 gives the extensions of its usage into mobile terminals, including PDA. Finally, Section 5 concludes this paper.

Motivation: Telemedicine

Shimane Prefecture: Sparcely-populated District

Shimane prefecture is one of the most rural regions in Japan, whose location is shown in Fig. 1. The population is 761,503 in 2001, which is the last two prefectures in Japan (e.g., 46th of 47 prefectures) and smaller than one ward in Tokyo. The population consists of 111,982(less than 15), 460,103(15 to 64 years old) and 189,031 (larger than 65 years old), which shows 24.8 Concerning medical issues, Shimane has only

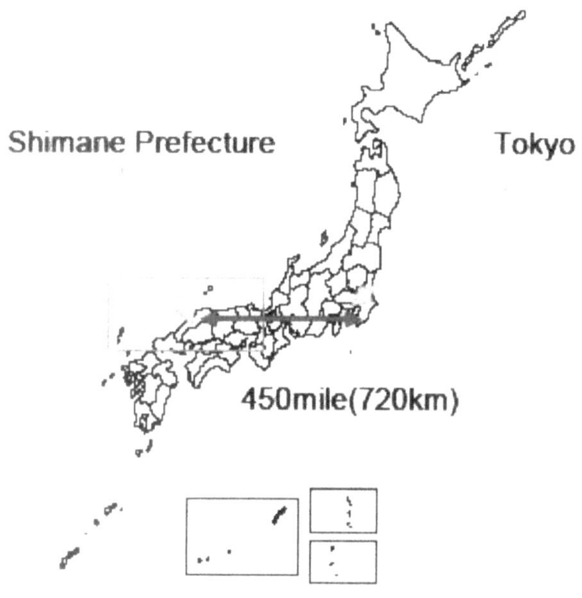

Fig. 1. Location of Shimane Prefecture

a few large hospitals, although it has very many home doctors (22th largest of 47 prefectures), which suggests that telemedicine is very important (Fig. 1). Especially, since most of the home doctors are very old person (the average age is 62.0), learning new medical knowledge is very difficult for them. Moreover, they have difficulties in learning computer literacy.

Shimane Community Healthcare Network System

Shimane community network system (Shimane CNS) is developed as a tool for healthcare communication between Shimane Medical University Hospital and home doctors in 1999. The system has two interfaces: one is for communication between hospital information systems and community network system, based on HL-7 which is a specific form of XML in medicine. Fig. 3 shows the architecture of Shimane CNS. Data which are allowed to be open for home doctors are transported from hospital information system to Patient Records DB Server.

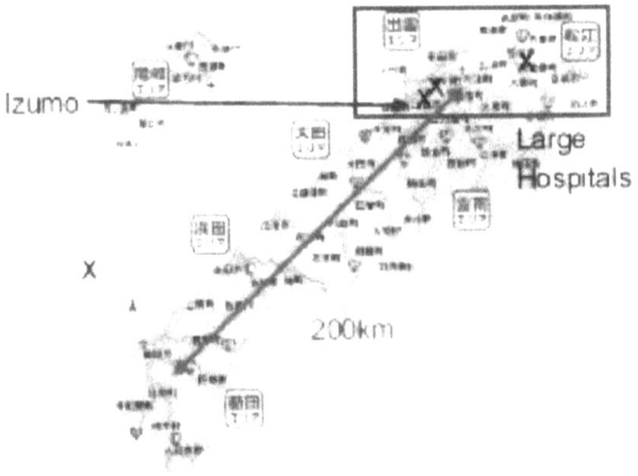

Fig. 2. Location of Large Hospitals

These data will be applied to application server for data visualization, whose results are sent to Web server. Web server provides an interface between hospital information systems and home doctors. This server has the following three functions: (1) exchange of e-mails on medical consultation, (2) the reports from the university hospital, (3) transportation of medical images and the results of laboratory examinations.

E-decision support

Basic Model: Focusing Mechanism

One of the characteristics in medical reasoning is a focusing mechanism, which is used to select the final diagnosis from many candidates [2,3,4]. For example, in differential diagnosis of headache, more than 60 diseases will be checked by present history, physical examinations and laboratory examinations. In diagnostic procedures, a candidate is excluded if a symptom necessary to diagnose is not observed. This style of reasoning consists of the following three kinds of reasoning processes:

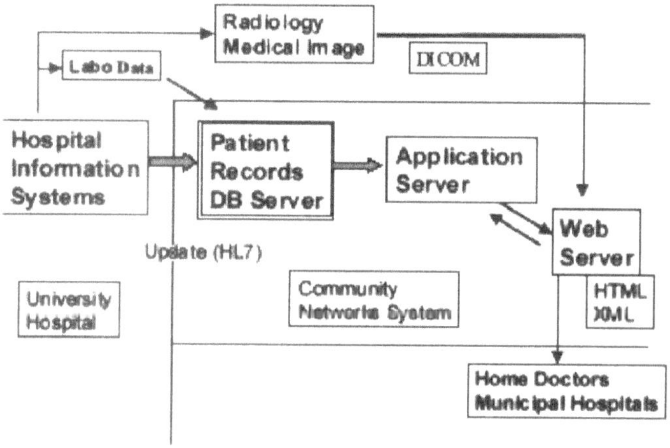

Fig. 3. Location of Large Hospitals

exclusive reasoning, inclusive reasoning and detection of complications, which corresponds to screening, differential diagnosis, and close examination of each case.

The diagnostic procedure will proceed as follows: first, exclusive reasoning excludes a disease from candidates when a patient does not have a symptom which is necessary to diagnose that disease. Secondly, inclusive reasoning suspects a disease in the output of the exclusive process when a patient has symptoms specific to a disease. Finally, it is checked whether all the input symptoms are explained by the final candidates. The symptoms which are not explained suggest the complications of other diseases. These three steps are modelled as usage of three kinds of rules, exclusive rules, inclusive rules and disease image, all of which corresponds to exclusive reasoning, inclusive reasoning and detection of complications.

Rules for Focusing Mechanism

Based on the focusing mechanism, we developed RHINOS, an expert system which diagnoses clinical cases on headache or facial pain from manifestations. In this system, a diagnostic model proposed by Matsumura [2] is applied to the domain, which consists of the following three kinds of reasoning processes: exclusive reasoning, inclusive reasoning, and reasoning about complications. First, exclusive reasoning excludes a disease from candidates when a patient does not have a symptom which is necessary to diagnose. Secondly, inclusive reasoning suspects a disease in the output of the exclusive process when a patient has symptoms specific to a disease. Finally, reasoning about complications suspects complications of other diseases when some symptoms which cannot be explained by the diagnostic conclusion obtained. Each reasoning style is rule-based and all the rules needed for diagnostic processes are acquired from medical experts in the following way.

Exclusive Rules

This rule corresponds to exclusive reasoning. In other words, the premise of this rule is equivalent to the necessity condition of a diagnostic conclusion. From the discussion with medical experts, we select the following six basic attributes which are minimally indispensable to defining the necessity condition: 1. Age, 2. Pain location, 3. Nature of the pain, 4. Severity of the pain, 5. History since onset, 6. Existence of jolt headache. For example, the exclusive rule of common migraine is defined as:

In order to suspect common migraine,
the following symptoms are required:
pain location: not eyes, nature :throbbing or persistent
or radiating,
history: paroxysmal or sudden and
jolt headache: positive

One of the reasons why we select the six attributes is to solve the interface problem of expert systems: if the whole attributes are considered, we also have to input all the symptoms which are not needed for diagnosis. To make exclusive reasoning compact, the only minimal requirements are chosen. It is notable that this kind of selection can be viewed as the ordering of given attributes, and it is expected that such ordering can be induced from databases. Therefore we intend to formulate induction of exclusive rules by using the whole given attributes. It is because we can acquire the minimal requirements for describing exclusive rules after all the exclusive rules are induced.

Inclusive Rules

The premises of inclusive rules are composed of a set of manifestations specific to a disease to be included. If a patient satisfies one set, we suspect this disease with some probability. This rule is derived by asking the following items for each disease to the medical experts: 1. a set of manifestations by which we strongly suspect a disease. 2. the probability that a patient has the disease with this set of manifestations: SI(Satisfactory Index) 3. the ratio of the patients who satisfy the set of all the patients of this disease: CI(Covering Index) 4. If the total sum of the derived CI(tCI) is equal to 1.0 then end. Otherwise, goto 5. 5. For the patients of this disease who do not satisfy all the collected set of manifestations, goto 1.

Therefore a positive rule is described by a set of manifestations, its satisfactory index (SI), which corresponds to it accuracy measure, and its covering index (CI), which corresponds to total positive rate. Note that SI and CI are given empirically by medical experts.

For example, one of three positive rules for common migraine is given as follows.

If history: paroxysmal,
jolt headache: yes,
nature: throbbing or persistent,
prodrome: no, intermittent symptom: no,
persistent time: more than 6 hours,
and location: not eye,
then common migraine is suspected with
accuracy 0.9 (SI=0.9) and this rule covers 60 percent
of the total cases(CI=0.6).

Disease Image

This rule is used to detect complications of multiple diseases, acquired by all the possible manifestations of the disease. Using this rule, we search for the manifestations which cannot be explained by the conclusions. Those symptoms suggest complications of other diseases. For example, the disease image of common migraine is:

The following symptoms can be explained
by common migraine: pain location: any
or depressing: not or jolt headache:
yes or ...

Therefore, when a patient who suffers from common migraine is depressing, it is suspected that he or she may also have other disease.

E-Decision Support System

Fig. 4 shows the architecture of expert system RH (Rule-based System for Headache) in Shimane CNS. All the systems are written in Java (EJB) and connected with Web server and DB server. In the new version, rule mining engine is implemented in the application sever.

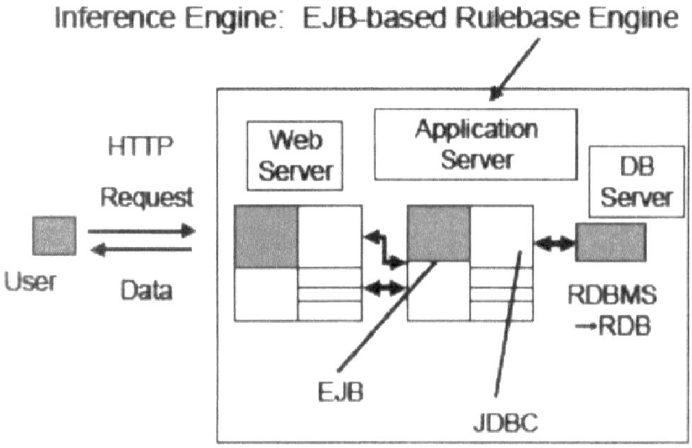

Fig. 4. Architecture of Expert System with Data Mining Engine in Application Server

Interface for E-Decision Support

Since most of the home doctors in Shimane are more than fifty years old, it is very difficult to expect that they will adapt to the advanced interface. Thus, we arranged that most of the inputs can be acquired by simple interfaces: radio button, except ofpersonal information of the patients, such as name and age. The other point is that all the inputs and outputs (diagnostic conclusions and explanations) are summarized into natural language and shown as texts to the doctors.

Fig. 5. Architecture of Expert System with Data Mining Engine in Application Server

For example, Fig. 5 show the final results with the explanations by exclusive and inclusive rules, respectively. These explanations are originally described by XML documents, and translated into natural language in the Web interface. Also,home doctors who are interested in the knowledge base and patient records, can access the data from the web interface. It is notable that these aspect shows that such data input system, results from clinical decision support system, and experts' feedbackcan generate a new type of data collection, which can be called intelligent data collection. All those data can be stored as XML documents, which can be applied to rule mining engine or accessed into databases.

Towards Mobile Clinical Decision Support

The Shimane CNS and E-clinical decision support system can be accessed from PDA and Mobile computers via PHS communication service. PDAs are expected to give more potential to the usability of clinical decision support: pen-touch interface is much easier to select inputs from lists. They are not heavy. The battery will retain much more than mobile computers. However, it is found the programming environment of PDA is very limited, and it is very difficult to implement a new intelligent interface, which suggests that the computing power is close to the personal computer in 1980's. Thus, we had to stop the development of intelligent parts in PDA and to decide to use the HTTP service as the interface for CNS and E-decision support system. It will be our future work to implement the intelligent program into PDA for more intelligent data collection.

Conclusion

This paper reports Shimane community network system and E-decision support system with intellgent web interface. The recent advances in web technologies were used for an efficient interface of medical expert system. Then, the system was put on the internet to provide an intelligent decision support in telemedicine and is now being evaluated by region medical home doctors. The results show that such an internet-based medical decision support enables home doctors to take a quick action to the applied domain. Moreover, those systems can be accessed from PDA, which enables us to expand the applicability of healthcare network system. However, the applicability of PDA is much more limited at this moment: more intelligent programming environments are required for PDA tools.

References

1. Buchnan BG and Shortliffe EH (1984) Rule-Based Expert System, Addison-Wesley, New York, USA.
2. Matsumura Y, et al. (1986) Consultation system for diagnoses of headache and facial pain: RHINOS. Medical Informatics, Vol. 11, pp. 145–157.
3. Tsumoto S and Tanaka H (1996) Automated Discovery of Medical Expert System Rules from Clinical Databases based on Rough Sets. Proceedings of the Second International Conference on Knowledge Discovery and Data Mining 96, AAAI Press, Palo AltoCA, pp.63–69.
4. Tsumoto S, 1998. Automated Extraction of Medical Expert System Rules from Clinical Databases based on Rough Set Theory. Information Sciences, Vol. 112, pp. 67–84.

Web Intelligence Meets Immunology

Jiming Liu[1], Ning Zhong[2], Yiyu Yao[3], and Jing-Long Wu[4]

[1]School of Computer Science, University of Windsor, Canada
[2]Dept. of Information Engineering, Maebashi Institute of Technology, Japan
[3]Dept. of Computer Science, Regina University, Canada
[4]Dept. of Intelligent Mechanical Systems, Kagawa University, Japan

In this article, we will discuss how the *Wisdom Web (WW)* from the next-generation *Web Intelligence (WI)* development contributes to the research studies in *immunology*, i.e., the scientific discovery of HIV-immune interaction dynamics. In particular, we will show that a new computing paradigm, called *Autonomy Oriented Computing (AOC)*, will be necessary to support the Wisdom Web to meet this challenge. Throughout our discussion, we will pose several *Research Questions*, through which the authors intend to highlight the technical issues involved and to stimulate thinking and innovations from the community of researchers pursuing this exciting, interdisciplinary endeavor.

Key Words. Web Intelligence, Immunology

The *Wisdom Web*

We envision that the next paradigm shift in Web Intelligence (WI) should go beyond the role of supporting tools and further incorporate the notion of *wisdom* to be embedded in the Web itself [6,13,17,19,20]. According to Webster Dictionary, the word *wisdom* means *the quality of being wise; knowledge, and the capacity to make due use of it; knowledge of the best means and the best ends.* We refer to the World Wide Wisdom Web as the Web that can autonomously

- discover the best means and ends,
- mobilize distributed resources,
- enrich social interaction, and
- enable users to gain practical wisdom of living, working and playing.

In short, the Wisdom Web will enable users to go beyond the existing on-line information search and knowledge queries functionalities and to gain, from the Web,[1] practical wisdom for problem solving.

To support such a Wisdom Web, we envision that a grid-like computing infrastructure with intelligent service agencies is needed, where these agencies can interact, self-organize, learn, and evolve their course of actions, identities and interrelationships for new knowledge creation, as well as scientific and social evolution [5,6].

Web Intelligence Meets *Immunology*

Discovering HIV-Immune Interaction: The *Open Problem*

As argued in [18], human immune system is a typical example of a highly sensitive, adaptive and self-regulated complex system involving numerous interactions among a vast number of cells that belong to different types. Immune response is an emergent phenomenon from the interactions of numerous entities, which protects the human body from invaders such as bacteria, virus and other parasites. Despite there are many clinical case studies and empirical findings [1,15,2,14,16], the working mechanism un-

[1]Hence, the notion of "Web" should be taken in a much broader sense.

derlying the complex process of HIV invasion, erosion and eventual crash on the immune system, such as how the local interactions at the level of HIV, T cells and B cells affect this process, remains to be fully understood (i.e., characterized and predicted).

Conventional modeling and simulation technologies are useful only to a very limited extent due to computational scale and cost involved. Hence, it would be most natural and desirable to make the best use of the above mentioned grid-like computational infrastructure with intelligence service agencies to carry out such a large-scale, distributed scientific knowledge discovery, i.e., characterization and prediction of the human immune responses [4,10,18].

The *Real* Challeng

This endeavor presents a tremendous challenge to the filed of computing, i.e., (1) the task of computing is seamlessly carried out in a variety of physical embodiments, and (2) there is no single multi-purpose or dedicated machine that can manage to accomplish a job of this nature. In other words, the key to success in such applications lies in a large-scale deployment of computational agents capable of autonomously making their localized decisions and achieving their collective goals.

Towards the *Autonomy Oriented Computing (AOC)* Paradigm

Recently, Liu et al. have been interested in the metaphors of autonomy as offered by nature and their roles in addressing our practical computing needs. His research agenda concerns the development of a new computing paradigm, called autonomy oriented computing (AOC)[7,10]. AOC makes use of autonomous entities in solving computational problems and in modeling complex systems [8,9,12].

Characteristics of AOC

AOC starts from the smallest and simplest element of a complex system based on the following characteristics of the entities in the system [10,11]:

- *Autonomous* The systems elements are rational individuals that will act independently. In other words, a central controller for directing and coordinating individual elements is absent.

Research Question 1: *How can the behavior and utilities of the entities be goal-directed? How would the entities acquire and improve their reactive behavior based on their local and/or shared utilities?*

- *Distributed* The entities of localized decision-making capabilities are distributed in a heterogeneous computing environment and are locally interacting among themselves, without too much communication overhead.

Research Question 2: *How would the entities measure, update, and share information among themselves?*

- *Emergent* They exhibit, often not simple, behaviors that are not present or predefined in the behavior of the autonomous entities within complex adaptive systems.

Research Question 3: *How would you formally model and quantitatively measure the inter-relationships between the local goals of the entities and the desired global goal(s) of the AOC system?*

- *Adaptive* They often change their behavior in response to changes in the environment in which they are situated.

Research Question 4: *How would you define (and update) the goals and on-going feedback (e.g., rewards) of individual entities?*

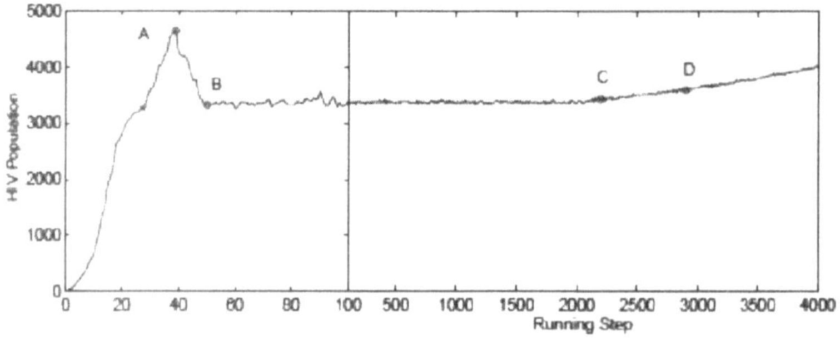

Fig. 1. The three-stage dynamics of HIV infection as generated from an AOC-based simulation [18]. This result consistent with the earlier empirical findings [3]

- *Self-organized* They are able to organize the elements to achieve the above behaviors.

Research Question 5: *How would the entities locally interact among themselves? What are the rules?*

The AOC Approach to Discovering HIV-Immune Interaction on the Wisdom Web

In order to deal with numerous interactions among a vast number of autonomous agents in AOC-based massive multi-agent (MMAS) simulation, we have studied an AOC approach that incorporates the characteristics of Cellular Automata (CA) and system-level mathematical equation modeling to simulate the complex interactions in the process of human immune response to HIV. In our AOC approach, the mathematical equations are used within a single site, which keeps the spatial characteristics of the system and reduces the heavy computational costs of the CA model.

Research Question 6: *How would you define and model (and let entities to efficiently explore and update) the large state space during their interaction?*

Figure 1 presents the temporal emergence of three-stage dynamics in HIV infection, which was generated from our earlier AOC-based massive multi-agent (MMAS) simulation [18]. The three stages are: (1) before B: the primary response, (2) $B \sim C$: the clinical latency, and (3) after D: the onset of AIDS. At A, the HIV population reaches a maximum. Starting from C, the mechanism that decreases the natural ability of an organism in producing T cells is incorporated. This AOC-generated result is consistent with empirically observed phenomena [3].

Furthermore, it also revealed from our experiments [18] that AIDS cannot break out if HIV only destroy T cells without weakening the T cell reproduction mechanism. The emergence in "shape space" indicates that it is because the HIV's fast mutation that the immune system cannot eradicate HIV as easily as it does to other invaders. The discoveries help researchers to understand HIV-immune interaction dynamics more comprehensively.

Research Question 7: *How would you design and implement the AOC system on the Wisdom Web, consisting of such multiple entities? In real time, millions of entities must concurrently operate and interact themselves.*

Discussion

The AOC paradigm for the Wisdom Web computing is different from the traditional imperative, logical, constraint, object oriented, or component based paradigms, not only in the characteristics of its fundamental constructs as mentioned in Section ? but also in the effectiveness and efficiency of computing that can be achieved through these special characteristics.

The Real power

Specifically speaking, the real power of AOC in the Wisdom Web manifests in dealing with the problems of the following nature:

1. The problems are of high complexity (e.g., large- scale, high-dimension, highly nonlinear relationship, and highly interrelated variables);
2. The problems are dynamically updated or changing in real time;
3. The problems are highly-distributed and locally- interacting in nature (i.e., not centralized, not ready/efficient for batch processing); or
4. The goal of AOC in data-mining and classification is not to extract some superficial patterns/relationships (i.e., data- transformation from one form to another), but really to discover and understand the "deep patterns" -- the underlying mechanisms/processes that produce the data (i.e., to provide an explanation of the cause/origin). As in real-world we always have certain ideas or knowledge about the general mechanisms/processes that generate the data, we are able to implement AOC to achieve this goal by means of developing entities that incorporate such ideas or knowledge.

Research Question 8: *What hard data-mining and classification problems of the above mentioned characteristics would you suggest to apply the AOC paradigm?*

Most of the presently available algorithms (e.g., competitive learning algorithms, bottom-up clustering, particle swarm, evolutionary/genetic algorithms, and neural networks – please survey them) that simply make use of some autonomy ideas are used to solve *common* sequentially batch-processing, data-mining problems. In other words, the AOC systems that successfully solve the above mentioned hard problems differ from these algorithms in both the nature of problems to be solved and the computational philosophy to be realized.

Research Question 9: *Try to make an in-depth survey and compare AOC with the existing approaches, as mentioned above. Would you be able to formulate the former as a theoretical generalization of the latter e.g., competitive learning, active learning, and particle swarm optimization.*

The *New Frontier* in Computing

As compared to other paradigms, such as centralized computation and top-down systems modeling, the AOC on the Wisdom Web can become extremely appealing in the following aspects:

1. To capture the essence of autonomy in natural and artificial systems;
2. To solve computationally hard problems, e.g., large- scale computation, distributed constraint satisfaction, and decentralized optimization, that are dynamically evolving and highly complex in terms of interaction and dimensionality;
3. To characterize complex phenomena or emergent behavior in natural and artificial systems that involve a large number of self-organizing, interacting entities;
4. To discover laws and mechanisms underlying complex phenomena or emergent behaviors.

The two most distinct applications of AOC for the Wisdom Web have been (1) designing and developing computational solutions to hard problems, and (2) characterizing and understanding complex phenomena or systems behavior.

Acknowledgment

We would like to acknowledge the support of the following research grants: (1) Hong Kong Research Grant Council (RGC) Central Allocation Grant (HKBU 2/03/C) and Earmarked Research Grants (HKBU 2121/03E)(HKBU 2040/02E), and (2) Major State Basic Research Development Program of China (973 Program) (2003CB317001).

References

1. Coffin J M (1995) HIV population dynamics in Vivo: Implications for genetic variation, pathogenesis, and therapy. Science 267:483–489
2. Fauci A S (1988) The immunodeficiency virus: Infectivity and mechanisms of pathogenesis. Science 239:617–622
3. Hershberg U, Louzoun Y, Atlan H, Solomon S (2000) HIV time hierarchy: Winning the war while loosing all the battles. Physica A 289:178–190
4. Jin X, Liu J (2004) From individual based modeling to autonomy oriented computation. In: Nickles M, Rovatsos M, Weiss, G. (eds) Agents and Computational Autonomy, LNAI 2969. Springer, pp. 151-169
5. Kauffman S (1993) Origins of Order: Self-Organization and Selection in Evolution. Oxford University Press
6. Liu J (2003) Web Intelligence (WI): What makes Wisdom Web? Proceedings of the Eighteenth International Joint Conference on Artificial Intelligence (IJCAI-03), Acapulco, Mexico, Aug. 9-15, 2003. Morgan Kaufmann Publishers, pp. 1596-1601 (invited talk)
7. Liu J (2001) Autonomous Agents and Multi-Agent Systems: Explorations in Learning, Self-Organization and Adaptive Computation. World Scientific.
8. Liu J, Han J, Tang Y Y (2002) Multi-agent oriented constraint satisfaction. Artificial Intelligence 136(1):101-144
9. Liu J, Jin X, Tang Y (2004) Multi-agent collaborative service and distributed problem solving. Cognitive Systems Research 5(3):191-206
10. Liu J, Jin X, Tsui K C (2005a) Autonomy Oriented Computing (AOC): From Problem Solving to Complex Systems Modeling. Springer/Kluwer Academic Publishers
11. Liu J, Jin X, Tsui K C (2005b) Autonomy Oriented Computing (AOC): Formulating computational systems with autonomous components. IEEE Transactions on Systems, Man and Cybernetics, Part A: Systems and Humans 35(6):879- 902
12. Liu J, Jin X, Wang S (2005c) Agent-based load balancing on homogeneous minigrids: Macroscopic modeling and characterization. IEEE Transactions on Parallel and Distributed Systems 16(7):586-598
13. Liu J, Zhong N, Yao Y Y, Ras Z W (2003) The Wisdom Web: New challenges for Web Intelligence (WI). Journal of Intelligent Information Systems 20(1)
14. McCune J M (2001) The dynamics of CD4+ T-cell depletion in HIV disease. Nature 410:974–979
15. Perelson A S, Newmann A U, Markowitz M, Leonard J M, Ho D D (1996) HIV-1 dynamics in Vivo: Virion clearance rate, infected cell life-span, and viral generation time. Science 271:1582–1586
16. Wei X, et al. (1995) Viral dynamics in Human Immunodeficiency Virus Type 1 Infection. Nature 373:117–122

17. Yao Y Y, Zhong N, Liu J, Ohsuga S (2001) Web Intelligence (WI): Research challenges and trends in the new information age. Web Intelligence: Research and Development, LNAI 2198. Springer, pp. 1-17
18. Zhang S, Liu J (2005) A massively multi-agent system for discovering HIV-immune interaction dynamics. In: Ishida T, Gasser L, Nakashima H (eds) Massively Multi-Agent Systems, LNCS. Springer, pp. 161-173
19. Zhong N, Liu J, Yao Y Y (2002) In search of the Wisdom Web. IEEE Computer 35:27-31
20. Zhong N, Liu J, Yao Y Y (2003) (eds) Web Intelligence. Springer

Part 4
Complex Technology in Rehabilitation

Hand Movement Compensation on Visual Target Tracking for Patients with Movement Disorders

Junko Ide[1], Takenao Sugi[2], Masatoshi Nakamura[3], and Hiroshi Shibasaki[4]

[1]Division of Social Welfare, Seinan Gakuin University, Fukuoka, Japan
[2]Department of Electrical and Electronic Engineering, Saga University, Saga, Japan
[3]Department of Advanced Systems Control Engineering, Saga University, Saga, Japan
[4]National Institute of Neurological Disorders and Stroke, National Institute of Health, Washington, USA

Chapter Overview. Skillful motor control of human is achieved by the appropriate motor commands generating from the central nerve system. Brain dysfunction in the cerebral cortex, cerebellum and/or basal ganglia causes serious movement disorders such as cerebellar ataxia, Parkinson disease and so on. Compensation of the hand movement by adding an external force is a way for recovering a motor function of human and will be helpful for improving the quality of patients' daily life. A test of visual target tracking is one of the effective methods for analyzing the human motor functions. We have previously examined a possibility for improving the hand movement on visual target tracking by additional assistant force through a simulation study.

In this study, a method for compensating the human hand movement on visual target tracking by adding an assistant force is proposed. Effectiveness of the compensation method was investigated through the simulation study for patients with movement disorders and the experiment for four healthy adults. Sufficient improvement of the hand movement such as reactivity, smoothness and so on was seen through both the simulation and experimental study.

The proposed method can reflects the properties of the hand movement for different subjects in the structure of the compensator. Therefore, the proposed method will be easily adjustable to the individual properties of patients with various movement disorders caused from brain dysfunctions.

Key Words. Visual target tracking test, Compensation of hand movement, Assistant force, Model-based compensator, Movement disorders

1. Introduction

Generation of appropriate motor commands by central nerve system practices the realization of dynamic and skillful motion of human. Motor and supplemental motor area in cerebral cortex, basal ganglia, cerebellum and so on have great contribution for realizing accurate motor control. Serious defects in motor control occur when the cerebellum or the basal ganglia has injured. Movement disorder originating from the cerebellum such as cerebellar ataxia deteriorates the accurate movement and the motor learning. Symptom of movement disorder in the basal ganglia such as Parkinson disease is the delay of reaction and no smooth change of the motion. A technique for modifying the hand movement by adding assistant force is effective for such patients with various movement disorders. Development of active orthosis for improving an accuracy of hand movement will be useful for recovering a quality of patients' daily life.

A test of visual target tracking is one of the effective methods for analyzing the human motor functions [1,2]. A large number of studies have been made on analysis and evaluation of motor function through mathematical models [3]. Some studies for compensating the human motion have also done [4].

We have previously reported the quantitative evaluation on the relationship between movement disorders and the characteristics of motor functions [2]. Then, the characteristics of hand movement of the patients with movement disorders were expressed by the model, and the differences among healthy adults, patients with cerebellar ataxia and patients with Parkinson disease could be found in the model parameters. A possibility on improving the hand movement by additional assistant force was also investigated through a simulation study [5].

In this paper, the method for compensating the hand movement on visual target tracking was developed and examined through the simulation and the experiments. First of all, a mathematical model for expressing the property of human hand movement was obtained from the measurement data of visual target tracking for each subject. Then, the compensator of hand movement was constructed by using the information of model structure.

Compensation of the hand movement on visual target tracking was examined through the simulation study for patients with movement disorders (cerebellar ataxia and Parkinson disease). Furthermore, the proposed method was applied to actual experiment for four healthy adults. Effectiveness of the hand movement compensation technique was quantitatively investigated.

2. Method of Hand Movement Compensation Based on Visual Target Tracking Model

2.1 Experimental equipment and visual target tracking

The experimental equipment for measuring the visual target tracking and for hand movement compensation used in this study consists of personal computer (NEC PC-9801Xa), monitor and XY-table (RIKEN DENSHI F45) with a handgrip. XY-table equips force sensors and position sensors in the base of handgrip and those values are transmitted to the computer through A/D converter. XY-table also has two servo motors for moving the base of handgrip both horizontally and vertically, and control inputs as compensation signal for hand movement compensation are given from the computer to the servo motor through D/A converter.

Subject holds on a handgrip by his/her hand and moves it, just then the corresponding green circle point (pursuit point) is moved on the monitor. Visual target with red rectangular is also displayed on the same monitor. Subject carries out the task to pursue the visual target as correctly and speedily as possible by moving a handgrip on XY-table.

Visual target used in the experiment moves 3 sec long from the upper part to the lower part of the monitor straightforwardly with a constant speed of 3 cm/s.

2.2 Modeling of human hand movement

Properties of the hand movement on visual target tracking was expressed by mathematical model. The model structure adopted in this study is shown in Figure 1 (a). $P_r(s)$ as input signal for the system means the position of visual target and $P_h(s)$ as output signal of the system denotes the hand position of subject. Model structure was based on the second order mechatoronic servo model with time delay [5]. Transfer function of the model shown in Figure 1 (a) can be expressed as

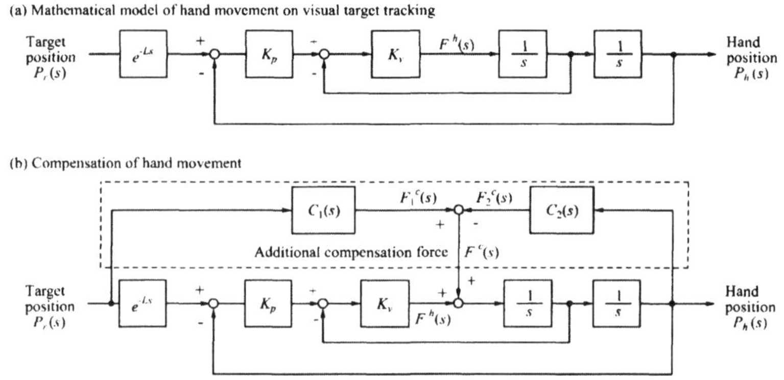

Fig. 1. The block diagram for hand movement compensation based on the model. (a)Mathematical model of hand movement on visual tracking, (b)compensation of hand movement. Compensator consists of two transfer functions, $C_1(s)$ and $C_2(s)$

$$G(s) = \frac{K_p K_v}{s^2 + K_v s + K_p K_v} e^{-Ls} \tag{1}$$

where L means the reaction delay, K_p is the position loop gain and K_v is the velocity loop gain. Model structure is simple, but the model can grasp the characteristic of hand movement for healthy adults, patients with cerebellar ataxia and patients with Parkinson disease sufficiently [5]. Model parameters L, K_p and K_v were determined by using the least squares method.

Control inputs as compensation signal for hand movement compensation are given from the computer to the servo motor through D/A converter. Signals for generating compensation force were target position $P_r(s)$, hand position $P_h(s)$ and contact force between subject's hand and handgrip. Compensator of hand movement on visual target tracking was then constructed by using the model information. Block diagram including the compensator is shown in Figure 1 (b).

Compensator is constructed by two blocks; feedforward compensation block $C_1(s)$ based on target position and feedback compensation block $C_2(s)$ based on hand position. The compensation signal $F^c(s)$ as assistant force is calculated by the summation of those outputs, and added as $F^c(s) = C_1(s)P_r(s) - C_2(s)P_h(s)$. Transfer function of the total system including compensator $C_1(s)$ and $C_2(s)$ shown in Figure 1 (b) is calculated as

$$G^a(s) = \frac{K_p K_v e^{-Ls} + C_1(s)}{s^2 + K_v s + K_p K_v - C_2(s)} \tag{2}$$

The structure of transfer functions $C_1(s)$ and $C_2(s)$ were designed in the following way. First of all, the referential model, which executed the ideal movement on visual target tracking, was established as same structure of equation 1.Transfer function of the referential model was $G^*(s)$, and the referential model parameters L^*, K_p^* and K_v^* can be adjusted freely. Transfer functions $C_1(s)$ and $C_2(s)$ were then determined by solving equation as $G^a(s) = G^*(s)$.

In this study, transfer functions $C_1(s)$ and $C_2(s)$ were settled as

$$C_1(s) = K_p^* K_v^* e^{-L^* s} - K_p K_v e^{-Ls} \tag{3}$$

$$C_2(s) = (K_v^* - K_v)s + K_p^* K_v^* - K_p K_v \tag{4}$$

Transfer function $C_1(s)$ includes the component of reaction delay, so the $C_1(s)$ mainly compensates the quickness of the reaction of the hand movement. Transfer function $C_2(s)$ compensates the other dynamical properties such as smoothness, changes of the motion and so on.

2.3 Evaluation of hand movement compensation

In order to evaluate the effectiveness of the hand movement compensation quantitatively, the following parameters were calculated from the experimental data. First, the reaction time was defined as the time difference of the beginning of movement between visual target and pursuit point. Secondly, the position error e_p was calculated the difference of the vertical movement of the visual target and the hand position in the whole time (from 0 s to 3 s). Thirdly, standard deviation of the velocity for vertical direction v_d was calculated in the latter interval (from 1.5 s to 3 s). The above three parameters adopted in this study were already established as useful ones for extracting the properties of the hand movement on visual target tracking in the previous study [2]. In addition to the above three parameters, the sum of force $f'(t)$ in the former interval (from 0 s to 1.5 s) was calculated as $f_i = \int_0^{1.5} |f'(t)| dt$.

3. Results

3.1 Simulation for patients with movement disorders

Measurement data of a patient with Parkinson disease, a patient with cerebellar ataxia and a healthy adult were used for evaluating the hand movement compensation through the simulation. Three models were constructed from the measurement data respectively. The referential model obtained from the healthy adult, and then the compensator was constructed by using the model for each patient and referential model.

Characteristics of three models were coincided with those of the measurement data for patients with cerebellar ataxia, patients with Parkinson disease and healthy adults respectively. Model parameters for each subject were shown in Table 1. Reaction delay L of healthy adult was smallest. The position loop gain K_p was smallest at patient with Parkinson disease.

Simulation result of the hand movement compensation for patient with Parkinson disease and that with cerebellar ataxia were shown in Figure 3, respectively. Gray thin line shows the target movement, solid line is subject's hand movement, broken line means model output and dotted line displays the simulation result of the compensation. Hand movement of patients with movement disorders were improved as same as that of healthy adult by the compensator.

Table 1. Model parameters for each subject on simulation study

Subject	L	K_p	K_v
Parkinson disease	0.18	3.5	6.2
Cerebellar ataxia	0.08	21.5	2.6
Healthy adult	0.03	24.5	12.8

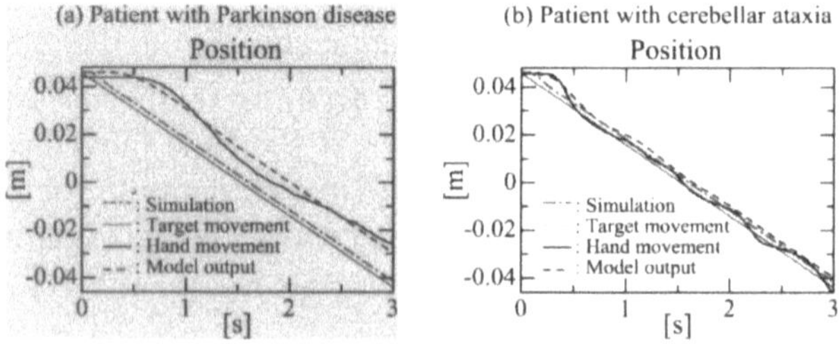

Fig. 2. Simulation of the hand movement compensation for patients

3.2 Experiments on healthy adults

Actual experiment on hand movement compensation was examined for four healthy adults. Continuous 20 trials of visual target tracking test was done for each subject. The mathematical model was obtained from the actual data for constructing the compensator of the hand movement. Then, the experiment was done with a same measurement condition. In this experiment, values of the parameters in the referential model were set at $K_p^* = 2.5$, $K_v^* = 12.5$ and $L^* = 0.1$, respectively. Experimental results were evaluated quantitatively by using the parameters described in the previous section.

Calculated model outputs were compared with the actual measurement data on visual target tracking, and almost all them except two cases grasped the characteristics of actual measurement data enough. Mean value and standard deviation of model parameters for each subject were shown in Table 2. Value of the model parameters were not so different for all subjects.

Experimental result of the hand movement compensation was shown in Figure 3. Left hand side shows the result of visual target tracking without adopting the hand movement compensation, and right hand side is that with compensation. Upper part means the hand position (a-1 and b-1), middle part is the velocity (a-2 and b-2), and lower part is the contact force between human hand and handgrip (a-3 and b-3). Force of human hand, additional assistant force by the compensator and total force were independently displayed in the figure concerning the result with compensation (b-3). Reaction time, position error and variability of the velocity of the

Table 2. Mean value and standard deviation of model parameters for each subject

	Subject	L	K_p	K_v
Without compensation	Subject 1	0.26 ± 0.02	7.0 ± 3.1	1.2 ± 0.8
	Subject 2	0.25 ± 0.01	8.0 ± 2.6	1.1 ± 0.4
	Subject 3	0.26 ± 0.02	7.9 ± 3.4	1.1 ± 0.6
	Subject 4	0.27 ± 0.02	8.1 ± 3.3	1.1 ± 0.6
	Average	0.26 ± 0.02	7.7 ± 3.1	1.1 ± 0.6
With compensation	Subject 1	0.10 ± 0.01	9.3 ± 2.0	2.0 ± 0.4
	Subject 2	0.10 ± 0.02	11.2 ± 3.7	1.7 ± 0.4
	Subject 3	0.09 ± 0.02	9.7 ± 2.1	1.9 ± 0.3
	Subject 4	0.08 ± 0.02	11.8 ± 3.2	1.4 ± 0.5
	Average	0.10 ± 0.02	10.5 ± 3.0	1.8 ± 0.5

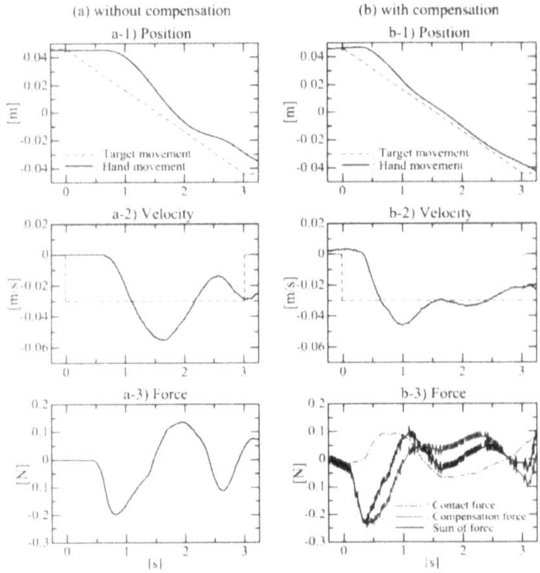

Fig. 3. Experimental results for hand movement compensation. (a) Without compensation, (b) with compensation. Effect of the compensation could be seen through the comparison

result with compensation were precisely reduced as compared with that without hand movement compensation. Assistant force had an effect especially in the beginning of hand movement. Almost all calculated models outputs with compensation were compared with the actual measurement data.

Figure 4 shows the comparison of the characteristic parameters defined in the previous section between the result of visual target tracking without compensation and that with compensation. Mean value and standard deviation of respective parameters for all data and those for each subject were displayed. Gray bar shows the parameter values obtained from the data without hand movement compensation and black bar denotes those with hand movement compensation. Reaction time decreased, position error became half value, SD of velocity decreased markedly and force of former interval increased by compensation.

From the comparison between experiments with hand movement compensation and those without compensation, sufficient improvement of the hand movement compensation was seen through the four characteristic parameters obtained from the experimental data.

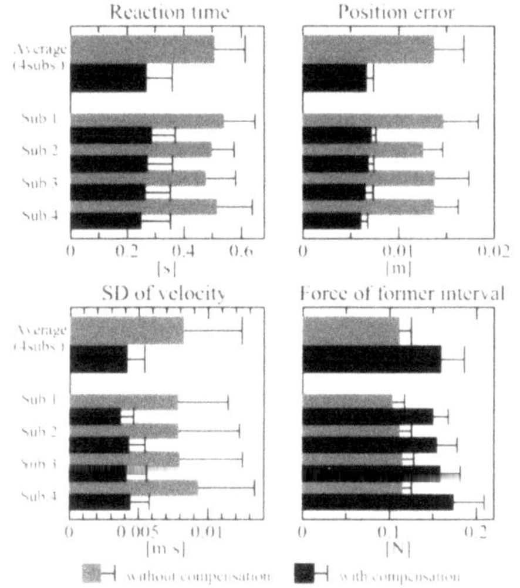

Fig. 4. Quantitative evaluation of hand movement compensation. Average and standard deviation of characteristic parameters; reaction time, position error, SD of velocity and force of former interval; for all subjects and each subject

4. Discussion

Effectiveness of the hand movement compensation for patients with movement disorders was verified through the simulation study, and the specific improvement was seen. Possibility for realizing the hand movement compensation was then investigated by the experimental study for healthy adults. Improvement of reactivity and smoothness of the hand movement were confirmed by comparing the experiment without compensator. Deterioration of the motor function is completely different for respective patients with movement disorders, and depends on a degree and/or a kind of the brain dysfunction. The proposed method is adjustable to the individual properties of patients with various movement disorders.

In the proposed method, a compensator was constructed by using the model information obtained from the measurement data of visual target tracking. The hand movement compensation method proposed in this study was constructed by using the measurement data on visual target tracking for each subject. So the properties of the hand movement for different subjects can be reflected in the structure of the compensator.

Generally, the obtained model often includes the modeling error, so the influence of modeling error should be taken into account. In this study, the stability of the compensator was proved by the Rauth method under the contamination with modeling error and unpredictable disturbance. In addition to the above result, the stability of the proposed compensator was also ascertained through the simulation study. At the experimental results for the total 80 trials on four subjects, unstable motion or continuous oscillation were not observed in the hand movement compensation at all. To consider the above results, the proposed hand movement compensator based on models has enough robustness and stability for the modeling error and the unpredictable disturbance.

5. Conclusion

This study proposed a method for improving the hand movement on visual target tracking by additional assistant force. Based on the experimental data of visual target tracking, characteristics of the hand movement was extracted by the mathematical model and the compensator was constructed. Sufficient improvement of hand movement was verified through the experiment for four subjects.

Proposed compensation technique will be effective for assisting the movement of patients with various movement disorders caused from brain dysfunction.

References

1. Vercher J -L, Gauthier GM (1992) Oculo-manual coordination control: Ocular and manual tracking of visual targets with delayed visual feedback of the hand motion, Experimental Brain Research, 90: 599-609
2. Ide J , Sugi T, et al (2000) Quantitative evaluation of learning effect for different movement disorders by use of manual tracking to moving visual target (in Japanese), The Commercial Review of Seinan Gakuin University, 46-3,4: 347-361
3. Karniel A, Inbar GF (2000) Human Motor Control: Learning to Control a Time-Varying, Nonlinear, Many-to-One System, IEEE Trans. System, Man, and Cybernetics, Part C: Applications and Reviews, 30-1: 1-11
4. Pledgie S, Barner KE, et al (2000) Tremor Suppression Through Impedance Control, IEEE Trans. Rehabilitation engineering, 8-1: 53-59
5. Bai O, Nakamura M, et al (2001) Compensation of hand movement for patients by assistant force: relationship between human hand movement and robot arm motion, IEEE Trans. NSRE, 9-3: 302-307

Approach Motion Generation of the Self-Aided Manipulator for Bed-ridden Patients

Akihiko Hanafusa[1], Hiroko Washida[2], Johta Sasaki[1], Teruhiko Fuwa[1], and Yasuhito Shiota[1]

[1]Department of Rehabilitation Engineering, Polytechnic University, Kanagawa, Japan
[2]Electronic Technology Group, Shizuoka Prefectural Numazu Vocational Training Center, Shizuoka, Japan

Chapter Overview. Rehabilitation robots that are used with humans should move naturally and should not cause discomfort. A system that can generate natural approach motion for a self-aided manipulator for bed-ridden patients based on the individual patient's characteristic motion has been developed. The self-aided manipulator is assumed to provide the function of bringing a drink from a side table to the patient.

Approach motions resembling those of individual subjects can be generated using six characteristic points of the path and spline interpolation. In order to calculate the position and time of characteristic points, linear approximation formulas derived from measured paths of each subject are used. In addition, the target lip position and direction were measured using a single video camera, and a manipulator drive test was performed in order to demonstrate the effectiveness of the system.

Key Words. Approach motion, Self-aided manipulator, Rehabilitation robot, Video camera.

1. Introduction

Rehabilitation robots [1] that aid to eating or drinking [2] and that can be mounted on the wheelchair [3] have been developed. Robots that interact with humans in rehabilitation roles should move naturally and should not cause discomfort. We have developed a system that can generate natural approach motion for a self-aided manipulator [4] for bed-ridden patients based on the individual patient's characteristic motion. The self-aided manipulator is assumed to provide the function of bringing a drink from a side table to the patient, and for this purpose, the newly developed system can detect the target lip position and direction using a single color video camera.

2. Approach Motion Generation

2.1 System configuration of self-aided manipulator system

Figure 1 shows the system configuration of the self-aided manipulator system. This system consists of a self-aided manipulator, a manipulator controller, a simulator and a central controller with a video camera. The central controller inputs images from the video camera and calculates the target position. Based on the target position, the approach motion of manipulator is generated and commands to move the manipulator are sent to the simulator and the manipulator controller. The manipulator controller controls the motors of manipulator, and the simulator can simulate the movement of the manipulator beforehand.

The self-aided manipulator, shown in Figure 1, is an arm-type robot that has shoulder, elbow and wrist joints. The total length of the self-aided manipulator is 900 mm. As each joint has rotation and flexion, there are six degrees of freedom in total, and the robot has a hand with three fingers that can move simultaneously. In order to move the manipulator, a series of move commands that specify the coordinates of the target position and the vectors of target posture of the hand are necessary. In addition, a speed command can be inserted before move commands in order to change the speed. The target time, motor rotating angles and motor angular velocities of each move command are calculated according to the specified speed and coordinates. Since motor angular velocities are feedback-controlled to finish the move command by the target time, the coordinates between commands are not interpolated linearly.

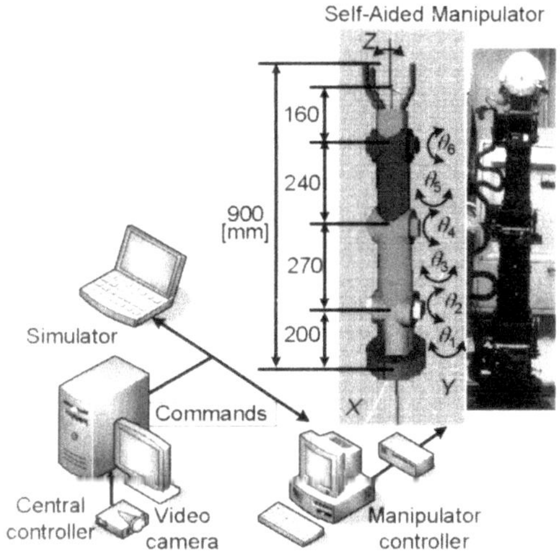

Fig. 1. System configuration of self-aided manipulator system.

2.2 Approach motion measurement

In order to derive the individual characteristics of the approach motion generated by each subject, we used a dummy to represent the bed-ridden patient and measured the approach motion of the subject's hand grasping a cup with a straw from the side table to the dummy head. The reclining angle of the bed was set to 30, 45 and 60 degrees. In addition, the face angle of the dummy head was set to –60, –30, 0, 30 and 60 degrees in 30-degree increments. The experiment considered the approach motion of five subjects, denoted as A, B, C, D and E, and the approach motion of each subject was measured using a three-dimensional motion capture system MA-6250 (Anima Corp.) that uses infrared reflective markers. The results of horizontal and vertical movement and velocity variation for a

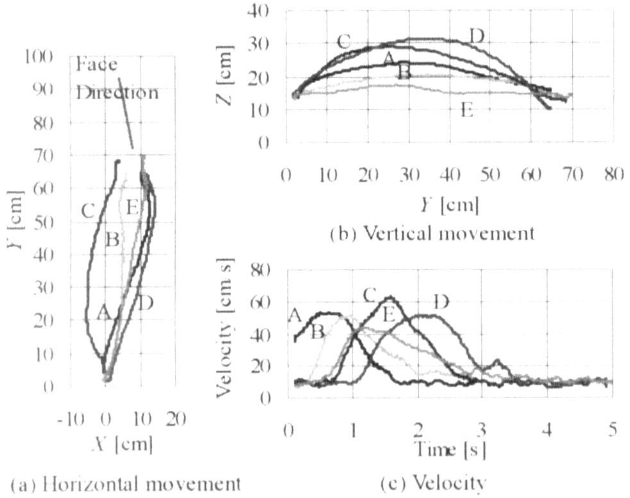

Fig. 2. Approach motion toward dummy head on a bed having a reclining angle of 30° and a face angle of 30°.

bed reclining angle of 30 degrees and a face angle of 30 degrees are shown in Figure 2. All subjects sat at approximately the same position (X = –50 cm, Y = 20 cm). The horizontal detour movement varied according to the subject. Subjects A, D and E approached from the front of the face, subject B approached directly, and subject C approached from the near side. In addition, the detour distance differed according to the patient face angle and the subject. On the other hand, the hand vertical movement and velocity variation were similar in that each had a single peak. However, the value and position of the peak differed by subject.

2.3 Generation of move data

In order to generate an approach motion resembling that of a specific subject, characteristic points representing the position along the path were derived from the measured path. The six characteristic points used for this purpose, shown in Figure 3, are start, end, most distant point from the direct straight path, point of maximum speed, and the first and last points at which the speed reaches half of maximum. Three points related to speed are necessary in order to approximate the variation of velocity. Next, the parameters of the linear approximation formula that are necessary in order to generate the coordinates of the characteristic points based on the start and end point coordinates and the face angle are derived separately using the measured data for each subject. A motion that resembles that of a specific subject is generated as follows. First, the start and end point coordinates and the face direction are input. Next, the time and position of the six characteristic points are calculated using the approximation

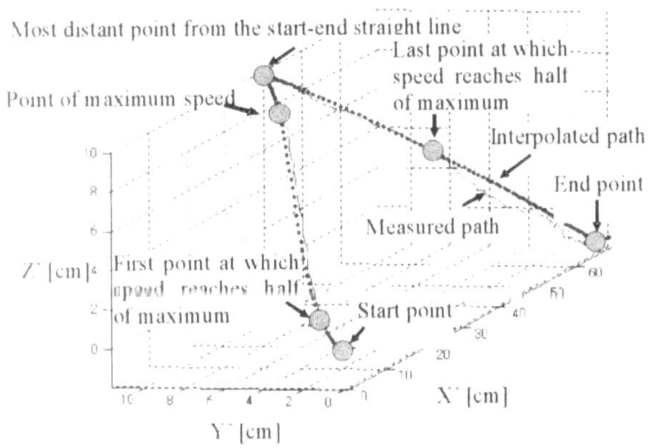

Fig. 3. Characteristic points and path obtained by cubic spline interpolation.

formula. Finally, path data for the approach motion are generated by cubic spline interpolation of the calculated characteristic points.

2.4 Experimental motion test by the manipulator

An experimental motion test is performed using the manipulator for two cases of data generated by the system. In the first case, the move commands generated from the six characteristic points were used. In the second case, in addition to the six characteristic points, 12 spline-interpolated points were used to generate the move commands. Interpolation points were generated when either the distance from the previous point exceeded 4 cm or the elapsed time from the previous point exceeded 0.7 sec. The movement of the manipulator hand was measured using the three-dimensional motion capture system. The results are shown in Figure 4. The maximum distance between the spline interpolated path and the measured path using commands generated without interpolation, indicated by the gray line, was 6.6 cm, and the maximum distance of the measured path with interpolation, indicated by the black line, was 1.8 cm and was approximately consistent with the spline interpolated path.

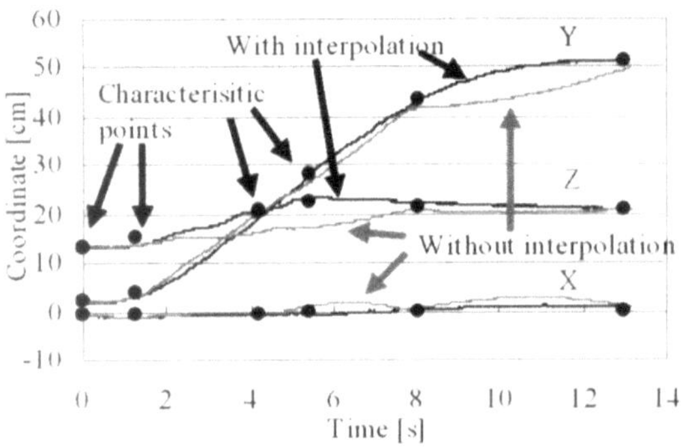

Fig. 4. Comparison of measured path with and without interpolation.

3. Target Measurement by Video Camera

3.1 Calculation of distance from the camera

A single-color video camera EVI-D30 (SONY Corp.) was used to measure the position of the end target point. The pan, tilt and zoom of the camera were controlled by computer, and the status, including the current focus value, could be obtained. The relation of focus value (F) and distance (zc) of the target object is shown in equation (1). Values of coefficients differ by the zoom value. By changing the distance between the dummy and the camera, focus values are measured for zoom values of 850 and 1,250. The results are approximated by equation (1), and distances are recalculated by the equation using measured focus values. Actual and calculated distances are compared to check the accuracy. Absolute errors are shown in Table 1. Better accuracy could be obtained when the maximum zoom value of 1,023 was used, as compared to a zoom value of 850.

The distance is the Z coordinate of camera coordinate system (x_c, y_c, z_c). In addition, the X and Y coordinates of the camera coordinate system are calculated using the target pixel position (u, v) of the image and z_c obtained experimentally by equations (2) and (3):

$$z_c = 1/\left(\alpha_F e^{\beta_F/F} + \gamma_F\right) \tag{1}$$

$$x_c = (\alpha_c z_c + \beta_c)(u - u_{max}/2) \tag{2}$$

$$y_c = (\alpha_c z_c + \beta_c)(v - v_{max}/2). \tag{3}$$

Table 1. Absolute errors of calculated distance using focus value.

Zoom value	Maximum [cm]	Average [cm]	SD [cm]
850	7.89	2.56	3.10
1,023	3.59	1.91	1.03

3.2 Calculation of the coordinates of the target

The target coordinates were calculated based on the position and posture of the camera, the zoom and focus value and the pixel position of the target. The target point used in the present experiment was located on the lips of a dummy head. In order to recognize the lip area and calculate the lip position, the flesh-colored area is first extracted from the captured image, and the lip area is then determined by template matching [5] within the flesh-colored area. The pre-captured lip image was used as a template. Then the camera is moved using pan and tilt functions, in an attempt to center the lips in the image. Finally, in order to calculate the coordinates of the target, the camera is zoomed until the maximum value of 1,023. In order to move the manipulator, the target position in the camera coordinate system (x_c, y_c, z_c) should be converted to the manipulator coordinate system (x_m, y_m, z_m). The conversion is performed using the following equation:

$$(x_m, y_m, z_m, 1)^T = \mathbf{T}_c\mathbf{R}_c \cdot \mathbf{R}_p \cdot \mathbf{T}_t\mathbf{R}_t \cdot \mathbf{T}_{-t} \cdot \mathbf{T}_p (x_c, y_c, z_c, 1)^T \tag{4}$$

where $\mathbf{T}_c\mathbf{R}_c$ is the translation and rotation matrix to the pan coordinate system of the camera as defined by the manipulator coordinate system, $\mathbf{T}_t$ is the position of the tilt axis in the pan coordinates, $\mathbf{R}_t$ is the rotation of tilt, $\mathbf{T}_p$ is the position of the pan axis in the camera coordinates, and $\mathbf{R}_p$ is pan rotation.

3.3 Calculation of face angle

If the same feature points can be extracted from a series of images, then the three-dimensional positions of feature points can be reconstructed by factorization [6]. In addition, the position and rotation of the camera used to capture image can be calculated. In other words, when the camera is fixed, the position and rotation of a target can be calculated. If we define the face coordinate system and a feature point coordinate (x_f, y_f, z_f), then the relationship between the face coordinate system and the camera coordinate system (x_c, y_c, z_c) is as follows:

$$(x_c, y_c, z_c, 1)^T = \mathbf{T}_f \mathbf{R}_f (x_f, y_f, z_f, 1)^T \quad (5)$$

where $\mathbf{T}_f$ and $\mathbf{R}_f$ are the translation and rotation matrices that indicate the position and rotation of the face, respectively, and can be calculated by factorization. Thus, the face angle can be calculated from multiple images and feature points.

We used the right and left edges of the eyes and mouth, the upper and lower edges of the mouth and the nostrils as feature points. Feature points extraction was performed primarily by the template matching method. Therefore, templates generated from directly in front of the face were registered beforehand. These include a coarse template of the face area, including the eye, the nose and the mouth, and fine templates of the edges of the eyes and mouth. Feature points are extracted by following process:

1. The flesh-colored area is extracted from the captured image.
2. The face area is searched within the flesh-colored area using the coarse template of the face.
3. Feature points assumed areas are defined using the face area information.
4. The nostrils are extracted as feature points using the brightness data of the assumed area.
5. The feature points of the edges of the eyes and mouth are searched by the fine templates within the assumed area.

Figure 5 shows the results of feature extraction when the face angle is from –25 to 25 [deg]. In the case of 10 [deg] and above, feature point extraction of the upper or lower edge of the mouth failed. In the cases of either less than –25 [deg] or more than 25 [deg], feature point extraction of the endmost feature points was not possible, as the feature points were hidden.

A face angle calculation program using factorization was developed based on the literature [7]. The results of calculation using all 10 feature points were –26.7, –20.0, –14.0, –9.3, –5.9, –1.5, 6.7, 10.0, 16.3, 18.1, and 24.7 [deg], respectively, and the average error was 1.0 [deg]. By preparing all of these patterns beforehand, it is possible to calculate the face angle when a new image is input. At present, this program has not been integrated into the system, and the calculated face angle should be input manually into the manipulator path generation program.

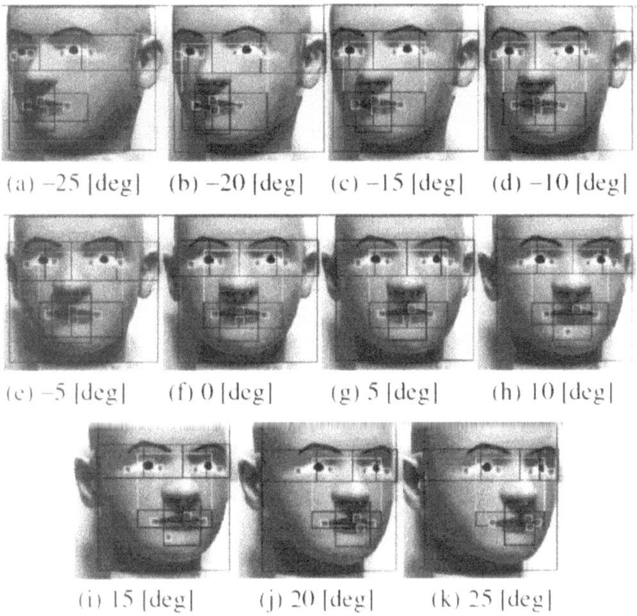

Fig. 5. Extracted feature points of face angles from −25 to 25 [deg].

3.4 Integral experiment

An integral experiment was performed in order to check whether the total system functioned effectively. The target lip position of the dummy head was measured using a video camera. The approach motion was generated by the measured target position, and the manipulator was moved. In the experiment, the reclining angle of the bed was changed from 45 degrees to 60 degrees in five-degree increments. In addition, the target lip position and the movement of the manipulator hand were measured using a three-dimensional motion capture system. The error distances of the target between coordinates calculated by the camera and that of three-dimensional motion capture system were calculated. The average error was 2.4 cm, and the maximum error was 4.7 cm.

Figure 6 shows the attitude of the self-aided manipulator at the end of the path for bed reclining angles of 45 degrees and 60 degrees. The XYZ

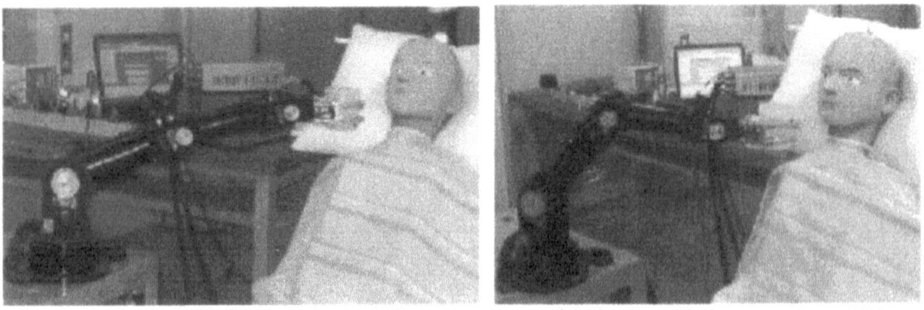

(a) Bed reclining angle = 45° (b) Bed reclining angle = 60°

Fig. 6. Posture of the manipulator at the end point of the path.

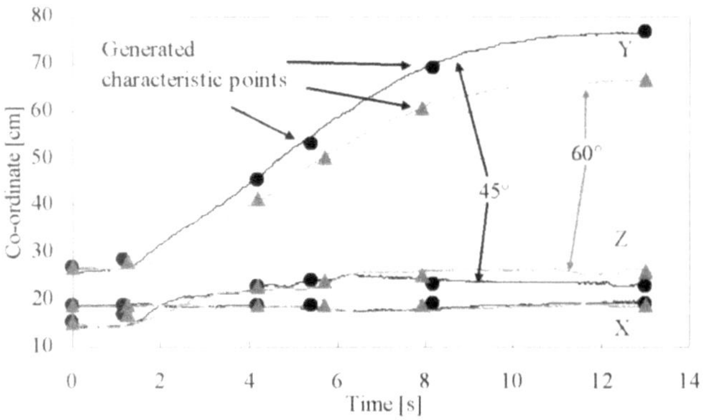

Fig. 7. Comparison of XYZ coordinates of characteristic points and the measured path between bed reclining angles of 45° and 60°.

coordinates of generated characteristic points and the measured path are shown in Figure 7. The paths passed over characteristic points, except for second point. The end point of the path for the bed reclining angle of 45 degrees was 10 cm longer in the Y direction and 3 cm lower in the Z direction, compared to that at the bed reclining angle of 60 degrees. This result confirms that the manipulator moved in accordance with the position of the lips for different reclining angles.

4. Conclusion

Approach motions resembling those of individual subjects can be generated using six characteristic points of the path and spline interpolation. In order to calculate the position and time of characteristic points, linear approximation formulas derived from measured paths of each subject are used. In addition, the target lip position and direction were measured using a single video camera, and a manipulator drive test was performed in order to demonstrate the effectiveness of the system.

In order to perform the test using human patients, the accuracy of the position calculation by the video camera and the accuracy of robot motion control should be improved along with the overall safety and reliability.

References

1. Mahony RM (1994) Robotic products for rehabilitation : Status and strategy. Proc 4th Int. Conf. Rehab. Robotics : 12.
2. Topping M, Smith J (1999) The development of Handy 1, a robotic system to assist the severely disabled. Proc 6th Int. Conf. Rehab. Robotics : 244-249.
3. Verburg G, Milner M, Naumann S, Bishop J, Sas O (1992) Evaluation of the manus wheelchair mounted manipulator. Proc 16th RESNA Conf. : 602-604.
4. Ikeda T, Washida H, Manan ZBAb, Yoda M, Kakimoto A, Hanafusa A, Nanao K, Shiota Y (2002) Development of self-help robot using biological signals. Proc 18th LST Conf. : 36.
5. Barnea DI, Silverman HF (1972) A class of algorithms for fast digital image registration. IEEE Trans. Computer, vol.c-21, no.2 : 179-186.
6. Tomasi C, Kanade K (1992) Shape and motion from image streams under orthography – A factorization method. Int. J. Comput. Vision, vol. 9, no.2 : 137-154.
7. Kanatani K, Suguya Y (2004) Complete recipe for factorization. http://www.suri.it.okayama-u.ac.jp

Lower-limb Joint Torque and Position Controls by Functional Electrical Stimulation (FES)

Koji Ito[1], Takahiro Shioyama[2], and Toshiyuki Kondo[1]

[1]Department of Computational Intelligence and Systems Science, Tokyo Institute of Technology, Japan
[2]Nihon Kohden Corporation, Japan

Chapter Overview. The present paper discusses a FES control method for promoting the functional recovery of voluntary movements in the early rehabilitation of hemiplegia patients or imperfect paralysis patients by apoplexy etc. If the patient's muscles could be activated through the external stimulation synchronizing with his/her motor intentional images, the feedback impulses from the sensory receptors within the limb must stimulate the sensory and motor areas, which results in helping to rebuild a new motor map in the cerebral cortex.

We propose a method to control the joint torque and position coordinated with the patient's motor intention by FES including the compensation for the muscle fatigues.

Key Words. Joint Torque, Functional Electrical Stimulation (FES), and Position Control.

1. Introduction

With the progress of the coming aging society, it gains in more importance to develop the efficient rehabilitation method for various motor paralytic patients. Primary concern in the traditional rehabilitation has been directed to what such a physically handicapped person can carry out by using the remained functions. Then, it has been mainly considered to maximize the patient's residual motor capability. However, attention is recently focused on the rehabilitation method, which aims to recover his/her motor functions by making active use of neuroplasticity in the brain [1, 2]. Then, from the initial stage of rehabilitation process, the patient's paralyzed limb is repeatedly moved by the physical therapist. This aims to prevent the joint contracture and also rebuild the lost motor map in the brain by stimulating the sensory and motor areas through the proprioceptive feedbacks.

Under the above background, the present paper proposes to apply Functional Electrical Stimulation (FES) to the rehabilitation process for the physically handicapped [3, 4, 5, 6]. Much effort of FES has been directed toward developing the joint control systems of upper and lower limbs. For

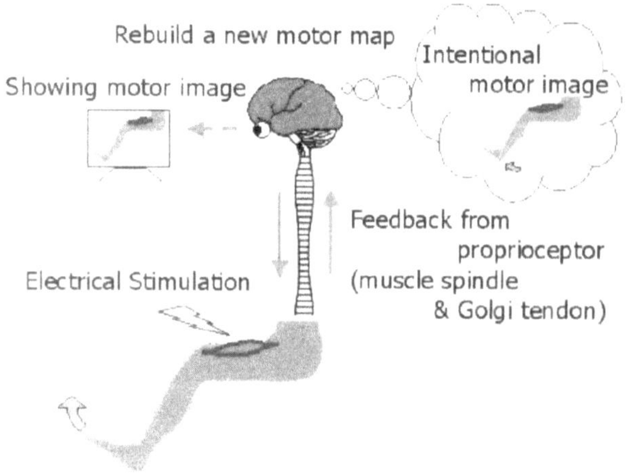

Fig.1. Schematic flow chart of the application of FES control to rehabilitation.

designing the controller, many studies have centered on identifying the dynamic behaviors of electrically activated musculoskeletal systems considering both linear and nonlinear models in isometric and isotonic conditions [7,8,9,10,11], and further to estimate the internal parameters of muscle systems, e.g. fatigue, habituation, spasticity [12,13]. In spite of the above studies, the complex mathematical representation and unknown parameters make difficult to apply them to individual subjects in the clinical setting.

The aim of the present study is to develop a simple procedure useful for the clinical sites that would enable us to apply FES to the early rehabilitation of persons injured by the cerebral infarction or hemorrhage.

The schematic flow chart is shown in Fig.1. A target motor pattern such as the joint rotation or torque is indicated on the TV display, and then the patient is required to imagine the visually captured target motion within the brain. At the same time, the patient's paralyzed leg or arm is activated by FES to generate the target motor pattern. The corresponding feedback impulses from the sensory receptors in the limb (muscle spindles, Golgi tendon organs etc.) must reach to the sensory and motor areas in the brain. Then, the synchronization between the feedforward motor signals created from his/her intentional motor images and the proprioceptive sensory feedback signals must promote to rebuild a new motor map in the cerebral cortex.

The knee joint muscles were chosen because of their fundamental role in the main activities (standing up, sitting down, walking etc.) and to the relative simplicity of the knee joint compared to the other joints.

Now, in order to apply FES to the functional recovery training, it is required to satisfy the following requirements.

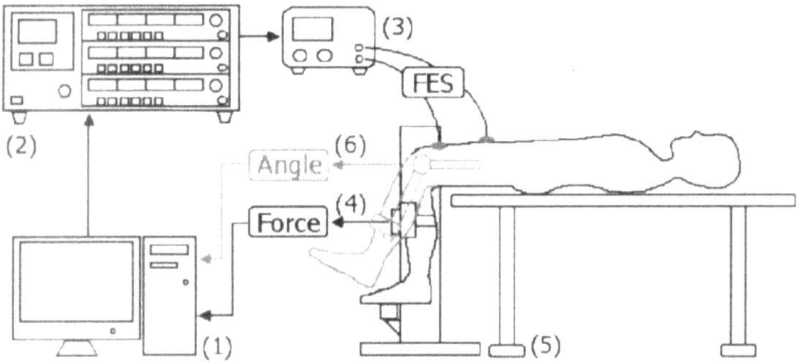

Fig.2. Experimental apparatus

(1)PC for control and measurement, (2) Electrical stimulator, (3) Isolator, (4) Strain gauges and transducer, (5) Bed for subject, (6) Goniometer

1) Various motor patterns must be presented to the patient repeatedly.
2) It is required to compensate for the change of motor pattern due to the muscle fatigues.
3) No sensor should be attached to the patient.

We propose a method to control the lower-limb joint torque and position by FES satisfying the above.

2. Experimental Environments

The overview of the experimental apparatus for the torque and position control by FES is shown in Fig.2. It consists of Electrical stimulator (SEN-7203, Nihon Kohden), Isolator (SS-104J, Nihon Kohden), Strain gauges and transducer (DPM-6K, Kyowa Electric Instruments), A-D/D-A converters (PCI-3135, PCI-3325, Interface), and computer. The muscle is stimulated through the isolator by the pulses of electric current, which is transformed from the voltage pulse sequence produced in the computer. The stimulation parameters (frequency: 25 Hz, pulse width: 0.2 msec) were determined by trial and error with due consideration for the muscle fatigues and joint controllability.

The subject for the experiments was lied face up on the bed. The surface electrodes for the stimulation are made of silicon rubber (EW0601P, National). The stimulation positions were chosen near each motor point of the medial vastus muscle and lateral vastus muscle. Before the measurement, the stimulation intensity was gradually increased from 0 to 30 mA in order to confirm that the subject feels no discomfort and pain.

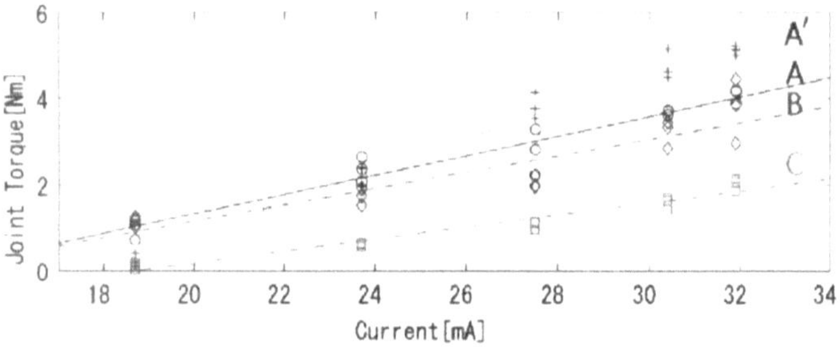

Fig.3. Relation between the electrical stimulation current and joint torque

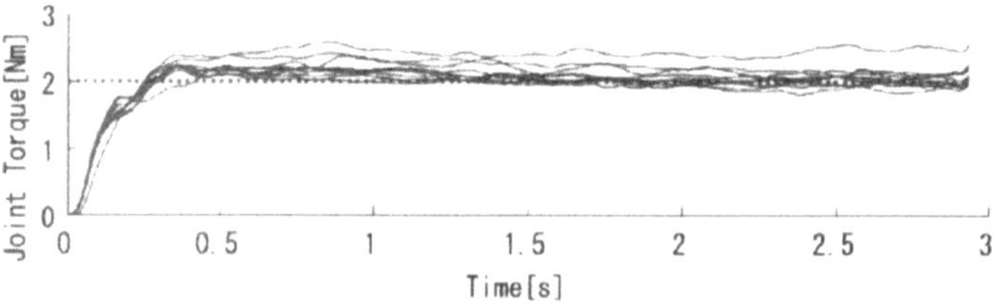

Fig.4. Joint torque control (Subject A: reference torque 2 Nm)

For the torque control, the lower-limb is belted at the knee flexion angle of 90 degrees. Then a couple of muscles were contracted by the electrical stimulation and the extension torque of knee joint was measured by the strain gauge. For the joint angle control, the knee and ankle joints are free and the angle of knee joint was measured by the potentiometer type Goniometer. The joint torque and angle were inputted into the computer at the sampling rate 2 kHz. The subjects were three healthy students.

3. Joint Torque Control

3.1 Electrical stimulation current and joint torque

For the joint torque control by FES, it is necessary to identify the relation between the electrical stimulation current and joint torque. The electrical currents for the stimulation were changed at five levels (18.7, 23.7, 27.5, 30.4, 31.9 mA). The duration of stimulation is 3 sec and the rest period between trials is 60 sec. The identification experiments were performed three times for each subject. The results for three subjects (A, B, C) are shown in Fig.3. The dotted line A' shows the result measured at another day for the

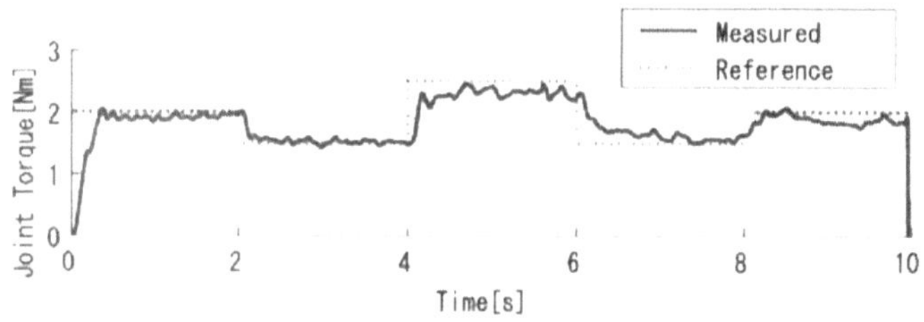

(a) First trial

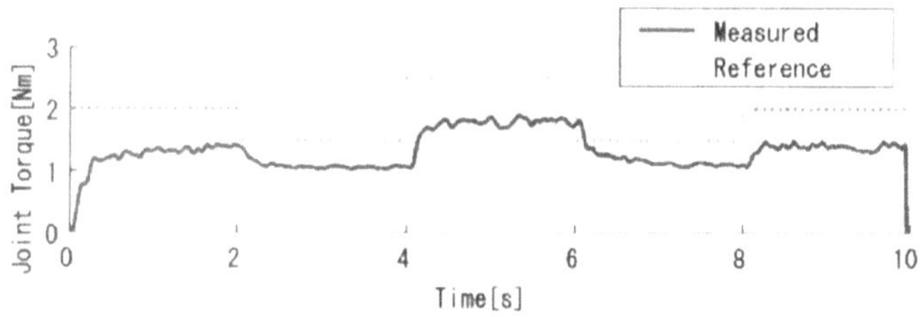

(b) 20th trial

Fig.5. Joint torque control in stepwise patterns for every 2sec during 10sec

subject A. It is seen that, though the gradients of lines are different among the subject and date, the relation between the electrical current and joint torque is approximated by a linear line for each subject.

3.2 Joint torque control experiment

By using the linear line for each subject in Fig.3, it is possible to determine the stimulation current value necessary to generate a desired joint torque. The stimulation experiments of 10 trials were performed for a reference joint torque (2 Nm). The duration of stimulation is 3sec, and the rest period between trials is 60 sec. Fig.4 shows the joint torques of ten trials for the subject A. It is seen that the reference joint torque is precisely generated for every subject.

Next, the reference torque was changed in a stepwise pattern for every 2 sec during 10sec. Fig.5 (a) is the result of the first trial, and Fig.5 (b) is the 20th trial. The rest period between trials was taken for 60 sec. The joint torque is able to generate the reference torque at the first trial. It is, however, found that it is unable to reach to the reference level at the 20th trial. This comes from the muscle fatigue.

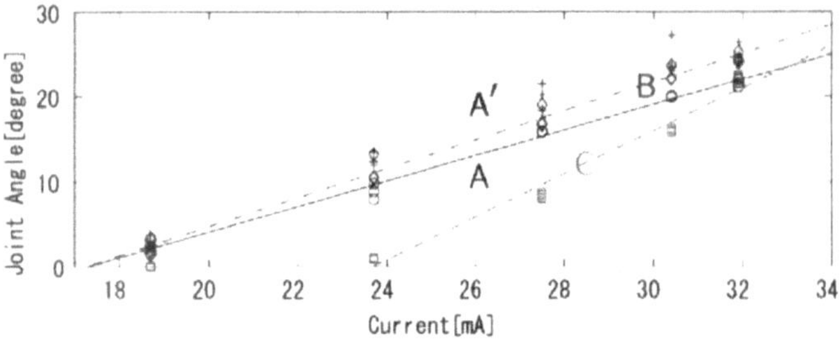

Fig.6. Relation between the electrical stimulation currents and joint angles

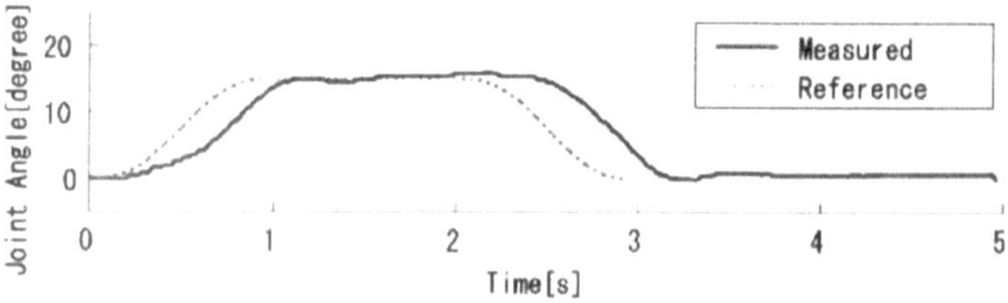

Fig.7. Joint angle control following the minimum jerk trajectory (15 degree)

4. Joint Angle Control

4.1 Electrical stimulation current and joint angle

For the joint angle control by FES, we have to identify the relation between the electrical stimulation current and joint angle. The electrical currents were changed at five levels (18.7, 23.7, 27.5, 30.4, 31.9mA) for each subject in the same way as the joint torque.

The electrical currents and joint angles at 1 sec from the beginning of stimulation are plotted for three subjects (A, B, C) in Fig.6. Though the gradients of lines are different among the subjects and dates, the relation between the electrical current and joint angle is approximated by a linear line for each subject.4.2 Joint angle tracking control
It is required to generate an arbitrary joint angle pattern by FES. Here, the desired joint trajectory is planned as follows. The joint angle is reached to

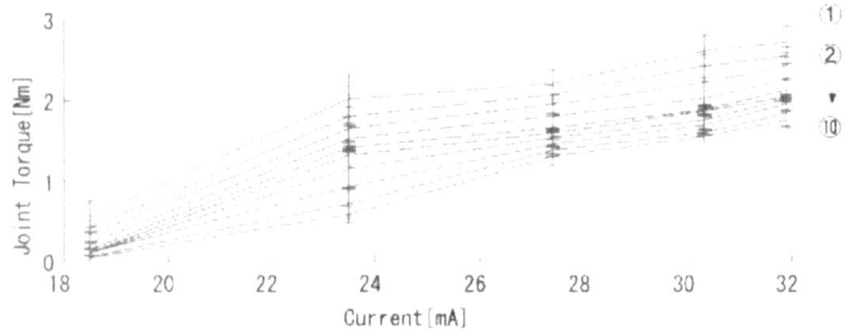

(a) Average value and standard deviation for every three sets

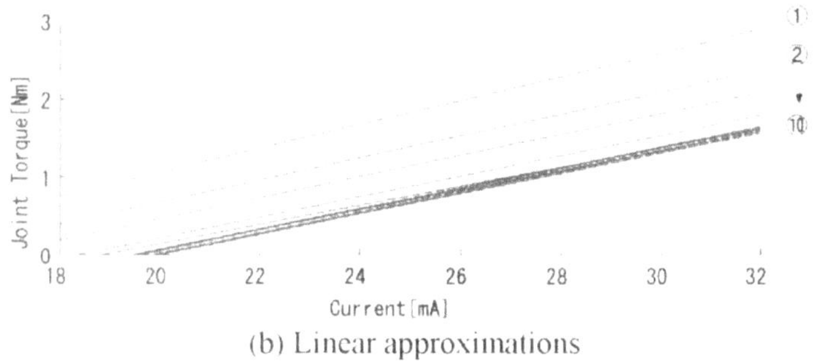

(b) Linear approximations

Fig.8. Joint torque decline due to the muscle fatigue

the reference angle by 1sec following the minimum jerk trajectory [14], and the angle is maintained for 1 sec. Then it returns to the initial angle by 1sec in the same way. The reference angle is 15 degree.

The joint angle trajectory of Subject A is shown in Fig.7. It is seen that the knee joint angle follows the minimum jerk trajectory with no oscillation. There is, however, the time delay of about 0.25 sec for the reference angle. So, when the desired motion pattern is presented to the patient, it should be delayed for 0.25 sec from the FES current command.

5. Compensation for Muscle Fatigue

5.1 Muscle fatigue in the joint torque control

As seen in Fig.5, the relation between the electrical stimulation current and joint torque has been changed with the muscle fatigue.

First, we performed the same experiments as Fig.4, but the rest period between trials is 15 sec. One trial set consists of five electrical currents

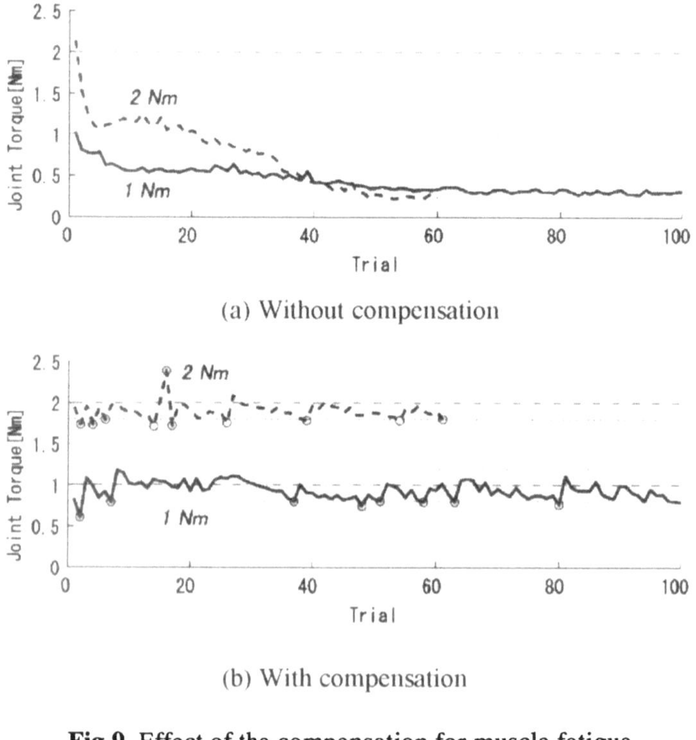

Fig.9. Effect of the compensation for muscle fatigue

(18.7, 23.7, 27.5, 30.4, 31.9 mA). Thirty sets (total of 150 trials) were stimulated successively. The result is shown in Fig.8, where Fig.8 (a) is the average value and standard deviation for every three sets. The joint torque decreases gradually by repetitions of the electrical stimulation, which is due to the muscle fatigue. Fig.8 (b) shows the linear approximation for each three set. Note that the gradients of line are same. From the above result, if the stimulation currents are updated according to the approximated linear lines, it is possible to compensate for the muscle fatigue.

5.2 Compensation for the muscle fatigue

Fig.9 shows the experimental results on the compensation for the muscle fatigue. The muscle fatigue is not compensated for in Fig.9 (a), but compensated for in Fig.9 (b). In both cases, the duration of stimulation is 3 sec and the rest period between trials is 10 sec. Two kinds of reference torques are set up, which are 1 Nm (solid line) and 2 Nm (broken line). The small round mark in Fig.9 (b) denotes that the stimulation current was updated to the new value at each point. As seen in Fig.9 (a), in the case of no compensation, there is a marked decline in the joint torques due to the muscle fatigue with increase in the trial number. On the other hand, Fig.9 (b) shows no decline and the desired joint torque are still generated even more than 50 trials.

6. Conclusion

The present paper proposed the method to control the lower-limb joint torque and position by FES compensating for the muscle fatigue with the view of applying to rehabilitation. In the both cases of the joint torque and position controls, it was possible to identify the relation between the electrical stimulation current and joint torque and position after pasting the stimulation electrodes on the skin. In addition, it was verified that the reference level in the joint torque control could be maintained under the muscle fatigue by updating the FES currents.

At the present, the standard time for rehabilitation is 40 minutes in Japan. It takes no more than 20 minutes from pasting the stimulation electrodes to identifying the personal parameters. Therefore, it will be possible to practice 80-90 trainings for FES control under the condition of 3 sec stimulation and 10 sec rest for the joint torque and position controls. Since the patient's lower limb is moved synchronizing with the motor images, it is expected that the synergistic effects between the feedforward motor signals generated from his/her intentional motor images and the proprioceptive sensory feedback signals could promote to rebuild a new motor map in the cerebral cortex.

When applying the FES control to the patient, we have to consider various problems such as the difference among individuals for the electrical stimulation, joint contracture, etc. Also when a part of motor function is remained, we have to find how to combine FES control with the patient's motor control? In addition, it is desired to investigate the multi-joint FES control and the impedance control of joint by the concurrent stimulations of agonist and antagonist muscles.

Acknowledgment

This research was supported in part by Grants from the Japanese Ministry of Education, Culture, Sports, Science and Technology (No. 14350227, 16760337).

References

1. Merzenich M.M. et al (1984) Somatosensory cortical map changes following digit amputation in adult monkeys. J. Comp Neurol., 224(4), 591-605.
2. Wolfgang H.R. et al (1999) Effects of constraint-induced movement therapy on patients with chronic motor deficits after stroke, Stroke, 30(3), 586-592.
3. Marsolais E.B., Kobetic R (1983) Functional walking in paralyzed patients by means of electrical stimulation, Clin. Orthop, 175, 30-36.
4. Handa Y. (1997) Current topics in clinical functional electrical stimulation in Japan, J. Dlectromyogr. Kinesiol, 7(4), 269-274.
5. Pedotti A. et al (1996) Neuroprosthetics: from basic research to clinical applications, Springer-Verlag.
6. Levin O., Mizrahi J., Isakov E. (2000) Transcutaneous FES of the paralyzed quadriceps: Is knee torque affected by unintended activation of the hamstrings? J. of Electromyography and Kinesiology, 10, 47-58.
7. Bernotas L., Cargo P et al (1986) A discrete-time model of electrically stimulated muscle, IEEE Trans. Biomed. Eng., 33, 829-838.

8. Veltink P.H., Chizeck H.J. et al (1992) Nonlinear joint angle control for artificially stimulated muscle, IEEE Trans. Biomed. Eng., 39, 368-380.
9. Dorgan S. J. Malley M.J.O' (1997) A nonlinear mathematical model of electrically stimulated skeletal muscles, IEEE Trans. Rehab. Eng., 5, 179-194.
10. Chizeck H.J., Chang S. et al (1999) Identification of electrically stimulated quadriceps muscles in paraplegic subjects, IEEE Trans. Biomed. Eng., 46, 51-61.
11. Ferrarin M., Pedotti A. (2000) The relationship between electrical stimulation and joint torque: A dynamic model, IEEE Transactions on Rehabilitation Engineering, 8(3), 342-352.
12. Boom H.B.K., et al (1993) Fatigue during functional neuromuscular stimulation, Progr. Brain Res. 97, 409-418.
13. Stefanovska A. et al (1989): FES and spasticity, IEEE Trans. Biomed. Eng., 36, 738-745.
14. Flash T., Hogan N. (1985) The coordination of arm movements: An experimentally confirmed mathematical model, J. of Neuroscience, 5(7), 1688-1703.

Pattern Recognition of EEG Signals During Right and Left Motor Imagery
~ Learning Effects of the Subjects ~

Katsuhiro Inoue[1], Daiki Mori[1], Gert Pfurtscheller[2],
and Kousuke Kumamaru[1]

[1] Department of Systems Innovation and Informatics, Factory of Computer Science and Systems Engineering, Kyushu Institute of Technology, Iizuka, Fukuoka, Japan
[2] Department of Medical Informatics, Institute of Biomedical Engineering, Graz University of Technology, Graz, Austria

Chapter Overview. Electroencephalograph (EEG) recordings during right and left motor imagery can be used to move a cursor to a target on a computer screen. Such an EEG-based brain-computer interface (BCI) can provide a new communication channel to replace an impaired motor function. It can be used by e.g., handicap users with amyotrophic lateral sclerosis (ALS). In this study, statistical pattern recognition method based on AR model was introduced to discriminate the EEG signals recorded during right and left motor imagery. And the learning effects of the subjects are investigated.

Key Words. Pattern Recognition, EEG (electroencephalogram), Motor Imagery, and BCI (Brain Computer Interface).

1. Introduction

Classification of EEG signals is a difficult task, especially when the derived classification result is to be used to control an electronic device, because in this case the classification has to be performed on a single-trial basis (i.e. not averaged). Such a system which transforms signals from the brain into control signals is known as a brain-computer interface (BCI) [1,

2]. For the single-trial signal recognition method, there are many methods were proposed. For examples, artificial neural network (ANN) with frequency components as a feature vectors [3], and linear discrimination method based on adaptive auto-regressive (AR) parameter [5]. We also have proposed such a system based on AR model [8].

In this paper, statistical pattern recognition method based on AR model was introduced to discriminate the EEG signals recorded during right and left motor imagery. And the learning effects of the subjects are investigated.

2. Method and Materials

Six subjects (22-24 years old) participated at this study. All were right-handed and free of medication and central nervous system abnormality.

2.1 Experimental paradigm, training and testing sessions

During the experiment, the subject fixated a computer monitor 100 cm in front of him. Each trial was 8 sec long and started with the presentation of a fixation cross at the center of the monitor, followed by a short warning tone ('beep') at 2000 ms (see Fig.1). At 3000 ms, the fixation cross was overlaid with an arrow at the center of the monitor for 1250 ms, pointing either to the left or to the right. Depending on the direction of the arrow, the subject was instructed to imagine a movement of the left or the right hand. Prior to the experiment, each subject was given the opportunity to practice and perform actual movements of the left and right hand according to the arrow direction displayed on the monitor. Feedback consisted of a bar-graph is presented in the center of the monitor since 4250msec to 8000msec; the length of the bar-graph depending on how well a subject-specific classifier could recognize movement-dependent EEG characteristics. There were two types of sessions: in the training sessions, data were collected for the creation of a subject-specific classifier and, therefore, no feedback was provided. In the following test sessions, the classifier was then used to classify the subject's EEG on-line while he imagined the requested kind of hand movement, and feedback was given to the subject as described above. Each of the initial subjects participated in 10 or 4 sessions, all on different days. Each session consisted of 3 experimental runs of 60 trials (30 'left' and 30 'right' trials) and lasted for about 1 hour. The sequence of 'left' and 'right' trials, as well as the duration of the breaks between consecutive trials (ranging between 500 and 2500 ms), were randomized throughout each experimental run.

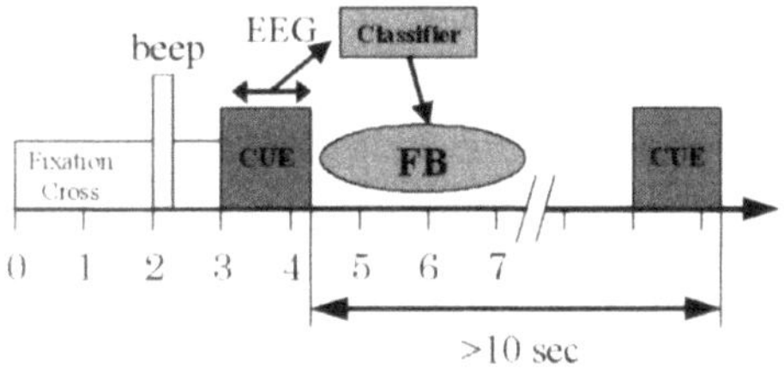

Fig.1. Timing chart of experiment

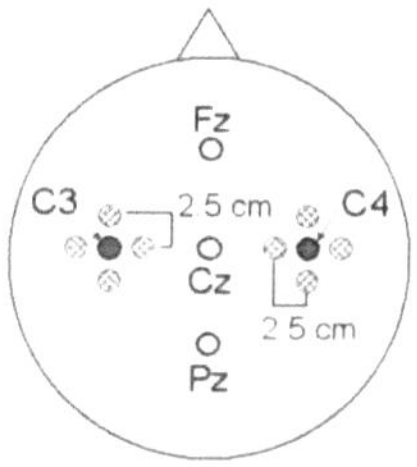

Fig.2. Position of the electrodes

The subjects are grouped into two groups (Group A, B). Subjects who belong in Group A participate in 10 sessions. And in 2~10 session, the feedbacks are executed based on the parameter estimated from the previous session's data. On the other hand, subjects who belong in Group B participate in 4 sessions. And in 2~4 session, the feedbacks are executed based on the parameter estimated from another subject's data.

2.2 EEG recording and data acquisition

Small Laplacian filtered signals are used as EEG signals. Channel C3 (or C4) was derived from following 5 electrodes. (See Fig.2) (y_0 : electrode placed C3(C4), y_A : electrode placed 2.5cm anterior to C3(C4), y_P : electrode placed 2.5cm posterior to C3(C4), y_L : electrode placed 2.5cm left to C3(C4), y_R : electrode placed 2.5cm right to C3(C4).
Channel C3 (or C4) is derived by following equation.

$$y = y_0 - (y_A + y_P + y_L + y_R)/4 \tag{1}$$

The EEG signals were amplified and band-pass filtered between 0.5 and 33Hz by a Nihon Khoden amplifier and then sampled at 128Hz. EOG was derived from two electrodes, one placed medially just above the right eye and the other laterally just below the right eye, in order to detect vertical as well as horizontal eye movements. These signals were used to screen the EEG recordings for eye movement artefacts.

3. Pattern Recognition Method Based on AR Model

The EEG signals are assumed to be generated from an AR model,

$$y_t = \sum_{j=1}^{m} \phi_j y_{t-j} + v_t = \Phi^T z_{t-1} + v_t \qquad \begin{cases} \Phi = [\phi_1, \phi_2, \cdots, \phi_m]^T \\ z_{t-1} = [y_{t-1}, y_{t-2}, \cdots, y_{t-m}]^T \end{cases} \tag{2}$$

Where, y_t is the observed EEG signal at time t, and v_t is the independent random variable with normal distribution, respectively.

$$E[v_t v_s] = \rho \delta_{t,s}, \qquad v_t = N[v_t; 0, \rho]$$

A feature vector θ is defined as follows

$$\theta = \left[\Phi^T, \rho\right]^T$$

The first m observations $y_1, y_2, \cdots, y_m$ serve as initial conditions for (2), following equation is assumed.

$$p(y_1, y_2, \cdots, y_m, \theta_k) = p(y_1, y_2, \cdots, y_m) \cdot p(\theta_k) \tag{3}$$

And it is assumed that the feature parameter θ in the class ω_k is deterministic. This assumption is expressed as follows

$$p(Z_N / \omega_k) = p(Z_N / \theta_k), \qquad Z_N = \{y_1, y_2, \cdots, y_N\} \tag{4}$$

These assumptions yield the following explicit expression for the conditional probability density functions.

$$p(Z_N / \omega_k) = P(y_1, y_2, \cdots, y_m) \cdot (1/2\pi\rho_k)^{\frac{N-m}{2}} \cdot \exp\left[-\frac{1}{2\rho_k} \sum_{t=m+1}^{N} \left(y_t - \Phi_k^T \cdot z_{t-1}\right)^2\right] \tag{5}$$

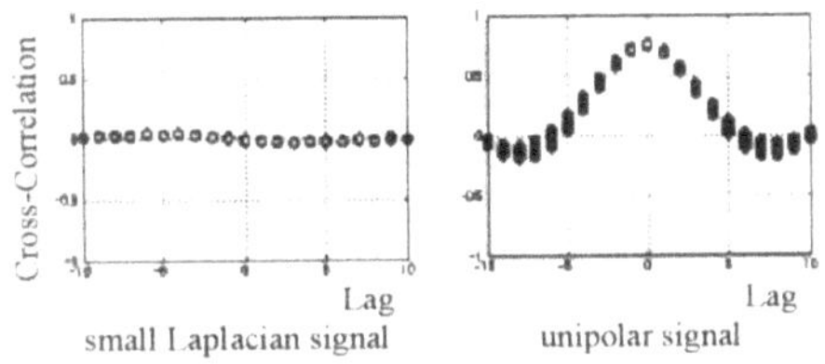

Fig.3. Cross-Correlation between C3 and C4 signals.

The purpose of this paper is to identify the human's will (right and left motor imagery) by using 2 channels EEG signals (C3 and C4). Therefore, following Bayes decision rule is adopted.

$$k^* = Arg \underset{k}{Max} \Pr(\omega_k / Z_{DN}), \qquad Z_{DN} = \{ Y_{1N}, Y_{2N} \} \tag{6}$$

$$Y_{1N} = \{y_{11}, y_{12}, \cdots, y_{1N}\}, \quad Y_{2N} = \{y_{21}, y_{22}, \cdots, y_{2N}\}$$

Where, Z_{DN} is the two channels signal, Y_{1N} is the C3 signal, and Y_{2N} is the C4 signal and ω_k's are the class (k=1: Left Movement Class, k=2: Right Movement Class), respectively.

In this study, EEG signals (electrode C3 and C4) are pre-processed by spatial filter. Then, it is conceivable that these signals are mutually independent. The validity of this assumption is confirmed from Fig.3. Therefore, following assumption is adopted.

$$p(Z_{DN}, \omega_k) = p(Y_{1N}, Y_{2N} / \omega_k) = p(Y_{1N} / \omega_k) \cdot p(Y_{2N} / \omega_k) \tag{7}$$

3.1 Parameter Estimation

For the estimation of parameters, following likelihood function is adopted.

$$L_i = \sum_{j=1}^{N_p} \ln p(Y_{iN}^j / \omega_k) \tag{8}$$

where, N_P is the number of samples of given class k, and Y_{iN}^j is j-th sample of channel i (C3 or C4). To maximize the criterion function L_i, partial differentiation of L_i with respect to each parameter is done and then equated to zero. Then the obtained simultaneous equations are solved to give rise the below equations.

$$\hat{\Phi}_{ik} = \left[\sum_{j=1}^{N_P} \sum_{t=m+1}^{N} z_{i,t-1}^{j} z_{i,t-1}^{jT} \right]^{-1} \cdot \left[\sum_{j=1}^{N_P} \sum_{t=m+1}^{N} y_{i,t}^{j} z_{i,t-1}^{jT} \right] \tag{9}$$

$$\hat{\rho}_{ik} = \frac{1}{N_p(N-m)} \sum_{j=1}^{N_P} \sum_{t=m+1}^{N} \left(y_{i,t}^{j} - \hat{\Phi}_k^{T} z_{i,t-1}^{j} \right)^2 \tag{10}$$

Now, applying Bayes rule to (6) and substituting (7), the concrete decision rule is obtained.

3.2 Decision rule

$$k^* = Arg \underset{k}{Max} \ [\ln(\Pr(\omega_k)) - \frac{N-m}{2}\ln(2\pi\rho_{1k}) - \frac{N-m}{2}\ln(2\pi\rho_{2k}) - \frac{1}{2\rho_{1k}} \sum_{t=m+1}^{N} \left(y_{1,t} - \Phi_{1k}^{T} \cdot z_{1,t-1} \right)^2 - \frac{1}{2\rho_{2k}} \sum_{t=m+1}^{N} \left(y_{2,t} - \Phi_{2k}^{T} \cdot z_{2,t-1} \right)^2] \tag{11}$$

In this study, $\Pr(\omega_1) = \Pr(\omega_2) = 0.5$ is assumed.

4. Pattern Recognition of the EEG

In this study, 112 kinds of period (Data Length: 7 kinds, 0.5～2.0 [sec], Decision Time: 13～19 kinds, 3.5～8.0 [sec]) are considered. And the order of the AR model is set to ten based on the previous study [8]. To obtain a more general view of the ability of classification, a 10 times 10 fold cross validation is performed.

The 10 times 10 fold cross validation mixes the data set randomly and divides it into 10 equally sized disjunct partitions. Each partition is then used once for testing, the other partitions are used for training. This results in 10 different accuracies (the percent ratios of the number of the samples classified correctly and the total number of the testing samples), which are averaged. This is the accuracies of a 10 fold cross validation. To further improve the estimate the procedure is repeated 10 time and again all accuracies are averaged.

4.1 Pattern recognition results in Group A

The recognition results (Group A) are shown in Fig.4. These results are obtained based on 10 times 10 fold cross validation. Testing data are classified based on the parameter which is estimated by using the same period of training data. Table 1 shows the best decision point and data length among 112 kind of period from view point of recognition results in each subject. And Fig.5 shows the transition state of these accuracies. This table suggests that the data length should be short when the best accuracy is low. This means that EEG activity is not stable when the learning is not advanced. And, the best decision point is late when the learning advances. It seems that this means the subject obtained the ability to adjust the stable imagination method according to the feedback signals by repeating ex-

Table 1. The Best Decision Point and Data Length

Subject	SeN	1	2	3	4	5	6	7	8	9	10
	DcP.	7.75	3.75	3.70	5.25	4.00	4.25	7.75	6.75	3.75	7.75
f1	DtL.	0.5	0.5	0.75	0.5	0.5	0.5	0.75	0.5	0.75	0.75
	Ac.	67	72	70	74	72	69	69	65	68	76
	DcP.	7.75	4.25	6.75	7.75	7.75	7.25	7.00	7.50	8.00	8.00
m2	DtL.	0.5	0.75	0.5	1.5	0.5	0.5	1.0	2.0	2.0	2.0
	Ac.	71	66	73	64	75	70	75	78	90	98
	DcP.	7.50	7.50	6.50	6.50	8.00	6.50	8.00	7.75	8.00	7.50
k3	DtL.	0.5	2.0	1.5	0.5	1.5	1.0	1.5	2.0	2.0	2.0
	Ac.	66	68	71	75	73	86	72	90	80	90

SeN: Session Number, Dcp: Decision Point, DtL: Data Length, Ac.: Accuracy

iments. Fig.5 suggests that the number of sessions necessary for learning depends on the subject. The number of sessions (10 sessions) is enough for subject m2. It is insufficient for subject k3, and, is too short for subject f1. But, for all subjects, it is understood that accuracy is improved by doing the experiment repeatedly.

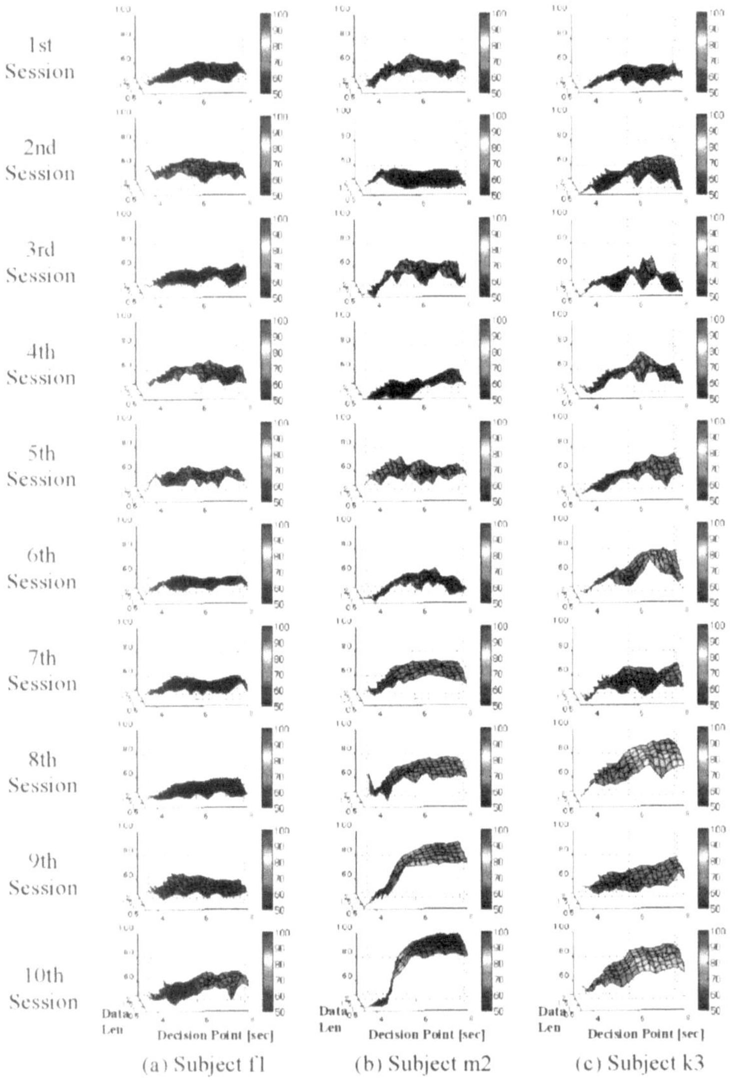

Fig.4. Pattern Recognition Results (Group A)
Accuracies vs. Data Length and Decision Point. (AR model order: 10)

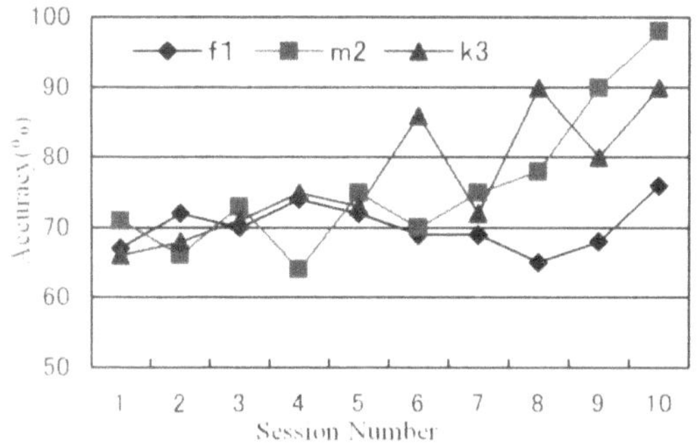

Fig.5. Transition state of accuracies (Subject: f1, m2, k3)

4.2 Parameter estimation results in Group A

The transition states of the AR parameters ($\phi_1 \sim \phi_5$) are shown in Fig.6 together with accuracies. It is found that the parameters do not converge in case of the subject who was not able to obtain high accuracy. In case of the subject who was able to obtain high accuracy, the parameters are relatively stable, but they do not converge yet except parameter ϕ_1. On the other hand, parameter ϕ_1 becomes about 0.8 in the sessions in which the accuracies become over 90%, (Subject m2 and k3). This phenomenon suggests the following facts.

- The learning process might still not end, although the accuracy becomes satisfied level in ten sessions.
- The optimal parameters common to everyone might exist.

4.3 Pattern recognition results in Group B

The recognition results (Group B) are shown in Fig.7. The form of this figure is equal to Fig.5 in group A. In 2~4 session in Group B, the feedbacks are executed based on the parameter estimated from another subject's data. In case of two subjects (o5, m6), there is no improvement. But, in case of subject h4, the accuracy becomes 86% at the 3rd session. This is the fastest learning speed among six subjects (Group A and Group B). This phenomenon suggests that humans are able to adjust to others' parameters, and also that the use of the optimal parameter according to the subject may help the initial learning.

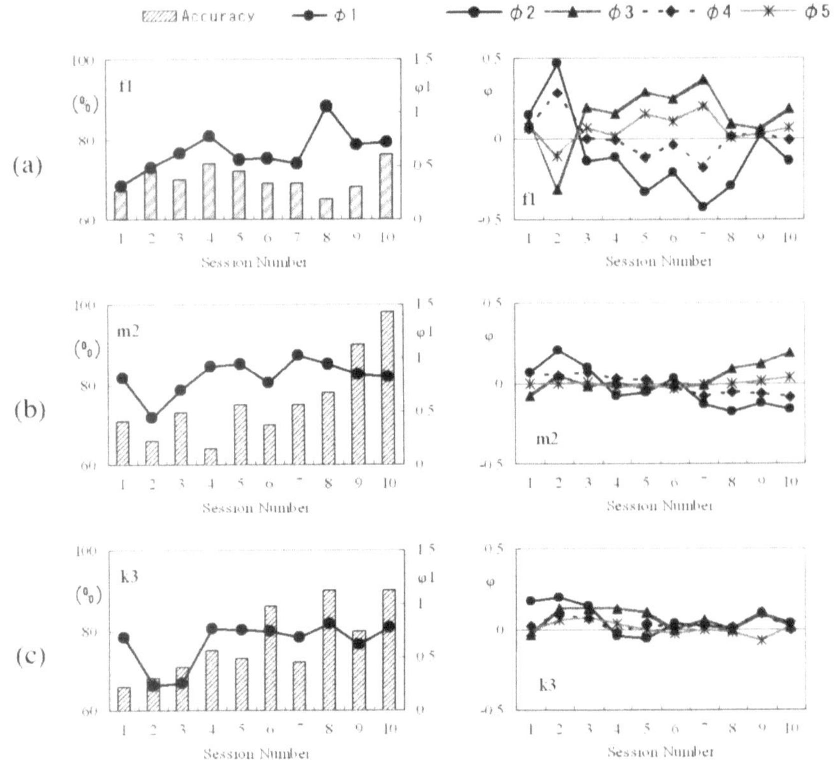

Fig.6. Transition state of accuracies and parameter $\varphi_1 \sim \varphi_5$
AR model oder: 10 (a) Subject f1, (b) Subject m2, (c) Subject k3

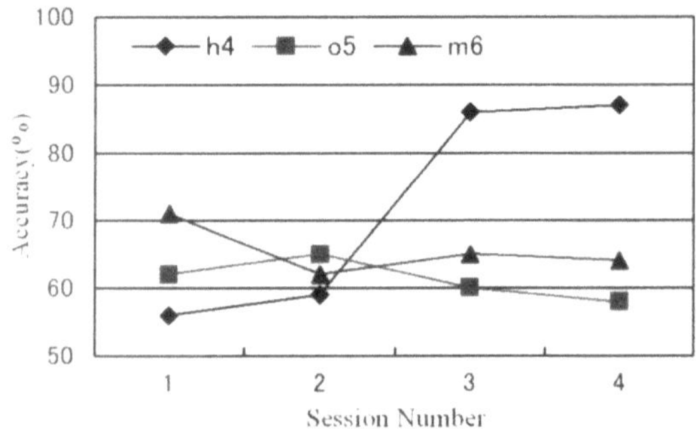

Fig.7. Transition state of accuracies (Subject: h4, o5, m6).

5. Conclusion

In this paper, the learning effects of the subjects were investigated under the situation in which a statistical pattern recognition method based on AR model is applied. And it was confirmed that some objects acquire ability to take communications by using EEG by learning for about ten days, although there are the individual variation. And it was also confirmed that subject can adjust to even others' parameter although it is easy to adjust to his own parameter. This suggests that the learning speed may be influenced from the setting method of the initial parameter. The research on the optimal setting method is under consideration.

References

1. Vidal J (1973) Toward direct brain-computer communication, Annual Rev. Biophysics Bio-engineering. pp.157-180
2. Wolpaw J and McFarland D (1997) Multichannel EEG-based brain-computer communications. Electroencephalography and clinical Neuro-phyiology. 90, pp.444-449
3. Björn P O, et al. (1997) Prompt Recognition of Brain States by their EEG Signals. Theory Bioscience, 116, pp.290-301
4. Pfurtscheller G, Neuper Ch, Flotzinger D, et al. (1997) EEG-based discrimination between imagination of right and left hand movement. Electroencephalography and clinical Neurophyiology, 103, pp.642-651
5. Pfurthscheller G, Neuper Ch, Schlögl A, et al. (1998) Separability of EEG Signals Recorded During Right and Left Motor Imagery Using Adaptive Autoregressive Parameters. IEEE Trans. on Rehabilitation Engineering, 6-3, pp.316-325
6. Pfurtscheller G, et al. (1999) EEG-based discrimination between imagination of right and left hand movement. Electroencephalography and clinical Neurophyiology, 103, pp.642-651
7. Inoue K, Rao G N, Mizutori A, et al. (1991) Classification of Gas-Liquid Two-Phase Flow Regime by an AR-Model. Proc. of IFAC Symposium on Identification and System Parameter Estimation, pp.597-602
8. Inoue K, et al., (2003) Pattern Recognition of EEE Signals During Right And Left Motor Imagery. Proc. of IFAC Symposium on System Identification, pp.138-143

A Hierarchical Interaction in Musical Ensemble Performance: Analysis of 1-bar Rhythm and Respiration Rhythm

Tomohito Yamamoto[1] and Yoshihiro Miyake[2]

[1]Department of Information and Computer Science, Kanazawa Institute of Technology
[2]Department of Computational Intelligence and Systems Science, Interdisciplinary Graduate School of Science and Engineering, Tokyo Institute of Technology

Chapter Overview. In recent years, music therapy has been focused on as a method to cure patients who have a physical or mental illness. Music therapy is a kind of musical communication between a patient and a therapist. Clarifying the mechanism of musical communication is the effective way to conduct this type of therapy. In this study, an ensemble performance was analyzed with musical and respiration rhythms using a dual task method to clarify the mechanism. The results showed that there was no effect of a subtask on musical and respiration rhythms in playing simple music. However, when playing complex music, musical and respiration rhythms were changed by a subtask. From these results, we proposed a hierarchal communication model of an ensemble performance.

Key Words. Communication, Music, Cognitive system, and Respiration.

1. Introduction

In recent years, music therapy has been focused on as a method to cure patients who has physical or metal illness. Music therapy is a kind of musical communication between a patient and a therapist. For example, in improvisational music therapy, patients move their body to a sound, while the therapist makes sounds that are synchronized to the patient's movement.

Although effectiveness of such therapy is often reported, the mechanism of musical communication is not analyzed enough. To clarify the mechanism, it is effective to conduct therapy more efficiently. In this study, we clarify the mechanism by analyzing a musical ensemble performance. Based on the result, we propose a musical communication model between players.

Musical performance is divided into two types. One is a solo performance, and the other is an ensemble performance. Although there is a lot of research about a solo musical performance, there is little research concerning an ensemble performance. One of a few examples of the research on an ensemble performance is an analysis of synchronization between players [1]. That research has revealed that a player of melody part precedes the ones who take other musical parts by 10msec on average and that there is always 30-50msec asynchrony when the musicians play simultaneously.

Research that clarifies the mechanism of a solo musical performance analyzes the musical tempo (speed of musical performance), ago-gics (temporal development of musical rhythm) and so on. Of these elements, ago-gics have most often been analyzed. Representative research about ago-gics illustrates that ago-gigs are not changed when the players play the same music [2] [3], or players use more ago-gics as a musical expression than other musical techniques [4].

In addition to such research about ago-gics, there is some research that investigates player's physiological aspects [5]. In this research, the relationship between 1-bar rhythm and respiration rhythm was analyzed when pianists played the same music in different meters. The results showed that the pianists' respiration period during a performance was shorter than that in normal breathing circumstances, and the coupling between 1-bar and respiration rhythm while playing in an unconventional meter of 7/4 and 5/4 was stronger than that while playing in a conventional meter of 3/4 and 4/4. The relationship between respiration and listening to music has often been analyzed [6, 7]. Such research has showed that human respiration was changed when listening to music and the difference depended on musical rhythm. The fact that the respiration changes depending on musical rhythm suggests that respiration have relation with the musical cognitive system, which is an effective element to analyze.

The research reviewed suggests that a temporal development of musical rhythm is an important element for an analysis of musical performance, and that it has relation with the respiratory rhythm. As for musical communication between players, only synchronization of musical ensemble performance has been analyzed. This study has analyzed an ensemble performance with a musical and respiration rhythm, and also investigates the cognitive aspect of ensemble performance by using a dual task method.

With the results of these analyses, we propose a musical communication model of an ensemble performance.

2. Method

2.1 Task and subjects

In this experiment, two subjects played face to face on two electric pianos at the same time. The subjects were told only to play the music cooperatively. The subjects were three graduate students (24-31) who had at least 12 years of experience playing the piano. The music for the experiment was Music A and B (Fig.1a, b). Music A was composed with only quarter notes. Music B was composed with 8th and 16th notes. Music B was more difficult to play than Music A because it required fast finger movement to express fast pitch change.

The subjects repeated the music 7 times (= 28 bars) in each trial of two experimental conditions. One was under normal condition (N condition) in which they played the music as usual. The other condition was a dual task condition (D condition) in which they played the music while performing a word-memory task.

In the dual task condition, subjects had to perform a main task and a sub task simultaneously. If subjects could perform each task smoothly, these tasks would be processed with different cognitive systems. If the subjects could not perform the main task, these tasks shared the same cognitive system. Using this method, it was possible to determine the cognitive system used for the main task, to some extent. In this study, the main task was playing the piano, and the subtask is memorizing 5 words which are presented as visual information. The word-memory task was performed by using laptop computer placed in front of each subject. The subjects had to memorize 5 unrelated words (Example: Children, Apple, Pen, Tokyo, Bed) before playing the piano and they had to answer the words immediately after playing the piano. The following is the experimental procedures:

1-1. Playing Music A (4 bars * 7 = 28 bars) 5 times in N condition
1-2. Playing Music A (28 bars) 5 times in D condition
2-1. Playing Music B (28 bars) 5 times in N condition
2-2. Playing Music B (28 bars) 5 times in D condition

Pianists were asked to practice the music until playing it smoothly and rehearsed together to adjust the tempo a few times before the experiment.

Fig. 1a, b. Music for the experiment, a: Music A, b: Music B.

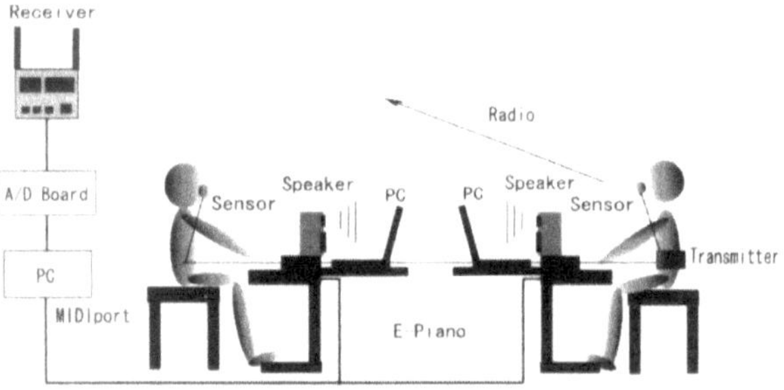

Fig. 2. Experimental system

2.2 Experimental system

Fig.2 shows experimental system. Two electric pianos (Roland: RD-600) and speakers (ONKYO: GX-R3) were used. Subjects sat 2.7m apart face to face. The musical performance was recorded when music sequence software in MIDI format. Time resolution of MIDI instruments was 0.96 msec. Player's respiration was measured by a thermistor sensor (NIHON KODEN: TR-511G) attached in a nostril cavity. All the measured respiration data was sent from the transceiver to a receiver (NIHON KODEN: Multi Telemeter System WEB-5000), and the data was digitalized by A/D board (ADTEK: AXP-AD02) at 256 Hz sampling rate with 12bit resolution. The digitalized data finally stored in a computer. Time resolution of respiration measurement is 7.8msec. A laptop computer was placed in front of the subjects for word-memory task. The entire experimental system was synchronized by MIDI signal, which made it possible to compare the musical data and respiration data.

2.3 Data analysis

To analyze the musical synchronization between players, the absolute value of time differences between 1-bar rhythms was as on index. 1-bar rhythm was defined as the time interval between the first note onset of a bar and that of a next bar (see Fig.3). To analyze the rate of the 1-bar rhythm, the period of 1-bar rhythm was used. To analyze the rate of respiration rhythm, the period of respiration rhythm was used. Respiration rhythm was defined as the time interval between inspiration peaks. In this analysis, 24 bars of the 28 bars are analyzed and the last 4 bars were deleted because it is impossible to calculate the last 1-bar period.

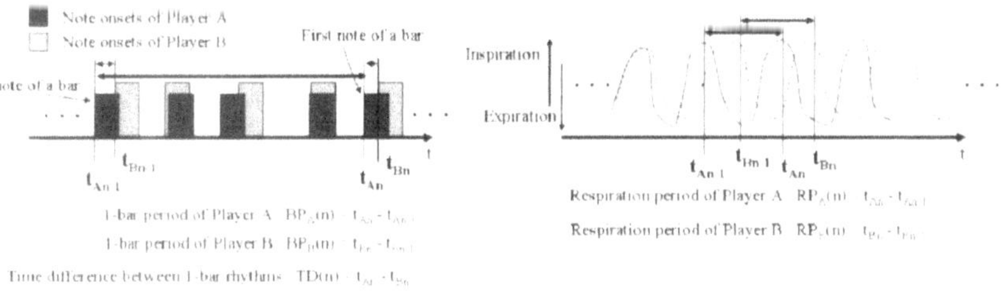

Fig. 3. 1-bar rhythm and respiration rhythm and indices for analysis

3. Results

The results of word-memory tasks were as follows: the correct answer rate while playing Music A was 92.0% (138/150), and while playing Music B was 85.3% (127/150). These results indicated that in D condition, players performed the word-memory task properly.

3.1 Change of time course of 1-bar and respiration period

We first analyzed the temporal development of 1-bar and respiration periods and the relationship between them qualitatively in all conditions.

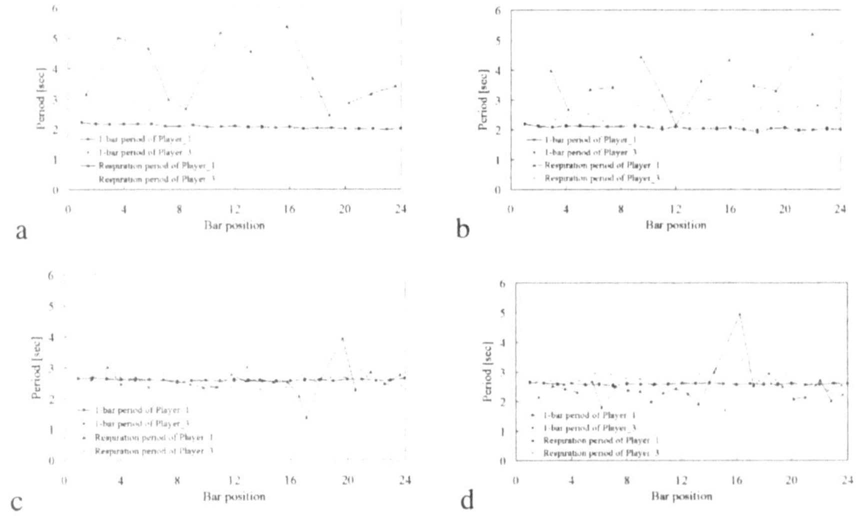

Fig. 4. Time course of 1-bar and respiration period (Player_1 and Player_3). a: Music A and N-condition, b: Music A and D-condition, c: Music B and N-condition, d: Music B and D-condition

Fig.4a, b, c, d are typical results of Player_1 and Player_3 which show the temporal development for 1-bar and respiration period in each condition. The respiration period shows large fluctuations in Fig.4a, but in Fig.4c, the fluctuation is very small. In all figures, 1-bar period shows little fluctuation. Comparing temporal developments of Fig.4a to that of Fig.4b, there are no obvious differences. Comparing temporal developments of Fig.4c to that of Fig.4d, respiration period of Fig.4d shows somewhat larger fluctuation. In the following section, these tendencies are analyzed quantitatively.

3.2 Relation between 1-bar rhythm of players and relation between respiration rhythm of players

Fig.5a shows the mean value of absolute time differences between 1-bar rhythms (the mean value of (Combination of subjects = 3)*(5 trials)*(24 bars) = 360 data in each condition), and error bars show standard deviation.

There was no significant difference between Music A and Music B in the N-condition (t (718) =0.0262, $p>0.25$). These results indicate that there is some asynchrony between players in Music A and Music B. Moreover, there was no effect of the condition for Music A (t (718) =1.71, $p>0.25$), however effect of the condition was significant for Music B (t (718) =3.49, p<0.001). These results suggest that there was no effect of the subtask on musical synchronization in Music A; however, in Music B, synchronization was changed by the subtask.

Fig.5b shows the mean value of difference between the respiration periods of players (the mean value of (Combination of subjects = 3)*(5 trail) = 15 data). The difference between the respiration periods of players is calculated by the absolute difference between mean periods of one trial. There was a significant difference between Music A and B in N condition (t (28) =3.03, p<0.001). These results indicate that the mean difference of respiration period in Music B become smaller than that in Music A.

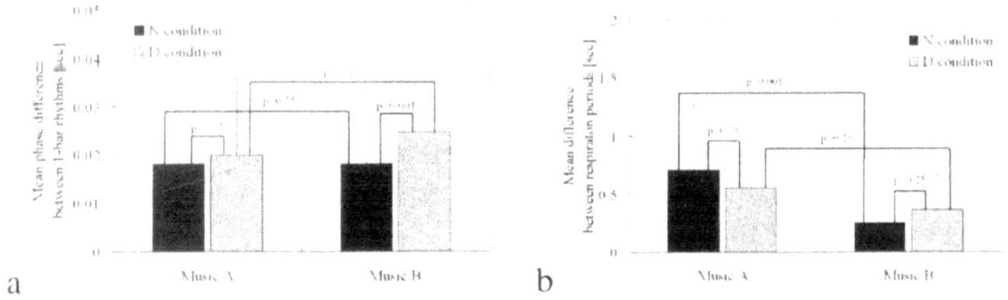

Fig. 5a, b. a: Mean absolute time difference between 1-bar rhythms, b: Mean difference between respiration periods

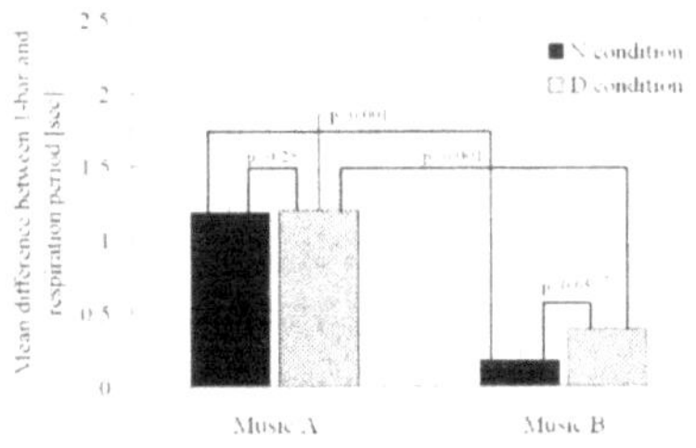

Fig. 6. Mean difference between 1-bar and respiration period

3.3 Relation between 1-bar and respiration rhythm of a player

Fig.6 shows the mean value of the difference between the 1-bar and respiration period of a player (the mean value of (Combination of subjects = 3)*(5 trail)*(Number of subjects = 2) = 30 data in each condition). The difference between the 1-bar and respiration period of a player was calculated by the absolute difference between mean periods of one trial.

There was a large difference between the value of Music A and Music B in the N-condition (t (58) =7.07, p<0.0001), the condition had no effect on

Music A ($t(58)=0.117$, $p>0.25$), but it affected Music B ($t(58)=3.05$, $p<0.05$). These results suggest that in Music B, the mean difference between the 1-bar and respiration period was increased by the subtask.

4. Discussion

Result summary:

1. The analysis of the relationship between the 1-bar rhythms of players revealed that there had no effect of the subtask on musical synchronization for Music A; but for Music B, synchronization was changed by the subtask.
2. The analysis of the relationship between the respiratory rhythms of players revealed that the mean difference of respiration period of Music B was smaller than that for Music A in the N condition but became closer to that of D condition.
3. The analysis of the differnce between the 1-bar and respiration rhythm of each player revealed that it was much smaller in Music B than in Music A, and for Music B only, it was increased by the subtask.

The respiration period was effected by the kind of music played. This result suggests that the coupling between 1-bar and respiration rhythm depends on type of music as Ebert showed in his work [5]. It is indicated that there is an interaction between a physical and respiration rhythm, and a finger movement involves the coordination of physical and respiration rhythms [8, 9]. These results suggest that the interaction between players in a cooperative performance is composed of not only a sound element but also a physiological element such as respiration.

In this study, a visually presented word-memory task is selected as the subtask in the dual task condition. When performing this subtask, broca area, supramarginal gyrus and supplementary motor area are activated (e.g. [10, 11, 12]). These areas are indicated to have relation with speech production. In this study, an effect of the subtask appeared only in the condition of Music B. This result suggests that when playing Music B, the areas of speech production are more active than when playing Music A.

The difference between Music A and Music B is difficulty and speed of finger movement and changing speed of pitch. The research about brain activity in musical perception revealed that music having slow pitch change and slow rhythmical change is processed in an auditory area of the right hemisphere [13, 14]. On the other hand, music having the fast pitch change and fast rhythmical change is processed in not only the auditory area of right brain but also a language processing area of left hemisphere [13, 14, 15].

This research can explain the increased asynchrony of 1-bar rhythms in Music B in the D condition. When playing music such as Music A, the musical signal is mainly processed in the auditory system of right hemisphere. Therefore, if the subtask occupied language system, there was no effect on the musical synchronization. However, when playing music such as Music B, if the subtask occupied the language system, players could not control the 1-bar rhythm as well as under normal circumstances.

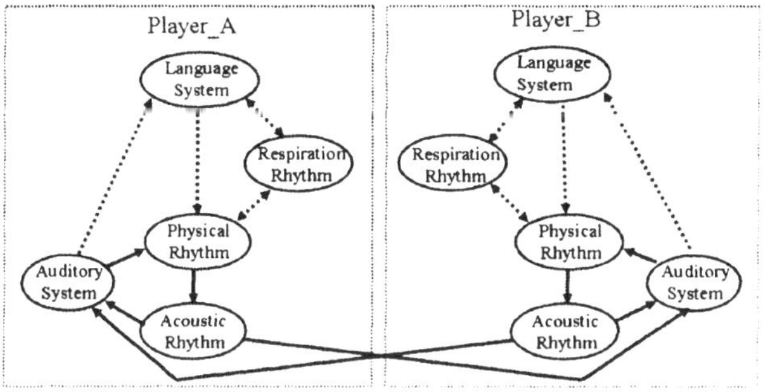

Fig. 7. Proposed communication model of a musical cooperative performance

This activation of the language system can also explain the change of respiration period. Language system is an area related to speech production. Therefore it has a strong relationship to respiration system. When playing Music B, the language system is more active than when playing Music A. It is speculated that the change of respiration rhythm is generated by this activation. In particular, the result that the difference between 1-bar and respiration period was very small in Music B, suggest that the melody of that music was generated as if producing speech. When two players play Music B, their respirations tend to synchronize.

We propose a communication model based on these ideas (Fig.7). This hierarchal model is composed of two layers: the lower layer comprises the auditory and physical feedback system, and the higher layer comprises the language and respiration systems. In this model, when music of type A is played, the musical signal is processed in the auditory system and the processed signal is fed back to the physical system (solid arrows in Fig.7). As a result, musical performance is controlled. This control system is considered as a simple sensori motor synchronization [16]. On the other hand, when music of type B is played, the musical signal is processed in not only the auditory system but also in the language system (dashed arrows in Fig.7). The signal processed in the language system is sent to the physical system, and at the same time the signal is sent to the respiratory system. Finally, musical performance is generated as a coordination of both physical and respiratory systems.

In general, the lower layer and higher layers work synergistically in musical performance. In future research, it will be necessary to verify this model and to analyze the temporal development of 1-bar and respiratory rhythm further to propose a model which takes the dynamics of each system into account.

5. Conclusion

In this study, a musical performance was analyzed in terms of 1-bar and respiration rhythms, and the dual task method was used to clarify the communication mechanism of a musical ensemble performance. Our results showed that there are two types of music. One is music of slow pitch and rhythm change which is not affected by the word-memory task. The second is music of fast pitch and rhythm change which is affected by the word-memory task. From these results, the hierarchical communication model of a cooperative performance is proposed. In future work, it is necessary to analyze the temporal development of the 1-bar and respiration rhythm to propose a model which considers the dynamics of each system.

References

1. Rasch R A (1979) Synchronization in performed ensemble Music. ACUSTICA 43:121-131
2. Palmer C (1992) The role of interpretive preferences in music performance. In: Jones M R, Holleran S (eds) Cognitive Bases of Musical Communication APA Washington D.C. pp249-262
3. Shaffer L H, Clarke E F, Todd N P (1985) Metre and rhythm in piano playing. Cognition 20:61-77
4. Shaffer L H (1981) Performances of Chopin, Bach and Bartok: Studies in motor programming. Cognitive Psychology 13:326-376
5. Ebert D (2002) Coordination between breathing and mental grouping of pianistic finger movements. Perceptual and Motor Skills 95:339-353
6. Diserens C M (1923) Reactions to musical stimuli. Psychological Bulletin 20:173-199
7. Haas F, Distenfeld S, Axen K (1986) Effects of perceived musical rhythm on respiratory pattern. J Appl Physiol 61:1185-1191
8. Wilke J, Lansing R W, Rogers C A (1975) Entrainment of respiration to repetitive finger tapping. Physiological Psychology 3:345-349

9. Rasler B (2000) Mutual nervous influences between breathing and precision finger movements. Eur J Appl Physiol 81:479-485
10. Grasby P M, Frith C D, Friston K J, et al (1993) Functional mapping of brain areas implicated in auditory-verbal memory function. Brain 116:1-20
11. Petrides M, Alivisatos B, Evans A C (1995) Functional activation of human ventrolateral frontal cortex during mnemonic retrieval of verbal information. Proc Nat Acad Sci 92:5803-5807
12. Fiez J A, Raife E A, Balota DA, et al (1996) A positron emission tomography study of the short-term maintenance of verbal information. J Neurosci, 16:808-822
13. Zatorre R J, Belin P, Penhune V B (2002) Structure and function of auditory cortex: music and speech. TRENDS in Cognitive Science, 6:37-46
14. Tervaniemi M, Hugdahl K (2003) Lateralization of auditory-cortex functions. Brain Research Reviews 43:231 246
15. Koelsch S, Gunter T C, Cramon D Y, et al (2002) Bach speaks: A cortical "language-network" serves the processing of music. NeuroImage, 17:956-966
16. Fraisse P (1980) The sensorimotor synchronization of rhythms. In J.Requin (Ed.) Anticipation et comportement Centre National pp233-257

Comparison of the Reaction Time Measurement System for Evaluating Robot Assisted Activities

Tomomi Hashimoto[1], Kunio Sugaya[1], Toshimitsu Hamada[2], Toshiko Akazawa[3], Yoshihito Kagawa[4], Yasuyuki Takakura[5], Yoshie Takahashi[5], Shusuke Kusano[6], Mitsuru Naganuma[6] , and Ryuhei Kimura[6]

[1] Saitama Institute of Technology
[2]Tsukubagakuin University
[3]Kitasato University
[4]Takusyoku University
[5]Saitama medical center, Saitama medical school
[6]Teikyo University of Science and Technology

Chapter Overview. RAA (Robot Assisted Activities) allow people to feel joy and pleasure and recover through contact with a robot. Considerable research has reported that RAA is useful for elderly persons, children hospitalized in pediatric wards and so on. However no specific evaluation method concerning RAA has yet been established. In this paper, we propose a reaction time measurement system for evaluating RAA. The level of caution or concentration of a subject can be estimated in terms of the reaction time based on visual or acoustic stimuli. We try to evaluate RAA using the variance in the reaction time before and after RAA respectively. We have developed a reaction time measurement system with hardware, following which we compared several reaction time measurement systems concerning performance and operability. We reported the results of comparisons featuring such measurement systems and the basic results of the RAA evaluation experiment using reaction time measurement. The results suggested that selection of the optimal enforcement form in RAA has been enabled through application of this reaction time measurement system.

Key Words. Robot assisted activity, Reaction time measurement system

1. Introduction

In Japan, due to consideration for the "aging society," there are high expectations for mechanical systems that will provide support for senior citizens.

Considerable research and development has been undertaken in the area of "welfare devices" that provide physical support. Some examples are robots that support senior citizens during mealtimes [1], automated robot systems that carry meals [2], and ambulatory support robots [3], and welfare support systems [4]. It is unclear, however, whether the users gain a sense of emotional satisfaction from the use of such devices.

Senior citizens' homes have implemented methods of providing emotional support to their occupants through the presence of animals, as exemplified in Animal Assisted Therapy (AAT) and Animal Assisted Activity (AAA). It has been reported that thanks to the application of these methods, subjects have developed more energetic lifestyles, and have demonstrated a more positive attitude in their day-to-day lives. However, this method also incurs various problems, such as "pet loss," the transmission of diseases, and the effort involved in maintaining such pets.

Shibata proposed a "Mental Commitment Robot" able to provide emotional comfort to users [5, 6]. Shibata created robots in the form of a seal and a cat, and, through experiments, demonstrated the potential for robots to have emotional effects on humans, such as enjoyment and a sense of well-being. However, since these robots were not designed to offer physical support, it is necessary to consider other methods for doing so.

Hashimoto et. al. proposed Robot Assisted Activity (RAA) as a means of providing physical support, for example by bringing objects, opening windows, and providing the emotional support which would otherwise be gained from keeping a pet [7, 8]. By using the RAA method described here, senior citizens can expect to enjoy the advantages of support offered by machines and by pets, while also avoiding the disadvantages of both. RAA involves carrying certain pet type robots into a care house etc, where occupants would play with them for 30 - 60 minutes. However, the subject became tired of RAA during extended periods and attention was focused elsewhere when we enforced RAA. Therefore, an appropriate means of classification (for example a short RAA enforcement time) for an appropriate means of enforcing RAA was desired.

By the way, Okada et. al. proposed a system that measured reactive time to the higher-order brain function problem of the figure acknowledgment etc [9]. The proposal system presented the problem requiring the subject to push the mouse only when various figures were displayed on the monitor, including the specified figure. The system presumed the ability of registering, acknowledgment, and judgment to reflect the higher-order brain function by measuring reactive time.

Hashimoto et. al. developed the reaction time measurement system for RAA [10]. It applies the Okada system in estimating the level of proficiency in the subject's attention and concentration through use of aural and visual stimulation. However, there was a considerable margin for error in the measurement since the system proposing it was real-time software.

In this paper, 1) we established a hardware measurement system, and 2) we performed a comparative study between the hardware and software systems.

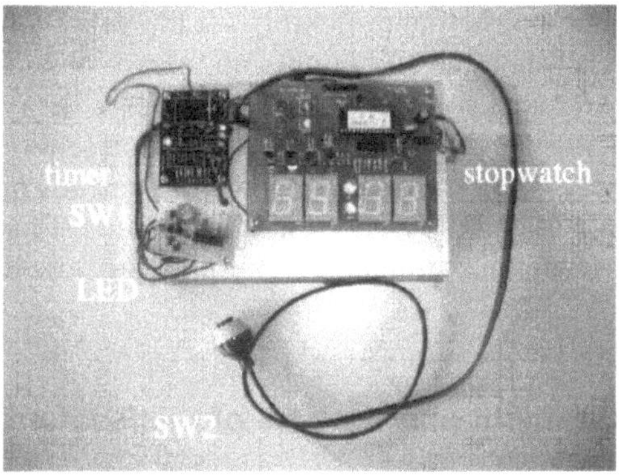

Fig. 1. Composition of the reaction time measurement system

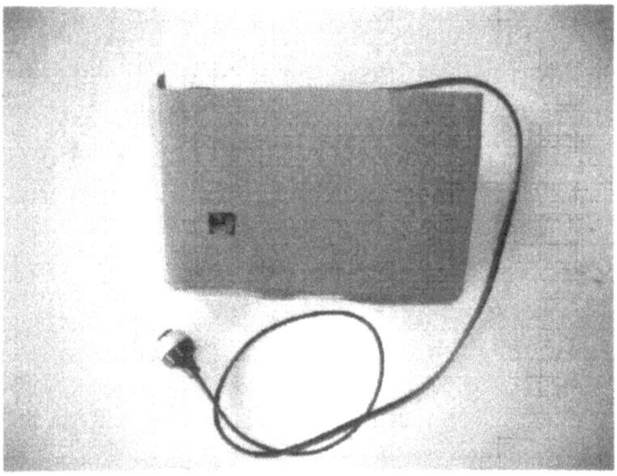

Fig. 2. Reactive time measurement system with the cover in place

2. Outline of the Reaction Time Measurement System

2.1 Composition of the reaction time measurement system

The proposed measurement system consists of a delay time outbreak circuit (a timer), a measurement circuit (a stopwatch), a stimulation presentation circuit (a LED lighting circuit) (Figure 1). Through use of a variable resister, the operator can set a delay time of about 1-7 seconds. To decrease stimulation other than that of a visual nature (LED lighting) for the subject, the measurement system covers the actual use of the measurement system at (Figure 2).

The system operational procedure is as follows.

Initially, an operator is free to flexibly set the delay time. Next, they instruct the subject "when the LED lights up, please push a switch". Subsequently, the operator pushes SW1 and generates a delay time. The LED lights up automatically after this delay time, representing visual stimulation for the subject. The system then starts simultaneous measurement. The subject pushes SW2, and the measurement system is stopped. It is reactive time that time from this stimulation presentation to pushing SW2 was measured.

2.2 The measurement of error time of this system

As for this system, the reactive time of the subject is that which elapses from the LED lighting up to the subject pushing SW2. In this chapter, we examined the error time associated with this measurement circuit (Figure 3). We used a digital oscilloscope (IWATSU Test Instruments Co., DS-8812) and measured it with a 1ms unit.

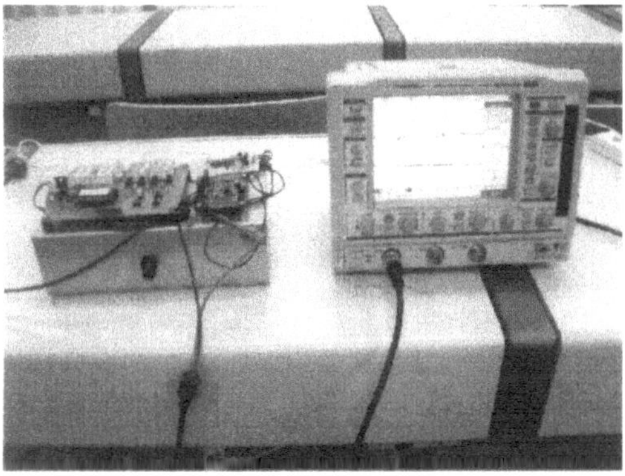

Fig. 3. Measurement of the error time

Following ten measurements of the error margin time, we decided on an absolute value to represent the error time of measurement systems (table 1). It reflects the fact that the system is slower than the time displayed by the oscilloscope as a plus error margin (+2ms) in table 1. On the other hand, the system is rapid in the event of a negative error margin (-6ms). For instance, when this system displayed 170ms, and the oscilloscope displayed 176ms, the error margin time of the measurement system would become -6ms.

Finally, the error time of this system is maximum of 9ms and an average of 3ms.

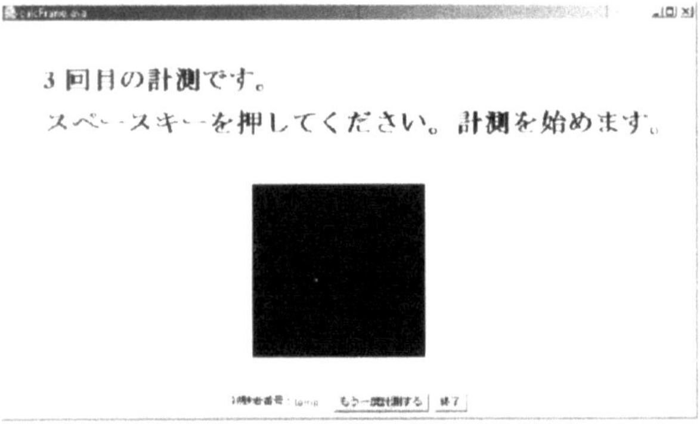

Fig. 4. Visual stimulation using software

Table 1. Error time

Number	Error time (ms)
1	-6
2	-4
3	+2
4	-1
5	-1
6	-2
7	+2
8	-1
9	+9
10	-2

Average (absolute) is 3 ms

3. Comparison of Hardware and Software Systems

3.1 Comparison method

From December, 2004 to January, 2005, we executed an evaluation assessing the psychological impression of hardware and software (using notebook and desktop PCs) with 21 male subjects aged 20 and 21 years old (table 2 and Figure 4).

The subjects registered their impressions of subjectivity using seven

Table 2. Part of the impression evaluation list

Hardware Evaluation								
	very	considerably	little	0	little	considerably	very	
bright								dark
interesting								dull
complicated								simple
vulgar								refined
hard								soft
safe								dangerous
warm								cool
artificial								natural
light								heavy
delayed								rapid

phases and ten adjective pairs such as "bright - dark," "interesting - dull," "complicated - simple," "vulgar - refined," "hard - soft," "safe - dangerous," "warm - cool," "artificial - natural," "light - heavy," and "delayed - rapid".

The order of the experiment involved three kinds such as "hardware → notebook PC → desktop PC," "notebook PC → desktop → hardware PC" and "desktop PC → hardware → notebook PC", in order to eliminate any influence of the experiment order on the evaluation. We also ensured the experiment would be balanced by ensuring the inclusion of positive and negative factors.

3.2 Evaluation result

Figure 5 shows the mean value for an evaluation involving 21 subjects. Here, the higher the numerical value, the stronger the impression of the item. Conversely, smaller numerical values indicate that the reverse impression is intensified. Finally, the numerical value of three is considered neutral. For instance, if the numerical value of "bright" were five, then that would mean "considerably bright".

Next, we checked whether the statistical work was possible, through the use of Mauchly's test of sphericity. We conducted the test as follows for values of 0.05% or more (table 3 (a)(b)). After passing Mauchly's test, we used repeated measures to examine whether there was a predominant sta-

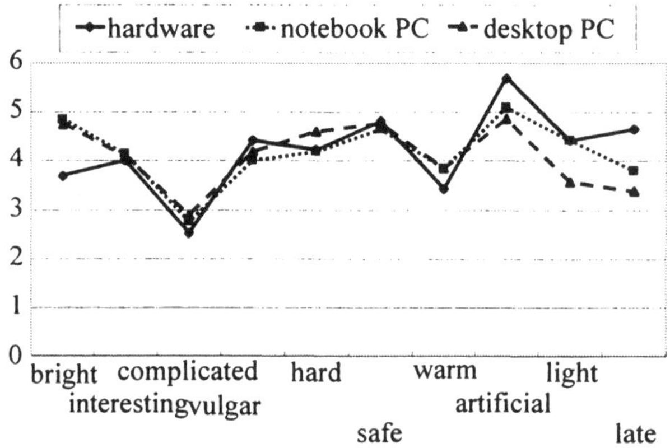

Fig. 5. Evaluation of Average Impression Values

tistical difference between the machines (hardware, and notebook / desktop PCs). During the repeated measures, we assumed the value of 0.05% or more to represent a significant difference. Moreover, we examined the nature of the difference between specific machines when such a significant difference between machines was revealed using the Bonferroni method.

Finally, significant differences were apparent concerning the following three paired categories: "artificial - natural," "light - heavy," and "delayed - rapid".

For the pair of "artificial - natural," a meaningful difference was detected between the hardware and desktop PCs respectively and the same was true for the "light - heavy" and "delayed - rapid" pairs respectively.

3.3 Discussion

We considered the subject to have a similar overall impression for each machine. However, there was a stronger impression that the hardware was

Table 3. Results of test (a)

Item	Mauchly's test	Repeated measures
artificial	0.835	0.008
delayed	0.304	0.007
light	0.096	0.035
bright	0.000	0.014
interesting	0.761	0.688
complicated	0.011	0.452
vulgar	0.012	0.051
hard	0.904	0.377
safe	0.986	0.884
warm	0.302	0.146

(b)

Item	Bonferroni
artificial	between Hardware and desktop PCs : 0.022
delayed	between Hardware and desktop PCs : 0.028
light	between Notebook and desktop PCs : 0.036

"artificial" and "slow", and also an increased feeling that the impression of the notebook PC was "lighter".

It seemed that the impression of the hardware was of one of "slowness" since the SW2 button was hard to use for subject. Moreover, it seemed difficult for the subject to see, since the hardware used both LED and visual stimulation, although software displayed the figure on the display.

We therefore considered it important to change the shape of SW2, and to change from an LED into a flashing display system.

4. Conclusion

In this paper, 1) we established a hardware measurement system, and 2) we performed a comparative study between hardware and software systems.

The suggested measurement system consists of delay time outbreak, measurement and stimulation presentation circuits respectively. Through an operator operated variable resister, the delay time can be set within a range of about 1-7 seconds, with an associated error time for this system reaching a maximum of 9 ms and an average of 3 ms.

We executed an evaluation of the psychological impression concerning hardware and software (using notebook and desk-top PCs) and 21 male subjects aged 20 and 21 years old. Finally, there were significant differences revealed in three pairs, namely "artificial - natural," "light - heavy," and "delayed - rapid". We considered the subject to have a similar impression overall for each machine. However, the impression of the hardware being "artificial" and "slow" was stronger, as was the impression of the notebook PC being "lighter". Moreover, it seemed to be difficult for the subject to see, since the hardware used LED sight stimulation, although software displayed figures on the display.

References

1. Sumio Ishii (1998) Meal Assistance Robot for Disabled. Journal of the Robotics Society of Japan, Vol.16, No.3, :306-308,
2. Toshiyuki Kouno and Shinji Kanda (1998) Robot for Carrying Trays to the Aged and Disabled. Journal of the Robotics Society of Japan, Vol.16, No.3, :317-320

3. Masakatsu Fujie (1996) Walk Supporting Robot. Journal of the Robotics Society of Japan, Vol.14, No.5, :628-631
4. Tomomi Hashimoto, et. al. (1998) Emotion-oriented Man-machine Interface for Welfare Intelligent Robot. Journal of the Robotics Society of Japan, Vol.16, No.7, :993-1000
5. Takanori Shibata (1999) Mental Commit Robot for Healing Human Mind. Journal of the Robotics Society of Japan, Vol.17, No.7, :943-946
6. Takanori Shibata (2000) Affective Artifacts. Jouranl of Japan Society for Fuzzy Theory and Systems, Vol.12, No.6, :752-761
7. Tomomi Hashimoto (2002) Enforcement Progress Report of Robot Assisted Activities. SCIS and ISIS 2002 :23Q1-6
8. Toshimitsu Hamada, et. al. (2001) Pet Robot Application to Welfare and Support of the Aged Study on Robot Assisted Therapy (1). The 19nd Annual Conference of the Robotics Society of Japan :149-150
9. Yoshikazu Okada, et. al. (2000) Validity of reaction times of various recognitive tasks for evaluating higher cortical functions. JPN J Stroke, Vol.22, No.2, :307-312
10. Tomomi Hashimoto, et. al. (2004) Development of the Reaction Time Measurement System and Application of it. The 22nd Annual Conference of the Robotics Society of Japan :3E15

Part 5
Cognitive Neuroscience and Technology

Influence of Interhemispheric Interactions on Paretic Hand Movement in Chronic Subcortical Stroke

Nagako Murase[1,2], Julie Duque[1,3], Riccardo Mazzocchio[1,4] ,
and Leonardo G. Cohen[1]

[1]Human Cortical Physiology Section, National Institute of Neurological Disorders and Stroke, National Institutes of Health, USA
[2]Department of Neurology, National Hospital Organization, Utano National Hospital, Kyoto, Japan
[3]Laboratory of Neurophysiology, University of Louvain, Brussels, Belgium [4]Neurophysiologic Clinic Section, Department of Neurological Sciences, University of Siena, Italy.

Chapter Overview. In patients with chronic stroke, the primary motor cortex of the intact hemisphere ($M1_{intact}$) may influence functional recovery, possibly through transcallosal effects exerted over M1 in the lesioned hemisphere ($M1_{lesioned}$). Here, we studied interhemispheric inhibition (IHI) between $M1_{intact}$ and $M1_{lesioned}$ in the process of generation of a voluntary movement by the paretic hand in patients with chronic subcortical stroke and in healthy volunteers. IHI was evaluated in both hands preceding the onset of unilateral voluntary index finger movements (paretic hand in patients, right hand in controls) in a simple reaction time paradigm. IHI at rest and shortly after the Go signal were comparable in patients and controls. Closer to movement onset, IHI targeting the moving index finger turned into facilitation in controls but inhibition still continued in patients, a finding that correlated with poor motor performance. IHI targeting the resting finger remained deep all through the reaction time. In case of intact hand movement, IHI targeting the moving fingers showed similar facilitation with that in control subjects. These results suggest an abnormally high interhemispheric inhibitory drive from $M1_{intact}$ to $M1_{lesioned}$ in the paretic hand movement. It is conceivable that this abnormality could adversely influence motor recovery in some patients with subcortical stroke.

Key Words. Interhemispheric inhibition, Subcortical stroke, Motor recovery, Magnetic stimulation, and Reaction time

1. Introduction

Functional recovery of motor function from stroke is incompletely understood. The motor cortex (M1) in the intact hemisphere ($M1_{intact}$) was activated in association with movements of the paretic hand in chronic stroke [1-3], suggesting that $M1_{intact}$ plays a crucial role to the functional recovery process [4, 5]. There are, however, several considerations that weaken this idea. The magnitude of activation (functional magnetic resonance imaging or positron emission tomography) of the $M1_{intact}$ with movements of the paretic hand does not correlate with functional recovery, [6,7] and (2) functional ipsilateral corticomotoneuronal connections from the intact hemisphere to the paretic hand are more commonly detected in patients with poor motor recovery [6]. Up to now, there is no direct study of interhemispheric interaction in the control of the paretic hand movements. Boroojerdi and colleagues [8] showed that interhemispheric inhibition (IHI) from $M1_{lesioned}$ to $M1_{intact}$ in patients with chronic subcortical stroke does not differ from healthy controls. However, control of motor activity in the paretic hand is likely to be influenced by IHI exerted from $M1_{intact}$ hemisphere to $M1_{lesioned}$ hemisphere. Thus we tested IHI using a previously described protocol in association with voluntary movements of the paretic hand in a RT paradigm in patients with chronic subcortical infarcts, intact corpus callosum, and moderate to good motor recovery. We hypothesized that the $M1_{lesioned}$ which controls movements in the paretic hand, receives abnormally high task-related inhibitory influences originating in the $M1_{intact}$ in patients relative to healthy age-matched controls.

2. Methods

A. Subjects and Experimental Procedure

Nine patients (5 men and 4 women, aged 65 ±13 [SD] years old; Table 1) and 8 age- and sex-matched healthy volunteers (5 men and 3 women, aged 62±13 years old) participated in the study. Patients were included if they had (1) history of a single ischemic subcortical infarction (with intact corpus callosum) more than 6 months preceding the study, (2) a score of 3 or higher in the Medical Research Council (MRC) scale and preserved movement of the paretic index finger (see Table 1).

Table 1. Clinical Features

Patient (sex)	Age (yr)	Lesion Description	Paretic Hand	Handedness	Disease Course (yr)	Modified Rankin Scale	MRC Scale	Modified Ashworth Scale	Finger Tap* (in 10 sec) (paretic/nonparetic)
1 (F)	74	R BG	L	R	1.6	1	5	0	28/27
2 (F)	36	L DWM	R	R	1.8	1	5	0	36/37
3 (M)	66	L BG, DWM	R	R	8	2	4	2	17/34
4 (M)	61	L BG, DWM	R	R	6	3	4	1+	20/40
5 (M)	59	L BG	R	L	3	3	4	4	16/38
6 (F)	78	L Thalamus	R	R	3	4	4	3	16/37
7 (M)	76	L BG, DWM	R	R	7	4	4	2	24/40
8 (M)	65	R BG	L	R	11	4	3	4	10/43
9 (F)	69	R BG, DWM	L	R	1.8	4	3	4	3/39
Mean ± SD	64.8 ± 12.7		6R/3L	8R/1L	4.8 ± 3.3	2.9 ± 1.3	4 ± 0.7	2.4 ± 1.7	

*Speed of finger tapping as measured by counting the number of movements in 10 seconds.

R = right; L = left; BG = basal ganglia; DWM = deep white matter; MRC = Medical Research Council; SD = standard deviation.

Each subject performed voluntary index finger movements in response to a Go signal in a simple reaction time (RT) paradigm (Fig 1); patients used the paretic hand and healthy volunteers used the right hand. Trials with mirror or background EMG activity were excluded from the analysis.

B. Measurement of Task-related Interhemispheric Inhibition

The interval between TS application and EMG onset, commonly used in healthy volunteers to study premovement modulations in M1 excitability, is not a good indicator of comparable premovement intervals in patients and controls because TMS delays RT more in stroke patients than in controls, precluding a direct comparison. To get around this problem, we defined three different measures; (1) maximum IHI (**IHImax**), defined as the maximal (deepest) IHI identified in a given subject in any of the timings preceding EMG onset in the RT paradigm; (2) IHI around movement onset (**IHImvt-onset**), defined as IHI determined at the time interval immediately preceding the mean RT in unstimulated trials. All trials at that particular interval were included for analysis. A drawback of this method is that the natural variability of RT results in the inclusion of trials in which TS fell before EMG onset and others in which TS fell after EMG onset. To avoid this problem, we also defined a third measure; (3) IHI immediately before movement onset (**IHIbef-mvt**). We first identified the timings at which the TS fell after EMG onset in a proportion of up to 50% of the trials. We then included in the analysis only the trials in which TS fell before EMG onset at the defined intervals. Because this method does not include in the analysis trials in which TS fell after EMG onset, it offers a cleaner reflection of premovement IHI.

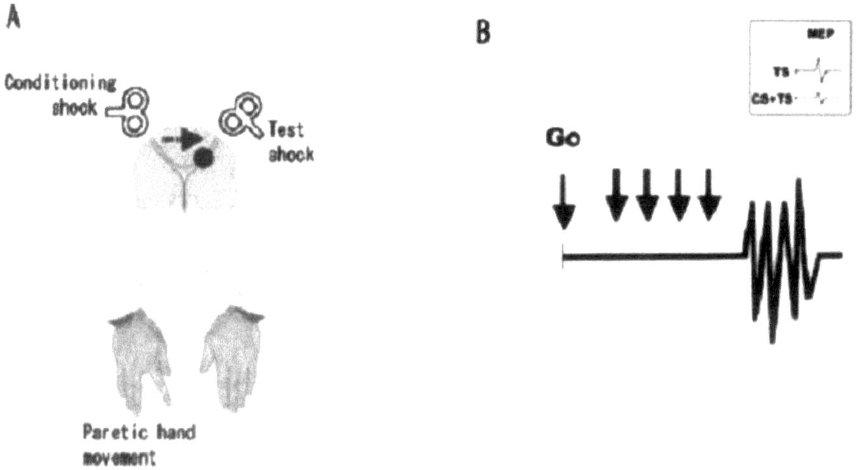

Fig.1. (A) Diagram illustrating the proposed hypothesis: interhemispheric inhibition (IHI) from the intact hemisphere to the M1 in the lesioned hemisphere may be enhanced in patients with subcortical stroke relative to healthy controls. IHI was evaluated with two double coils giving conditioning (CS) and test shocks (TS) in each. Interstimulus interval between CS and TS was 10 msec. (B) IHI was evaluated in the process of generation of a voluntary movement in a simple reaction time (RT) paradigm. Subjects responded with a brisk voluntary index finger movement to a Go signal. For measurement of IHI, motor-evoked potentials (MEPs) were always recorded from the hand contralateral to the TS. IHI was expressed as the amplitude of the conditioned MEPs in the double-pulse trials (CS-TS) relative to the amplitude of the test MEP when TS was delivered alone (CS-TS)/TS (see inset). IHI was measured at 10 different timings during the RT period. Because reaction time varies across subjects and IHI was determined at regular timings, the time interval between IHI determinations differed slightly across subjects (by 20 to 40 msec).

3. Results

A. Reaction Time, Corticomotor Excitability, and Interhemispheric Inhibition at Rest

RTs in unstimulated trials tended to be longer in patients than in normal volunteers (220 ±44 msec and 176 ±39 msec, respectively, p=0.07). Absolute motor thresholds were 50 ±12% and 47±10% in M1 of the affected and intact hemispheres, respectively, and 48±8% and 45±8% in the left and right M1 of controls. There was no significant difference between those values.

B. Interhemispheric Inhibition preceding Unilateral Index Finger Movements

There was a significant effect of the following factors: GROUP (F=18.76, p=0.003), TIMING (F=8.77, p=0.003), and the interaction GROUP x TIMING (F=4.57, p=0.03) on IHI.

$IHI_{Healthy\ volunteers}$ targeting the moving index finger decreased progressively as testing approached movement onset to become facilitation with movement (Figs 2 and 3). In the patient group, $IHI_{Paretic\ hand}$ targeting the paretic moving index finger also decreased progressively with proximity to the movement onset but did not reach facilitatory levels (see Figs 2 and 3).

IHImvt-onset, which occurred at 161±38 msec and at 206±40 milliseconds after the Go signal in healthy volunteers and patients, respectively (not significant), was significantly different in magnitude in both groups, showing persistent inhibition in the patient group (0.73±0.33) and facilitation in the healthy controls (1.15±0.23; p<0.01; see Fig 3). Exclusion of the three patients with left hand paresis (see Table 1) did not modify this result (n =6, P= 0.008).

The deeper IHImvt-onset from M1intact hemisphere to M1lesioned hemisphere, the lower the MRC scale scores (n =9; r=0.7, p=0.02) and the slower performance in the finger tapping task (n=9; r=0.9, p=0.001), a result also detected when excluding the three patients with left hand paresis (n=6; r=0.9, p=0.02).

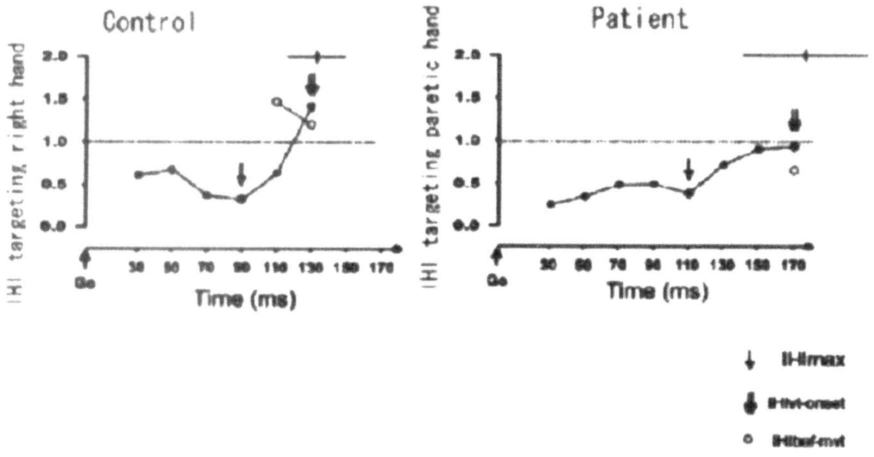

Fig.2. Interhemispheric inhibition (IHI) targeting the moving hand in a reaction time (RT) paradigm in a healthy volunteer and a stroke patient (Patient2 inTable1). The abscissa shows the timing (msec) of application of the test stimulus (TS) relative to the Go signal. The ordinate shows the magnitude of IHI targeting the moving hand (1 indicates absence of facilitation or inhibition; <1indicates facilitation and >1 indicates inhibition). Note the deep maximum IHI (IHImax, single arrow, <1) that progressively became less prominent in intervals close to movement onset in both subjects. Around movement onset, IHImvt-onset (double arrow) and IHIbef-mvt (open circles) turned to facilitation in the control subject but remained inhibited in the patient. The gray lines in the top right corner indicate mean±1SD of RTs in unstimulated trials.

IHIbef-mvt, which occurred at 166±33 and 193±44 msec after the Go signal in healthy volunteers and patients, respectively, showed clear inhibition in the patient group (0.71±0.19) and facilitation in the healthy controls (1.26±0.35; p=0.001; see Fig 3). Exclusion of the three patients with left hand paresis (Table 1) did not modify this result (n=6, p=0.003).

In contrast, $IHI_{Healthy\ hand}$ targeting the homonymous resting index finger remained largely unchanged regardless of movement of the other index finger in either patients or controls.

In the generation of intact hand movement, IHIbef-mvt and IHImvt-onset were similar in stroke patients with those of healthy volunteers.

4. Discussion

We tested IHI at different time intervals after a Go signal in a simple RT paradigm. This design allowed us to study IHI at different timings relative to the onset of a voluntary movement by the paretic hand and to compare the results with those obtained in a group of age-matched healthy volunteers. We used parameters of TMS that elicited comparable IHI in both groups at rest (see Subjects and Methods). Other experimental approaches for the evaluation of IHI, such as ipsilateral silent periods, require a constant background contraction and are suboptimal for the evaluation of premovement modulation in a RT paradigm.

IHI in controls was profound close to the Go signal, but it decreased progressively as movement approached to turn into facilitation at movement onset. This time dependent modulation of IHI by voluntary drive in normal volunteers may support accurate motor control for unilateral hand movements. Stroke patients on the other hand failed to show this modulation of IHI by voluntary movements. The most important finding was that despite comparable IHImax soon after the Go signal, IHImvt-onset and IHIbef-mvt were more pronounced in the paretic hand of patients than in controls. The movement-related modulation of IHI in the paretic hand selectively affected IHI from $M1_{intact}$ to $M1_{lesioned}$ because IHI targeting the

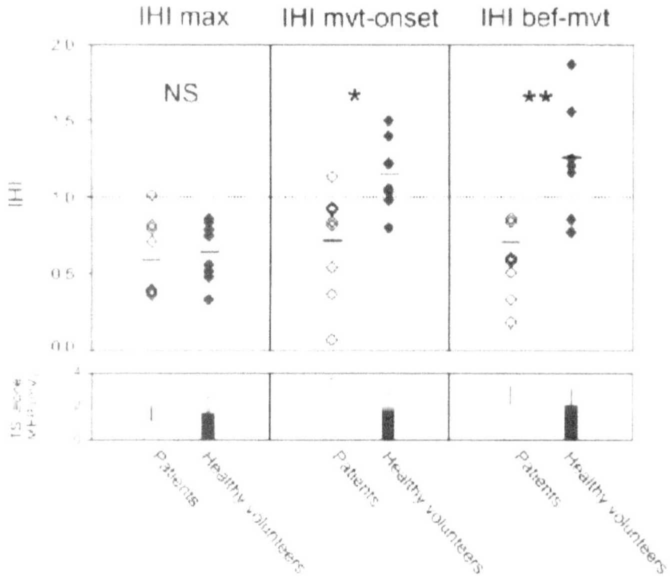

Fig.3. Interhemispheric inhibition (IHI) targeting the moving hand in stroke patients and in healthy volunteers. Note that maximum IHI (IHImax) was comparable in both groups, whereas IHI around movement onset (IHImvt-onset) and IHI immediately before movement (IHIbef-mvt) were significantly deeper in the stroke patients than in healthy volunteers (*$p<0.01$and**$p<0.001$ for IHImvt-onset and IHIbef-mvt, respectively). MEP amplitudes evoked by the test stimulation (TS) were comparable across groups (bottom graphs). NS=not significant.

resting hand was comparable in patients and controls.

We found that the deeper IHImvt-onset from M1intact hemisphere to M1lesioned hemisphere, the lower the MRC scale scores and the slower performance in the finger tapping task in the patient group. Speed of finger tapping is a motor skill that correlates well with the achievement of functional goals in patients with brain lesions undergoing rehabilitative treatments, and the MRC score is an accepted measure of motor function. These results suggest a link between high interhemispheric inhibitory drive from M1intact hemisphere to M1lesioned hemisphere and poor motor recovery in some patients with chronic stroke. In addition, they indicate a substantial difference in the way interhemispheric interactions operate in both groups in the process of generation of a voluntary movement, raising the hypothesis that stronger IHI from $M1_{intact}$ to $M1_{lesioned}$ could contribute to the motor disability still present in some patients with stroke. Longitudinal studies could provide additional information on the timing of development of these changes. This interpretation is consistent with the finding of abnormalities in interhemispheric processing of sensory input after stroke in patients with unilateral extinction. The mechanisms that mediate functional recovery after stroke may differ depending on such factors as time from the insult, degree of recovery, or lesion site.

In summary, our results indicate that an abnormal interhemispheric interaction operate in the process of generation of voluntary movements, and probably influences functional recovery in patients with chronic subcortical stroke [9]. It is conceivable that modulation of such interactions might influence motor disability in these patients, as documented in sensory systems, an issue for future investigation.

References

1. Chollet F, DiPiero V, Wise J (1991) The functional anatomy of motor recovery after stroke in humans: a study with positron emission tomography. Ann Neurol 29: 63–71
2. Cao Y, D'Olhaberriague, Vikingstad M (1998) Pilot study of functional MRI to assess cerebral activation of motor function after post stroke hemiparesis. Stroke 29: 112–122
3. Marshall R, Perera M, Lazar M (2000) Evolution of cortical activation during recovery from corticospinal tract infarction. Stroke 31: 656–661
4. Traversa R, Cicinelli P, Pasqualetti P (1998) Follow-up of interhemispheric differences of motor evoked potentials from the "affected" and "unaffected" hemispheres in human stroke. Brain Res 803: 1–8
5. Trompetto C, Assini A, Buccolieri A (2000) Motor recovery following stroke: a transcranial magnetic stimulation study. Clin Neurophysiol 111: 1860–1867
6. Netz J, Lammers T, Homberg V (1997) Reorganization of motor output in the non-affected hemisphere after stroke. Brain 120: 1579–1586
7. Young M, The Technical Writer's Handbook, Mill Valley, CA: University Science, 1989.
8. Boroojerdi B, Diefenbach K, Ferbert A (1996) Transcallosal inhibition in cortical and subcortical cerebral vascular lesions. J Neurol Sci 144: 160–170
9. Murase N, Duque J, Mazzocchio R (2004) Influence of Interhemispheric In teractions on Motor function in Chronic Subcortical Stroke. Ann Neurol 55: 400-409

BOLD Contrast fMRI as a Tool for Imaging Neuroscience

Robert Turner

Wellcome Department of Imaging Neuroscience, Institute of Neurology, Queen Square, London.

Chapter Overview. Since its discovery in 1991 the technique of mapping human brain activity by observing changes in cortical blood oxygenation with MRI has blossomed into a methodology of major importance for brain science and medicine. This paper summarizes recent improvements in the use of fMRI as a tool for imaging neuroscience.

Key Words. BOLD, brain, fMRI, neuroimaging

1. Introduction

Ogawa [1] and Turner [2] showed in 1990-1991 that MR images could be made sensitive to the level of oxygenation of cerebral blood, and thus to changes in regional CBF, indicating localized brain activity. This led to the development of the technique of Functional MRI (fMRI), which generally uses Blood Oxygenation Level Dependent (BOLD) contrast to depict brain areas showing changes in neuronal activity associated with specific brain tasks or conditions.

The topic of fMRI has been reviewed several times [e.g. 3], but new insights continue to be developed, and improvements in MRI and computing hardware have enabled a number of earlier problems to be successfully tackled. This paper updates the progress made in dealing with the technical challenges that have arisen in its implementation as a tool for imaging neuroscience.

2. Biophysics of Blood Oxygenation Level Dependent Contrast

2.1 Mechanisms

The relationships between changes in neuronal electrical activity, in cellular metabolism, in oxygen extraction and blood flow, and hence changes in blood oxygenation, are still not fully understood. However, recent research in several laboratories has clarified several important issues.

There is now a consensus that the BOLD signal provides a time-delayed index of changes in Local Field Potential (LFP), rather than in neuronal spike discharges [4]. Most of the energy expenditure of neuronal tissue arises from post-synaptic activity [5], and it is this activity that is manifested in the LFP. The BOLD signal represents a balance between the increased blood flow triggered by chemical messengers produced during increased neuronal activity, and the increased oxygen consumption required to support the increased local brain metabolism. Because oxygen transport from red cells to tissue is limited primarily by diffusion, provision of adequate oxygen concentrations at the mitochondria requires an excess of oxygen in the bloodstream [3]. This rise in blood oxygenation gives an observable localized rise in MR image intensity (Figure 1), because deoxyhaemoglobin acts as an intravascular paramagnetic contrast agent [1].

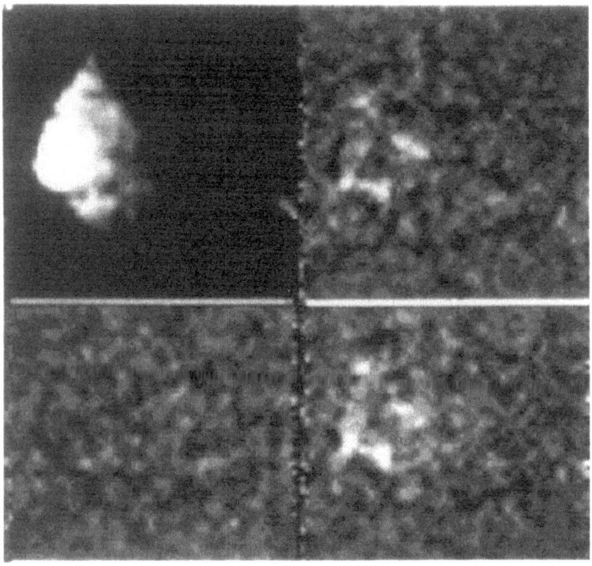

Fig.1. Sagittal images obtained at 4 T showing the effect of visual stimulation on image intensity in human occipital cortex. 64 x 64 pixel matrix EPI images interpolated to 128 x 128. Field of view 160 mm, TE 25 ms. Top left: source image, obtained using 20 cm transmit-receive surface coil at back of head. The other images are single-subtraction images. Top and bottom right: visual stimulation on - visual stimulation off. Bottom left: visual stimulation off - visual stimulation off. Differences in image intensity are as much as 20%. Images are windowed to show background noise (Reprinted from Turner [6], with permission).

2.2 Spatial localization

It is apparent that the spatial matching of cerebral blood flow to neuronal activity cannot be perfect, given the 5 - 10 micron spacing of neurons, the 50 micron spacing of capillaries, and more importantly, the 0.5 - 1 mm spacing of resistance arterioles, the proximal locus of control of flow [7]. Thus the changes in blood oxygenation detected in BOLD contrast cannot be expected to mirror electrical data at a scale of microns. Model calculations [8,9] and some observations [10] indicate that at field strengths below 3 T the signal is dominated by changes of oxygenation in small draining venules, likely for geometrical reasons to be within a few millimetres of the site of neuronal activity, rather than in the parenchymal capillaries actually feeding the active neurons.

After leaving the cortical capillaries the blood drains into the jugular veins via a network of intracortical veins and subpial veins. Thus changes in blood oxygenation will be propagated downstream and away from sites of increased neuronal activity, to give a spurious BOLD signal apparently arising from non-activated brain areas. Turner [11] provides estimates of the extent of this compromise of fMRI's spatial resolution, which strongly depends on the spatial extent of the changes in neuronal activity (Figure 2).

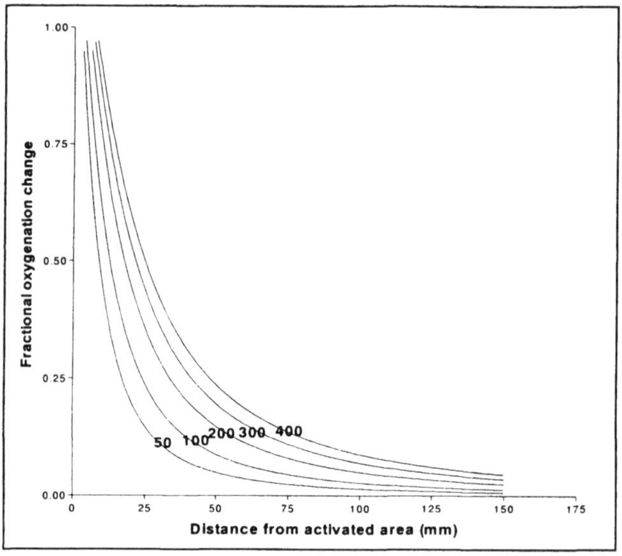

Fig.2. Fractional oxygenation change in a draining vein as a function of linear distance from activated area, for 50% change of blood flow, 25% change in oxygen uptake. Numbers marked on curves refer to the area (in square mm) of increased neuronal activation

3. fMRI Technical Requirements

The key technical requirements in the use of fMRI to study brain function are a) equipment to implement the brain task in question while the subject lies in an MRI scanner with head immobilized, and b) use of an MR imaging sequence made sensitive to changes in T2*, with good image quality, SNR and stability.

3.1 Task presentation

For maximum flexibility and control over software management, the Functional Imaging Laboratory created a task presentation computer language, named Cogent2000, based mainly on the commercial language MATLAB, and which is thus compatible with the statistical analysis package, SPM. For convenience, Cogent2000 can be run on a standard PC (the Stimulus PC), which drives all peripherals, mostly using the USB port. The shareware package Cogent2000 can be downloaded freely from the FIL website: www.fil.ion.ucl.ac.uk.

The standard requirements for any task presentation set-up—which we call the Cognitive Interface--are as follows:

a) Task presentation synchronization with scanner data acquisition. We use optical serial signals from the scanner, which provide a slice-by-slice image count, to drive the Stimulus PC output.

b) Visual presentation. We use a good quality LCD projector. The criteria for projector selection are: high refresh rate, high pixel resolution, and good colour fidelity.

c) Graphics library. Cogent2000 contains Cogent Graphics, a library of simple graphics primitives written in the fast low-level language Direct-X. This allows very rapid and precise visual stimuli to be delivered.

d) Video capability. Many paradigms require video presentations. This can be provided easily within Cogent2000.

e) Keypad. Several nonmagnetic keypads are now commercially available, with foolproof optical coupling to the task presentation PC. Joysticks and manipulanda can also be obtained and interfaced easily if required.

f) Data Logging. It is essential to be able to monitor subject performance during scanning. The Stimulus PC automatically logs key presses and timings.

g) Auditory presentation. Many studies require auditory presentation. Use of pneumatic headphones is not recommended because of their very low auditory fidelity. Electrostatic headphones are available giving excellent sound power and fidelity, which are compatible with magnetic fields. Appropriate filters must be installed in the leads to avoid rf interference which can cause image artefact. To obtain good sound fidelity without the intrusion of scanner acoustic noise, the strategy of sparse sampling is often used [12].

h) Eye-tracking. Even for paradigms that do not explicitly study neural control of eye movements, use of an eye-tracker is advisable, to ensure eye direction fixation as required, and to monitor attention to the task. In our experience the best option, in terms of flexibility, freedom from image artefact, and reliability, is a commercial remote infrared source and CCD

camera with telescope to view the eye's pupil. This system includes efficient computer analysis and display software.
i) Additional features included in the Cognitive Interface are: an olfactometer, which presents a selection of odours; devices for delivery of taste stimuli; devices for somatosensory stimuli, by pneumatic, electrical, or mechanical means; hardware and software for combined simultaneous EEG and fMRI [13]; and combined simultaneous TMS and fMRI [14].

3.2 FMRI Sequences and image reconstruction

In general, the important sequence requirements continue to be sensitivity, spatial resolution, speed, complete brain coverage, freedom from motion artefact, and good image quality. Echo-planar imaging (EPI) and its variants [15] maintain good SNR over the whole brain, consisting of 30 or more slices, down to TR times of 1 s. Given the widespread requirement for whole-brain coverage in a few seconds, only this type of sequence is seriously considered for use in imaging neuroscience. EPI is relatively simple to programme, and is time-efficient and gives good SNR per unit time. While it is prone to image artefacts--ghosting, distortion and dropout--these are easily characterized and can mostly be corrected without a severe penalty in processing time. Typically, to provide sufficient T2* weighting, an echo time of 50 ms is used at 1.5 T, the lowest field strength normally used for fMRI. Shorter echo times are used with scanners at 3 T (30 ms) and 4 T (25 ms). Spatial resolution of 3 mm x 3 mm x 3 mm is readily achievable, and in-plane resolution of 1.5 mm is possible with scanners having powerful gradient hardware. To obtain higher spatial resolution, segmented EPI sequences can be used. Submillimeter resolution has been achieved [16]. Because images requiring several rf pulses to collect the complete set of data in k-space are prone to motion artefact, it is essential to use navigator echoes for such acquisitions.

3.3 Image quality issues

While EPI sequences have so many benefits that they are selected by most researchers for fMRI research, echo-planar images have several particular difficulties regarding image quality. These can be classified under the headings of image artefacts and field inhomogeneity effects.

3.4 Image artefacts: ghosting

The trajectory traversed during EPI data acquisition reverses in direction between each echo. Before image reconstruction can take place, the data for alternate echoes must be time-reversed. If there is the slightest difference in the gradient waveform between the two trajectory directions, or if the noise-limiting band pass filters in the receiver have any phase roll, the echoes show a double-period modulation, which appears as a half-field or Nyquist ghost in the reconstructed image. In practice it is extremely difficult to eradicate all sources of asymmetry in k-space traversals, and a low intensity ghost is almost always present in EPI images

Recently Josephs [17] has developed an image reconstruction methodology, used routinely at the Functional Imaging Laboratory, which reduces image ghosting to negligible levels while maintaining real-time image reconstruction speed. The method is known as Trajectory-Based Reconstruction (TBR).

Conventional MR image reconstruction, which uses fast Fourier transforms (FFT), assumes perfect gradient performance, with no eddy currents, and a precise temporal relationship between gradient switching and sampling comb. In reality, these assumptions are incorrect, leading to image ghosting, as just mentioned. In order to provide a more precise image reconstruction, TBR uses a reference prescan, with a modified EPI sequence that allows the sampling trajectory in k-space to be precisely measured. The trajectory prescan needs to be performed only once after each scanner service, and a phantom may be used. The additional use of navigator echo data acquired at each imaging slice, with no added cost in scan time, allows the phase referencing of each scan to the first scan of series, with further improvements to image quality and stability. In reconstructing the images, a generalised reconstruction replaces one of the FFT steps by a matrix multiplication. With a standard PC this can be performed offline in real time, and allows a continuous monitoring of image quality to take place. The result of the TBR method is much reduced image ghosting (Figure 3) and greater image stability.

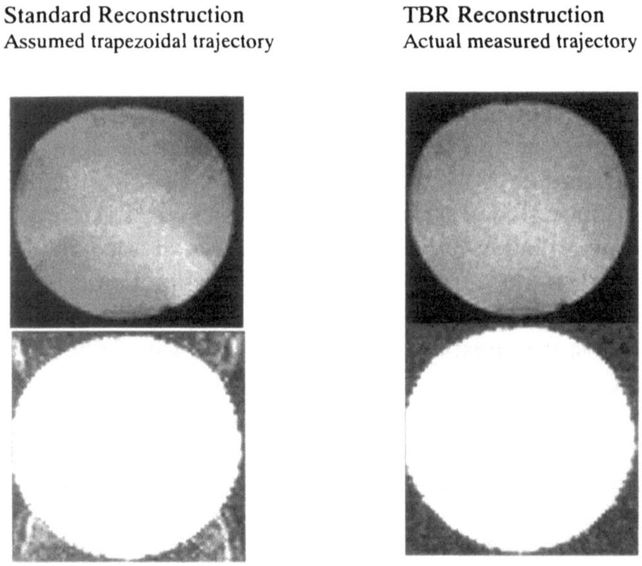

Fig.3. Comparison of results of image reconstruction using the same EPI data set of a gel-filled phantom, showing the striking improvement of ghost artefact. Left: Standard FFT reconstruction. Right: Trajectory-based reconstruction. Lower images are windowed to show more clearly background noise and ghosting levels.

3.5 Field inhomogeneity effects: distortion and dropout

The use of gradient echo techniques to detect changes in deoxyhaemoglobin concentration means that the MR image is also sensitive to bulk variations in static magnetic field. In practice, at the field strengths of 1.5 T and above that are useful for BOLD fMRI, the major source of field inhomogeneity is the endogenous variations in magnetic susceptibility of the head itself. Simple MRI techniques can map these field variations very accurately [18]. Because they typically have rapid spatial variation, they can be impossible to correct by means of shim coils mounted in the magnet's bore tube.

The effects of susceptibility-induced field variations are more complex than might first appear, and can be classified into the following types: a) image distortion, b) slice refocusing dropout, c) transverse dropout. These will be discussed in turn.

a) Image distortion

MRI works by mapping frequency offsets caused by known applied linear gradients onto spatial co-ordinates. Fast imaging sequences such as EPI produce pixels with a small frequency range per pixel in the phase-encoding direction, say 25 Hz. Thus the effect of field inhomogeneities becomes important, and distortions of 3 pixels or more in the phase-encoding direction are quite typical in ventral brain regions. It is simple and quick to create magnetic field maps of the entire volume of tissue in the head. Such maps can be easily transformed into tables of displacements in the phase-encode direction required to eliminate the distortion [19], which can be applied at the same post-processing stage as the removal of head motion effects. As with motion correction, care must be taken with the resampling of the image data to ensure that interpolation is done correctly [20].

A significant improvement in dealing with distortion, in the context of a time series of EPI brain volumes as used for fMRI, arises from considering the effects of head motion on distortion. Because the distortion-producing magnetic fields do not move with the head for rotations about the two transverse co-ordinate axes, such motions, even of very small amplitude, provide an additional source of variance in image intensity that is not corrected by simple rigid-body realignment algorithms. Andersson et al [21] discovered a method for removing this interaction between head movement and distortion, which is compatible with the image analysis package SPM.

b) Slice refocusing dropout

To achieve the desired T2* weighting for BOLD contrast to be observed, a relatively long echo time is required, giving through-slice dephasing more time to evolve and hence to reduce the net signal. The effect is that regions of the head with strong through-slice inhomogeneity gradients disappear, or drop out, from the image.

c) Transverse dropout

If strong endogenous field gradients exist in some part of the object, the trajectory applying to that part of the object is distorted. The trajectory may not cross the k-space axis, and the large-amplitude data at this location will then not be sampled. Furthermore, for a finite echo time, the acquisition period may not even begin at the correct location in k-space, and the effective trajectory may not cross either axis. Thus spatial frequency content of the signal from such regions will be poorly sampled, and image drop-out will be observed. This effect may be caused by transverse gradients in the phase-encode direction, moving the k-space coverage sideways in the phase-encode direction, or in the read direction, imparting a shear to the k-space coverage. If the ensuing k-space coverage fails to include the centre of k-space, there will be drop-out in the final image.

To date, the most comprehensive and practical solution to this problem has been developed recently by Deichmann [22], who proposed obtaining just three images per time point, each with different optimized compensation gradient pulses applied before data acquisition. These pulses compensate for the effect of endogenous gradients in slice, phase-encode, or read directions, each of which can cause dropout, and also impaired BOLD contrast, which arises when an endogenous gradient in the phase-encode direction changes the effective echo-time. The three images are combined using a weighted sum-of-squares method. A further method, which involves tilting the slice direction out of the axial plane by an angle of 30 degrees, allows good images of ventral frontal regions to be obtained using only one set of images, with a single compensation pulse in the through-slice direction [23].

One final effective stratagem to reduce dropout is to reduce the echo time. This has the inevitable effect of reducing the magnitude of BOLD contrast changes [24], but with scanners of sufficient sensitivity and stability, and with sufficiently long data acquisition runs, it is still possible to obtain useful results with echo times as short as 25 ms at 2 T.

3.6 High field fMRI

The most radically effective means of improving BOLD contrast is to use higher magnetic field strength [25, 26]. The sensitivity is improved in two ways. Firstly, the increased magnetic field gives higher image SNR, especially if fast imaging sequences are used. Secondly, the effects of a given susceptibility difference between oxygenated and deoxygenated blood also increase with field strength. Furthermore, the BOLD signal is more focal, because proportionately more of the signal derives from the capillary bed local to the neuronal activity, as compared with pial venules [27].

As previously mentioned, one major drawback of high field fMRI is the increased artefact caused by bulk susceptibility differences between brain and re-entrant air-filled regions near the brain, such as the frontal and ethmoid sinuses, and the ear canal, Most parts of the brain are unaffected by this problem, however, and many important results have come from scanners working at field strengths of 3 T and higher [28, 29].

While there have been a number of studies comparing BOLD fMRI at different field strengths, few have made a comparison using appropriate statistical methods, with an adequate number of human subjects. Krasnow et al [30] have made such a study, showing that for a task giving highly consistent activation across a group of subjects, 3 T field strength gives clearly better results than 1.5 T. On the other hand, Turner et al [31] with a different paradigm, show that intra-subject differences can easily outweigh the increased sensitivity found at higher field. The question arises, what are the most reliable benefits of using higher field strength magnets?

An interesting debate has arisen recently in this regard. There are two main sources of noise affecting the MRI signal: thermal noise, mostly arising from the tissue itself, and physiological noise, caused by processes associated with live tissue homeostasis, such as slow fluctuations of blood oxygenation [32], and from cardiac and respiratory pulsa pulsations. Thermal SNR increases linearly with field strength, given comparable receiver performance. Krueger [33] showed that if optimal TE is used, the size of the BOLD response MRI signal increases with field strength, but the physiological noise associated with the BOLD signal also increases with field strength.

When physiological noise dominates thermal noise, BOLD contrast-to-noise ratio related to functional activation does not increase very strongly with field strength. Such conditions prevail when the voxel size is above 7 mm^3 at 1.5 T. At progressively higher field strengths, as shown by Triantafyllou et al [34] (Figure 4), smaller voxels can be used before these diminishing returns are encountered. It is likely that an optimum is obtained for any given field strength when the physiological noise for a typical grey matter voxel is equal to the thermal noise.

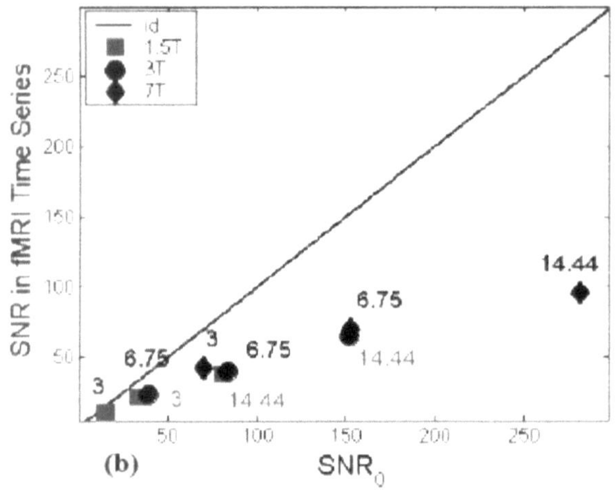

Fig.4. Variations of image SNR modulated by altering the image resolution, plotted against thermal image SNR, for BOLD EPI images obtained at 1.5 T, 3 T and 7 T. Voxel volume is given in mm3, where 3, 6.75 and 14.44 mm3 equate to voxel dimensions of 1x1x3, 1.5x1.5x3 and 1.9x1.9x4 mm3 respectively. (Adapted from Triantafyllou et al, ISMRM 2004)

3.7 Speed and brain coverage

The use of several rf receiver coils surrounding the head, each connected to an independent receiver, with its own preamplifier and digitizer, allows so-called 'parallel imaging' techniques to be employed, the best known named SENSE [35] and SMASH [36]. These allow images to be acquired still faster, with speed-up factors equal to the number of coils used in parallel. There is a consequent drop of SNR in such images, which can be compensated to some extent by acquiring more images in the same scanning time. Shortening the acquisition time means a reduction in image distortion and the possibility of event-related fMRI with volume acquisition more closely resembling a simultaneous whole-head data capture.

4. Applications of FMRI

The field of Imaging Neuroscience is already wide and is rapidly expanding. A comprehensive review of fMRI applications is already impossible.

In general terms, the following areas are being explored: a) Elucidation of the functional organization of the human brain down to submillimetre distance scales. b) Exploration of changes in cortical organization during development and learning. c) Localization of eloquent cortical areas in planning for neurosurgery. d) Detection of functional abnormalities in developmental disorders. e) Monitoring of recovery after brain damage caused by trauma or stroke, allowing management optimization. f) Investigation of the cerebral correlates of mental illness such as schizophrenia, mood disorders, eating disorders, and addiction.

5. Conclusions

The technique of BOLD contrast fMRI has now developed to the point at which it is in routine use as an important tool in imaging neuroscience. The improvements to image quality developed in the FIL have been outlined in this summary.

Acknowledgment

The work described here was supported by the Wellcome Trust.

References

1. S. Ogawa, T.M. Lee, A.R. Kay and D.W. Tank. Brain magnetic resonance imaging with contrast dependent on blood oxygenation. Proc. Natl. Acad. Sci. U S A vol. 87, pp. 9868-72, 1990
2. R. Turner, D. Le Bihan, C.T. Moonen, D. Despres and J. Frank. Echo-planar time course MRI of cat brain oxygenation changes. Magn. Reson. Med. vol. 22, pp.159-66, 1991
3. R. B. Buxton. Introduction to Functional Magnetic Resonance Imaging: Principles and Techniques, Cambridge: Cambridge University Press 2001
4. N.K. Logothetis and J.Pfeuffer.On the nature of the BOLD fMRI contrast mechanism. Magn. Reson. Imaging, vol 22, pp. 1517-31, 2004
5. D. Attwell and C.Iadecola. The neural basis of functional brain imaging signals. Trends Neurosci., vol. 25, pp. 621-5, 2002
6. R. Turner. Magnetic resonance imaging of brain function. Am J Physiol Imaging, vol. 7, pp.136-45, 1992

7. S.S. Segal and B.R. Duling. Conduction of vasomotor responses in arterioles: a role for cell-to-cell coupling? Am. J. Physiol. vol. 256, pp. H838-45, 1989
8. S. Ogawa, R.S. Menon, D.W. Tank, S.G. Kim, H. Merkle, J.M. Ellermann and K. Ugurbil.Functional brain mapping by blood oxygenation level-dependent contrast magnetic-resonance-imaging - a comparison of signal characteristics with a biophysical model. Biophys. J. vol. 64, pp. 803-812, 1993.
9. R.M. Weisskoff, C.S. Zuo, J.L. Boxerman and B.R. Rosen. Microscopic susceptibility variation and transverse relaxation - theory and experiment. Magn. Reson. Med., vol. 31, pp. 601-610, 1994.
10. R.S. Menon, J.S. Gati, B.G. Goodyear, D.C. Luknowsky and C.G. Thomas. Spatial and temporal resolution of functional magnetic resonance imaging. Biochem. Cell Biol. 76: 560-71, 1998.
11. R. Turner How much cortex can a vein drain? Downstream dilution of activation-related cerebral blood oxygenation changes. NeuroImage, vol.16, pp.1062-7, 2002.
12. D.A. Hall, M.P. Haggard, M.A. Akeroyd, A.R. Palmer, A.Q. Summerfield, M.R. Elliott, E.M. Gurney and R.W. Bowtell. "Sparse" temporal sampling in auditory fMRI. Hum. Brain Mapp., vol. 7, pp. 213-23, 1999.
13. P.J. Allen, O. Josephs and R. Turner. A method for removing imaging artifact from continuous EEG recorded during functional MRI. Neuro-Image, vol.12, pp. 230-9, 2000.
14. D.E. Bohning, A. Shastri, Z. Nahas, J.P. Lorberbaum, S.W. Andersen, W.R. Dannels, E.U. Haxthausen, D.J. Vincent and M.S. George Echoplanar BOLD fMRI of brain activation induced by concurrent transcranial magnetic stimulation. Invest. Radiol. vol. 33, pp. 336-40, 1998
15. F. Schmitt, M.K. Stehling and R. Turner (eds). Echo-Planar Imaging, Berlin: Springer, 1998
16. R.S. Menon and B.G. Goodyear. Submillimeter functional localization in human striate cortex using BOLD contrast at 4 Tesla: implications for the vascular point-spread function. Magn. Reson. Med., vol. 41, pp. 230-5, 1999
17. O. Josephs, R. Deichmann and R. Turner, Trajectory measurement and generalised reconstruction in rectilinear EPI. NeuroImage vol.11, p S543, 2000
18. S. Li, B.J. Dardzinski, C.M. Collins, Q.X. Yang and M.B. Smith. Three-dimensional mapping of the static magnetic field inside the human head. Magn. Reson. Med., vol. 36, pp 705-14, 1996
19. P. Jezzard and R.S. Balaban. Correction for geometric distortion in echo planar images from B0 field variations. Magn. Reson.Med., vol. 34, pp. 65-73, 1995
20. C. Hutton, A. Bork, O. Josephs, R. Deichmann, J. Ashburner and R. Turner. Image distortion correction in fMRI: A quantitative evaluation. NeuroImage, vol.16, pp. 217-40, 2002
21. J.L. Andersson, C. Hutton, J. Ashburner, R. Turner and K. Friston. Modeling geometric deformations in EPI time series. NeuroImage, vol. 13, pp. 903-19, 2001

22. R. Deichmann, O. Josephs, C. Hutton, D.R. Corfield, R. Turner. Compensation of susceptibility-induced BOLD sensitivity losses in echo-planar fMRI imaging. NeuroImage, vol.15, pp.120-35, 2002
23. R. Deichmann, J.A. Gottfried, C. Hutton and R. Turner. Optimized EPI for fMRI studies of the orbitofrontal cortex. NeuroImage, vol 19,pp. 430-41, 2003
24. M.L. Gorno-Tempini, C. Hutton, O. Josephs, R. Deichmann, C. Price and R. Turner. Echo time dependence of BOLD contrast and susceptibility artifacts. *NeuroImage*, vol.15, pp.136-42, 2002
25. R. Turner, P. Jezzard, H. Wen, K.K. Kwong, D. Le Bihan, T. Zeffiro and R.S. Balaban. Functional mapping of the human visual cortex at 4 tesla and 1.5 tesla using deoxygenation contrast EPI. *Magn. Reson. Med.*, vol. 29, pp. 281-283, 1993
26. Y. Yang, H. Wen, V.S. Mattay, R.S. Balaban, J.A. Frank and J.H. Duyn. Comparison of 3D BOLD functional MRI with spiral acquisition at 1.5 and 4.0 T. *NeuroImage*, vol. 9, pp. 446-51, 1999
27. J.S. Gati, R.S. Menon, K. Ugurbil and B.K. Rutt. Experimental determination of the BOLD field strength dependence in vessels and tissue. *Magn. Reson. Med.*, vol. 38, pp. 296-302, 1997
28. A. Karni, G. Meyer, P. Jezzard, M.M. Adams, R. Turner and L.G. Ungerleider. Functional MRI evidence for adult motor plasticity during motor skill learning. *Nature* vol. 377, pp. 155-158, 1995
29. H.J. Neville, D. Bavelier, D. Corina, J.P. Rauschecker, A. Karni, A. Lalwani, A. Braun, V. Clark, P. Jezzard and R. Turner. Cerebral organization for language in deaf and hearing subjects: biological constraints and effects of experience. *Proc. Natl. Acad. Sci. USA vol.* 95, pp. 922-929,1998
30. B. Krasnow, L. Tamm, M.D. Greicius, T.T. Yang, G.H. Glover, A.L. Reiss and V. Menon. Comparison of fMRI activation at 3 and 1.5 T during perceptual, cognitive, and affective processing. *NeuroImage*, vol.18, pp. 813-26, 2003
31. R. Turner, A. Mechelli, U. Noppeney, C. Price, J. Glensman, S. Rombouts and D.Veltman. A comparison of the effectiveness of 3 T and 1.5 T MRI scanners in a multi-subject fMRI study of visual memory. *Proc. ISMRM*, 13th Annual Meeting, 2005
32. R.G. Wise, K. Ide, M.J. Poulin and I. Tracey. Resting fluctuations in arterial carbon dioxide induce significant low frequency variations in BOLD signal. *NeuroImage*, vol. 21, pp.1652-64, 2004
33. G. Kruger and G.H. Glover. Physiological noise in oxygenation-sensitive magnetic resonance imaging. *Magn. Reson. Med.*, vol. 46, pp. 631-7, 2001
34. C. Triantafyllou, L. L. Wald, C. J. Wiggins, A. Potthast, G. C. Wiggins, G. Krueger and R. D. Hoge. Physiological noise in fMRI: Comparison at 1.5T, 3T and 7T and dependence on image resolution. *Proc. ISMRM*, 12th Annual Meeting, 2004
35. K.P. Pruessmann, M.Weiger, M.B. Scheidegger and P. Boesiger. SENSE: sensitivity encoding for fast MRI. *Magn. Reson. Med.*, vol. 42, pp. 952-62, 1999

36. D.K. Sodickson and W.J. Manning. Simultaneous acquisition of spatial harmonics (SMASH): fast imaging with radiofrequency coil arrays. *Magn. Reson. Med.*, vol. 38, pp. 591-603, 1997

What can be Observed from Functional Neuroimaging?

Jorge Riera

New Industry Creation Hatchery Center (NICHe) Tohoku University, Sendai, Japan

Chapter Overview. In this paper, I will discuss some of the physics and physiological principles underlying the current development of neuroimaging based on EEG/MEG and fMRI techniques. I will highlight the non-uniqueness of the respective inverse problems to determine the implicit neuronal activity. The advantages of introducing compartmental models of the V-layer pyramidal cells and the balloon model for post-capillary venous compartments to constrain the temporal dynamics of particular physiological variables will also be evaluated.

Key Words. EEG and MEG Inverse Problems, Nonlinear Dynamic Systems, Electro-Vascular Coupling

1. Introduction

The use of several types of neuroimaging modalities (i.e. EEG/MEG and fMRI) to study the human brain is standard today in the neuroscience community. A buzz of excitement and fashionable interest has typically followed the introduction of each of these techniques, and after a massive volume of papers has been published, once their limitations have been revealed the excitement gives way to something more akin to a motivational crisis. Berger [1] is considered the pioneer, introducing the EEG (Electro-EncephaloGram) technique to study alpha rhythms in humans. The EEG data represents voltage differences at a number of sites on the scalp recorded with the temporal resolution of just milliseconds.

After producing a profusion of fruitful results over a long period, the EEG was strongly criticized for its susceptibility to the conductive profile of the head tissues. To overcome such limitations, a more expensive technique, the MEG (Magneto-EncephaloGram), came to light in the second half of the last century as a new prospect with huge potential. Cohen [2] was the first to measure using external coils the weak magnetic field produced by the brain activity. At that time, based on the apparent transparence of head tissues to the quasi-static magnetic field, the MEG was proclaimed as the ultimate technique for determining the "*brain current sources*" $\vec{J}_p$ with an acceptable spatial resolution and a similar temporal resolution as the EEG.

From the time it was possible to record local field potentials in animals using linear arrays of microelectrodes implanted perpendicular to the cortex, it has been recognized that brain current sources may result from unbalanced extra-cellular ionic gradients found between superficial and deep layers of the neocortex. These gradients originate mostly from the differentiated synaptic activity at diverse locations of the V-layer pyramidal cells (i.e. the apical and basal dendrites, the soma, etc). The V-layer pyramidal cells are symmetrically organized along the whole cortical surface. This fact, in addition to the special geometrical form of this type of neuron, makes them the best candidate for the electromotive force, of extra-cellular character, that appears perpendicular to the cortical surface when neuronal activity increases. The intensity of this force is a sign of the level of synchronization in the post-synaptic potentials of the V-layer pyramidal cells within an adjacent cortical area. That is why the EEG and MEG signals are thought to correlate well to the synaptic activation of this type of cortical neuron, although the possible contributions of others brain structures (i.e. brainstem, thalamus, cerebellum, etc) have not been disregarded.

Efforts to reconstruct $\vec{J}_p$ in the whole brain volume from EEG and MEG data gave birth to a new modality of functional neuroimaging. This 3D-reconstruction, which requires that the respective forward and inverse problems be solved, still constitutes a mammoth theoretical subject. Innumerable reports appeared after the MEG started to be used, and its popularity increased considerably when simultaneous EEG and MEG recordings using a huge number of sensors became realizable. In that golden era, neuroscientists indiscriminately reported to have found not only the regions where a lot of brain functions were carried out but also how these regions interacted. Even so, researchers have always been familiar with the ill-posed nature of the EEG and MEG inverse problems; the brain current sources cannot be determined uniquely. Hence, any inverse solution would depend on the particularities of the a priori information utilized to constrain $\vec{J}_P$ (e.g. **dynamics** [3], **anatomy** [4], **biophysics** [5], etc).

In the last years of the twenty-century, feelings of disheartenment in the neuroscience community due to the real limits of EEG and MEG techniques gave way to a new burst of excitement with the introduction of a new and promising technique, functional Magnetic Resonance Imaging (fMRI), thought to have excellent spatial resolution for localizing slow temporal fluctuations in the Blood Oxygen Level Dependent (BOLD) signals [6]. It was claimed that BOLD signals somehow reflected basic neuronal activity. This historical rebound of optimism has remained till the present. Thousands of papers have been published since that time, each of them reporting to have discovered regions associated with several cognitive functions in humans (i.e. language, attention, perception, etc), even though the limitations of the BOLD signals in determining the fast dynamics of neuronal activity as a consequence of the low-pass filter of vascular origin are well recognized. Hence, the main efforts in recent years have been on the development of high technology systems for recording EEG and fMRI data simultaneously [7].

At this point, the most attractive theoretical subject seems to be the determination from fused data of the *neuronal activity* in those brain regions where specific functions are executed and, additionally, the evaluation of how these regions are *effectively connected* [8]. However, there is no consensus among researchers using different techniques about **what these fundamental concepts mean** and it is not clear thus far **how researchers should best make use of data to investigate them**. In my personal opinion, neuroimagers must be more conservative when discussing their results and they should limit themselves to clearly state what has been observed with the technique utilized. Additionally, I strongly believe that the issues of "neuronal activation" or "effective connectivity" must be investigated on the basis of a neuronal substrate.

In this report, I would like to make a revision of the following topics:

1- What kind of physiological phenomena are observed with EEG/MEG and fMRI techniques?

2- What are the limitations and capabilities of each of these neuroimaging modalities?

In the discussion I have emphasized how the neuronal activation and the effective connectivity should be examined.

2. Revision of the Physical Problems

2.1. The maxwell equations

The existence of ions makes the head a good conductor which remains an equilibrium state of null net-charge. The fluctuations of charges that affect the EEG and MEG occur mainly in the extra-cellular medium, and they are represented by a physical magnitude denominating the brain current sources. These fluctuations are associated with a misbalance in the concentrations of the ions in the neighborhood of neuronal assemblies containing cells of special geometries (i.e. the V-layer pyramidal cells in the neocortex) temporally synchronized; hence, they are indisputably local in character. This fact ensures that sources and sinks are closed to each other; hence, no net charge is created. In the rare case where it happens, the ions will move out to the conductor boundary. As a second effect, the local brain current sources $\vec{J}_P$ induce a global movement of ions as a consequence the conductor being exposed to an electric field. However, both effects are slow enough to warrant a quasi-static approach of the Maxwell equations for the electric $\vec{E}$ and magnetic $\vec{B}$ fields (Eqs. 1-3)

$$\nabla \times \vec{E} = 0 \quad \Rightarrow \quad \vec{E} = -\nabla \varphi \tag{1}$$

$$\nabla \cdot \vec{B} = 0 \tag{2}$$

$$\nabla \times \vec{B} = \mu_0 \vec{J} \quad \Rightarrow \quad \nabla \cdot \vec{J} = 0 \tag{3}$$

The scalar field φ represents the electric potential. The non-divergent total current has two additive terms, one caused by the ohmic effect and the other due to the brain current source: $\vec{J} = \vec{\vec{\sigma}}\vec{E} + \vec{J}_P$, with $\vec{\vec{\sigma}}$ representing the tensor of conductivity.

The head is surrounded by a non-conductive medium (air); hence, the normal component of the total current at any position on the scalp will be always zero. In order to solve the partial differential equations (1)-(3), finite element methods must be used. However, a very well established volume conductor model is to assume the head is both isotropic and piecewise homogeneous (Fig. 1). In this model, the head is constituted by a set of enclosed and non-intersected compartments R_k, which are assumed to have scalar and constant conductivity value σ_k. It is assumed that the surface S_k separates compartments R_k and R_{k+1}. It is frequently hypothesized that $\vec{J}_p$ is different from zero only in an inner region R enclosing the brain tissues. The brain current sources, as a vector field, must be as general as possible; therefore, its normal $\vec{J}_p \cdot \vec{n}$ and tangential $\vec{J}_p \times \vec{n}$ components on the surface S of region R may take any value (i.e. free boundary conditions).

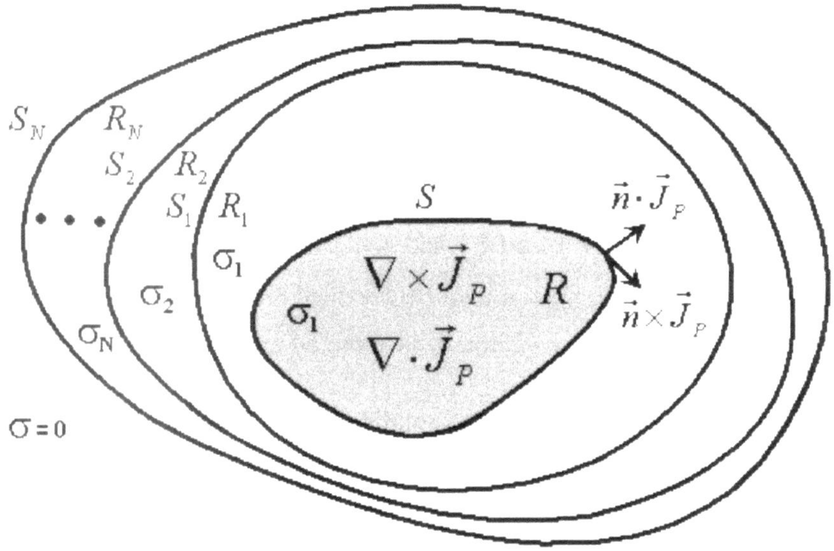

Fig.1. The isotropic and piecewise homogeneous volume conductor model of the head

The electric potential must satisfy additional linking conditions in the surfaces S_k:

$$\varphi_k\big|_{S_k} - \varphi_{k+1}\big|_{S_k} = 0 \tag{4}$$

$$\sigma_k \, \partial\varphi_k/\partial n\big|_{S_k} - \sigma_{k+1} \, \partial\varphi_{k+1}/\partial n\big|_{S_k} = 0 \tag{5}$$

In this case, the electric potential φ and magnetic field $\vec{B}$ at any place are represented by scalar (6) and vector (7) inhomogeneous Fredholm integral equations of the second kind respectively, with the secondary ohmic currents $\vec{J}_k(\vec{J}_P) = (\sigma_{k+1} - \sigma_k)\varphi_k(\vec{J}_P)\vec{n}_k/\Delta l$ being defined for each elemental volume Ω_k (i.e. the surface S_k of thickness Δl).

$$4\pi\sigma\varphi = \int_R \vec{J}_P \cdot \nabla g dv + \sum_k \int_{\Omega_k} \vec{J}_k(\vec{J}_P) \cdot \nabla g dv \tag{6}$$

$$4\pi \vec{B}/\mu_0 = \int_R \vec{J}_P \times \nabla g dv + \sum_k \int_{\Omega_k} \vec{J}_k(\vec{J}_P) \times \nabla g dv \tag{7}$$

The scalar field g is the Green function for the infinite space.

The EEG $o_l^E = \varphi(\vec{r}_l) - \varphi(\vec{r}_r)$ and the MEG $o_l^M = \vec{B}(\vec{r}_l) \cdot \vec{o}_l$ at the l-th sensor (i.e. located at $\vec{r}_l$) can then be obtained from equations (6) and (7), respectively. The unitary vector $\vec{o}_l$ is directed perpendicular to the MEG coil.

The use of the Lead Field (LF) formalism permits us to reformulate both EEG and MEG inverse problems by using vector Fredholm integral equations of the first kind.

$$o_l^T = \int_R \vec{K}_l^T \cdot \vec{J}_P dv \tag{8}$$

The kernels $\vec{K}_l^T$, representing the LFs of type T (i.e. electric "E" and magnetic "M"), have general properties in terms of their flux and vortex sources, as well as the specific boundary conditions (see Table I). A relevant issue is that both LFs have zero laplacian inside the brain; hence, they are harmonic vector functions.

Table 1. Properties of the LFs

LF	$\nabla^2_{\vec{r}}$	$\nabla_{\vec{r}}\cdot$	$\nabla_{\vec{r}}\times$	$\vec{K}_l(\vec{r})\cdot\vec{n}\big\|_S$
Electric	**Zero**	**Zero**	**Zero**	**Not zero**
Magnetic	**Zero**	**Zero**	**Not zero**	**Zero**

The LFs are proportional to the induced ohmic current $\vec{J}$ inside the conductor after energizing a particular sensor (i.e. the reciprocity theorem is illustrated in Fig. 2).

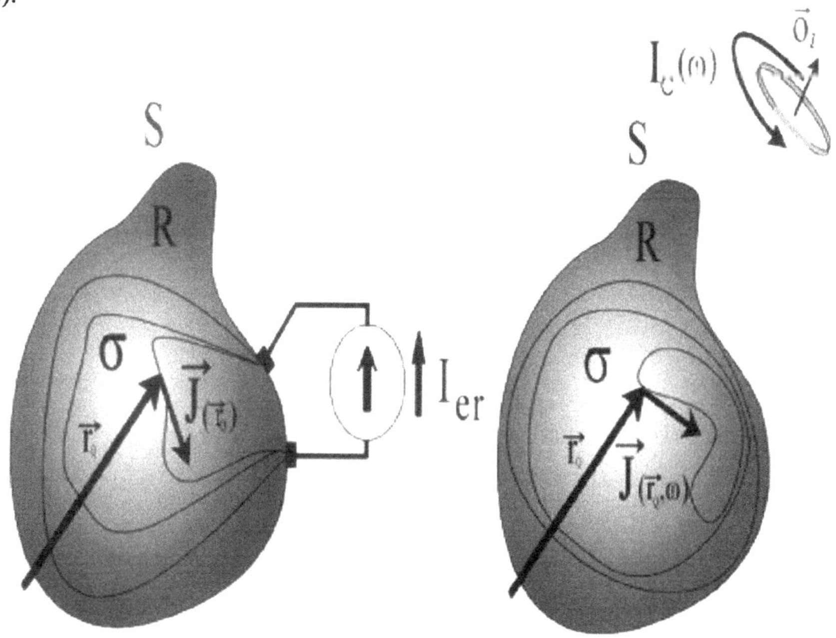

Fig.2. The reciprocity theorem

$$\vec{K}^E = -\vec{J}/I_{er}\sigma \qquad\qquad \vec{K}^M = -\vec{J}/i\omega\Delta S_c I_c\sigma$$

2.2. The vector field characterizing the brain current sources

The use of Hilbert spaces H, with convenient scalar product $\langle x,x\rangle_H$ permits the reduction of the ambiguity of $\vec{J}_p$ by considering only those belonging to H. A general class of H is formed by $\vec{J}_p$ having absolutely continuous derivatives and satisfying the condition $\nabla^2 \vec{J}_P(\vec{r}) \in \mathrm{L}_2(R)$. Within this general class of Hilbert spaces, a frequently used H (i.e. very convenient in Physics) is one endowed with a unique positive-definite second order tensor function, $\mathrm{R}: R \times R \to \Re^{3\times3}$, called the reproducing kernel. If R is continuous and also satisfies a bound condition $\iint_R (\mathrm{R}(\vec{r},\vec{r}'))^2 dv dv' < \infty$, then there is an orthogonal basis F of continuous eigenfunctions $\vec{\phi}_1, \vec{\phi}_2, \cdots$ in $\mathrm{L}_2(R)$ with eigenvalues $\lambda_1 \geq \lambda_2 \geq \cdots > 0$ (i.e. satisfying $\langle \vec{\phi}_i, \vec{\phi}_j \rangle_H = f_i^2 \delta_{ij} / \lambda_i$ and $\int_R \vec{\phi}_i \cdot \vec{\phi}_j dv = f_i^2 \delta_{ij}$) such that:

$$\int_R \mathrm{R}(\vec{r},\vec{r}') \cdot \vec{\phi}_i(\vec{r}') dv' = \lambda_i \vec{\phi}_i(\vec{r}) \tag{9}$$

The reproducing kernels can be evaluated by using the expression: $\mathrm{R}(\vec{r},\vec{r}') = \sum_i \lambda_i \vec{\phi}_i(\vec{r}) \circ \vec{\phi}_i(\vec{r}') / f_i^2$, $i = 1, 2, \cdots$. The symbol "$\circ$" represents the dyadic product of vectors. Furthermore, for any function $\vec{J}_P$ 朩 H, a Fourier series (or spectral decomposition) can be defined as $\vec{J}_P = \sum_i \alpha_i \vec{\phi}_i$, holding the following relationships $\alpha_i = \lambda_i \langle \vec{J}_P, \vec{\phi}_i \rangle_H / f_i^2$ and $\|\vec{J}_P\|_H^2 = \sum_i f_i^2 \alpha_i^2 / \lambda_i$. However, the most significant property of reproducing kernels is that for each $\vec{r} \in R$, the evaluation functional $L_{\vec{r}}$ is linear and bounded. The Riesz representation theorem guarantees that any function $\vec{J}_P$ belonging to H can be represented by:

$$\vec{J}_P(\vec{r}) = L_{\vec{r}} \vec{J}_P = \sum_l \langle \vec{R}_{\vec{r}}^l, \vec{J}_P \rangle_H \hat{e}_l \qquad l = \{x, y, z\} \tag{10}$$

Note that $\vec{R}_{\vec{r}}^l(\vec{r}') = \mathrm{R}(\vec{r},\vec{r}')\hat{e}_l$; hence, the reproducing kernel is

$$\mathrm{R}(\vec{r},\vec{r}') = \sum_{ll'} \langle \vec{R}_{\vec{r}}^l, \vec{R}_{\vec{r}'}^{l'} \rangle_H e_{ll'} \tag{11}$$

In this context, it is very convenient to characterize any linear functional (i.e. the integral operator in Eq. 8) defined into H by the use of its representer $o_l^T = \langle \vec{\psi}_l^T, \vec{J}_P \rangle_H$. The representers $\vec{\psi}_l^T$ is a well-defined element of H if, and only if, the linear functional is bounded. Therefore, from a physical point of view, the representers $\vec{\psi}_l^T$ of the LFs establish particular directions in H onto which $\vec{J}_P$ will be projected during observation. The representer for the linear functional in Eq. (8) can be obtained from the reproducing kernel by the relationship:

$$\vec{\psi}_l^T(\vec{r}) = \int_R \mathrm{R}(\vec{r},\vec{r}')\cdot\vec{K}_l^T(\vec{r}')dv' \tag{12}$$

The first term in (6) and (7) implicitly summarize both the volumetric and the surface contributions of $\vec{J}_P$ to the observations. In terms of the flux and vortex sources in the volume R, the $\vec{J}_P$ can be separated in longitudinal $\vec{J}_P^L$ and transversal $\vec{J}_P^T$ parts. The EEG is produced by $\vec{J}_P^L$ while MEG contains only contributions from $\vec{J}_P^T$. The longitudinal part has an arbitrary direction and satisfies $\nabla\times\vec{J}_P^L = 0$; therefore, it can be expressed as the gradient of a scalar potential $\vec{J}_P^L = \nabla\phi$. The transversal part normally has two components and satisfies $\nabla\cdot\vec{J}_P^T = 0$; hence, it can be expressed as a rotational of a vector potential $\vec{J}_P^T = \nabla\times\vec{A}$ (with the additional gauge $\nabla\cdot\vec{A} = 0$). In six of the eleven separable coordinates systems, it is possible to set up, by using an additional potential χ, a transversal component $\vec{J}_P^{T\perp} = \nabla\times(a_1\omega(\xi_1)\chi)$ tangentially to one of the coordinate surfaces [9]. The other transversal component $\vec{J}_P^{T\|}$ is defined such as $(\nabla\times J_P^{T\|}) \| a_l$.
In terms of the value of $\vec{J}_P$ on the surface S, it can be split into harmonic and non-harmonic components. For example, the physical implication of imposing $\vec{J}_P$ zero on S will be the exclusion of the harmonic components of $\vec{J}_P$ in R. Therefore, it is convenient to decompose the Hilbert space H in the sum of two orthogonal sub-spaces $H = H_0 \oplus H_1$ (i.e. $H_0 \perp H_1$). The sub-space H_0 represents the null space of the Laplace operator. The harmonic vector fields, with a scalar product $\langle\vec{f},\vec{g}\rangle_{H_0} = \oint \vec{f}\cdot\vec{g}ds$,

constitute H_0. Those $\vec{J}_P$ satisfying the homogeneous boundary condition $\vec{J}_P\big|_S = 0$ constitute the sub-space H_1 (i.e. non-harmonic vector fields). The use of a scalar product $\langle\vec{f},\vec{g}\rangle_{H_1} = \int_R \nabla^2\vec{f}\cdot\nabla^2\vec{g}dv$ involving derivatives will permit us to construct a reproducing kernel for this sub-space from the knowledge of its Green function:

$$4\pi R_1(\vec{r},\vec{r}') = \int_R \mathrm{G}(\vec{r},\vec{r}'')\cdot\mathrm{G}(\vec{r}',\vec{r}'')dv'' \tag{13}$$

By definition, the dyadic Green function relates to the reproducing kernel as $\mathrm{G} = -\nabla^2 R_1$.

Hence, the reproducing kernel of H is defined by the sum of the reproducing kernels of both harmonic H_0 and non-harmonic H_1 sub-spaces: $\mathrm{R}=\mathrm{R}_0+\mathrm{R}_1$. A superscript in the eigenfunctions $\vec{\phi}_i$ will be used henceforth to specify the projection of a vector-valued field in both sub-spaces (i.e. harmonic "0" and non-harmonic "1").

2.3 An example: the spherical model for both the head volume conductor and the brain region

The Hilbert space H will be specified for the particular case of spherical symmetry (i.e. R is a sphere of radius a) of the brain region. In this case, the basis F can be clearly identified for both harmonic and non-harmonic sub-spaces. The set of orthogonal eigenfunctions $\vec{\phi}_i^0(\vec{r}) = r^n \vec{P}_{nms}^k(\Omega)$ constitutes a basis for the harmonic sub-space, with:

$$f_i^0 = \sqrt{4\pi a^{2n+3}/(2n+1)(2n+3)\gamma_{nm}} \qquad \lambda_i^0 = a/2n+3 \qquad (14)$$

The subscript $i=\{k,n,m,s\}$ symbolizes, in the vector spherical harmonic (i.e. $\vec{P}_{nms}^k(\Omega_l)$), the kind, degree, order and parity, respectively. The Neumann factors ε_m are used to define quantities $\gamma_{nm} = \varepsilon_m (n-m)!/(n+m)!$.

The dyadic Green function for H_1 has been presented in a previous paper [10] as an expansion in terms of a dyadic product of the orthogonal basis: $\vec{\phi}_i^1(\vec{r}) = j_n(\kappa_p^n r/a)\vec{P}_{nms}^k(\Omega)$, with:

$$f_i^1 = \sqrt{2\pi a^3/(2n+1)\gamma_{nm}}\left|j_{n-1}(\kappa_p^n)\right| \qquad \lambda_i^1 = (a/\kappa_p^n)^4 \qquad (15)$$

The vector laplacian eigenfunctions inside the sphere constitute this basis, which is formed by the tensor product of Bessel spherical functions of the first kind $j_n(\kappa_p^n r/a)$ and $P_{nms}^k(\Omega)$. These two sub-spaces are orthogonal. The eigenvalues $\lambda_i^{0,1}$ are decreasing functions of the spherical harmonic degree, which correspond to reproducing kernels with the property of being low-pass band spatial filters. In this case, the subscript $i=\{k,p,n,m,s\}$ is extended with a new index "p" to symbolize the p-th zero κ_p^n of the spherical Bessel functions of the first kind.

By using properties of the LFs for volume conductors with spherical symmetry, the magnitudes o_l^T (for EEG average referenced and for MEG obtained using radial coils) can be represented by expansions in terms of current multipoles (i.e. multipole moments q_{nms}^T) at angular sensor position $\Omega_l = (\theta_l, \varphi_l)$.

$o_l^T = \sum_{n=1,ms} q_{nms}^T Y_{nm}^s(\Omega_l)$, with:

$$q_{nms}^E = \frac{2A_n\left[n, \sqrt{n(n+1)}\right]}{\sigma_N r_E^{n+1}(2n+1)} \left(\frac{a^{2(n+1)}}{2(n+1)} \begin{bmatrix} \alpha_{1nms}^0 \\ \alpha_{2nms}^0 \end{bmatrix} + \sum_p Q_{pn}^E \begin{bmatrix} \alpha_{1pnms}^1 \\ \alpha_{2pnms}^1 \end{bmatrix} \right)$$

$$Q_{pn}^E = \int_0^u j_n\left(\lambda_p^n \frac{r}{a}\right) r^{n+1} dr$$

$$q_{nms}^M = \frac{\mu_0 \sqrt{n(n+1)}}{r_M^{n+2}(2n+1)} \left(\frac{a^{2n+3}}{2n+3} \alpha_{3nms}^0 + \sum_p Q_{pn}^M \alpha_{3pnms}^1 \right) \qquad (16)$$

$$Q_{pn}^M = \int_0^a j_n\left(\lambda_p^n \frac{r}{a}\right) r^{n+2} dr$$

The coefficients A_n were explicitly reported in Rush and Driscoll [11]. The magnitudes r_E and r_M represent the radius of the surface containing the sensors.

1- It should be noted that the coefficients α_{3nms}^0 and a_{3pnms}^1 with $n > 0$ do not contribute to o_s^E. Hence, there is a sub-space of H that does not produce EEG, spanned by the following subset of orthogonal eigenfunctions:

$$H_1^E = \left\{ r^n \vec{P}_{nms}^3(\Omega); j_n\left(\kappa_p^n r/a\right) \vec{P}_{nms}^3(\Omega) \right\}$$
$$(n = 1, \cdots, \infty; m = 0, \cdots, n; p = 1, \cdots, \infty; s = e, o) \quad (17)$$

An additional sub-space H_2^E does not produce EEG, and is constituted by $\vec{J}_p$ with spectral coefficients related by the equations $q_{nms}^E = 0$. Let the whole-extended sub-space H^E be defined as $H^E = H_1^E \oplus H_2^E$. The symbol $\oplus$ denotes the direct sum due to the fact that these sub-spaces are orthogonal.

2- The coefficients α_{knms}^0 and α_{kpnms}^1 with $k = 1,2$ and $n > 0$ do not contribute to o_s^M. Hence, there is a sub-space of H which does not produce MEG.

This sub-space is spanned by the following subset of orthogonal eigenfunctions:

$$H_1^M = \left\{ r^n \vec{P}_{nms}^k(\Omega); j_n\left(\kappa_p^n r/a\right) \vec{P}_{nms}^k(\Omega) \right\}$$

$$\left(n = 1,\cdots,\infty; m = 0,\cdots,n; p = 1,\cdots,\infty; s = e,o; k = 1,2\right) \tag{18}$$

Similar to the electric case, there is an additional sub-space H_2^M that does not produce MEG, and is constituted by those $\vec{J}_P$ whose spectral coefficients are related to each other by the equations $q_{nms}^M = 0$. Let the whole-extended sub-space H^M be defined by $H^M = H_1^M \oplus H_2^M$. Furthermore, there are no monopolar contributions to either o_s^E or o_s^M, i.e. the spectral coefficients α_{100e}^0 and α_{1p00e}^1 do not contribute to the observations. Therefore, there is a common null sub-space $M = \left\{ \vec{0}; \hat{e}_r; j_0\left(p\pi\, r/a\right)\hat{e}_r \right\}$, with $p = 1,\cdots,\cdot$, for both LFs. Here, we write $\vec{P}_{00e}^{2,3}(\mathrm{W}) = \vec{0}$. It can be easily verified that all $\vec{J}_P$, with the exception of a central force field, belong to the sub-space $H_P = H \setminus M = H_l^E \oplus H_l^M$. The $\vec{J}_P$ in the central field force sub-space satisfy $\vec{J}_P(\vec{r}) = \left(\alpha_{100e}^0 + \sum_p \alpha_{1p00e}^1 j_0\left(p\pi\, r/a\right)\right)\hat{e}_r$. The null spaces of the electric $H^E \oplus M$ and magnetic $H^M \oplus M$ LFs are illustrated in Fig. 3. The null space of the combined electric and magnetic LFs will therefore be represented by the following sub-space $\left(H_2^E \oplus H_2^M\right) \oplus M$.

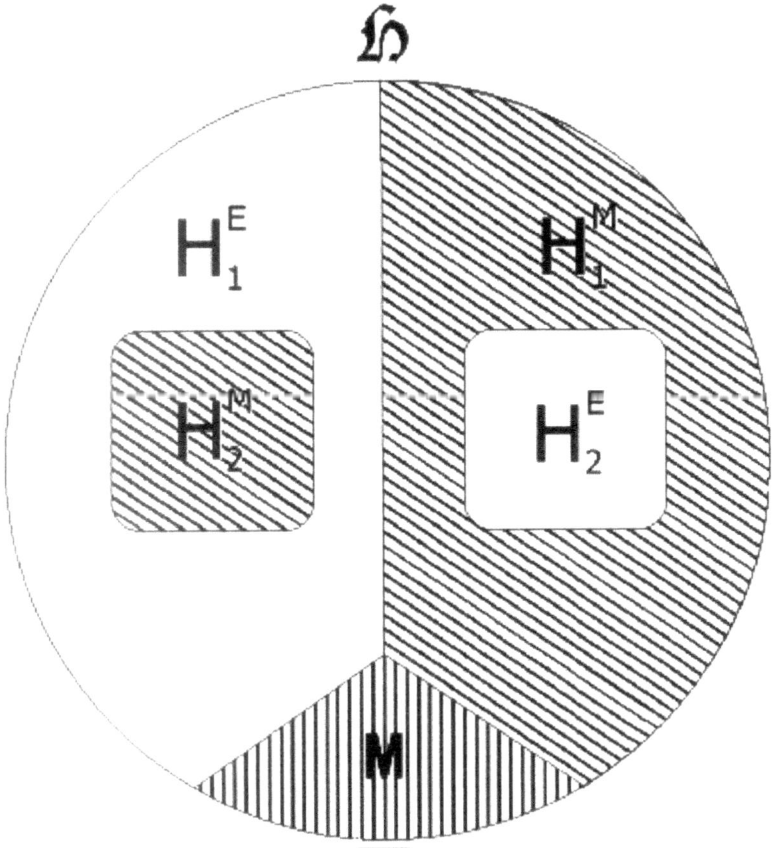

Fig.3. The representation of the null spaces of the LFs

2.4. The temporal dynamics of the brain current sources

The soma and the dendrite trees take part in the spatio-temporal integration of post-synaptic potentials in the V-layer pyramidal cells (see Fig. 4, left). According to where the main stems of these trees connect to the soma, the dendrite can be decomposed into basal and apical paths. The basal path is interpreted as a virtual compartment that summarizes a huge number of synapses arriving at multiple-bifurcated branches of the basal dendrite tree, and it is connected with an equivalent electric compartment of the soma by a small stem. It is assumed that the basal stem originates at the site where the basal dendrite tree reaches the maximal number of intersections. On the basis of morphological and electrotonic properties, the apical path is classified into three compartments: oblique, tuft and trunk. However, in this dynamic model, contributions of synaptic activity in the oblique and trunk compartments have been neglected (or they have been merged, in the equivalent electrical sense, with the synaptic activity flowing into the basal and apical tuft compartments, respectively). Therefore, I assumed that the apical tuft compartment is connected to the soma by a long isolated stem. The final configuration of two electric compartments with differentiated input effective resistances R_m^k and capacitances C_m^k ($k=1$ basal and $k=2$ apical tuft) can be interpreted as a parallel circuit (Fig. 4, right). These compartments are connected to the soma by two stems of different lengths; hence, they have distinct effective longitudinal resistance. The somatic compartment has a comparable but different effective resistance R_m and capacitance C_m.

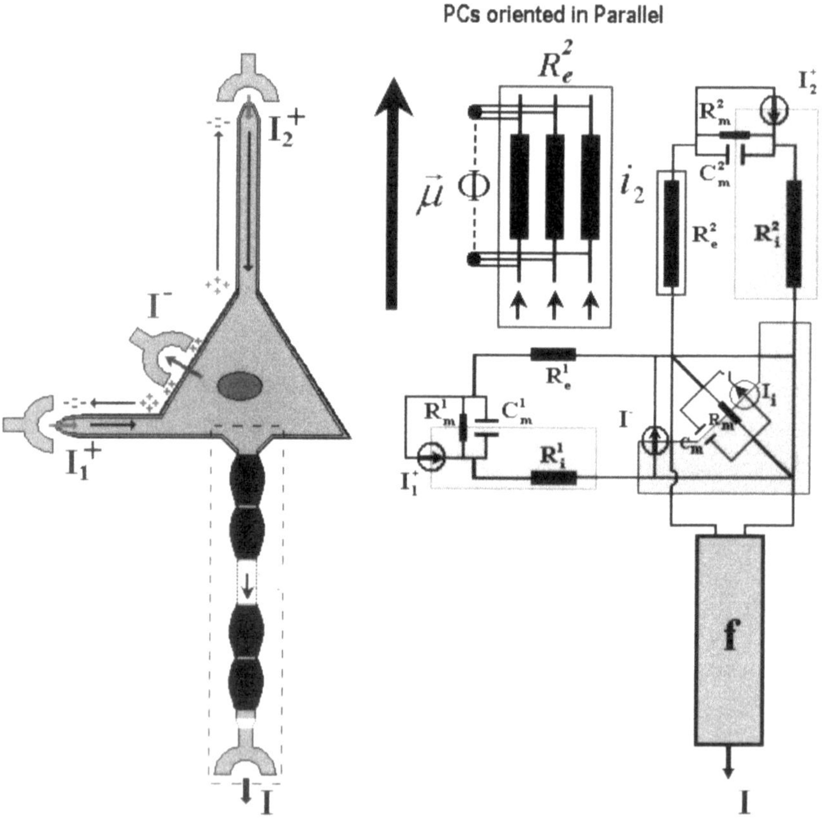

Fig.4. The electric circuit for the V-layer pyramidal cell

The dynamics equations for the membrane potentials at each compartment in this model are:

$$\tau_m \dot{V} + \left(\alpha_0 + \sum_k \frac{1}{\beta_k}\right) V + \frac{\Omega}{\prod_k \beta_k} + R_m I^- = \sum_k \left[\frac{R_m v_k}{\left(R_i^k + R_e^k\right)} - \frac{v_-}{\beta_k}\right] \tag{19a}$$

$$\tau_m \dot{\Omega} + \Omega = R_m \sum_k \frac{\beta_k \left(V - v_k\right)}{\left(R_i^k + R_n^k\right)} + V + v_- \tag{19b}$$

$$\tau_m \dot{v}_k + v_k = R_m I_k^+ \qquad k = \{1,2\} \tag{19c}$$

$$\tau_m \dot{v}_- + v_- = R_m I^- \tag{19d}$$

The coefficients $\beta_k = \left(R_i^k + R_e^k\right)/R_m^k$ play the role of equivalent voltage divisors in the circuit of each path. The coefficient $\alpha_0 = 1 + R_m \sum_k \left(R_i^k + R_e^k\right)^{-1}$ mixes the contributions coming from both basal and apical paths. The magnitude τ_m is the membrane time constant. There is a sole output I for the V-layer pyramidal cells while three inputs I_1^+, I_2^+ and I^- reaching these neurons at dissimilar layers are introduced.
Therefore, the brain current sources on the cortical surface will have fixed orientation $\vec{\mu}$, but its amplitude will vary with time proportional to the extra-cellular current $i(t)$.

$$\vec{J}_P(t) = i(t)\vec{\mu} \tag{20}$$

The normalized orientations $\vec{\mu}$ are defined by the curvature of the cortex, ensuring that they are always perpendicular to its surface. The dynamics of the extra-cellular currents $i = \Phi/R_e^2$ is represented by the equations:

$$\tau_m\dot{\Phi}+\left(\alpha_0+\sum_k\frac{1}{\beta_k}\right)\Phi+\frac{\Theta}{\prod_k\beta_k}=\frac{R_e^2}{\left(R_i^2+R_e^2\right)}\left(R_mI^-+R_m^2I_2^+\right)+\frac{\left[R_m^1\left(v_-+v_2\right)+R_m\left(v_2-v_1\right)\right]R_e^2}{\prod_k\left(R_i^k+R_e^k\right)} \tag{21a}$$

$$\tau_m\dot{\Theta}+\Theta=\left[1+R_m\sum_k\frac{1}{R_m^k}\right]\Phi \tag{21b}$$

2.5. The mechanical deformations of post-capillary venous

The BOLD signal reflects changes in the effective content of de-oxy hemoglobin within a small brain region originated due to two facts: i) neurons consume oxygen during activation and ii) post-capillary venules stretch out when the Cerebral Blood Flow (CBF) increases considerably. The BOLD effect has been accurately replicated using the balloon model, which has its foundation in the mechanically compelling representation of an expandable venule [12], and the standard Windkessel theory [13]. The balloon model, as originally proposed, includes a passive oxygen extraction effect occurring at the level of capillaries, although Zheng et al. [14] have generalized it to account for a vascular-metabolic crosstalk by introducing a dynamic relationship between neuronal activation and the oxygen extraction fraction. Friston et al. [15] put forward a complete hemodynamic approach by introducing a blood flow-inducing signal into the original balloon model relating the stimulus input u and CBF. The hemodynamic approach comprises four nonlinear and non-autonomous ordinary differential equations (14) which govern the dynamics of some "*intrinsic*" vascular variables (Fig. 5): the flow-inducing signal s, the CBF f, the blood volume v, and the effective content of de-oxy hemoglobin q. The whole dynamic system is, in effect, non-autonomous due to the time varying dependence of the stimulus input u.

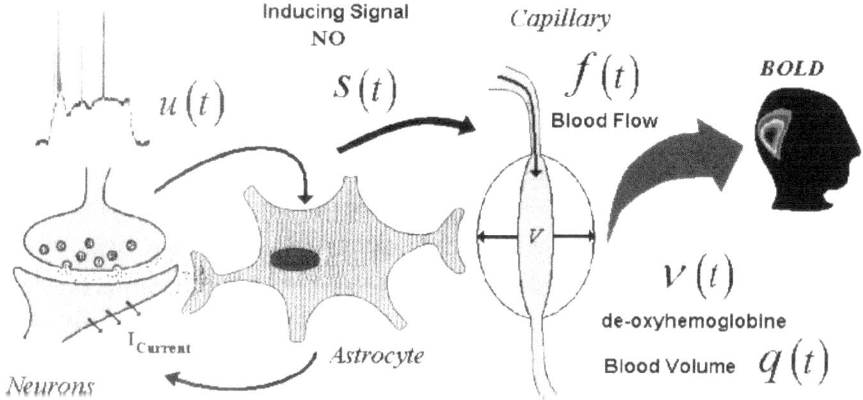

Fig.5. The schematic representation of the hemodynamic approach

$$\dot{s} = u - s/\tau_s - (f-1)/\tau_f \tag{22a}$$

$$\dot{f} = s \tag{22b}$$

$$\tau_0 \dot{v} = f - v^{1/\alpha} \tag{22c}$$

$$\tau_0 \dot{q} = \frac{f}{E_0}\left[1-(1-E_0)^{1/f}\right] - q v^{(1-\alpha)/\alpha} \tag{22d}$$

The BOLD effect is finally represented by the evaluation of vascular variables v and q in a non-linear static function.

$$BOLD = V_0\left[k_1(1-q) + k_2(1-q/v) + k_3(1-v)\right] \tag{23}$$

The constants k_1, k_2 and k_3 depend on the MRI scanner.

2.6. The electro-vascular coupling

The physiological mechanisms underlying the relationship between neuronal activation and vascular/metabolic controlling systems have been widely reported in the literature. In this work, it is hypothesized that instantaneously the stimulus input u to the hemodynamic approach is proportional to a low-pass filtered version of the total concentration of Nitric Oxide (NO) C_{NO} in a brain area containing a large number of post-capillary venules (16). Mechanisms associated with NO synthetase are in the neurons (i.e. dendrite and soma) and also in the glia cells. After transient stimulation, the energy required to reestablish the transmembrane ionic gradients in the neurons and also that used to reuptake neurotransmitters from the synaptic cleft could trigger the synthesis of NO. The transmembrane capacitive current summarized three different ionic flows that change these gradients: the purely postsynaptic currents (neurotransmitter-receptor gated ionic channels), the intrinsic electrophysiological currents (voltage gated ionic channels, e.g. sodium, potassium, calcium, etc) and the flow of ions through the membrane by means of passive channels (leakage currents). Hence, transmembrane capacitive currents also indirectly contain information about the concentration of the released neurotransmitters. We propose that the synthesis of NO can be explained by a nonlinear function of the transmembrane capacitive currents. This function must be symmetric around zero to take account of both inward and outward ionic currents and it must also include a saturation effect.

The nonlinear function $g_k(x) = \rho_k\left(1 - \exp\left(-x^2/\omega_k\right)\right)$, with positive constants ρ_k and ω_k have the above-referred properties. The subscript "*k*" is used to represent different type of neurons.

$$u(t) = \int_W \psi(t-\tau)C_{NO}(\tau)d\tau \tag{24}$$

$$C_{NO}(t) = \sum_k \chi_k g_k\left(I_C^k\right) \tag{25}$$

The magnitude $I_C^k = C_m^k A_k \dot{V}_k$ corresponds to the transmembrane capacitive current at the *k*-th type of neuron. For each type of neuron, the constants

C_m^k and A_k symbolize its specific membrane capacitance and the surface area of its whole cell membrane, respectively. The factors χ_k are introduced in order to make a distinction in the relative metabolic demand among neurons of different types. The function ψ describes a low-pass filter originated due to the diffusion phenomenon of NO within a small brain region. Also, it imitates a much required down sampling because of the slow temporal scale of the BOLD signals. In Riera et al. [16], the function ψ comprises a Kaiser class filter with windows parameter <1 and a boxcar of around 3 s in length. In this paper, we propose a low-pass filter defined by the impulse response function in the Laplace domain $\psi(s) = A\omega_0^2 / (s^2 + 2\delta\omega_0 s + \omega_0^2)$ with gain A, angular high cut frequency ω_0 and damping factor δ. Hence, the linear convolution (16) is transformed into a system of dynamic equations (18).

$$\dot{r} = -2\delta\omega_0 r - \omega_0^2 u + \omega_0^2 A C_{NO} \quad (26a)$$

$$\dot{u} = r \quad (26b)$$

3. Discussion

The post-synaptic potentials could be a sound indicator of the neuronal activity in a population of neurons. However, the characteristics of the neurons producing these post-synaptic potentials as well as the locations where they originate inside each neuron are particulars relevant when studying the neuronal activity. Therefore, the estimation of neuronal activity from EEG/MEG data is a difficult inverse problem not only because of the intrinsic ill-posedness which appears when trying to reconstruct brain currents sources solving the Maxwell equations but also due to the uncertainty that appear after that in the estimations of the dynamics of the post-synaptic potentials. For example, three inputs I_1^+, I_2^+ and I^- reach the V-layer pyramidal cell at dissimilar levels, giving possibility to an electrotonic canceling effect. Depending on the de-synchronization of the synaptic activity in these three inputs, a resulting extra-cellular ionic current will flow along the apical trunks of the V-layer pyramidal cells, i.e., originating a brain current source. Likewise, the brain vasculature acts as a low-pass filter of any temporal electrophysiological index of the neuronal activity. Hence, in spite of its excellent spatial resolution, BOLD signals must be used in combination with theoretical forward/generative models (e.g. balloon model) when trying to study the dynamics of cortical neural masses. In this sense, the combination of electrophysiological (e.g. EEG and MEG) and vascular/metabolic (fMRI) measurements will offer promising means to reconstruct the actual brain current sources. Finally, the analysis of the effective connectivity among neuronal populations must be performed taking into consideration that the nature of the connections can be of three types (i.e. excitatory at the basal/apical dendrites trees and inhibitory at the soma).

Acknowledgment

This study has been supported by a R&D promotion scheme for regional proposals promoted by TAO, the 21st Century COE Program entitled "A Strategic Research and Education Center for an Integrated Approach to Language, Brain and Cognition" (Tohoku University), as well as RISTEX and CREST from JST.

References

1. H Berger. On the electroencephalogram of man. *Journal fur Psychologie und Neurologie 40*, 160-179, 1930.
2. D Cohen. Magnetoencephalography: detection of the brain's electrical activity with a superconducting magnetometer. *Science* 175, 664-666, 1972.
3. O Yamashita, A Galka, T Ozaki, R Biscay, P Valdes-Sosa. Recursive penalized least squares solution for dynamical inverse problems of EEG generation. *Human Brain Mapping* 21, 221-235, 2004.
4. N Trujillo, E Aubert, P Valdes. Bayesian model averaging in EEG/MEG imaging. *Neuroimage* 21, 1300-1319, 2004.
5. R Grave de Peralta., MM Murray, CM Michel, R Martuzzi, SL Gonzalez. Electrical neuroimaging based on biophysical constraints. *Neuroimage* 21, 527-539, 2004.
6. S Ogawa, TM Lee, AR Kay, DW Tank. Brain magnetic resonance imaging with contrast dependent on blood oxygenation. *Proceedings of the National Academy of Sciences USA* 87, 9868-9872, 1990.

7. RI Goldman, JM Stern, JrJ Engel, MS Cohen. Acquiring simultaneous EEG and functional MRI. *Clinical Neurophysiology* 111, 11, 1974-1980, 2000.
8. The Brain Connectivity Workshop. A theme issue of the *Phil. Trans. R. Soc. B.* 360, 1457, 2005.
9. FM Morse and H Feshbach. *Methods of Theoretical Physics*. McGraw-Hill Book Company, Inc., 1953.
10. J Riera, ME Fuentes, P Valdés, Y Ohárriz. EEG Distributed inverse solutions for a spherical head model. *Inverse Problems* 14, 4, 1009-1019, 1998.
11. S Rush and DA Driscoll. EEG electrode sensitivity-an application of reciprocity. *IEEE Trans. Biomed. Eng.* 16, 15-22, 1969. see Eq. 25.
12. RB Buxton, EC Wong, LR Frank. Dynamics of blood flow and oxygenation changes during brain activation: the balloon model. *Magn. Reson. Med.* 39, 855-864, 1998.
13. JD Mandeville, JJA Marota, C Ayata, G Zaharchuk, MA Moskowitz, BR Rosen, RM Weisskoff. Evidence of cerebrovascular postarteriole windkessel with delayed compliance. *J. Cereb. Blood Flow Metab.* 19, 6, 679-689, 1999.
14. Y Zheng, J Martindale, D Johnston, M Jones, J Berwick, J Mayhew. A model of hemodynamic response and oxygen delivery to brain. *Neuroimage* 16, 617-637, 2002.
15. KJ Friston, A Mechelli, R Turner, CJ Price. Nonlinear responses in fMRI: the balloon model, volterra kernels, and other hemodynamics. *Neuroimage* 12, 466-477, 2000.
16. J Riera, E Aubert, K Iwata, R Kawashima, X Wan, T Ozaki. Fusing EEG and fMRI based on a bottom-up model: Inferring activation and effective connectivity in neural masses. *Phil. Trans. R. Soc. Lond. B.* 360, 1457, pp. 1025-1041, 2005.

Human Brain Atlases in Education, Research and Clinical Applications

Wieslaw L. Nowinski

Biomedical Imaging Lab; Agency for Science, Technology and Research

Chapter Overview. This paper overviews human brain atlases and addresses their use in neuroscience education, research, and clinical applications. It focuses on the electronic anatomical, functional, and vascular brain atlases as well as atlas-assisted applications being developed in our Biomedical Imaging Lab for over more than one decade.

Key Words. Brain atlas, functional neurosurgery, human brain mapping, stroke, neuroradiology, neuroscience education.

1. Introduction

For millennia people have been using atlases to navigate the globe. Navigating the human brain, the most complex living organ containing more than 100 billion neurons with one million billion interconnections between them, can also be facilitated by means of the atlas. Then, the atlas must capture not only the complexity of the brain but also its variability.

Drawings of the human brain have been in use for half a millennium to capture and communicate anatomical knowledge. The life span of anatomical print brain atlases is of an order shorter. Introduced in the 1950s they have found applications in stereotactic and functional neurosurgery as well as neuroscience education. With time, they also are being employed in human brain mapping and the interest in them is growing in the neuroradiological community.

Modern diagnostic imaging offers unprecedented views into human body's structure, function, metabolism, hemodynamics, and pathology. Our ability to generate images is much greater than that of understanding and interpreting them. Deformable body models with warping techniques facilitate analysis of medical images. Deformable brain atlases, particularly, are useful for analysis and interpretation of brain scans.

This work provides an overview of brain atlases in education, research and clinics, with the emphasis on the anatomical, functional, and vascular atlases and atlas-assisted applications developed in our lab.

2. Human Brain Atlases

Human brain atlases can be classified from various view points including: medium (print, electronic), type of source material (e.g., cryosections, radiologic images, Visible Human Data), population of source material (low (deterministic atlas), high (probabilistic atlas)), and content (anatomy, function, vasculature). Similarly, an atlas-based application can be considered in terms of: field (education, research, clinical), functionality (atlas-specific, problem specific), cost (e.g., a low cost CD versus a high end virtual reality solution), accessibility (web-based, stand-alone, plug-in library).

2.1 Print atlases

Numerous print brain atlases are published including [7,9,11,24,30,31,34]. In addition, many stereotactic brain atlases have been constructed since the 1950s, such as [2,3,63,65,66,72,73,74,75,80].

2.2 Electronic stereotactic atlases

Deformable electronic atlases overcome some shortcomings of the print atlases and open new avenues. In addition to atlas warping, they offer new features not available in print atlases, such as interactive labeling of scans, flexible ways of presentation in 2D and 3D (and generally in nD space),

mensuration, searching, integration of knowledge from multiple sources, and aggregation of information from numerous cases.
To combine the widely accepted stereotactic print atlases with the capabilities offered by electronic atlases, several print atlases have been converted into electronic form, including: Schaltenbrand-Bailey [16,22,84]; Schaltenbrand-Wahren [21,33,47,70]; Talairach-Tournoux [16,47]; Referentially Oriented Talairach-Tournoux [47]; Ono et al [48]; Afshar et al [32]; and Van Buren-Borke [16]. Computerized versions of print atlases may vary substantially from a simple, direct digitization of the original material to a sophisticated, fully segmented, labeled, enhanced, and 3D extended deformable atlas.

2.3 Other electronic brain atlases

Many other types of brain atlases have been developed. They include MRI-based atlases [23,26,68]; cryosection-based atlases [6,8,15,78]; multimodal Visible Human-derived atlases [18,67]; brain animations [71]; probabilistic anatomical atlases [10,29,77,79]; surface-based atlases [81]; surface-based probabilistic atlases [82]; and probabilistic functional atlases constructed from neuroimaging and electrophysiological data [46]. A substantial effort in building digital brain atlases has been made by the International Consortium for Brain Mapping [28].

2.4 Cerefy electronic brain atlas database

The Cerefy electronic brain atlas database [38, 48, 59] contains complementary atlases with gross anatomy, subcortical structures, brain connections, and sulcal patterns. It was derived from the classic print brain atlases edited by Thieme:

1) Atlas for Stereotaxy of the Human Brain by Schaltenbrand and Wahren [66];
2) Co-Planar Stereotactic Atlas of the Human Brain by Talairach and Tournoux [74];
3) Atlas of the Cerebral Sulci by Ono, Kubik, and Abernathey [63];
4) Referentially Oriented Cerebral MRI Anatomy: Atlas of Stereotaxic Anatomical Correlations for Gray and White Matter by Talairach and Tournoux [75].

These atlases, spatially co-registered, and available in 2D and 3D. The anatomical index has about 1000 structures per hemisphere and more than 400 sulcal patterns. The Cerefy brain atlas database is suitable for neurosurgery [37,46,57,58]; neuroradiology [36,59]; human brain mapping [35,55]; and neuroscience education [53]. Figure 1 shows some of the Cerefy brain atlases on CD-ROMs.

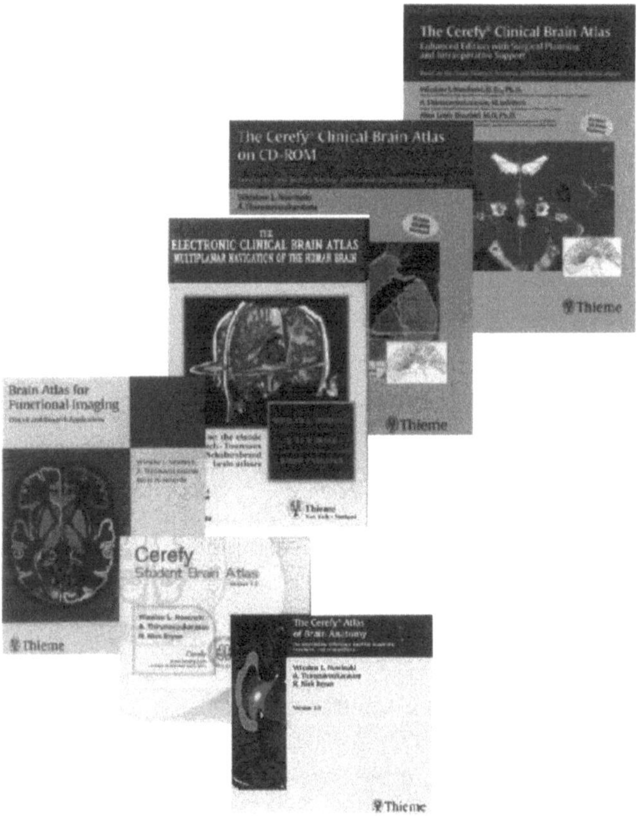

Fig.1. Cerefy brain atlases on CD-ROMs.

3. Brain Atlases in Neuroscience Education

Electronic human brain atlases are widely accepted in neuroscience education because of their high interactivity, visualization and manipulation of 3D models, multi-media nature, and friendly user interfaces. Pioneering applications include ADAM [1], BrainStorm [8], The Digital Anatomist [71], Voxel-Man [17]; several others are reviewed in [48]. Most of the existing atlases are teaching programs. They are typically HyperCard type systems with planar images. The Digital Anatomist, as opposed to the 2D systems, contains 3D brain animations. Most of atlases use predefined images and animations, and offer limited visualization capabilities. Those limitations are overcome by constructing 3D brain atlases based on radiological images or Visible Human Data, such as VOXEL-MAN, Anatomy Browser [14], and others listed above. The limitations of these 3D atlases include lack of registration capabilities and a coarse anatomical parcellation of cerebral structures.

The Cerefy Atlas of Brain Anatomy (CABA) [53] is a user-friendly application for medical students, residents, and teachers. It contains MRI and atlas images of gross anatomy as well as related textual materials. It also provides testing and scoring capabilities for exam preparation. Its novelty includes:

- atlas-assisted localization of cerebral structures on radiological images;
- atlas-assisted interactive labeling simultaneously on axial, coronal, and sagittal planes;
- automatic atlas-assisted testing against location and name of cerebral structures along with scoring;
- saving of the labeled images suitable for preparing teaching materials.

Our recent development is a cerebrovascular atlas combined with a 3D anatomical atlas [39] as well as a Chinese version of the CABA [53].

4. Brain Atlases in Research

The usefulness of electronic brain atlases in medical research is growing, particularly in medical image analysis, human brain mapping, and disease study.

4.1 Medical image analysis

The gold standard software package in medical image analysis is ANALYZE from Mayo [64]: a powerful, comprehensive visualization tool for multi-dimensional display, processing, and analysis of biomedical images from multiple imaging modalities. To facilitate analysis of brain images, the Cerefy brain atlas has been integrated with ANALYZE.

4.2 Human brain mapping

Functional neuroimaging (fMRI, PET, MEG) is used extensively to identify brain regions associated with motor, sensory, visual, auditory, and cognitive tasks. Functional brain mapping has tremendous potential as a tool for basic neuroscientific investigation (but also for the diagnosis of stroke and tumors as well as presurgical planning).

There are numerous software packages for functional image generation [13], [55]. None of them, however, provides atlas-assisted analysis of functional images. BrainMap [12] and the Brain Atlas for Functional Imaging [55] allow for labeling of functional images by means of the Talairach-Tournoux brain atlas (of gross anatomy). BrainMap, a web-based application, does not provide direct access to the atlas. To label the loci located beyond the atlas structures, BrainMap uses the Talairach Daemon [25] that looks for the cortical area closest to the activation locus. The Brain Atlas for Functional Imaging (BAFI) provides direct access to the atlas and the user can display the activation regions superimposed on the atlas, and subsequently place and edit the marks corresponding to the activation regions. Besides labeling, the BAFI supports localization analysis of functional images [35]. To our best knowledge, besides the BAFI, none of the existing software packages for functional image generation contains the Talairach-Tournoux atlas, which is the gold standard atlas in human brain mapping research. The BAFI also provides numerous functions such as fast data normalization, readout of Talairach coordinates, and data-atlas display. It has also several unique features including interactive warping facilitating fine tuning of the data-to-atlas fit, multiple labeling, navigation on the triplanar formed by the data and the atlas, multiple-images-in-one display with atlas-anatomy-function blending, loci editing, fast locus-controlled generation of results [62], and reading and saving of the loci list.

4.3 Disease study

Probabilistic brain atlases constructed from large populations capture brain variability and change over time. Their taxonomy was proposed in [44]. These atlases facilitate studying various processes in the healthy brain (such as development or aging) as well as in disease, for instance, schizophrenia and Alzheimer's disease [76].

5. Brain Atlases in Clinical Applications

Brain atlases are prevalent in functional neurosurgery and the interest in them is growing in neuroradiology for computer-aided diagnosis.

5.1 Computer-aided diagnosis

Two trends are currently observed in radiology: 1) the number of scans to be interpreted is growing about three times faster than the number of radiologists, and 2) about 65% of the world is radiologically void while another 30% has only basic technologies. Therefore, new approaches for facilitating and speeding up interpretation of scans as well as reducing the learning curve of radiologists are needed. In our opinion, model-enhanced radiology may be one of solutions. The advantages of atlas use in neuroradiology were addressed in [59]; namely, the atlas can potentially:

- reduce time in image interpretation by providing interactive labeling, triplanar display, higher parcellation than the scan itself, multi-modal fusion and displaying the underlying anatomy for functional and molecular images;
- facilitate communication about interpreted scans from the neuroradiologist to other clinicians, such as neurosurgeon, neurologist, or general practitioner, and medical students;
- increase the neuroradiologist's confidence in terms of anatomy and spatial relationships by providing interactive labelling of the scan on the orthogonal planes and triplanar display;
- reduce the cost by providing some information from mutually co-registered atlases which otherwise has to be obtained from other modalities acquired;
- reduce the time in learning neuroanatomy and scan interpretation by providing 3D and triplanar displays and labeling of multi-modal scans.

To our best knowledge the only model-enhanced applications for neuroradiology is our web-enabled Cerefy Neuroradiology Atlas (CNA) [42] with more than 1,100 users registered (available from www.cerefy.com). The CNA assists in speeding up scan interpretation by rapid labeling of morphological and/or functional scans, displaying underlying anatomy for functional studies, and facilitating multi-modal fusion. The labeled and annotated (with text and graphics) scan can be saved in Dicom and/or XML formats, giving the neuroradiologist possibility to store the atlas-enhanced scan in a PACS and to use it in web-enabled applications. In this way, the scan interpretation done by the neuroradiologist can easily be communicated to other clinicians and medical students.

While the CNA employs interactive atlas-to-data registration, our recent development [49] provides fully automatic warping of the atlas into morphological images in about 5 seconds. For atlas warping, we use the Talairach transformation [74] which scales the Cerefy atlas [56] piecewise linearly based on point landmarks whose fast, robust and accurate identification in the data is the core of our approach. We use the modified Talairach landmarks [61]: anterior commissure (AC) and posterior commissure (PC) located on the midsagittal plane (MSP), and 6 cortical landmarks determining the extent of brain. Three component algorithms calculate these landmarks automatically based on anatomy and radiologic properties of the scan. The identification of the MSP is detailed in [19], that of AC and PC in [5], and calculation of cortical landmarks in [20] . We have extended this concept for atlas to data warping in the presence of a brain tumor causing a mass effect. An automatic algorithm segments the tumor and warps the atlas non-linearly in 10-15 seconds [45].

We have also applied the atlas for interpretation of stoke images. Magnetic resonance (MR) diffusion and perfusion are key modalities for interpretation of stroke images. They provide information about the infarcted region and that at risk with high sensitivity and specificity. Their major drawbacks are the lack of underlying anatomy and blood supply territories as well as low resolution (for instance, a typical slice thickness of diffusion images is 5 mm while that of perfusion is 7.5 mm). We have developed a fast algorithm for overlapping the anatomical atlas as well as an atlas of blood supply territories on MR perfusion and diffusion images [50]. In addition, our solution allows for a simultaneous display of the atlas, diffusion image and a selected perfusion map (out of CBF, CBV, MTT, PKHT, TTP) with the user-controlled levels of transparency.

The key limitation of PET imaging is the lack of underlying anatomy. A combined PET-CT overcomes this shortcoming, making it one of the fastest growing modality. However, a PET-CT scanner is expensive. We have developed a more affordable solution by getting the underlying anatomy from the anatomical atlas warped non-linearly onto a PET scan [50].

5.2 Stereotactic and functional neurosurgery

From the 1950s, it took about two decades to have first brain atlases in electronic form available in clinical setting [4]. After another two decades the electronic brain atlases have become commonplace in stereotactic and functional neurosurgery [16,46,57]. The Cerefy brain atlas database

[38,48,59] has become the standard in stereotactic and functional neurosurgery. It has already been integrated with major image guided surgery systems including the StealthStation (Medtronic/Sofamor-Danek), Target (BrainLab), SurgiPlan (Elekta), SNN 3 Image Guided Surgery System (Surgical Navigation Network), a neurosurgical robot NeuroMate (Integrated Surgical Systems/IMMI), and Z-kat products.

One of the simplest atlas-assisted approaches is to use The Electronic Clinical Brain Atlas on CD-ROM [47]. It allows an individualized atlas to be generated without loading the patient-specific data which ability is useful for anatomical targeting. The planning procedure for stereotactic and functional neurosurgery by using this CD-ROM [58] employs a 2D local deformation done by matching the atlas rectangular region of interest to the corresponding data region of interest.

The NeuroPlanner is another application developed by us, useful for preoperative planning and training, intraoperative procedures, and postoperative follow-up [57]. It comprises all, mutually co-registered atlases from the Cerefy brain atlas database including their 3D extensions [48]. The NeuroPlanner provides four groups of functions: data-related (data interpolation, reformatting, image processing); atlas-related (atlas-to-data interactive 3D warping, 2D and 3D interactive multiple labeling); atlas-data exploration-related (interaction in three orthogonal and one 3D views, continuous data-atlas exploration); and neurosurgery-related (targeting, path planning, mensuration, simulating the insertion of a microelectrode, simulating therapeutic lesioning).

The BrainBench [69] is a virtual reality-based surgical planning system containing a suite of neurosurgery supporting tools and the Cerefy atlas. It facilitates preparing faster plans; have a more accurate anatomical targeting; improve the avoidance of critical structures; have fewer sub-optimal frame attachments and faster, more effective planning and training.

The advantages of using the Cerefy brain atlases for stereotactic and functional neurosurgery are summarized in [38].

The limitations of the Cerefy anatomical atlases include sparse data derived from a few specimens only and inconsistency in three dimensions. More importantly, these atlases are anatomical while the stereotactic target structures are functional. To overcome these limitations, we have developed the probabilistic functional atlas (PFA) generated from pre-, intra-, and post-operative data collected during the surgical treatment of Parkinson's disease patients [44]. So far, we have constructed PFA for the subthalamic nucleus [40] and ventrointermediate nucleus of the thalamus [41].

The PFA and the Schaltenbrand-Wahren atlas were registered spatially [51] and a dedicated application is developed for a combined anatomical-functional planning of functional neurosurgery [52]. Moreover, this application provides intraoperative support and serves as a personal archive.
The construction of PFA and development of PFA-based applications result in two conceptual breakthroughs. Firstly, the PFA is dynamic and can be updated with new cases, allowing knowledge from the currently operated cases to be accumulated. Secondly, the atlas can be built over the internet by the community by using a public-domain portal [43], [83] changing in this way the paradigm from the manufacturer-centric to community-centric.

Acknowledgment

The key contributors to the atlases and applications described here include A Thirunavuukarasuu, D Belov, A Fang, A Ananthasubramaniam, G Qian, GL Yang, QM Hu, KN Bhanu Prakash, I Volkau, and L. Serra, among many others.
This research is funded by the Biomedical Research Council; Agency for Science, Technology and Research; Singapore.

References

1. A.D.A.M. A.D.A.M Animated Dissection of Anatomy for Medicine. User's Guide, A.D.A.M.
2. Afshar E, Watkins ES, Yap JC: Stereotactic Atlas of the Human Brainstem and Cerebellar Nuclei. Raven Press, New York, 1978.
3. Andrew J, Watkins ES: A Stereotaxic Atlas of the Human Thalamus and Adjacent Structures. A Variability Study. Williams and Wilkins, Baltimore, 1969.
4. Bertrand G, Olivier A, Thompson CJ: Computer display of stereotaxic brain maps and probe tracts. Acta Neurochir Suppl 1974:21;235-243.
5. Bhanu Prakash KN, Hu Q, Volkau I, Aziz A, Nowinski WL: Rapid and automatic localization of the anterior and posterior commissure point landmarks in MR volumetric neuroimages. Academic Radiology (in press).
6. Bohm C, Greitz T, Eriksson L: A computerized adjustable brain atlas. European Journal of Nuclear Medicine, 1983:15;687-689.

7. DeArmond S. J., Fusco MM, Dewey MM: Structure of the Human Brain. A Photographics Atlas (3rd ed). Oxford University Press, New York, 1989.
8. Dev P, Coppa GP, Tancred E: BrainStorm: desiging in interactive neuroanatomy atlas. Radiology, 1992:185(P); 413.
9. Duvernoy HM: The Human Hippocampus: an Atlas of Applied Anatomy. Bergman, Munch, 1988.
10. Evans AC, Collins DL, Milner B: An MRI-based stereotactic brain atlas from 300 young normal subjects. Proceedings of the 22nd Symposium of the Society for Neuroscience, Anaheim, 1992; 408.
11. Fix JD: Atlas of the Human Brain and Spinal Cord. Aspen, Rockville 1987.
12. Fox PT, Mikiten S, Davis G, Lancaster JL: BrainMap: a database of human functional brain mapping. In: Functional Neuroimaging: Technical Foundations, Thatcher RW, Hallett M, Zeffiro T, John ER, Huerta M (eds.), 1994;95-106.
13. Gold S, Christian B, Arndt S, Zeien G, Cizadlo T, Johnson DL, Flaum M, Andreasen NC: Functional MRI statistical software packages: a comparative analysis. Human Brain Mapping, 1998:6; 73-84.
14. Golland P, Kikinis R, Halle M, Umans C, Grimson WE, Shenton ME, Richolt JA: Anatomy Browser: a novel approach to visualization and integration of medical information. Comput Aided Surg 1999:4; 129-43.
15. Greitz T, Bohm C, Holte S et al: A computerized brain atlas: construction, anatomical content, and some applications. Journal of Computer Assisted Tomography, 1991, 15(1); 26-38.
16. Hardy TL, Deming LR, Harris-Collazo R: Computerized stereotactic atlases. In: Advanced Neurosurgical Navigation (eds. Alexander III E and Maciunas RJ), Thieme, New York, 1999;115-124.
17. Hoehne KH. VOXEL-MAN 3-D Navigator: Brain and Skull. 2nd edition, Springer Verlag, Heidelberg 1997.
18. Hohne KH, Pflesser B, Pommert A et al: A realistic model of human structure from the Visible Human data, Method Inform Med:2001;83-9.
19. Hu Q, Nowinski WL: A rapid algorithm for robust and automatic extraction of the midsagittal plane of the human cerebrum from neuroimages based on local symmetry and outlier removal. NeuroImage 2003;20(4):2154-2166.
20. Hu Q, Qian G, Nowinski WL: Fast, accurate and automatic extraction of the modified Talairach cortical landmarks from MR images. Magnetic Resonance in Medicine, 2005; 53:970-976.
21. Kall BA. Computer-assisted stereotactic functional neurosurgery. In: Computers in Stereotactic Neurosurgery (eds. Kelly PJ, Kall BA), Blackwell, Boston, 1992; 134-142.
22. Kazarnovskaya MI, Borodkin SM, Shabalov VA, Krivosheina VY, Golanov AV: 3-D computer model of subcortical structures of human brain. Comput Biol Med. 1991:21(6);451-7.
23. Kikinis R, Shenton ME, Iosifescu DV et al : A digital brain atlas for surgical planning, model-driven segmentation, and teaching. IEEE Trans. on Visualization and Comp. Graphics, 1996:2(3);232-241.

24. Kraus G E ,Bailey G J: Microsurgical Anatomy of the Brain. A Stereo Atlas. Wiliams & Wilkins, Baltimore, 1994.
25. Lancaster JL, Woldorff MG, Parsons LM, Liotti M, Freitas CS, Rainey L, Kochunov PV, Nickerson D, Mikiten SA, Fox PT: Automated Talairach atlas labels for functional brain mapping. Hum Brain Mapp, 2000:10(3); 120-31.
26. Lehmann ED, Hawkes D, Hill D et al: Computer-aided interpretation of SPECT images of the brain using an MRI-derived neuroanatomic atlas. Med. Informatics 1991:16; 151-166.
27. Mapping cortical change in Alzheimer's disease, brain development, and schizophrenia, Neuroimage. 2004;23 Suppl 1:S2-18.
28. Mazziotta J: The International Consortium for Brain Mapping: a probabilistic atlas and reference system for the human brain. In: Brain Mapping. The Methods. Second edition (eds. Toga AW, Mazziotta JC). Academic Press, Amsterdam, 2002:727-755.
29. Mazziotta JC, Toga AW, Evans AC et al: A probabilistic atlas of the human brain: theory and rationale for its development. NeuroImage 1995:2;89-101.
30. McMinn R M H, Hutchings RT, Pegington J, Abrahams P: Color Atlas of Human Anatomy (3rd ed), Mosby Year Book, St. Louis, 1993.
31. Netter F H: The CIBA Collection of Medical Illustrations. Vol. 1 Nervous System. CIBA, 1991.
32. Niemann K, van den Boom R, Haeselbarth K, Afshar F: A brainstem stereotactic atlas in a three-dimensional magnetic resonance imaging navigation system: first experiences with atlas-to-patient registration. J Neurosurg; 1999:90(5);891-901.
33. Niemann K, van Nieuwenhofen I: One atlas - three anatomies: relationships of the Schaltenbrand and Wahren microscopic data. Acta Neurochir (Wien);1999:141(10);1025-38.
34. Nieuwennhuys R, Voogd J, van Huijzen C: The Human Central Nervous System. A Synopsis and Atlas (2nd ed). Springer-Verlag, Berlin 1981.
35. Nowinski WL, Thirunavuukarasuu A: Atlas-assisted localization analysis of functional images. Medical Image Analysis, Sep. 2001; 5(3):207-220.
36. Nowinski WL: Thirunavuukarasuu A: Electronic atlases show value in brain studies. Diagnostic Imaging Asia Pacific 2001:8(2); 35-39.
37. Nowinski WL: Anatomical targeting in functional neurosurgery by the simultaneous use of multiple Schaltenbrand-Wahren brain atlas microseries. Stereotactic and Functional Neurosurgery, 1998:71;103-116.
38. Nowinski WL: Computerized brain atlases for surgery of movement disorders. Seminars in Neurosurgery 2001:12(2); 183-194.
39. Nowinski WL, A Thirunavuukarasuu, Volkau I, Baimuratov R, Hu Q, Aziz A, Huang S: Three-dimensional brain atlas of anatomy and vasculature. Radiographics, 2005;25(1):263-271.
40. Nowinski WL, Belov D, Pollak P, Benabid AL: A probabilistic functional atlas of the human subthalamic nucleus. Neuroinformatics 2004;2(4):381-398.
41. Nowinski WL, Belov D, Thirunavuukarasuu A, Benabid AL: A probabilistic functional atlas of the VIM nucleus. Proc. American Association of Neuro-

logical Surgeons AANS 2005, 16-21 April 2005, New Orleans, Louisiana, USA.
42. Nowinski WL, Belov D: The Cerefy Neuroradiology Atlas: a Talairach-Tournoux atlas-based tool for analysis of neuroimages. Neuroimage, 2003;20(1):50-57; available from www.cerefy.com .
43. Nowinski WL, Belov, Benabid AL: A community-centric Internet portal for stereotactic and functional neurosurgery with a probabilistic functional atlas. Stereotactic and Functional Neurosurgery 2002;79:1-12.
44. Nowinski WL, Belov, Benabid AL: An algorithm for rapid calculation of a probabilistic functional atlas of subcortical structures from electrophysiological data collected during functional neurosurgery procedures. NeuroImage 2003;18(1):143-155.
45. Nowinski WL, Belov D: Towards atlas-assisted automatic interpretation of MRI morphological brain scans in the presence of tumor. Academic Radiology 2005;12:1049–1057.
46. Nowinski WL, Benabid AL: New directions in atlas-assisted stereotactic functional neurosurgery. In: Advanced Techniques in Image-Guided Brain and Spine Surgery (ed Germano IM), Thieme, New York, 2001.
47. Nowinski WL, Bryan RN, Raghavan R: The Electronic Clinical Brain Atlas. Multiplanar Navigation of the Human Brain. Thieme, New York – Stuttgart, 1997.
48. Nowinski WL, Fang A, Nguyen BT, Raphel JK, Jagannathan L, Raghavan R, Bryan RN, Miller G: Multiple brain atlas database and atlas-based neuroimaging system. Computer Aided Surgery 1997:2(1);42-66.
49. Nowinski WL, Hu Q, Bhanu Prakash KN, Qian G, Thirunavuukarasuu A, Aziz A: Automatic interpretation of normal brain scans. The Radiological Society of North America, 90th Scientific Assembly & Annual Meeting Program; Chicago, Illinois, USA, 28 November - 3 December 2004:710.
50. Nowinski WL, Hu Q, Bhanu Prakash, Volkau I, Qian G, Thirunavuukarasuu A, Liu J, Aziz A, Baimouratov R, Hou Z, Huang S, Luo S, Minoshima S, Runge V, Beauchamp N: Atlas-assisted analysis of brain scans. Book of Abstracts of European Congress of Radiology ECR 2005, European Radiology, Supplement 1 to Volume 15, Feb 2005:572.
51. Nowinski WL: Co-registration of the Schaltenbrand-Wahren microseries with the probabilistic functional atlas. Stereotactic Functional Neurosurgery 2004; 82:142-146.
52. Nowinski WL, Thirunavuukarasuu A, Benabid AL: The Cerefy Clinical Brain Atlas. Extended Edition with Surgery Planning and Intraoperative Support. Thieme, New York, 2005.
53. Nowinski WL, Thirunavuukarasuu A, Fu Y, Ma X, Lin Z, Wang S: The Cerefy Clinical Brain Atlas. Chinese Edition. Harbin Institute of Technology, China, 2005.
54. Nowinski WL, Thirunavuukarasuu A, Bryan RN: The Cerefy Atlas of Brain Anatomy. An Interactive Reference Tool for Students, Teachers, and Researchers. Thieme, New York – Stuttgart, 2002.

55. Nowinski WL, Thirunavuukarasuu A, Kennedy DN: Brain Atlas for Functional Imaging. Clinical and Research Applications. Thieme, New York – Stuttgart, 2000.
56. Nowinski WL, Thirunavuukarasuu A: The Cerefy Clinical Brain Atlas on CD-ROM. Thieme, New York – Stuttgart, 2004.
57. Nowinski WL, Yang GL, Yeo TT: Computer-aided stereotactic functional neurosurgery enhanced by the use of the multiple brain atlas database. IEEE Transactions on Medical Imaging 2000:19 (1);62-69.
58. Nowinski WL, Yeo TT, Thirunavuukarasuu A: Microelectrode-guided functional neurosurgery assisted by Electronic Clinical Brain Atlas CD-ROM. Computer Aided Surgery 1998:3(3);115-122.
59. Nowinski WL: Electronic brain atlases: features and applications. In: 3D Image Processing: Techniques and Clinical Applications (eds. Caramella D., Bartolozzi C.). Medical Radiology series, Springer-Verlag, 2002.
60. Nowinski WL: Model enhanced neuroimaging: clinical, research, and educational applications. In: Yearbook of Medical Informatics 2002:132-144.
61. Nowinski WL: Modified Talairach landmarks. Acta Neurochirurgica, 2001; 143(10); 1045-1057.
62. Nowinski WL, Thirunavuukarasuu A: A locus-driven mechanism for rapid and automated atlas-assisted analysis of functional images by using the Brain Atlas for Functional Imaging. Neurosurg Focus Jul. 2003; 15(1): Article 3.
63. Ono M, Kubik S, Abernathey CD: Atlas of the Cerebral Sulci. Georg Thieme Verlag/Thieme Medical Publishers, Stuttgart - New York, 1990.
64. Robb RA, Hanson DP: The ANALYZE software system for visualization and analysis in surgery simulation. In: Lavallee S, Tayor R, Burdea G, Mosges R (eds) Computer Integrated Surgery. Cambridge MA, MIT Press:1995;175-190.
65. Schaltenbrand G, Bailey: Introduction to Stereotaxis with an Atlas of the Human Brain. Georg Thieme Verlag, Stuttgart, 1959.
66. Schaltenbrand G, Wahren W: Atlas for Stereotaxy of the Human Brain. Georg Thieme Verlag, Stuttgart, 1977.
67. Schiemann T, Freudenberg J, Pflesser B et al: Exploring the Visible Human using the VOXEL-MAN framework. Comput Med Imaging Graph. 2000:24(3);127-32.
68. Schmahmann JD, Doyon J, McDonald D et al: Three-dimensional MRI atlas of the human cerebellum in proportional stereotaxic space. Neuroimage 1999:10(3 Pt 1);233-60.
69. Serra L, Nowinski WL, Poston T et al : The Brain Bench: virtual tools for stereotactic frame neurosurgery. Medical Image Analysis 1997:1(4);317-329.
70. Sramka M, Ruzicky E, Novotny M: Computerized brain atlas in functional neurosurgery. Stereotact Funct Neurosurg;1997:69(1-4 Pt 2);93-8.
71. undsten J W, Brinkley J F, Eno K, Prothero J: The Digital Anatomist. Interactive Brain Atlas. CD ROM for the Macintosh, University of Washington, Seattle, 1994.
72. Szikla G, Bouvier G, Hori T, Petrov V: Angiography of the Human Brain Cortex: Atlas of Vascular Patterns and Stereotactic Localization. Springer-Verlag, Berlin 1977.

73. Talairach J, David M, Tournoux P, Corredor H, Kvasina T: Atlas d'Anatomie Stereotaxique des Noyaux Gris Centraux. Masson, Paris, 1957.
74. Talairach J, Tournoux P: Co-Planar Stereotactic Atlas of the Human Brain. Georg Thieme Verlag/Thieme Medical Publishers, Stuttgart - New York, 1988.
75. Talairach J, Tournoux P: Referentially Oriented Cerebral MRI Anatomy. Atlas of Stereotaxic Anatomical Correlations for Gray and White Matter. Georg Thieme Verlag/Thieme Medical Publishers, Stuttgart - New York, 1993.
76. Thompson PM, Hayashi KM, Sowell ER, Gogtay N, Giedd JN, Rapoport JL, de Zubicaray GI, Janke AL, Rose SE, Semple J, Doddrell DM, Wang Y, van Erp TG, Cannon TD, Toga AW: Mapping cortical change in Alzheimer's disease, brain development, and schizophrenia. Neuroimage 2004:23 Suppl 1:S2-18.
77. Thompson PM, Woods RP, Mega MS et al: Mathematical/computational challenges in creating deformable and probabilistic atlases of the human brain. Hum Brain Mapp. 2000:9(2);81-92.
78. Toga AW, Ambach KL, Schluender S: High-resolution anatomy from in situ human brain. Neuroimage 1994:1(4);334-44.
79. Toga AW, Thompson PM: Maps of the brain. Anat Rec 2001:265(2);37-53.
80. Van Buren JM, Borke RC: Variations and Connections of the Human Thalamus. Springer-Verlag, New York, 1972.
81. Van Essen DC, Drury HA: Structural and functional analyses of human cerebral cortex using a surface-based atlas. J Neurosci. 1997:17(18);7079-102.
82. Van Essen DC, Lewis JW, Drury HA, Hadjikhani N, Tootell RB, Bakircioglu M, Miller MI: Mapping visual cortex in monkeys and humans using surface-based atlases. Vision Res 2001:41(10-11);1359-78.
83. www.cerefy.com
84. Yoshida M: Three-dimensional maps by interpolation from the Schaltenbrand and Bailey atlas. In: Computers in Stereotactic Neurosurgery (eds Kelly PJ and Kall BA), Blackwell, Boston, 1992, 143-152.

Deploying Chinese Visible Human Data on Anatomical Exploration: From Western Medicine to Chinese Acupuncture

Pheng Ann Heng[1,2], Shao Xiang Zhang[3], Yongming Xie[1],
Tien Tsin Wong[1], Yim Pan Chui[1,2], and Jack Chun Yiu Cheng[4]

[1]Department of Computer Science and Engineering, the Chinese University of Hong Kong, Hong Kong SAR, China
[2]Shun Hing Institute of Advanced Engineering, the Chinese University of Hong Kong, Hong Kong SAR, China
[3]Department of Anatomy, College of Medicine, the Third Military Medical University, Chongqing, China
[4]Department of Orthopedics and Traumatology, the Chinese University of Hong Kong, Hong Kong SAR, China

Key Words. Visualization, virtual anatomry, and visible human research.

1. The Visible Human Data

The advent of Visible Human Project (VHP) by the National Library of Medicine of United States in 1994 [1] has revolutionized the research in medical imaging. This was the first time true tissue color can be revealed for in depth study by all scientists and researchers worldwide. The visible human dataset has been applied to many different aspects of medicine including medical image diagnosis, anatomical education, and surgical simulation etc. Since October 2002, the Chinese Visible Human Project [2, 3] research team from The Third Military Medical University has successfully collected visible human datasets from several Chinese donors. Deploying latest processing techniques, extra-fine anatomic details, which cannot be displayed from the long ago American dataset, can now be unveiled by the Chinese visible human datasets. Together with the Visible Korean Man [4], there are totally three countries in the world has published the establishment of her own visible human data.

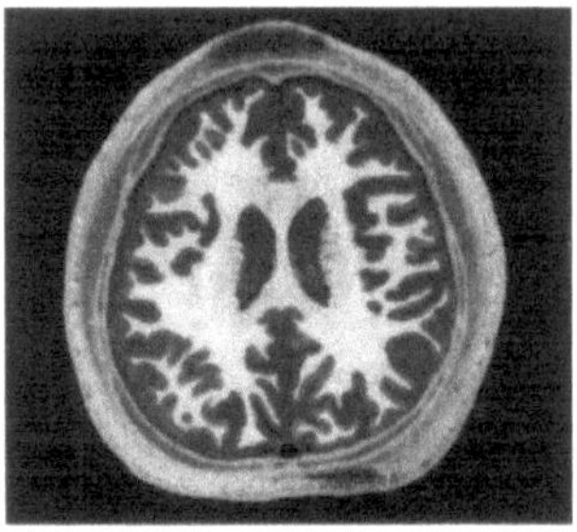

Fig. 1. An original slice from the Chinese Visible Human.

The first Chinese Visible Human (CVH) dataset was derived from a 35 years old male cadaver of medium build. No organic disease was observed macroscopically through naked eye observation [2, 3]. Figure 1 shows one of the original slices from this dataset. The latest male dataset comprises approximately 18,200 cross-sectional digital photographs acquired at 0.1 millimeter intervals (a pixel resolution of 4064×2704, 24 bits per pixel).

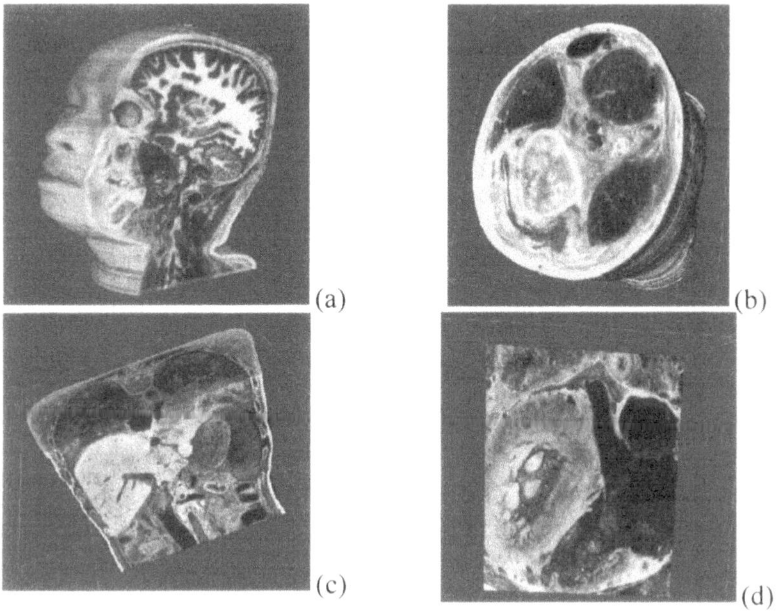

Fig. 2. Virtual anatomy based on Chinese Visible Human data. (a) eye, (b) knee, (c) liver, (d) heart

Figure 2 shows screenshots of various selected parts of the Chinese Visible Human data, e.g., eye, knee, liver, and heart. The smaller inter-slicing distance of the CVH project (0.1 mm thick), compared to the original American Visible Human sections (1.0 mm thick), has improved subsequent visualization of finer anatomical details. Figure 3 compares the cross-sectional dissection of the optic nerve and the rectus region from the VHP (Figs. 3 a, c, & e) and the CVH (Figs. 3 b, d, & f) datasets.

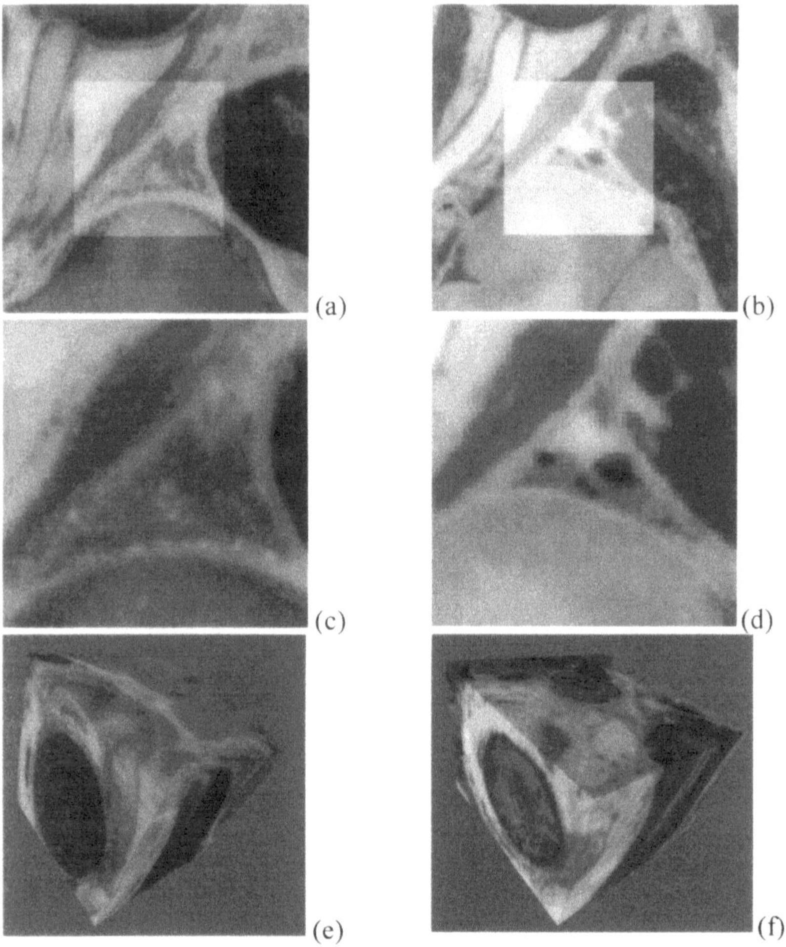

Fig. 3. Comparison of image slices and volume rendered data between the American Visible Human and CVH. Images of the lateral rectus region from the CVH and American datasets and corresponding enlarged views (a, c) and (b, d), respectively. Volume rendered views of the orbit from the CVH (e) and American (f) datasets.

2. Huge-Sized Volume Visualization

Due to the huge data size of the CVH datasets, we have developed various processing, compression and visualization techniques so that they can be displayed interactively on an ordinary personal computer with high fidelity to the original data. Delicate anatomic structures, previously difficult to resolve, can now be well displayed. With this advance, volume data can be observed from any orientation in real-time. Anatomic structures can be revealed dynamically by multi-phase and multi-temporal stages. In order to include more details in the 3D model, human data was processed by a special rebuilding method. By utilizing compressed texture images generated from the Visible Human data, greater realism of 3D reconstructed models has been accomplished.

Image data of cross-sectional volumes are processed according to the RGB components; by using the voxel as the basic modeling unit, the virtual body can be rendered directly. The dataset was post-processed so that irrelevant signals, such as pixels representing the frozen liquid, are removed. In addition, compression has to be performed so that data flow can be maintained in an interactive level on a single personal computer (PC). Using graphics accelerators and advanced visualization techniques, real-time visualization of the CVH dataset was accomplished in a 3D virtual environment. The virtual human data can be enlarged and navigated in any orientation so that internal structures can be clearly observed. These 3D reconstructed slices can also be stereoscopically viewed with no obvious loss in performance.

3. Virtual Anatomy

Backed by the advanced visualization engine, truly volumetric virtual anatomy applications is no longer impracticable. To develop virtual anatomical models, various structures including neural, musculoskeletal, and vascular tissues of the targeted body region were first identified and selected from the CVH dataset. Figure 4 illustrates a slice of the segmented brain. Up to this moment, the upper limb region, together with the head and neck region have been labeled in our first stage segmentation work.

By implementing efficient programmable Graphics Processing Unit (GPU) software, visualization of the segmented data can be performed interactively on a single PC. Our testing PC is equipped with Pentium 4 1.4G as the Central Processing Unit (CPU), 256M RAM, and Nvidia GeForce FX-grade graphics accelerator.

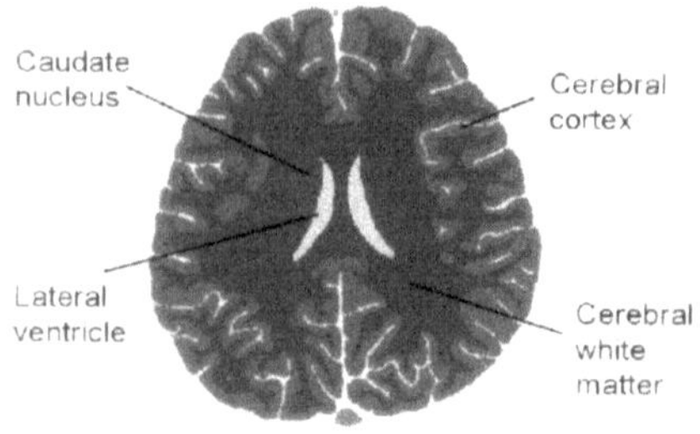

Fig. 4. Segmentation of the brain of the Chinese Visible Human.

We have constructed an intuitive navigational interface to facilitate user removal and enhancement of different tissue structures in the reconstructed anatomical model. Similarly, multi-planar reconstruction, zooming, rotation and flexible structure selections are supported. The user interface is shown in Figure 5. The user can freely assign any pseudo-color to various anatomical structures for the ease of tissue identification (Fig. 5b).Figure 8 shows two specific close-up views of the upper limb model. Muscular tissue (Fig. 5c) and the septum (Fig. 5d) can be visualized through different user manipulations.

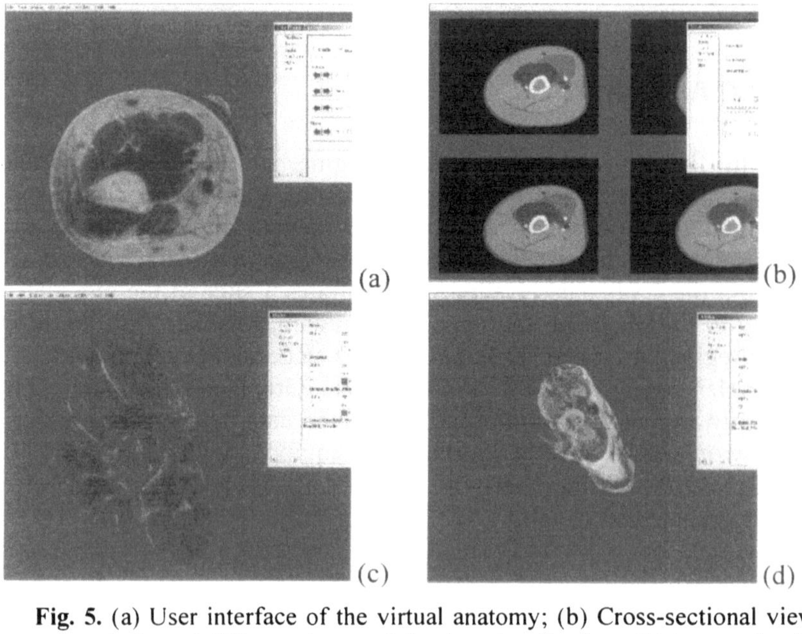

Fig. 5. (a) User interface of the virtual anatomy; (b) Cross-sectional views of pseudo-coloring of different tissues. Selective visualization of (c) muscular tissues and (d) the septum.

We have also deployed the CVH on virtual neuroanatomy. In addition to an opaque volume rendering, translucent rendering mode is also supported. By adjusting the alpha channel of different tissues interactively, the user can obtain their own desired translucent display of volume. Figure 6 shows a few volume-rendered brain structures through various alpha channel adjustments. We can also add pseudo-coloring onto different brain tissues, for example, in Figures 6c and 6d, we have dyed the limbic system and dural sinus in different pseudo-color.

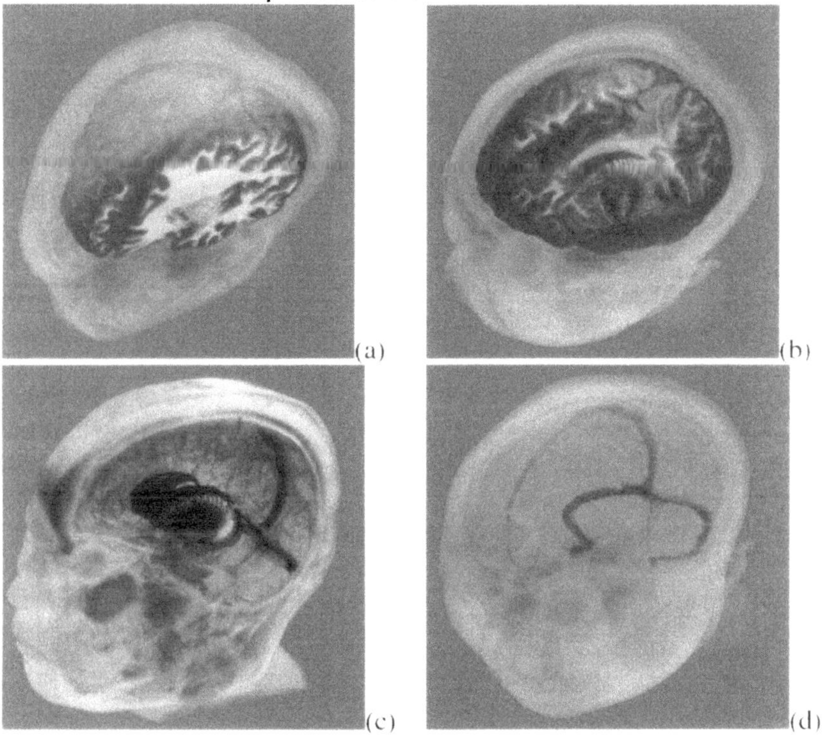

Fig. 6. Translucent volume rendering of brain and intracranial structures.

4. Virtual Acupuncture

In addition to the applications of the CVH dataset on western medicine, such as virtual arthroscopy [5], we have also explored the use of these photorealistic tissue depictions onto the field of Chinese medicine [6]. In traditional Chinese acupuncture, anatomical details has not yet fully investigated and deployed. The high complexity of the acupuncture system has imposed great difficulties in the learning process for novice acupuncturists. While the outstanding therapeutic performance of Chinese Acupuncture has aroused the interest of many medical professionals worldwide, we have developed advanced information technologies for the computer-assisted Chinese acupuncture learning and research. By integrating the virtual reality, visualization and imaging techniques, we have constructed a detailed virtual acupuncture digital human model based on the ultra-high resolution CVH dataset.

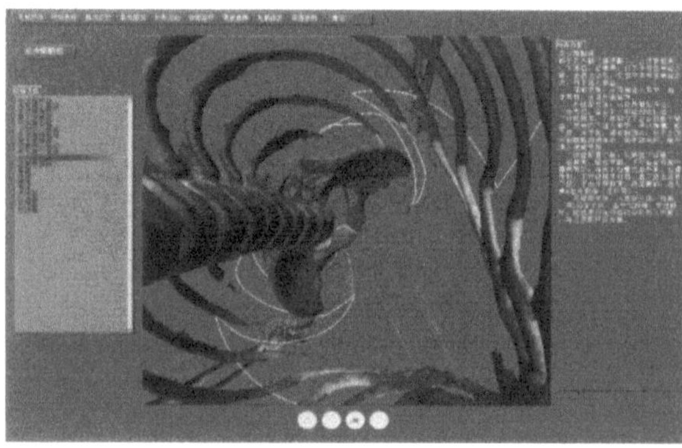

Fig. 7. Interactive visualization of the meridian system.

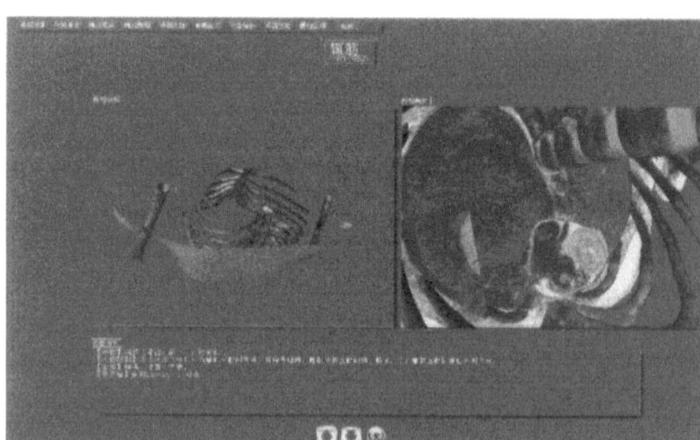

Fig. 8. Multi-layer dissection visualization.

To deliver a high-quality and precise virtual acupuncture system, we have combined both the volumetric and surface representations in our rendering kernel. In order to handle the huge CVH dataset in the rendering process, we have deployed the advanced GPU Shader Language programming techniques to generate realistic visual effects. In this sense, a systematic virtual acupuncture human digital model, which integrates the meridian system positioning (see Fig. 7), and 3-dimensional acupuncture point positioning, becomes feasible. This can shorten the learning path of acupuncture novices in acquiring the positioning concept. Our system serves as an interesting and interactive acupuncture learning platform.

In addition, arbitrary cutting-plane visualization, multi-layer dissection (see Fig. 8) is also supported. User can interactively zoom in or zoom out in order to view highly detailed dissective information. With an integration of anatomical information, traditional surface location can be related to the underlying anatomical structures. With these advances, further scientific research of traditional Chinese acupuncture theory can be carried out in depth. This is a noteworthy milestone on the integration of western medicine and traditional Chinese medicine.

5. Discussion

Compared with the American Visible Human data and the Korean Visible Man, the Chinese Visible Human dataset is able to present finer anatomical details. We have shown how this highly-detailed digital human data can be applied on virtual anatomy as well as Chinese acupuncture. The anatomical visualization for orthopedics training program is being used in Department of Orthopaedics and Traumatology, the Chinese University of Hong Kong. While the virtual acupuncture program is under intensive testing in the Institute of Chinese Medicine of the Chinese University of Hong Kong. Users generally give credit to the interactive visualization of highly-detailed anatomical structures. In the future, staged labeling work of the dissective details will continue; the advancement of anatomical exploration would definitely be beneficial to the scientific research of both western and Chinese medicine.

Acknowledgment

The work described in this paper was supported by a grant from the Research Grants Council of the Hong Kong Special Administrative Region, China (Project No. CUHK4223/04E) and CUHK Direct Grant for Research 2005/2006.

References

1. Spitzer VM, Whitlock DG (1998) The Visible Human dataset: The anatomical platform for human simulation. Anat. Rec., 253:49-57
2. Zhang SX, Heng PA, Liu ZJ, et al (2003) Creation of the Chinese Visible Human data set. Anatomical Record (Part B: New Anat.) 275B:190-195
3. Zhang SX, Heng PA, Liu ZJ, et al (2004) The Chinese Visible Human (CVH) Datasets incorporate technical and imaging advances on earlier digital humans. J. Anat. 204:165-173
4. Park JS, Chung MS, Hwang SB, et al (2005) Visible Korean Human: Improved serially sectioned images of the entire body. IEEE Trans. Med. Imaging, 24:32-360
5. Heng PA, Cheng CY, Wong TT, et al (2004) A virtual reality training system for knee arthroscopic surgery. IEEE Trans. Inform. Technol. Biomed., 8:2, pp 217-227
6. Heng PA, Wong TT, Yang R, et al (2006) Intelligent inferencing and haptic simulation for Chinese acupuncture learning and training. IEEE Trans. Inform. Technol. Biomed., 10:1, pp.28-41.

MEG Single-event Analysis: Networks for Normal Brain Function and Their Changes in Schizophrenia

Andreas A. Ioannides

Laboratory for Human Brain Dynamics, RIKEN Brain Science Institue (BSI), 2-1 Hirosawa, Wakoshi, Saitama 351-0198, Japan

Chapter Overview. Modern instrumentation offers the capability to capture in an instance the magnetic field around the head. There is agreement that the measurements are responsive to instantaneous changes of the current density inside the head, corresponding, at least partially, to neuronal events elicited by the stimuli and/or the task. There is however disagreement about the information that can be usefully extracted from the data. On the one hand theoretical arguments can be given why unambiguous reconstructions of generators from the data are impossible. On the other hand evidence has been gathering that surprisingly accurate information can be extracted from the data under minimal assumptions. Here we outline magnetic field tomography (MFT), a method for obtaining tomographic descriptions of brain activity from biomagnetic data. We show that if the underlying physics laws and biological complexity are to be respected then the non-linear reconstruction algorithm of MFT must be selected over the simpler minimum norm or other algorithms. Drawing on the enormous dynamic range of the measurements and the richness of information in the single trial data, MFT provides unique insights into brain function, as demonstrated by a detailed study of the brain activity and interactions elicited by a facial recognition task in healthy and schizophrenic subjects.

Key Words. Inverse Problem, Magnetic Field Tomography (MFT), Magnetoencephalography (MEG), Schizophrenia, Single trial analysis.

1. Introduction

The bioelectromagnetic inverse problem poses a paradox in modern neuroimaging. It has been known for nearly two centuries that the strict theoretical formulation of the problem leaves no room for a unique solution [1]. Many papers use non-uniqueness to justify the use of unrealistic models for the generators (e.g., point like descriptions) and fix their locations by appealing to evidence from totally different sets of measurements, e.g. fMRI. In contrast, magnetic field tomography (MFT) has been introduced over 15 years ago as a method capable of extracting tomographic description of activity in the brain from the magnetoencephalographic (MEG) signals [2]. In the last ten years the claim of accurate reconstructions has been extended to even single timeslices of single trial data [3]. Close examination of the mathematical properties of the lead fields provides the mathematical foundation of the MFT claim [4] and relates a generalized version of the method to most of the other popular distributed source methods [4,5]. From a purely pragmatic point of view applications of MFT to real MEG data have consistently produced solutions with surprisingly good accuracy. At the cortical level an accuracy of a few millimeters was demonstrated [6]. Recently, good localization was also demonstrated for deeper structures including the amygdala and brainstem [7] and cerebellum [8, 9]. In this paper we group together the main arguments and point out some misconceptions that prevail in the literature about the nature of distributed source solutions [10]. The key point is rather subtle. Although the "forward" relationship from the data to the generators is linear, a mathematically consistent formulation of the inverse relationship expressing the generators in terms of the data is non-linear [4]. It is the use of the optimal non-linearity that sets MFT apart from other methods [4, 5]. We demonstrate the amount and novelty of new information with MFT results from recent studies, emphasizing research on facial expression recognition in normal and schizophrenic subjects [11, 12, 13].

2. Methods

2.1 Mathematical preliminaries

A linear integral expression relates the signal d_m from the mth sensor (set of coils) to the primary current density vector $\mathbf{J}(\mathbf{r})$. The integration is over the source space, Q, which encompasses all regions with primary currents that can generate an MEG signal of interest, i.e., the entire brain,

$$d_m = \int_Q \mathbf{f}_m(\mathbf{r}) \cdot \mathbf{J}(\mathbf{r}) d\mathbf{r} \tag{1}$$

The lead field $\mathbf{f}_m(\mathbf{r})$ is a vector function that is completely determined by the geometric properties of the coils making up each sensor and the conductivity details of the biological medium. In the case of MEG a fairly simple model for the generators is often sufficient to produce accurate lead fields. The basic mathematical arguments are also applicable for EEG, but EEG will not be considered here to avoid sidetracking the discussion to details on accurate modeling of the head conductivity.

The fact that (1) is linear does not necessarily make the expression relating the generators to the data linear, as a naïve discretization of (1) will imply. If for example we replace the "naturally continuous current distribution" with a "large number of narrowly spaced dipole sources" $\mathbf{s}(\mathbf{r}_k)$, placed at source space points $\mathbf{r}_k$, then (1) reduces to

$$d_m = \sum_k \mathbf{f}_m(\mathbf{r}_k) \cdot \mathbf{s}(\mathbf{r}_k).$$

Recovering the sources is then simply a matter of choosing how the linear system of equations is to be solved, as pointed out in reference [10]. To understand the limitations of this seemingly innocent step we write explicitly the general form $\mathbf{J}(\mathbf{r})$ can take:

$$\mathbf{J}(\mathbf{r}) = \sum_m A_m \mathbf{f}_m(\mathbf{r})\,\omega(\mathbf{r},\mathbf{J}(\mathbf{r})) \tag{2}$$

The point emphasized here is that the weight factor ω can be a function of both the source space location and the unknown current density vector itself. We note that (2) contains no overlap with the null space, i.e. it contains no silent sources, but as can be easily demonstrated it is not a minimum norm (MN) solution [4]. The art of moving from the MEG measurements to the generators is how one chooses the weighting factor in a way that respects both the laws of physics and the biological reality. The rather obvious choice of setting $\omega(\mathbf{r},\mathbf{J}(\mathbf{r})) = \omega_{MN}(\mathbf{r})$ leads to the weighted minimum norm (wMN) ansatz,

$$\mathbf{J}(\mathbf{r}) = \sum_m A_m \mathbf{f}_m(\mathbf{r})\,\omega_{MN}(\mathbf{r}) \tag{3}$$

The wMN choice seems very natural and harmless. It has a huge advantage: it leads to a J-independent linear system of equations as can be readily verified by substituting (3) in (1). Many different methods for tackling the inverse problem can be reduced to different attempts to find an optimal choice for the J-independent *a priori* weight $\omega_{MN}(\mathbf{r})$. In recent years advances in source reconstruction have emphasized signal processing methods that one way or another implicitly or explicitly make use of (3). In the complete absence of any additional information all such methods end up with the minimum norm pseudoinverse (MNP) as correctly pointed out in reference [10]. However, the use of (3) leads to a garden-path that is only limited to the MN and wMN classes. It is not valid to generalize the limitations of MN or wMN methods to the parent recording modalities in general and particularly to MEG, as for example is made in [10] with the statement "the properties of the MNP, and in particular, its limitations like the inability to localize sources in depth, are not specific to the method but are fundamental limitations of the recording modalities".

Generalized MFT uses a power expansion of $\omega(\mathbf{r},\mathbf{J}(\mathbf{r}))$ in the modulus of the current density [4]:

$$\mathbf{J}(\mathbf{r}) = |\mathbf{J}(\mathbf{r})|^{p+1} \sum_m A_m \mathbf{f}_m(\mathbf{r}) \omega_p(\mathbf{r}) \tag{4}$$

This produces a family of methods [4-5]. The case with $(p = -1)$ includes the MN, wMN and LORETTA (where weights are introduced depending on the inverse square Laplacian operator). These methods lead to a linear system of equations and this is why they are popular. A version of the FOCUSS algorithms (corrected for gauge invariance) can also be put under the generalized MFT framework $(p = +1)$ [5]. Standard MFT corresponds to $(p = 0)$ and it was initially selected via a simulation study [2]. Details of the mathematical justification for standard MFT can be found in [4], and the main points are summarized next.

2.2 What is wrong with minimum norm? Why MFT?

The linearity of the reconstruction problem for the class of solutions corresponding to $p = -1$ makes possible the incorporation of signal processing methods into the reconstruction algorithm. This leads to powerful schemes of analysis for relatively small computations. Typically, complicated operations can be done at the level of signals, e.g., using the signal covariance matrix. Statistical inferences can then be directly applied to the MEG signal to produce the source distribution using J-independent matrix operations. In short the inverse problem is solved once for the entire data. Regional activations are then computed by matrix operations involving the signal and the local lead fields values. The problem is that the computational simplicity and efficiency is bought at the expense of physics. Equation (3) implies that the same single scalar function can be used for the recovery of any current density without further recourse to the data. The very nature of the non-uniqueness of the inverse problem points to the opposite conclusion. The linear methods (corresponding to $p = -1$) do not respect the symmetries implicit in the degrees of freedom for the current generators, because they assume that it is possible to express both direction and strength of the current density in one single linear expansion in lead fields. The blurred nature of the MN solutions reflects the over-reliance on the raw lead fields which are themselves very smooth functions.

Standard MFT (with $p = 0$) takes a different point of view. It assumes that a linear expansion in terms of lead fields can represent only the direction of the current density. This is as much as can be deduced from the underlying physics for fixed detectors and conductivity profile. The *a priori* weight, $\omega_0(\mathbf{r})$, can in fact be computed from simulations with computer generated data [2, 3] or in more general ways [4]. The full current density however must be obtained from a highly non-linear system of equations for each snapshot of data. Specifically, the strength must be determined more explicitly from the MEG signal itself. It is important to emphasize that the way standard MFT draws on the data is very appropriate for localization because it has two contrasting and highly desirable properties. First, the presence of the modulus of the current density on the right hand side of (4) (with $p = 0$) allows for sharp discontinuities in the current density vector with small values of the expansion coefficients. Second, standard MFT satisfies the principle of least sensitivity to both variations of the data and iterations of the non-linear norm constrains [4]. In summary, the standard MFT expansion (with $p = 0$) satisfies best the underlying physics and has the expected properties for localized distributed sources [4]. The penalty is however heavy in terms of computational resources. Because linearity is lost, the direct and safe appeal to the data must be made on every timeslice of the data, so a new non-linear system of equations must be solved each time. It is in this sense that MFT draws on all available information in the MEG signal.

2.3 Self-consistency checks and post-MFT options

The linearity of the forward problem implies that any linear combination of sources produces a total signal that is equal to the linear combination of the signals generated by each source. This is identically satisfied by linear methods, but it need not be true for a non-linear method like MFT, so it can be used to test the self-consistency of any non-linear solution. For example, the solutions derived from a non-linear method applied to the MEG signal after linear manipulations across trials, e.g. averaging, should be the same as the result of applying the same linear manipulations to the solutions derived from the single trial MEG signals. This has been indeed the case for MFT in all cases we have tested [3].

The availability of single trial tomographic solutions for each timeslice of data allows for a range of new options. We will distinguish three types of post-MFT analysis. The first, (post-MFT-Rt) uses regional timecourses derived from the tomographic solutions and can be thought of as the analysis of virtual electrodes, sensing the local non-silent primary current density in the brain. The variability of single trial solutions is quantified at a regional level and clusters of similar trials can be selected. Methods for such analysis have been introduced borrowing concepts from pattern analysis methods and graph theory [14] and non-linear timeseries analysis [15]. A second range of options makes use of statistical parametric mapping (SPM) and it is referred to hereafter as post-MFT-SPM. Post-MFT-SPM is similar to what is routinely used for fMRI and PET, but with windows of varying length, allowing the recovery of brief changes in activity (between active conditions, or between an active condition and a baseline) as a function of time [3, 6-8, 11-13]. A third range of options (post-MFT-c) focuses on the relationship between regional activations using information theoretic measures like mutual information to derive estimates of connectivity between brain regions [3,7,12-13].

3. Results

3.1 Localization accuracy and single trial variability

The question how well MFT can localize was for a long time difficult to answer. Known generators could be identified from computer generated data with very high accuracy. Until recently however there was no gold standard case with real MEG data to compare. Recently, we perform a combined MEG/fMRI experiment and used post-MFT analysis to test the localization accuracy attainable by MEG [6]. It was found that activity within V1 could be reliably identified within just a few millimeters, as good as the co-registration accuracy would allow.

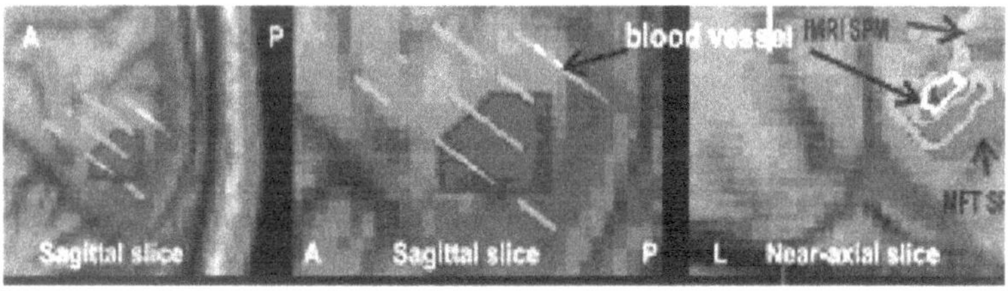

Fig. 1. Foci of statistical significant increase in activity within V1 elicited by a small checker board stimulus placed in the lower left quadrant of the visual field. The MFT SPM corresponds to a brief increase in activity 42 ms after stimulus onset and lasting for just a few milliseconds The fMRI SPM foci are shown as green outlines. They appear as straight lines in the sagittal slice (left) and in its zoomed version (middle) because they are confined in the near-axial slices used for fMRI acquisition.

Very similar results were obtained for each one of the four subjects studied in [6] so typical results from just one subject are shown in Fig. 1. The fMRI SPM foci are shown in green after strong activity due to blood vessels (marked in white) is eliminated by thresholding. No such thresholding is needed for the MFT results. The first statistically significant post-MFT-SPM focus (red blob) is identified in V1 at 42 milliseconds after stimulus onset. This was the only statistically significant SPM focus of activity at this timeslice throughout the brain. The red post-MFT-SPM foci do not extend into the blood vessel and they localize very close to the fMRI SPM foci. Focal V1 activations arrive in waves, each lasting just a few milliseconds before activity spreads to V2 and other extrastriate areas. Early brief V1 activations were missed in previous studies because the data were low passed filtered at 100 Hz, or lower frequencies and because activations were deemed significant only if they persisted for about 10 ms. Even the 200 Hz low pass filter used in [6] may be too low. A recent study showed that "MEG spikes", task relevant collective brain activity lasting about a millisecond, can be recorded with MEG [15].

The most incisive results are obtained when different post-MFT options are combined. Post-MFT-SPM can identify very focal regions, as demonstrated in the fMRI/MEG study [6]. These regions can be used as guides for the definition of regions of interest (ROIs). Post-MFT-Rt can then be applied to ROI timeseries of single trial activations to identify compact clusters of similar activity. The connectivity analysis (post-MFT-c) for the activations in different areas within each cluster leads to the definition of networks that are remarkably similar across subjects [13, 17, 18]. The application of such post-MFT analysis steps has demonstrated that even a simple stimulus elicits different responses in single trials which can be grouped into different clusters and shown to correspond to consistent network interactions across different subjects. This has been demonstrated for both the auditory [16] somatosensory [17] and visual systems [18]. For the remainder of the paper we outline results from our facial expression recognition studies with normal and schizophrenic subjects.

3.2 Recognizing facial expressions of emotion

MFT was applied to a series of experiments dealing with object recognition with particular emphasis on activations elicited while subjects attempted to recognize facial expressions. This research was chosen partly because of its intrinsic merit, including the relevance to pathology, especially schizophrenia. The topic was also an excellent vehicle for testing the MFT methodology because specific brain areas, widely spread in the cortical mantle and deep in the brain are known to play a role. Our studies [11-13] confirmed the accuracy of localization, identifying foci of activity with Talairach coordinates [20] in excellent agreement to what other studies with fMRI have demonstrated. In addition to the areas established by other methods, changes in activity were also identified in the right amygdala and cerebellum [8, 13].

The post-MFT analysis of single trial and average data produced new insights for both the processing of faces in normal subjects and changes in processing patterns in schizophrenia. The MFT analysis of average data showed reduced activity in a number of key areas in schizophrenic compared to control subjects, very much in agreement with previous studies [12]. Weaker activations were identified in patients in inferior prefrontal, temporal, occipital and inferior parietal areas at circumscribed latencies. Group differences did not occur in early visual areas during a first sensory related activation between 60 and 120 ms. Strength of activation was associated with behavioral performance in inferior prefrontal areas, in the region of right posterior fusiform gyrus, in right anterior temporal cortex and in right inferior parietal cortex.

The detailed single trial MFT analysis provided a new insight, forcing a re-interpretation of the earlier results [13]. The reduced activations identified in the MFT analysis of the average data were shown to be the result of higher variability in patients. In fact the activity in early visual cortices especially for early latencies was stronger in patients. This is demonstrated in Fig. 2.

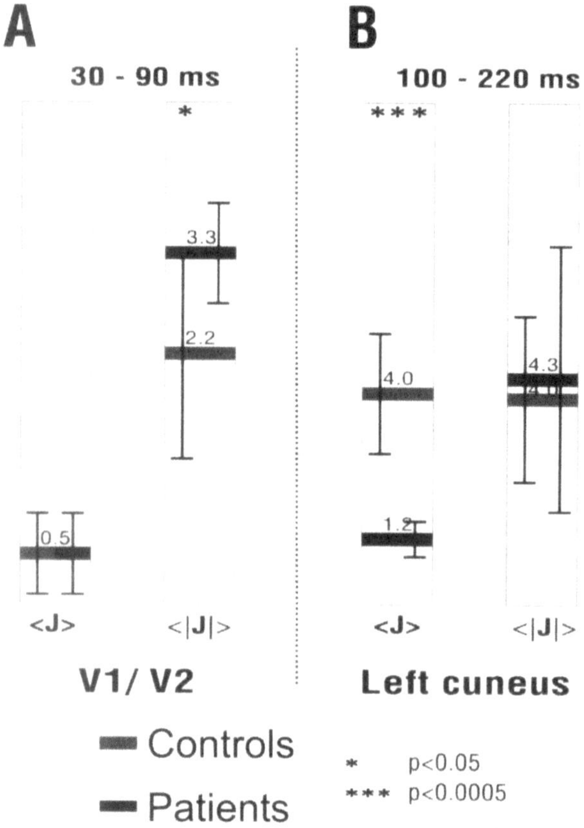

Fig. 2. Mean and standard deviation for the expectation values for the current density vector < **J** > and modulus < |**J**| > for normal and schizophrenic subjects in early visual areas.

The figure shows the mean and standard deviation for two expectation values. The first is computed from the current density vector of single trial MFT solutions (equivalent to the MFT solution derived from the average MEG signal). The second is computed from the moduli of the single trial MFT solutions (roughly expressing the power). In V1/V2 the power of activation is higher (at statistical significance $p < 0.05$) in patients while no difference is evident for the current density vector (or average signal) MFT solutions. For the cuneus the expectation values based on the vector current density shows highly statistically significant ($p < 0.0005$) reduction of activity in patients, but no such change is identified for the expectation values based on the power.

The results summarized above provide a powerful hint that the deficit in schizophrenia is associated with the organization rather than the strength of activations. This has been investigated further using mutual information to establish the connectivity between areas. Figure 3 summarizes the end product of a large volume of analysis. The mutual information analysis demonstrated that in normal subjects the activity is organized in relatively distinct stages of processing. Within 50 ms of stimulus onset a weak and labile activation is present in the right amygdala and it is linked to activity in V1/V2 40 ms later. This linked activity is not identified in patients, where instead a persistent activity is identified in the right amygdala. Later stages of activation are identified encompassing first extrastriate posterior areas and then frontal areas. In schizophrenic subjects no such distinct stages of processing are identified, not even for occipital visual areas where the patients show higher activity (as measured by the power of single trial activity).

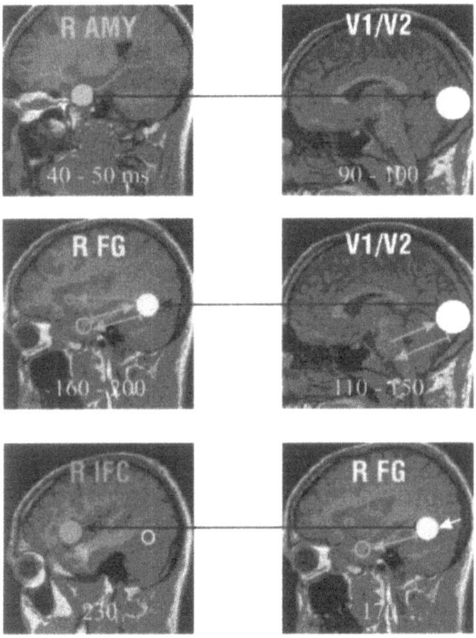

Fig. 3. Main result for MI analysis of single trial data of normal subjects, focusing on the right hemisphere. The activations are grouped into stages. The first stage (top) centers around processing in V1/V2 that is modulated by very early activity in the amygdala, possibly corresponding too ultra-fast magnocellular input bypassing V1. The second stage (middle) is centered around the FG and it is associated with interactions with both earlier visual areas and the amygdala. The third stage (bottom) is centered around the inferior frontal cortex and it involves interactions between all areas that have been activated previously. The numbers in the lower part of each panel show the latency range of strong activity for each highlighted area.

4. Discussion and Conclusions

Restricting the inverse problem algorithms to linear or non-linear forms is more than an academic exercise; it determines both the complexity of the computations and the nature of the questions that can be asked. Specifically, the explicit non-linearity of MFT allows the data of each single trial to influence the solutions directly without any assumption about stationarity. MFT solutions applied to single timeslices of single trial data leave the use of statistics open into the post-reconstruction stage and hence provide a source-space based statistical analysis that includes detailed connectivity mapping. This is in sharp contrast to other methods that rely on linearity and further simplifications (e.g. averaging) at the level of the signal.

MFT solutions extracted from average signals show smooth changes in source configuration, usually with an initial focal activation that is quickly followed by strong and extended activations. The application of MFT analysis to real time MEG data produced a view of the brain that is much more dynamic. Single trial MFT solutions show intermittent [3], extremely fast [15] and labile activity. Precise tests for localization confirmed that remarkable accuracy could be achieved in the cortex, typically a few millimeters [6] especially from post-MFT-SPM analysis of the single trial solutions. Even for generators at the level of the amygdala and the brainstem the accuracy is sufficiently good to distinguish activity from the left and right gaze centers and flocculi [7, 15]. The new insights are obtained when the information in the single trial data is really mined by the techniques described earlier.

For the case of processing of activity during facial expression recognition our MFT and post-MFT analysis showed that it is changes in the dynamics of the activations in specific areas and their interactions that distinguishes between processing in normal and schizophrenic subjects. In summary, the real time MFT view of brain activity provided direct evidence for speculations positing disturbances in functional connectivity in schizophrenia [21].

Acknowledgment

The work described here was developed over many years, at the Open University (UK), Jülich Research center (Germany) and RIKEN Brain Science Institute (Japan). Many colleagues contributed to the development of these ideas, particularly J.G Taylor for the theoretical work on lead fields, L.C. Liu and P.B.C Fenwick for many of the applications and V. Poghosyan and the late M. Streit for the study of facial affect recognition.

References

1. H. von Helmholtz,. ber einige Gesetze der Vertheilung elektrischer Ströme in körperlichen Leitern, mit Anwendung auf die thierisch-elektrischen Versuche. *Annals of Physics and Chemistry* 89, 211 – 233, 353– 377 (1853).
2. A.A. Ioannides, J.P.R. Bolton and C.J.S. Clarke, "Continuous Probabilistic Solutions to the Biomagnetic Inverse Problem", *Inverse Problems* 6. 523-542 (1990).
3. A.A. Ioannides, "Real Time Human Brain Function: Observations and Inferences from Single Trial Analysis of Magnetoencephalographic Signals", *Clinical EEG* 32: 98-111 (2001).
4. J.G. Taylor, A.A. Ioannides and H.-W.Mueller-Gaertner, "Mathematical Analysis of Lead Field Expansions", *IEEE Trans. Med. Imag.* 18: 151-163 (1999).
5. A.A. Ioannides and J.G. Taylor, "Minimum norm, Magnetic Field Tomography and FOCUSS", in Recent Advances in Biomagnetism, T. Yoshimoto, M. Kotani, S. Kuriki, H. Karibe and N. Nakasato Eds, Tohoku University Press, Sendai, 1999, pp 228-231.
6. F. Moradi *et al.*, "Consistent and precise localization of brain activity in human primary visual cortex by MEG and fMRI", *NeuroImage* 18: 595-609 (2003).
7. A.A. Ioannides *et al.*, "MEG tomography of human cortex and brainstem activity in waking and REM sleep saccades", *Cerebral Cortex* 14: 56-72 (2004).
8. A.A. Ioannides and P.B.C. Fenwick, "Imaging cerebellum activity in real time with magnetoencephalography", *Prog Brain Res.* 148: 139-150 (2005).
9. A.A. Ioannides. P.B.C. Fenwick and L.C. Liu "Widely distributed magnetoencephalography spikes related to the planning and execution of human saccades", J. Neurosc. 25(35): 7950-7967 (2005).
10. O. Hauk, "Keep it simple: a case for using classical minimum norm estimation in the analysis of EEG and MEG data", *NeuroImage* 21: 1612-1621 (2004).
11. L.C. Liu, A.A. Ioannides and M. Streit, "Single Trial Analysis of Neurophysiological Correlates of the Recognition of Complex Objects and Facial Expressions of Emotion", *Brain Topography* 11: 291 – 303 (1999).
12. M. Streit *et al.*, "Disturbed Facial Affect Recognition in Schizophrenia as a Result of Hypoactivity in Distributed Brain Regions: a Magnetoencephalographic Study", *American Journal of Psychiatry* 158: 1429-1436 (2001).
13. A.A. Ioannides, V. Poghosyan, J. Dammers and M. Streit, "Real-time neural activity and connectivity in healthy individuals and schizophrenia patients", *NeuroImage* 23: 473-482 (2004).

14. N. Laskaris and A.A. Ioannides, "Exploratory Data Analysis of Evoked Response Single Trials based on Minimal Spanning Tree", *Electroenceph. Clin. Neurophysiol* 112: 698-712 (2001).
15. N. Laskaris and A.A. Ioannides, "Semantic geodesic maps: a unifying geometrical approach for studying the structure and dynamics of Single trials evoked response", *Electroenceph. Clin. Neurophysiol* 113: 1209-1226 (2002).
16. L.C. Liu, A.A. Ioannides and H.-W. Mueller-Gaertner, "Bi-hemispheric Study of Single Trial MEG Signals of the Human Auditory Cortex", *Electroenceph. Clin. Neurophysiol.* 106:64-78 (1998).
17. A.A. Ioannides *et al.*, "Timing and connectivity in the human somatosensory cortex from single trial mass electrical activity", *Human Brain Mapping* 11: 77-92 (2000).
18. N. Laskaris, L.C. Liu and A.A. Ioannides, "Single-Trial Variability in Early Visual Neuromagnetic Responses: An Explorative Study Based on the Regional Activation Contributing to the N70m Peak", *NeuroImage* 20: 765-783 (2003).
19. A.A. Ioannides, V. Poghosyan, L.C. Liu and M. Streit, "Early amygdala activations in normal and schizophrenic subjects", *Soc. Neurosci. Abstr.* 28: 521.5 (2002).
20. J. Talairach, J. and P. Tournoux, Co-planar stereotaxic atlas of the human brain: 3-dimensional proportional system: an approach to cerebral imaging. Thieme, New York, 1988.
21. N.C. Andreasen, "Schizophrenia: the fundamental questions", *Brain Research Reviews* 31, 106-112 (2000).

MEG and Complex Systems

Gareth R.Barnes, Michael I.G.Simpson, Arjan Hillebrand, Avgis Hadjipapas, Caroline Witton, and Paul L.Furlong

Neurosciences Research Institute, Aston University, Birmingham, U.K.

Chapter Overview. MEG beamformer algorithms work by making the assumption that correlated and spatially distinct local field potentials do not develop in the human brain. Despite this assumption, images produced by such algorithms concur with those from other non-invasive and invasive estimates of brain function. In this paper we set out to develop a method that could be applied to raw MEG data to explicitly test his assumption. We show that a promax rotation of MEG channel data can be used as an approximate estimator of the number of spatially distinct correlated sources in any frequency band.

Key Words. MEG, beamformer, promax, factor.

1. Introduction

The aim of Magnetoencephalography (MEG) is to determine the spatio-temporal characteristics of neuronal sources based on extracranial recordings of magnetic fields. The entails an inverse problem in which a 3 dimensional current distribution has to be inferred from a 2 dimensional measurement plane. In order to solve such a problem it is necessary to make some a-priori assumptions about the structure of the underlying current distribution. Recent work has shown that a class of MEG inversion algorithms, known as beamformers, produce images which concur with those from other non-invasive imaging modalities [1,2,3,4] and invasive recordings from other species [5]. The fundamental assumption behind the beamformer is that no two cortically distinct electrical sources are linearly correlated in their activation timeseries [6]. The beamformer could be thought of as a class of algorithm best suited to measuring a complex system in which, locally (a few mm), cortical field potentials are integrated (long-term correlated) but globally (a few cm), they are segregated (long-term uncorrelated). In this study we seek to develop an algorithm that can directly test this hypothesis prior to any inversion step. We are specifically interested in the ongoing cortical dynamical processes such as alpha rhythm, rather than activity reflecting sensory afferent pathways such as the evoked response.

Anatomically, the case for local connectivity is strong: lateral connections between axonal and dendritic arbors are prevalent [7] and cortico-cortical mylenated projections have dense local yet sparse global connectivity [8,9,10].

Electrically, intra-cranial measurements suggest that cortical rhythms have limited domains [11,12]. For example, Bullock and coworkers [11] estimated average local coherence between ECoG electrodes at around 0.4 at 10 mm separation, dropping to <0.1 at 30 mm.

In contrast, using conservative Laplacian estimates to minimise volume conduction effects, some of the scalp EEG literature [13,14,15] is still at odds with the intra-cranial data, especially in the alpha frequency band where relatively high coherence estimates (0.4-0.5) have been shown at distances of 15-20 cm.

We set out to address the question: Do correlated, yet spatially distinct, local field potentials develop in the human brain? To do this we have attempted to define a simple procedure that does not rely on the strong assumptions associated with most inversion algorithms. The rationale of our approach is that the multi-channel data within any frequency band can be decomposed into groups of highly correlated measurement channels. If these channel groupings appear to map to more than a single source (or more than a single dipolar pattern) then there must be at least one other one correlated, and spatially distinct, generator in this frequency band.

2. Methods

We were interested in a decomposition (or rotation) of the 151 MEG channel data that would correctly identify the number of underlying correlated sources in a particular frequency band. We considered both principal component and factor analysis as possible methods to distinguish between temporally correlated and correlated dipole pairs. Principal component analysis (PCA) identifies orthogonal channel weightings which maximise the variance in each component. Factor analysis, in contrast, attempts to find components (or channel weightings) in which maximally covariant channels are grouped together.

2.1 Simulation

We simulated pairs of current dipoles on a realistic cortical surface within the configuration of a CTF systems 151 neuromagnetometer (as in [16]). System noise was set at 10ft/√Hz. Each dipole pair was strongly correlated at 30Hz but orthogonal at 10Hz. Dipoles had sinusoidal or co-sinusoidal timeseries, amplitude 2nAm and always placed at least 5cm from the head centre and 3cm distant from one another.

2.2 Recorded data

We used data from a steady state auditory stimulation paradigm. Epochs were 4.5 seconds long with 1s pre-stimulus baseline. Auditory stimulation was either pure tone (500Hz) or pure-tone amplitude modulated at 32Hz. We used all the data from 150 epochs, without regard to the stimulus type.

2.3 Algorithm

In the recorded data we used independent component analysis to extract eye blink and heartbeat artefacts [17]. We then calculated channel covariance matrix over a specific frequency band. In order to maximise signal to noise we then extracted all but the two largest principal components. We then compared the field maps produced from the rotations (or eigenvectors) due to the principal component decomposition to those obtained using the first two components as identified by factor analysis (the promax rotation).

3. Results

Fig.1 shows first and second (left, right) PCA and promax (top, bottom) components due to a pair of dipoles simulated in left and right auditory cortex with uncorrelated 10Hz time courses. The schematic shows the approximate location of the (simulated) head in the sensor array during recordings (nose at top). In an ideal decomposition the weightings on each component would have a single dipolar profile either to the left, or the right of the midline. The first two PCA components each contain large weightings for channels grouped around both sources. (top row). Although the time series of the simulated sources are orthogonal, their field maps are not. This means that some channels contain a strong signal from both sources. The first PCA component (top left), chosen to maximise the variance, therefore contains large weightings for both sources. In contrast, the promax rotation (bottom row) attempts to identify groupings of maximally covariant channels. In this case, each component maps spatially to a simple dipolar field pattern.

Fig.2 shows the case in which the simulated sources are correlated. In this case, both the first PCA component and promax weightings contain two dipolar field patterns. In other words, the underlying factor is no longer a single source but a correlated source pair.

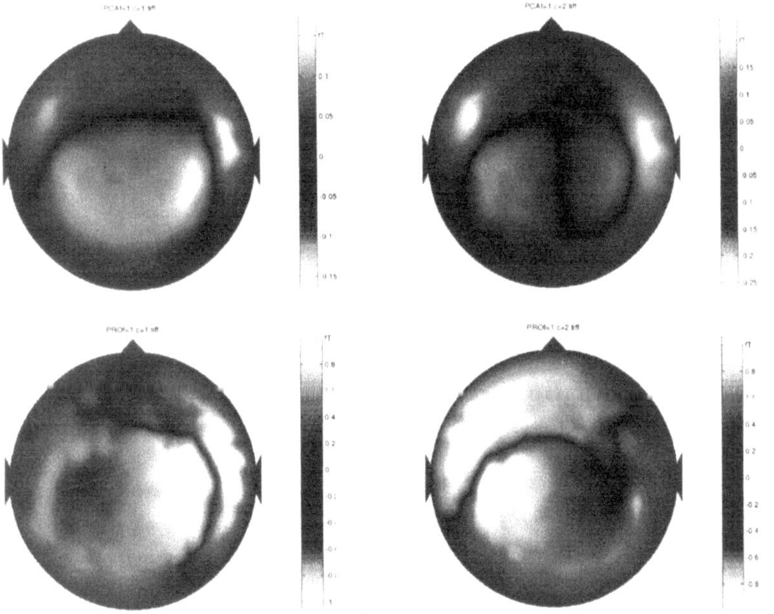

Fig.1. Field maps showing the weightings for first (left) and second (right) PCA (top row) and promax (bottom row) components for simulated uncorrelated current dipoles in left and right auditory cortices. The first two PCA components (top row) both map to two dipolar field patterns, even though their time series are orthogonal. The promax rotation (bottom row), which attempts to identify the maximally covariant channel groupings, correctly partitions the left and right dipoles into separate components.

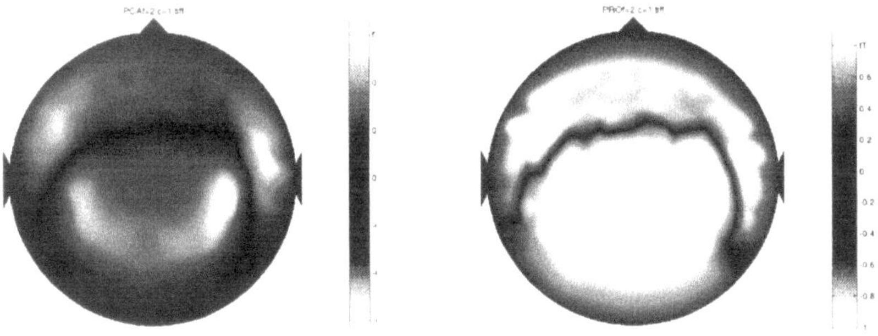

Fig.2. First PCA (left) and promax (right) components for the case of two correlated dipoles in left and right hemispheres. The complex pattern suggests that these weightings cannot be explained by single source.

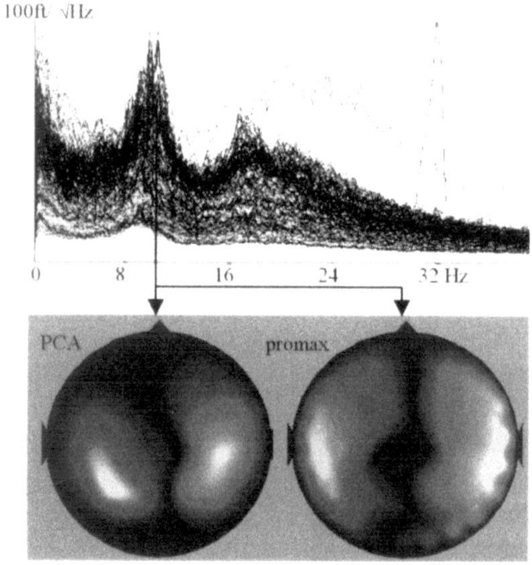

Fig.3. Overlaid power spectrum of 151 measurement channels for a single subject, note the 10Hz alpha peak. Below are the weightings for the first PCA (left) and promax (right) components in the 8-12Hz band. Both methods identify a single dipolar pattern, suggesting that the largest alpha component is due to a single underlying source.

4. Conclusions

We have presented a method to directly test the hypothesis that correlated and spatially distinct local field potentials do not develop in the human brain. Explicitly we are testing whether the factor weightings identifying the strongest oscillatory components within any frequency band are dipolar or not. The algorithm is simple and robust, its only aim to identify factor weightings that are not dipolar. We are currently testing the method, and a simple pattern recognition algorithm, on large sets of simulated correlated and uncorrelated source pairs. If this is successful it will allow us to directly test an assumption set which could greatly simplify non-invasive imaging of the human brain.

Acknowledgment

We would like to thank Jiri Vrba for his insightful comments on principal component analysis. Michael Simpson is supported by the Wellcome Trust.

References

1. Singh,K.D., Barnes,G.R., Hillebrand,A., Forde,E.M.E., Williams,A.L., Task-Related Changes in Cortical Synchronization Are Spatially Coincident with the Hemodynamic Response, NeuroImage, 16 (2002) 103-114.
2. Coppola,R., Calicotte,J.H., Holroyd,T., Verchinski,B.A., Sust,S., Weinberger,W.J. MEG activation comparison to fMRI BOLD for a working memory task. Proceedings Biomag 2004, Boston . 2004.
   Ref Type: Abstract
3. Taniguchi,M., Kato,A., Fujita,N., Hirata,M., Tanaka,H., Kihara,T., Ninomiya,H., Hirabuki,N., Nakamura,H., Robinson,S.E., Cheyne,D., Yoshimine,T., Movement-Related Desynchronization of the Cerebral Cortex Studied with Spatially Filtered Magnetoencephalography, NeuroImage, 12 (2000) 298-306.
4. Furlong,P.L., Hobson,A.R., Aziz,Q., Barnes,G.R., Singh,K.D., Hillebrand,A., Thompson,D.G., Hamdy,S., Dissociating the spatio-temporal characteristics of cortical neuronal activity associated with human volitional swallowing in the healthy adult brain, NeuroImage, 22 (2004) 1447-1455.
5. Hall,S.D., Holliday,I.E., Hillebrand,A., Singh,K.D., Furlong,P.L., Hadjipapas,A., Barnes,G.R., The missing link: concurrent human and primate cortical gamma oscillations, NeuroImage, (2005).
6. van Veen,B.D., van Drongelen,W., Yuchtman,M., Suzuki,A., Localization of Brain Electrical Activity via Linearly Constrained Minimum Variance Spatial Filtering, IEEE Trans. Biomed. Eng., 44 (1997) 867-880.
7. Holmgren,C., Harkany,T., Svennenfors,B., Zilberter,Y., Pyramidal cell communication within local networks in layer 2/3 of rat neocortex, J Physiol, 551 (2003) 139-153.
8. Cherniak,C., Mokhtarzada,Z., Rodriguez-Esteban,R., Changizi,K., Global optimization of cerebral cortex layout, Proc. Natl. Acad. Sci. U. S. A., 101 (2004) 1081-1086.
9. Passingham,R.E., Stephan,K.E., Kotter,R., The anatomical basis of functional localization in the cortex, Nat. Rev. Neurosci., 3 (2002) 606-616.
10. Sporns,O., Tononi,G., Edelman,G.M., Connectivity and complexity: the relationship between neuroanatomy and brain dynamics, Neural Netw., 13 (2000) 909-922.
11. Bullock,T.H., Mcclune,M.C., Achimowicz,J.Z., Iragui-Madoz,V.J., Duckrow,R.B., Spencer,S.S., EEG coherence has structure in the millimeter domain: subdural and hippocampal recordings from epileptic patients, Electroencephalogr. Clin. Neurophysiol., 95 (1995) 161-177.
12. Shen,B., Nadkarni,M., Zappulla,R.A., Spectral modulation of cortical connections measured by EEG coherence in humans, Clin Neurophysiol., 110 (1999) 115-125.
13. Nunez,P.L., Neocortical Dynamics and human EEG rhythms. In Oxford University Press, New York, 1995.

14. Nunez,P.L., Silberstein,R.B., On the Relationship of Synaptic Activity to Macroscopic Measurements: Does Co-Registration of EEG with fMRI Make Sense, Brain Topogr., 13 (2000) 79-96.
15. Srinivasan,R., Spatial structure of the human alpha rhythm: global correlation in adults and local correlation in children, Clin Neurophysiol., 110 (1999) 1351-1362.
16. Hillebrand,A., Barnes,G.R., A quantitative assessment of the sensitivity of whole-head meg to activity in the adult human cortex, NeuroImage, 16 (2002) 638-650.
17. Makeig,S., Bell,A.J., Jung,T.-P., Sejnowski,T.J., Independent Component Analysis of Electroencephalographic Data. In Touretzky,D, Mozer,M, Hasselmo,M (Eds.), Advances in Neural Information Processing Systems 8 MIT Press, Cambridge, MA, 1996, pp. 145-151.

MEG Source Localization under Multiple Constraints: An Extended Bayesian Framework

Jeremie Mattout[1], Christophe Phillips[2], Richard Henson[3] , and Karl Friston[1]

[1]Wellcome Department of Imaging Neuroscience, London, United Kingdom
[2]Centre de Recherches du Cyclotron, Liège, Belgium
[3]MRC Cognition & Brain Sciences Unit, Cambridge, United Kingdom

Chapter Overview. Reconstructing the sources of MagnetoEncephalographic (MEG) measurements is an ill-posed inverse problem which does not admit a unique solution. Forward modeling of the observed signal and prior information or constraints on the solution are needed. Many types of constraints have been considered. However, a general framework is required that can accommodate for multiple priors and enables the evaluation of their contribution. With this in mind, we developed an empirical Bayesian framework based on hierarchical linear models. Our approach entails the estimation of both the expected source distribution and its conditional variance, the latter being constrained by an empirically determined mixture of prior variance components corresponding to the constraints one wants to introduce. Finally, a second level of inference based on Bayesian model selection enables one to compare different combinations of priors. We describe here the methodological framework, the estimation scheme and illustrate its application on a single subject MEG data set.

Key Words. Empirical Bayes, Expectation Maximization, Hierarchical models, Inverse problem, MEG, Model selection, ReML, Source localization.

1. Introduction

Electro-encephalography (EEG) and Magneto-encephalography (MEG) provide non-invasive, passive and instantaneous measures of the whole brain activity. The recorded signal over the scalp reflects synchronous post-synaptic potentials of cortical neural populations [1]. However, mapping and characterizing these electromagnetic sources rests on solving a mathematically ill-posed inverse problem [2]. Priors are needed to define a unique solution. Various priors have been considered such as anatomical, mathematical, physiological and functional constraints. However, most approaches can only accommodate a single constraint and provide the user with a rather arbitrary (or no) estimation of the optimal contribution of the prior information within the implicit regularization.

Following imaging-like modelling, most recent advances have employed a voxel-wise representation of the solution space, either by covering the brain volume, or by restricting it to the cortical surface where the MEG sources are most likely to lie. Each voxel's putative activity is usually modelled by a current dipole with either a free or fixed orientation. This type of representation affords a hierarchical linear generative model of the observed signal Within that context, a general and convenient framework for incorporating priors on the source parameters and providing inference on brain regional activity is provided by Parametric Empirical Bayesian (PEB) methodology which has been recently introduced for the analysis of functional neuroimaging data [3].

In recent communications, we showed that PEB can be successfully applied to the MEG/EEG inverse problem, allowing for both multiple constraints and the data-driven estimation of their optimal contributions [4,5]. Moreover, an extension of this framework enables model selection based on the Log-evidence [5], where the model refers to all the unknown parameters that need to be estimated (e.g. the number of dipoles, the dipole physical parameters, the multiple constraints).

Relying upon a two-level hierarchical model, the PEB approach is probabilistic and subsumes random variables which we assume as Gaussian. Both measurement noise (first level) and source activity (second level) are then defined by their respective mean and variance. Multiple constraints on the sources can be incorporated in terms of variance components. The estimation of model parameters (mean values) and hyperparameters (variance components) obtains using Expectation-Maximization (EM). The latter yields the Restricted Maximum Likelihood (ReML) estimate of the hyperparameters and the Maximum A Posteriori (MAP) estimate of the source parameters.

The paper is organized as follows. In the methods section we reprise the so-called distributed (or imaging) model and its classical weighted Minimum Norm (MN) solution. We then describe the extended empirical Bayes approach, which generalizes the former. The hierarchical linear model, enabling the introduction of multiple priors, is presented together with the estimation procedure and Log-evidence. In the applications section, we evaluate the approach on a real MEG data set. Contrary to previous studies, we evaluate (i) different constraints on the sources, (ii) the effect of various variance components at the first level and (iii) the effect of different solution space. Our results are presented and discussed in the two last sections, respectively.

2. Methods

2.1 Forward model

Consider a t-sample-wide window of MEG measurements acquired on n sensors. To explain these data, we rely upon a distributed model with p dipoles, spread over the cortical surface, with fixed position and orientation [6]. This gives the following linear forward model to be inverted

$$Y = KJ + E \tag{1}$$

where Y is the $n \times t$ observation matrix, K is the $n \times p$ forward operator, J is the $p \times t$ matrix of unknown dipole magnitudes and E is additive measurement noise. Columns (resp. rows) of K describe the measurements observed across all sensors induced by a particular dipole (resp. the flow of current for a given sensor through each dipole location). K is fully determined by the dipole locations and orientations, the geometry and the conductivity of head tissues [7].

2.2 Classical weighted minimum norm approach

The distributed linear model represented by equation (1) is highly underdetermined since the number of dipoles needed to cover the whole cortical surface (several thousands) is much larger than the number of measures (a few hundred sensors). As a consequence, the priors are needed for a single solution for the parameter matrix J.

A common approach to this ill-posed MEG inverse problem is the Weighted Minimum Norm (WMN) solution [8]. Assuming the noise component is Gaussian $E \sim N(0,C_e)$ with known covariance C_e, WMN solutions minimize a linear mixture of some weighted norm $\|WJ\|$ and the residuals of the data fit

$$\begin{aligned} J_{WMN} &= \underset{J}{\operatorname{argmin}}\{\| C_e^{-1/2}(Y-KJ)\|^2 + \lambda \| WJ \|^2\} \\ &= \underset{J}{\operatorname{argmin}}\{(Y-KJ)^T C_e^{-1}(Y-KJ) + \lambda J^T (W^T W) J\} \end{aligned} \tag{2}$$

where W is the weighting matrix and λ is the hyperparameter which expresses the balance between fitting the data and minimizing the prior term. For a given λ, the solution of equation (2) is

$$\begin{aligned} J_{WMN} &= [K^T C_e^{-1} K + \lambda (W^T W)]^{-1} K^T C_e^{-1} Y \\ &= (W^T W)^{-1} K^T [K (W^T W)^{-1} K^T + \lambda C_e]^{-1} Y \end{aligned} \tag{3}$$

by the matrix inversion Lemma.

Importantly, the WMN solution corresponds to the Bayesian estimate of the source parameters under Gaussian assumptions. The conditional expectation or posterior mean of J is given by

$$\begin{aligned} \mathrm{E}(J\,|\,Y) &= [K^T C_e^{-1} K + (C_j^{-1})]^{-1} K^T C_e^{-1} Y \\ &= C_j K^T [K C_j K^T + C_e]^{-1} Y \end{aligned} \tag{4}$$

where C_j is the prior covariance of the source parameters. Comparing equations (3) and (4) shows that the weighting matrix W relates directly to the prior covariance of the sources

$$(\lambda W^T W)^{-1} = C_j \tag{5}$$

or, equivalently, to the precision (inverse of variance)

$$\lambda(W^T W) = C_j^{-1} \tag{6}$$

which motivates particular forms for W [9]. Note that when C_j^{-1} tends to zero (flat priors), solutions (3) and (4) correspond to the Maximum Likelihood estimate of the source parameters.

The WMN solution (3) depends non-linearly on the value of hyperparameter λ. Namely, the higher the noise in the data, the more one should regularize (increase λ). However, this is insufficient for estimating λ. A widely used heuristic for estimating λ consists of plotting the weighted norm $\|WJ_{WMN}\|^2$ of the regularized solution, versus the norm of the residuals $\|C_e^{-1/2}(Y-KJ_{WMN})\|^2$, for different values of λ. This yields a L-shape curve whose inflection point gives a satisfactory value for λ [10]. However, a major drawback of the L-curve approach is that the WMN solution has to be calculated for a large number of different hyperparameter values. Moreover, this heuristic is not extendable to the estimation of multiple hyperparameters that are needed for accommodating multiple priors.

The empirical Bayesian approach introduced in [4] enables us to cover multiple hyperparameters.

2.3 Parametric empirical bayesian approach

In the Parametric Empirical Bayesian (PEB) approach, forward model (1) is rewritten using a two-level hierarchical model

$$\begin{aligned} Y &= KJ + E_1 \\ J &= E_2 \end{aligned} \tag{7}$$

where errors at both level follow a Gaussian distribution with zero mean, $E_1 \sim N(0,C_e)$ and $E_2 \sim N(0,C_j)$. Contrary to the classical WMN approach, the two covariances are unknown. They are modelled as a linear combination of variance components Q, which embody, at the second level, the various priors one wants to introduce

$$\begin{aligned} C_e &= \mu_1 Q_{e,1} + \mu_2 Q_{e,2} + \ldots \\ C_j &= \lambda_1 Q_{j,1} + \lambda_2 Q_{j,2} + \ldots \end{aligned} \tag{8}$$

and relate to the spatial data covariance as follows

$$E\{YY^T\} = KC_jK^T + C_e \tag{9}$$

which can be estimated empirically.

We now denote by Λ the vector of unknown hyperparameters at both levels. Λ is estimated, together with parameters J, using an EM algorithm.

EM maximizes the likelihood of the data $p(Y|\Lambda)$ conditional on hyperparameters Λ, in the presence of unobserved variables J

$$p(Y \mid \Lambda) = \int_j p(Y,J \mid \Lambda) \tag{11}$$

which amounts to maximizing the log-likelihood

$$\log p(Y \mid \Lambda) = \log \int_j p(Y,J \mid \Lambda) \geq F(q(J),\Lambda) \tag{12}$$

where $q(J)$ is any distribution over the model parameters and F is the negative free energy. F is the sum of the energy and the entropy terms

$$F(q(J),\Lambda) = \int_j q(J)\log p(Y,J \mid \Lambda) - \int_j q(J)\log q(J) \tag{13}$$

The EM algorithm maximizes $F(q(J), \Lambda)$ by alternatively:

- maximizing $F(q(J),\Lambda)$ with respect to $q(J)$, while keeping Λ constant (E-step);
- maximizing $F(q(J),\Lambda)$ with respect to Λ, while keeping $q(J)$ constant (M-step).

In fact, the maximum in the E-step obtains when $q(J) = p(J|Y,\Lambda)$ at which point the inequality in (10) becomes an equality. In our case, the E-step solution corresponds to the MAP estimate of the source parameters whose analytic form is given by equation (4). Consequently, the E-step can be embedded in the M-step which then yields the Maximum Likelihood (ML) estimate of the hyperparameters by integrating $p(Y,J|\Lambda)$ over the parameters using the current estimate of their conditional distribution. It can be shown that this solution is equivalent to the ReML estimate of Λ which accounts for the loss of degrees of freedom in the model incurred from simultaneously estimating J [11].

Given the hierarchical model (7) and Gaussian assumptions, the negative free energy is

$$F = -\tfrac{1}{2}\log|C_e| - \tfrac{1}{2}(Y-KJ)^T C_e^{-1}(Y-KJ) - \tfrac{1}{2}tr\{C_{J/Y}K^T C_e^{-1}K\} + \tfrac{1}{2}\log|C_{J/Y}| + const \tag{14}$$

where the conditional covariance is

$$C_{J/Y} = (K^T C_e^{-1} K + C_j^{-1})^{-1} \tag{15}$$

The derivatives of F with respect to each hyperparameter can be calculated and the maximum found [3].

2.4 Bayesian model selection

In [5], we introduced model selection based on the model evidence or marginal likelihood. A model M is defined by the set of variance components $\{Q_{e,i};Q_{j,i}\}$, the forward operator K and the corresponding cortical mesh defining the dipole positions and orientations.

The log-evidence of M is defined by

$$\log p(Y\,|\,M) = \log\int_j p(Y,J\,|\,M)p(J\,|\,M) \tag{16}$$

For a given mesh and forward solution, the log-evidence reduces to log $p(Y|\Lambda)$ which corresponds to the objective function maximized by the ReML scheme. The higher the log-evidence, the better the model. Hence, given the same data, different models can be compared based on their log-evidence. Furthermore, the ratio of two model evidence defines the Bayes factor, which can be interpreted as described in [12].

Note that, contrary to the L-curve approach, the PEB approach maximizes the complete marginal likelihood which accounts for the accuracy as well as the complexity of the model.

3. Application

3.1 Previous studies

In recent communications [4, 5], we showed, using either simulated EEG or MEG data, that the PEB approach is useful as:

- the ReML solution was significantly better than a classic WMN solution under a single constraint;
- the regularization (the influence of the prior terms) increased as the noise level increased;
- invalid location priors did not significantly impair ReML performance when valid location priors were also considered.
- Bayesian model selection based on log-evidence indentified the optimal solution among those elicited with different prior sets.

3.2 Evaluation on real MEG data

We here further evaluate the PEB approach on a single subject MEG dataset and illustrate the usefulness of Bayesian model selection, when considering two different mesh sizes and different sets of variance components at both levels.

The subject made symmetry judgments on faces and scrambled faces (for a description of the paradigm, see [13]). The MEG data were acquired on a 151-channel CTF Omega system at the Wellcome Trust Laboratory for MEG Studies, Aston University, England. The epochs (67 face trials, collapsing across familiar and unfamiliar faces, and 84 scrambled trials) were baseline-corrected from -100ms to 0ms, low-pass filtered (cutoff at 35Hz) and averaged over trials. The two types of event-related fields were substracted (faces – scrambled) to study the face-specific perception effect characterized by the M170 signal component occurring around 170ms after stimulation. A data window centred on the signal peak, from 165 to 210ms, was used. A T1-weighted MRI was also obtained with a resolution of

1x1x1mm^3. Head-shape was digitized with a 3-D Polhemus Isotrak and used to coregister the MRI with the MEG data. A segmented cortical mesh was created using Anatomist [14], with approximately 7200 dipoles oriented normal to the grey matter. Finally, a single-shell spherical head model was constructed using BrainStorm [15] to compute the forward operator K.

At the first (sensor) level, we considered three variance components. $Q_{e,1}$ was a simple identity matrix, modelling an independent and identically distributed noise over sensors. $Q_{e,2}$ (resp. $Q_{e,3}$) was determined empirically by anti-averaging the face (resp. scrambled) trials.

At the second (source) level, we considered two different prior variance components, either separately or together. The first one was defined by $Q_{j,1}=DD^t$, where a spatial smoothness operator D was defined on the cortical mesh, using a Gaussian kernel with a standard deviation of 8 millimetres. The second prior, $Q_{j,2}$, was an intrinsic functional prior based on Multivariate Source Prelocalization (MSP), previously used in [5]. Comparing the observed scalp topography with the dipole forward field, MSP provides an activation coefficient for each dipole of the model. This coefficient can be used to both restrict the solution space to the dipoles that are more likely to be active, as well as to define a quantitative prior in terms of a variance component [16].

Here, we compared the ReML solutions obtained for different combinations of the above variance components and when restricting the solution space to, either 1500 or 3000 sources, that were most active according to MSP.

4. Results

Table 1 shows the log-evidence for the different sets of variance components and the two mesh sizes considered.

Table 1. Log-evidence obtained for the different sets of variance components and the two sizes of solution space: 3000 (in bold) and 1500 dipoles

	$Q_{j,1}$	$Q_{j,2}$	$Q_{j,1}$ and $Q_{j,2}$
$Q_{e,1}$	**433.24**	**415.31**	**410.81**
	409.7	403.49	388.39
$Q_{e,1}$, $Q_{e,2}$ and $Q_{e,3}$	**448.37**	**446.95**	**443.44**
	436.14	436.09	358.53

Whatever the set of variance components, the results in table 1 indicate that models based on the 3000 dipole mesh yield the highest model evidence. Furthermore, for this mesh size only, including the empirical estimates of the two trial-type noise covariances yields a better model, for all sets of priors on the sources. Finally, the highest log-evidence obtains when accounting only for the spatial smoothness constraint while the lowest one obtains when combining the smoothness and the intrinsic functional priors.

The best model is the one based on the 3000 dipole mesh and including the three variance components at the first level with a spatial smoothness prior at the second level. A probabilistic interpretation of Bayes factors, as explained in [12], further provides us with positive or even strong evidences in favour of this model.

Figure 1 shows the cortical activation map, estimated using the PEB approach with the selected model and its projection onto the subject's structural Magnetic Resonance Images (sMRI). The iterative computation required only a few seconds to converge, on a standard workstation. The cortical projection represents, for the most active regions, the absolute dipole amplitudes at signal peak normalized to the map maximum.

The brain area showing the main difference in evoked power, between face and scrambled trials, was the right Inferior Occipital Gyrus (IOG) (see Fig. 1.a). Weaker activations were also found in the horizontal segment of the right Superior Temporal Sulcus (STS) (see Fig 1.b), the bilateral precuneus and the left calcarine sulcus (see Fig. 1.c). The activation of these regions is consistent with the same comparison using fMRI data [13]. The activation of the ventral and lateral occipitotemporal regions is also consistent with recent localisations of the M170 component [17,18].

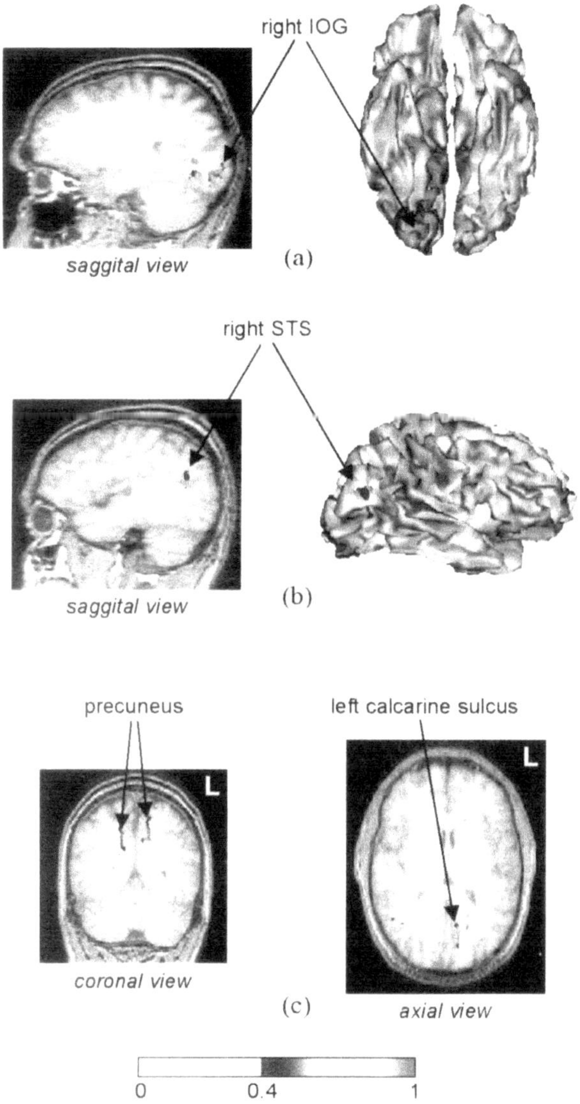

Fig. 1. PEB source localization obtained for the model with highest log-evidence: main active sources are shown using both the cortical mesh and projection onto subject's MRI.

5. Conclusion

The Parametric Empirical Bayesian approach to solve the MEG/EEG inverse problem affords an extended framework for incorporating multiple priors upon the source spatial distribution. Priors are introduced in terms of variance components and their optimal contribution is derived from the data via ReML estimation of hyperparameters. The latter provides a graceful generalization of the L-curve heuristic. The MAP estimate of the source amplitudes and ReML estimates of the hyperparameters are obtained using an EM algorithm. The EM scheme is not computationally demanding since only low dimension matrices needs to be inverted (n x n matrices see equation 9). Finally, Bayesian model selection can proceed using model evidence. Such an approach is of particular interest in the context of multimodal integration for neuroimaging data fusion.

In previous studies, the PEB approach was extensively evaluated on EEG and MEG synthetic data. Here, we evaluated the ReML solution and Bayesian model selection on a single subject MEG dataset. We showed that model selection can adjudicate among different sets of priors. Critically, it can also be used to compare models in terms of dipole mesh size or variance components at the sensor level.

We have recently extended this PEB approach to estimate the induced responses which, contrary to the evoked ones, are not phase-locked with the stimulus [19].

Acknowledgments

The authors are grateful to Pia Roshstein for useful discussion about the results. The Wellcome trust funded this work. A Marie Curie Fellowship from the EU funded JM.

References

1. P.L. Nunez and R.B. Silberstein "On the relationship of synaptic activity to macroscopic measurements: does co-registration of EEG with fMRI make sense?," *Brain Topogr.*, 1999, vol. 13, pp. 79-96.
2. S. Baillet, J.C. Mosher and R.M. Leahy "Electromagnetic brain mapping," *Sign. Proc. Mag.* 2001, vol 18, pp. 14-30.
3. K.J. Friston, W. Penny, C. Phillips, S. Kiebel, G. Hinton and J. Ashburner "Classical and Bayesian inference in neuroimaging: theory," *NeuroImage*, 2002, vol. 16, pp. 465-483.

4. C. Phillips, J. Mattout, M.D. Rugg, P. Maquet and K.J. Friston "An empirical Bayesian solution to the source reconstruction problem in EEG," *NeuroImage*, 2005, vol. 24, pp. 997-1011.
5. J. Mattout, C. Phillips, W. Penny, M.D. Rugg, and K.J. Friston "MEG source localization under multiple constraints: an extended Bayesian framework," *NeuroImage*, submitted.
6. A.M. Dale and M. Sereno "Improved localization cortical activity by combining EEG and MEG with MRI surface reconstruction: a linear approach," *J. Cognit. Neurosci*, 1993, vol. 5, pp. 162-176.
7. J.C. Mosher, R.M. Leahy and P.S. Lewis "EEG and MEG: Forward solutions for inverse methods," *IEEE Trans. Boimed. Eng.*, 1999, vol. 46, pp 245-259.
8. M.S. Hamalainen and R.J. Llmoniemi "Interpreting magnetic fields of the brain - minimum norm estimates," *Med. Biol. Eng. Comput.*, 1994, vol. 32, pp 35-42.
9. O. Hauk "Keep it simple: a case for using classical minimum norm estimation in the analysis of EEG and MEG data," *NeuroImage*, 2004, vol. 21, pp 1612-1621.
10. P.C. Hansen "Analysis of discrete Ill-posed problems by means of the L-curve," *SIAM Rev.*, 1992, vol. 34, pp 561-580.
11. D.A Harville "Maximum likelihood approaches to variance component estimation and related problems," *J. Am. Stat. Assoc.*, 1977, vol. 72, pp. 320-338.
12. W.D. Penny, K.E. Stephan, A. Mechelli and K.J. Friston "Comparing Dynamic causal models," *NeuroImage*, 2004, vol. 22, pp. 1157-1172.
13. R.N. Henson, Y. Goshen-Gottstein, T. Ganel, L.J. Otten, A. Quayle and M.D. Rugg "Electrophysiological and haemodynamic correlates of face perception, recognition and priming," *Cereb. Cortex.*, 2003, vol. 13, pp. 793-805.
14. J.F. Mangin, D. Riviere, A. Cachia, E. Duchesnay, Y. Cointepas, D. Papadopoulos-Orfanos, P. Scifo, T. Ochiai, F. Brunelle and J. Regis "A framework to study the cortical folding patterns," *NeuroImage*, 2004, vol. 23, pp. 129-138.
15. S. Baillet, J.C. Mosher and R.M. Leahy "BrainStorm beta release: a Matlab software package for MEG signal processing and source localization and visualization" *Neuroimage,* 2000, vol. 11, pp. S915.
16. J. Mattout, M. Pélégrini-Issac, L. Garnero and H. Benali "Multivariate Source Prelocalization: use of functional Informed Basis Functions to better condition the MEG inverse problem," *NeuroImage*, in press.
17. R.N. Henson, J. Mattout, K. Friston, S.Hassel, A. Hillebrand, G.R. Barnes and K.D. Singh "Distributed source localization of the M170 using multiple constraints," *Human Brain Mapping Conrefence 2005*, in press.
18. T. Tanskanen, R. Nasanen, T. Montez, J. Paallysaho, R. Hari "Face recognition and cortical rcsponses show similar sensitivity to noise spatial frequency" *Cereb. Cortex*, in press.
19. K.J. Friston, R. Henson, C. Phillips and J. Mattout "Bayesian estimation of evoked and induced responses," submitted.

Differential Contribution of Early Visual Areas to Perception of Contextual Effects: fMRI Studies

Yoshimichi Ejima

Kyoto Institute of Technology

Chapter Overview. Perception of a visual target and the responses of cortical neurons can be strongly influenced by the context surrounding the target. The specific layout of the surrounding features can produce a suppressive or facilitatory effect. We investigated the contextual effects of metacontrast masking, chromatic induction, and amodal completion of contours of a partially occluded object on the brain activity in human early visual areas using functional magnetic resonance imaging (fMRI). Signal modulation corresponding to the perceptual effects of contextual stimuli was detected in the retinotopic areas, including V1, V2, and V3. Brain activity in higher order visual areas showed more prominent specificity to stimulus attributes and contexts. Such a contextual effect is discussed in the light of different contextual modulation roles at different stages.

Key Words. Contextual effect, visual cortex, fMRI, retinotopy, spatio-temporal interaction.

1. Introduction

A fundamental question in visual perception is how the visual cortex integrates information from local image features into global configurations and yields percepts such as contours, surfaces, visual shapes and objects. We do not perceive a collection of isolated visual features but structured ensembles, where the perception of the characteristics of the features depends on the context in which it is presented. This perceptual organization and the contextual sensitivity of object perception have been difficult to reconcile with our current understanding of visual processing in the brain. Traditionally, the process of complex global shapes has been ascribed to higher-order cortical areas in the hierarchically organized visual system [1, 2]. However, recently, a growing body of evidence suggests that early visual areas (e.g., V1, V2) may respond to global, rather than simple local, features and that the primary visual cortex (V1) may play an important role [3]. It was found that stimuli outside what is called the classical receptive field (CRF) of a visual neuron in V1 could influence the magnitude of the excitatory response to an appropriate stimulus presented inside the CRF: maximal modulations of facilitator interactions are generally observed when center and surround stimuli are iso-oriented, co-aligned and the same spatial frequency. A simple, conceptual model of how the observed pattern of contextual modulations might form the neural basis of contour integration and surface segmentation has been proposed [4]. The contextual modulation may be of fundamental importance in understanding visual processing in the brain.

In human vision, the contextual effect manifests itself in various perceptual phenomena, where a context is defined by the spatio-temporal configuration of perceptual attributes such as orientation, luminance and color contrast, spatial frequency, geometric elements and so on. Whether the perceptually observed contextual effects are correlated to the contextual modulatory activity in the early visual cortex is an open question. Here, brain activities during the perception of three types of perceptual phenomena, meta-contrast, chromatic induction and amodal completion were measured to elucidate the role of early visual areas in the integration of local features into global shapes, by functional magnetic resonance imaging (fMRI). FMRI allows simultaneous measurement of responses to the same stimulus set from multiple visual areas that is not possible with standard recording techniques. We performed fMRI studies in the human visual cortex, where recent fMRI studies provided evidence for functional topology of striate and extrastriate visual areas [5]. The correlation between the fMRI signal and the underlying neural responses [6] emphasizes the importance of fMRI studies for bridging the gap between the neurophysiological findings in monkeys and those reported in combined psychophysical and imaging investigations with humans. FMRI can reveal the overall activity patterns seen in large neuron populations, while still allowing different visual areas to be studied separately.

2. Experimental Protocol

The primate visual cortex contains multiple topographic maps of the visual hemifield, known as retinotopy. Roughly speaking, the angular positions in the visual field are represented along the posterior-anterior dimension on the cortical surface, and the eccentricity in the visual field is represented along the lateral-medial dimension on the cortical surface, as shown in Fig.1a [7]. In an fMRI study, we can determine the retinotpic areas in the human visual cortex using phase encoding stimuli [5, 8]: an expanding checkered ring was used to map eccentricity in the visual field, and a rotating checkered wedge was used to map the angular positions in the visual field. Functional homology between the human and the macaque visual cortices has provided an important cue to functional subdivisions of the human visual cortex [9, 10].

Retinotopy in the visual cortex enables us to measure spatial interaction revealed by brain activity during the perception of contextual effect. To this end, the experimental design and the analysis of fMRI data consisted of two stages, shown in Fig. 1b: one was the localization of a target sub area within individual retinotopic areas on the cortical surface; the other was the measurement of the modulatory activity in the target sub area induced by a contextual stimulus. When a target stimulus was alternated by a uniform stimulus in a block design paradigm, a cortical region representing the target stimulus was identified. The shape and the position of the target stimulus were designed by taking into account the retinotopy. This target region was the ROI (region of interest). Using localizers and retinotopic mapping stimuli we were able to subdivide each early visual area into regions corresponding to the target and surrounding features. To measure contextual modulation activity, a block design paradigm, in which the target stimulus was continuously presented and the surrounding stimulus was alternated between what produced a perceptual contextual effect and what did not produce it, was used. In each ROI we investigated the hemodynamic responses and compared responses between with and without the contextual effect.

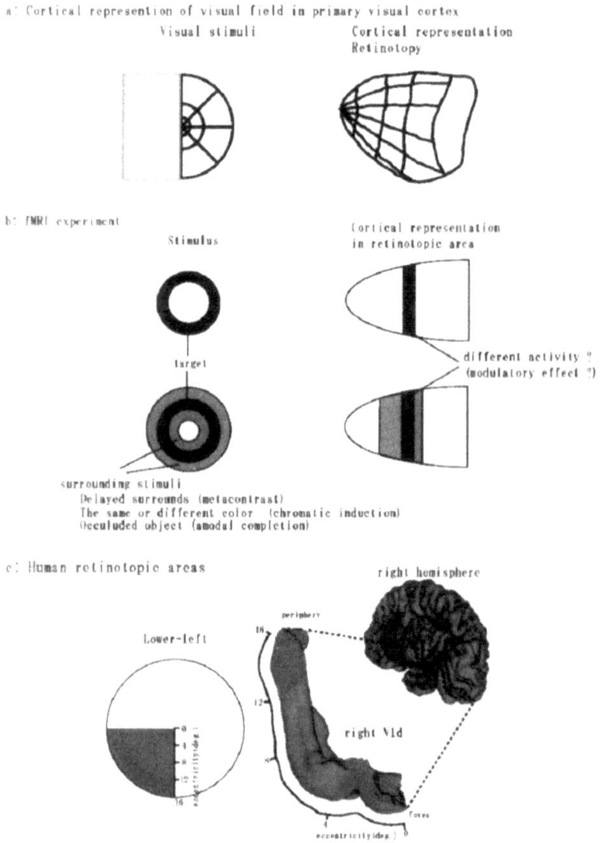

Figure 1 Experimental protocol
a:Retinotopic organization in the primary visual cortex is shown schematically based on the macaque data [7]
b:Visual stimulation and cortical representation in fMRI study of contextual modulation effect
c:Human retinotopic area (V1d)

Fig. 1.

3. Contextual Modulation of Brain Activity

3.1 Metacontrast masking

Metacontrast masking is a type of backward masking that occurs when a target is followed by a non-overlapping mask which shares a contour. The visibility of a target stimulus can be strongly reduced when it is followed in time by a spatially non-overlapping mask stimulus. Visual masking has been, and continues to be, a powerful psychophysical tool for investigating the properties of spatial-processing mechanisms. A clear understanding of the underlying mechanisms by which such a masking effect can occur is required for the informed use of masking as a methodological tool to investigate the temporal sequencing and various levels of information in the visual system.

We investigated the responses of human visual cortex to metacontrast stimuli. Perceptually, metacontrast masking strongly depends on the color of the stimuli: the perceptual visibility of the target stimulus was much more reduced (masking effect) when the same color is used for a target and a mask stimulus than when different colors were used for a target and a mask stimulus. By alternating between the same color and different color conditions, the modulation effect of metacontast was measured in human visual cortical areas. Retinotopically mapped visual areas showed responses that varied in a qualitatively similar way to the perceptually observed metacontrast masking. Whereas brain activity increased while the subject perceived the target (the different color condition), it decreased while the visibility of a target stimulus was strongly reduced by the mask stimulus (the same color condition). Although there were individual differences, the group analysis showed that the brain activations in V1, V2 and V3 were modulated by the masking stimulus. The magnitude of signal modulation in V2 and V3 was larger than that in V1. This indicates that the response corresponding to the masking effect may be more robust in V2 and V3 than V1. The visual cortical areas higher than V3 also showed contextual modulation of the fMRI signal elicited by masking.

Psychophysical models of the metacontrast masking assume that the after-coming mask acts to erase visual information or to interrupt its further processing. The present results indicate that the reduction in brain activity in early visual areas elicited by the masking stimulus may be a neural correlate to erasure or interruption. It has been considered that higher level cognitive processes can modulate backward masking. For instance, the magnitude of backward masking is affected by perceptual grouping and segmentation and by the development of selective visual attention. The present finding suggests that the brain activities of early visual areas are profoundly influenced by the masking effect and that these influences may be of fundamental importance in understanding perceptual grouping and segmentation. Backward masking has been used to study certain clinical anomalies related to vision and brain function, such as amblyopia, closed head injury, developmental dyslexia, mania, and schizophrenia. Furthering our understanding of backward visual masking may, therefore, provide better clinical markers for these clinical conditions [11, 12].

3.2 Chromatic induction

When humans perceive the color of a region, their perception is influenced not only by the local distribution of wavelengths from the region, but also by longer distance color brightness contrast effects at the boundary of the region. When two circles have identical wavelength distributions, but the adjacent surrounding regions are different colors, the circles appear different in color, demonstrating that a region's color appearance is influenced by the surrounding color. The perception of the hue in a region in the visual field can be derived from color-opponent mechanisms like those observed in the LGN and V1 that respond to local color contrast [13]. However, the chromatic induction effect requires further cortical spatial processing. It requires neuronal elements that respond preferentially to color boundaries and that send this signal around the cortex to propagate the boundary's effect into perceived regions. It is unclear how the early visual cortex contributes to the color boundary effect and its propagation.

We investigated brain activities of the visual areas in the human visual cortex to following chromatic induction stimuli, by alternating the condition: where different hues were adjacent (chromatic induction) and where the same hue with different saturations was adjacent (no induction). In order to examine the color boundary effect and propagation processing, a gap between two color regions was manipulated. Visual areas showed modulatory activation depending on the surrounding color: the activation of visual areas increased when different hues were adjacent and the chromatic induction occurred, whereas it decreased when the same hue with different saturations was adjacent. The dependence of the brain activation on the spatial property of the stimuli (gap) varied among visual cortical areas. The brain activations in V1, V2 and V3 were strongly influenced by the gap between two color regions: a significant activation was observed for the stimulus in which the two color regions were adjacent without a gap, but it reduced when a gap was introduced between the two color regions. On the other hand, the brain activation in V4v was not sensitive to the spatial property of the boundary: significant activation was observed after introducing a gap between the test and inducing stimuli.

The present result provides evidence regarding the existence of at least two different spatial processing stages in the chromatic induction effect. One is the processing of local color boundaries, which may be achieved by the color opponent mechanisms in early visual areas that respond preferentially to color boundaries. The other is additional neuronal elements at a higher level (V4) that send the signal of color boundaries around the cortex to propagate the boundary's effect to perceived regions. We propose that these different color mechanisms may contribute separately to perception of color boundaries and colored regions.

3.3 Contour interpolation

Human observers can perceive a shape partially specified by image contours by overcoming the fragmentation of the optic input. There are three types of perceptual integration of optic fragments: 1) virtual lines perceived as the most natural chaining of dots or oriented elements; 2) amodal contours of partially occluded shapes, typically defined by T-junction stems; 3) modal contours perceived as illusory contours for Kaniza's figures. In virtual grouping, implicit lines are perceived as connecting separate elements. In amodal completion, image-specified contours continue behind occluders along trajectories that bind a partially occluded surface. Modal completions are characterized by the visibility of illusory contours that bind a perceptually integrated surface. Concerning the modal (illusory) contours, neurophysiological experiments in cats and monkeys suggest that neurons in at least two visual areas, V1 and V2, carry signals related to illusory contours and that signals in V2 are more robust than in V1, whereas functional magnetic resonance image experiments on the human visual cortex suggest that illusory contours are processed throughout the visual pathway, but that signals are strongest in higher-order areas, V3A, V7, V4v and V8 [14, 15]. It is of interest to compare the cortical circuits activated by shapes defined by amodal contours and by real contours. For this reason, we collected functional magnetic resonance images of the human visual cortex during the perception of amodal and real contours.

Perceptually, amodal completion is yielded by various cues. We distinguished two types of cues. With an implicit cue of occlusion, amodal completions were supported by the segregation of T-junction stems from

T-junction tops: paired stems became the visible portions of the partially occluded contour, whereas tops belonged to the contour of the occluding surface. With an explicit cue of occlusion, the stimulus further contained a global context of occlusion presented by a moving stimulus. With the implicit cue of occlusion, the contextual modulation effect of amodal completion was not observed in early visual areas of V1, V2 and V3. On the other hand, with the explicit cue of occlusion, significant contextual modulation effect was observed in V1, V2 and V3: the brain activation in the target area increased while the subject perceived the amodal contours, but no significant activation was observed when the subject did not perceive amodal contours. The present result suggests that neurons in early visual areas, V1, V2 and V3, carry signals related to the completion of amodal contours based on the dynamic and global context. Higher order visual areas, V4d (V3B) or LOC, showed robust activity while perceiving amodal completion. This suggests that the activation of higher order visual areas may be responsible for the modulation of brain activity in the early visual areas.

The present results have demonstrated that the global and dynamic context provided by stimulus sequence-depicting dynamic changes in a stimulus configuration, rather than the local context of the T-junction, affects the brain activity in early visual cortical areas. Most visual interpolation models adopt a two-stage approach [16], which states that the decision to interpolate is taken at the end of the first stage in which the input is analyzed to evaluate its compatibility with geometric constraints: if the input is geometrically compatible, optic fragments are fed to a shape integrator that generates the interpolated trajectory in the second stage. Other models [17] adopt a dynamic approach, which states that optic fragments represent the stimulus conditions for a completion process whose final state can achieve a variable degree of stability: both salience and compatibility derive from the dynamic constraints of the interpolation process. Our results are consistent with the dynamic approach and with the idea that visual interpolation reflects organization according to the global and dynamic context, which strongly influences brain activity in early visual areas.

4. Conclusion

The early visual system is a significant part of vision proper, although many processes other than those of early vision enter into the construction of visual representations of the perceived world. I have demonstrated that the early visual system carries out complex computations, showing evidence for the contextual modulation in early visual areas in the human brain. With the advent of brain imaging techniques, the neuroanatomical locus of early vision in the human brain has been charted with precision and its functional properties have been articulated in detail over the years. I suggest that many contextual effects in vision may come about after the processing of the early visual areas. Contextual modulators act as a kind of gain control that affects the salience of features detected at each level of the hierarchy: dynamic grouping combines the contrasts in each of the separate feature dimensions into a single, generalized contour-boundary map.

Acknowledgments

I appreciate the assistance and co-work of H. Fukuyama, C.Tanaka, H.Yamamoto, T. Hanakawa, K. Maeda, H. Ban, A. Kondo, S. Takahashi, & S. Urayama. This work was supported by "T&D support for funding selected IT proposals" of the Ministry of Public Management, Post and Telecommunications.

References

1. D.J.Felleman & D.C. Van Essen, "Distributed hierarchical processing in the primate cerebral cortex", Cerebral Cortex, vol.1, pp.1-47, 1991
2. J.H.R. Maunsell & W.T. Newsome, "Visual processing in monkey extrastriate cortex", Ann. Rev. Neurosci, vol.10, pp.363-401, 1987
3. D.Fitzpatrick, "Seeing beyond the receptive field in primary visual cortex", Curr. Opin. Neurobiol. Vol.10, pp.438-443, 2000
4. M.K.Kapadia, G. Westheimer & D. Gilbert, "Spatial distribution of contextual interactions in primary visual cortex and in visual perception", J. Physiology, vol.84, pp. 2048-2062, 2000
5. B.A. Wandell, "Computational neuroimaging of human visual cortex", Ann, Rev. Neurosci., vol.22, pp.145-173, 1999
6. N.K. Logothetis et al., "Neurophysiological investigation of the basis of the fMRI signal", Nature, vol.412, pp.150-157, 2001
7. R.B.Tootell et al., "Functional anatomy of macaque striate cortex II. Retinotopic organization", J. Nuerosci., vol.8, pp.1531-1568, 1988.
8. M.I. Sereno et al., "Borders of multiple visual areas in humans revealed by functional magnetic resonance imaging", Science, vol.268, pp.889-893, 1995.

9. Y. Ejima et al, "Interindividual and interspecies vatiaions of the extrastriate visual cortex", Neuroreport, vol.14,pp.1579-1583, 2003
10. Y.Ejima & S.Takahashi, Positioning of retinotopic areas and patterning of cerebral cortex layout", Neuroreport, vol.17, pp. 9-12, 2005.
11. M.F. Green, H.H. Nuechterlein, & B. Breitmeyer, "Backward masking oreformance in unaffected siblings of schizophrenic patients", Arch. General Psychiatry, vol.54, pp.465-472, 1997
12. M. Williams, K. LeCluyse, & N.Bologna, "Masking by light as a measure if visual integration time in normal and disable readers", Cilin. Vision Sci, vol.5, pp.335-343, 1990
13. R. Shapley & M. Hawken, "Neural mechanisms for color perception in the primary visual cortex", Curr. Opin. Neurobiol. Vol.12, pp.426-432, 2002.
14. R. von der Heydt & E. Peterhans, "illusory contours and cortical neuron responses", Science, vol.224, pp.1260-1262, 1984.
15. J.M. Mendola et al., "The representation of illusory cne real contours in human cortical visual areas revealed by functional magnetic resonance imaging", J. Neurosci. Vol.19, pp.8560-8672, 1999.
16. P.J. Kellman & T.F. Shipley, "A theiry of visual interpolation in object perception", Cognitive Psychol. Vol.23, pp.141-221, 1991.
17. C.Fantoni & W.Gerbino, "Contour interpolation by vector-field combination", J. Vision, vol.3, pp.281-303, 2003

Brain-machine Interface to Detect Real Dynamics of Neuronal Assemblies in the Working Brain

Yoshio Sakurai

Department of Psychology, Graduate School of Letters, Kyoto University, Kyoto, Japan CREST, Japan Science and Technology Agency, Kawaguchi, Japan

Chapter Overview. In awaking and working brains, it is still not clear how dynamic the neuronal activity and their functional connectivity are and how high potentials for learning they have. Our research project will detect actual features of the dynamics and learning potentials of the working brains by constructing and applying a high-performance brain-machine interface (BMI). With the BMI system, neuronal activities of animals learning behavioral tasks directly control machine-outputs instead of animal's body-outputs and learn to perform behavioral tasks correctly. The keys to construct such a successful BMI are training animals in appropriate behavioral tasks, recording multi-neuronal activities from the behaving animals for long periods, and decoding neuronal code representing valid information in the working brain. The neuronal code might be expressed in synchronized activity among neighboring neurons which could consist a cell assembly. We, therefore, have developed a method to detect precise sub-millisecond activity interactions among closely neighboring neurons in the brains of behaving animals. The system uses a combination of independent component analysis (ICA) and newly developed multi-electrodes.

Key Words. Brain-machine interface, behavioral task, synchronized correlation, cell assembly, Independent component analysis.

1. Introduction

This paper exemplifies a research area dedicated to connect brain activity directly with computer-based devices. It is called brain-machine interface (BMI). For millions of individuals who have lost to disease or trauma an ability to control their limbs, this provocative new paradigm of a direct link between brains and computers attempts to reestablish the ability and so to interact autonomously with their environments. Therefore biomedical engineering has a central role to play in this exciting new area [1]. In the last decade, however, advances in extracellular and multi-neuronal recording techniques enable us to use BMI not only to aid injured motor systems but to investigate neuronal information process and plasticity in the normal and working brain. The advance introduces new methods for recording and analyzing large-scale brain activity in behaving animals and new signal processing algorithms. These methods with sophisticated sensors and robotics have the potential to coalesce into a new technology devoted to creating reliable and accurate BMI and could show us real dynamics and learning ability of the brain detected with the BMI. In other words, the bottlenecks in this endeavor are not only in the computer technology side but in systems neuroscience field, which is studying how to best record and analyze information in neuronal activity to be sent to computer based information systems. The keys to the success are appropriate behavioral tasks for animals, multi-neuronal recording, spike-sorting and analysis, and neuronal modeling to detect neuronal codes of information in the working brain.

2. Behavioral Task

What behavioral tasks we should use depends on aims and recording areas of the BMI study. BMIs to control the intention of arm movement record neuronal firing in pre and motor cortex and often employ target-hitting tasks to move robotic devices. However, BMIs to detect cognitive and learning-related signals in the brain should employ cognitive behavioral tasks [2,3]. Fig.1 illustrates an example, i.e., a conditional discrimination task we are using. Our behavioral study shows that the task has some advantages to detect learning-related neuronal signals and construct BMI which learns the task with the rat.

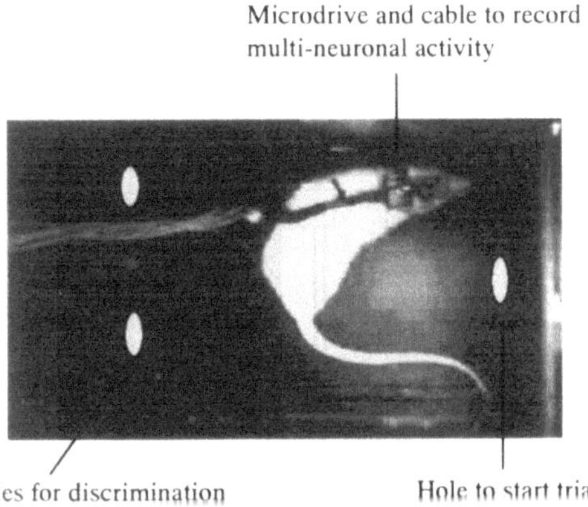

Fig.1. A rat performing the conditional discrimination task.Nose poke to the hole on the right wall start a trial and present high or low tones. The high tone and low tone require the rat to make a nose poke to the right-side hole and left-side hole on the left wall respectively. A microdrive to record multi-neuronal activity is mounted on the head and connected to the cable.

3. Multi-neuronal Recording

BMI needs long-term chronically implantation of microelectrode arrays in targeted brain areas of animals. Though recent results from several research groups raise the optimistic perspective that BMIs for humans will be feasible in the near future, the technical difficulties still lie in better electrodes, ultra low power electronics for amplification and signal processing algorithms for detection and coding in order to reduce the bandwidth of the raw signal, and enable low power transcutaneous transmission of the information to receivers installed in external robotic devices [1]. Fig. 2 is one of our standard microdrives with an array of 6 tetrode electrodes. This enables long-term and stable recording of multi-neuronal activity from behaving rats [4,5,6] and monkeys [7].

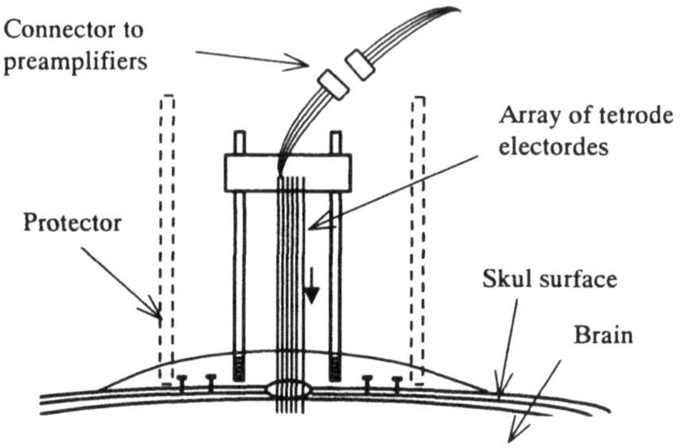

Fig. 2. A microdrive with an array of tetrode electrodes.

4. Spike-sorting and Analysis

Simultaneous and real-time multi-site recording of multi-neuronal activity plays an important role in the BMI. However, current major systems for BMI can only record summated activity of many neurons and cannot precisely isolate activities of individual neurons, especially when they fire simultaneously. Consequently they cannot use neuronal information revealed by interactions among the individual neurons, though one of the most fundamental functioning of the brain involves highly dynamic interactions among many neurons. Our former method with independent component analysis (ICA) [8,9] has solved this problem of spike-sorting, but a few problems have remained: the computational load is heavy, the method can be used only for off-line, not real-time, processing, and the electrode-neuron drift problem remains unsolved. Our recently-developed method [10] solves all these problems and construct a novel system consisting of automatic and real-time spike sorting with ICA in combination with a newly developed multi-electrode. The system has the potential to answer how sub-millisecond interactions among closely neighboring neurons act in information processing in freely behaving animals and therefore can be used to give the additional and useful neural information to the BMI.

5. Neuronal Modeling

Recent system neuroscience studies have suggested that the most essential feature of the brain is highly dynamic and distributed information processing by activity of functional neuron groups. D.O.Hebb called such a functional neuron group "cell assembly" and some recent electrophysiological experiments and theoretical consideration have indicated that cell assemblies surely represent information in the working brain [11,12]. Therefore highly adequate neuronal models involving cell assembly model should be investigated and used to construct high-performance BMI.

Acknowledgments

The research is supported by 21st Century COE Program, Grand-in-Aid for Priority Research and Development in Information Science, and CREST program to Y.S.

References

1. M.A.L. Nicolelis, N. Birbaumer, and K.-L. Muller, "Editorial", IEEE Trans. Biomed., vol.51, 877-880, 2005.
2. Y. Sakurai, "Elemental, configural, and sequential memory processes in the rat can be tested in a single situation in one day", Psychobiol., vol.27, pp.486-490, 1999.
3. Y. Sakurai, "Working memory for temporal and nontemporal events in monkeys", Learn. & Mem., vol.8, pp.309-316, 2001.
4. Y. Sakurai, "Involvement of auditory cortical and hippocampal neurons in auditory working memory and reference memory in the rat", J. Neurosci., vol.14, pp.2606-2623, 1994.
5. Y. Sakurai, "Hippocampal and neocortical cell assemblies encode memory processes for different types of stimuli in the rat", J. Neurosci., vol.16, pp.2809-2819, 1996.
6. Y. Sakurai, "Coding of temporal information by hippocampal individual cells and cell assemblies in the rat", Neurosci., vol.115, pp.1153-1163, 2002.
7. Y. Sakurai, S. Takahashi, and M. Inoue, "Stimulus duration in working memory is represented by neuronal activity in the monkey prefrontal cortex", Eur. J. Neurosci., vol.20, pp.1069-1080, 2004.

8. S. Takahashi, Y. Anzai, and Y. Sakurai, "Automatic sorting for multi-neuronal activity recorded with tetrodes in the presence of overlapping spikes", J. Neurophysiol., vol.89, pp.2245-2258, 2003.
9. S. Takahashi, Y. Anzai, and Y. Sakurai, "A new approach to spike sorting for multi-neuronal activities recorded with a tetrode - how ICA can be practical", Neurosci. Res., vol.46, pp.265-272, 2003.
10. S. Takahashi, and Y. Sakurai, "Real-time and automatic sorting of multi-neuronal activity for sub-millisecond interactions in vivo", Neurosci., 2005 In press.
11. Y. Sakurai, "Population coidng by cell assemblies - what it really is in the brain ?", Neurosci. Res., vol.26, pp.1-16, 1996.
12. Y. Sakurai, "How do cell assemblies encode information in the brain ?" Neurosci. Biobehav. Rev., vol.23, pp.785-796, 1999.

Prism-adaptation Therapy for Unilateral Neglect: from the Facts to a Bottom-up Explanatory Model

Revol Patrice [1,2,3], Pisella Laure[1,2], Luauté Jacques[3], Jacquin-Courtois Sophie[3], Farné Alessandro[1,2], Ota Hisaaki[1], Rode Gilles[1,2,3], Boisson Dominique[1,2,3] , and Rossetti Yves[1,2,3]

[1] Espace et Action, INSERM UMR-S 534, Institut National de la Santé et de la Recherche Médicale, and Université Claude Bernard - Lyon, 16 avenue Lépine, CP12, 69676 Bron, France.
[2]'Mouvement et Handicap', 'Institut Fédératif des Neurosciences de Lyon', and 'Hospices Civils de Lyon', France.
[3]Service de Rééducation Neurologique, Hopital H Gabrielle, Hospices civils de Lyon, Saint-Genis Laval, France

Chapter Overview. Unilateral neglect is defined as the patient's failure to report, respond to, or orient toward novel and/or meaningful stimuli presented to the side opposite to the brain lesion [1-3]. This syndrome is frequently consecutive to the damage of the right brain hemisphere. It often is in association with contralesional hemiplegia, hemianesthesia and hemianopia. Several types of motor deficits have been described in unilateral neglect, ranging from motor neglect or extinction to perturbations of intentional aspects of action.

Key Word. Prism adaptation, Unilateral Neglect, neuropsychological therapy, rehabilitation

1. Ecological Significance of Unilateral Neglect

The presence of unilateral neglect worsen the severity of associated motor or sensory deficit inducing many functionally debilitating effects on everyday life, and responsible for poor functional recovery and ability to benefit from treatment [4-6]. For example, neglect patients can forget to read the left part of a journal or a book; omit to eat the left half of a plate or forget to sheave the left hemi-face. Everyday activities are frequently perturbed in neglect patients and can be investigated by both self-rating and third-person rating (Bergero scale; [7]). As typical action deficit they may distribute fruits on the only right side of an apple tart (ex: [8, 9], exhibit difficulties in finding their room when on the left side of the corridor and in driving a wheel-chair [9], show a biased postural balance [10], slow down actions directed to their left or hit left-sided obstacles.

The history of neglect modulation by a specific intervention started with Diller and Weinberg [11] 25 years ago. It was the study by Rubens [12] that renewed interest in rehabilitation methods for unilateral neglect and a renewal of both theoretical questions and experimental approaches. The experimental neuropsychology approach to unilateral neglect brought in clearer constraints for the precise quantification of the patients' performance and stimulated interest in focal or partial aspects of this syndrome. At this stage local interventions with the patients were shown to alleviate most of the symptoms of unilateral neglect (review in [13]). More recently the idea that specific learning could improve patients has come back and incorporated the physiological aspects of the sensory stimulation [14].

2. Top-down vs Bottom-up Approach

Many attempts have been made in the last 20 years to rehabilitate neglect. Different approaches have been proposed relying mainly on passive physiological stimulations or active training (see review in: [15]). The main goal of these methods is to favor the re-orientation of the motor behavior toward the neglect side and the first difficulty to obtain a generalization of effects at a functional level.

The various manifestations of unilateral neglect share one major feature: patients remain unaware of the deficit they exhibit or at least fail to fully consciously attend to these deficits. This deficit in awareness is dramatically expressed in anosognosia and hemiasomatognosia [16]. It is therefore astonishing that the first methods proposed for the rehabilitation of neglect were mainly based on a voluntary orientation of attention to the left. This paradox was already underlined by Diller and Weinberg themselves: "The first step in the treatment of hemi-inattention is to make the patient aware of the problem. This is particularly difficult in hemi-inattention since this failure in awareness appears to be at the heart of the patient's difficulty" ([11], p.67). It may indeed appear paradoxical to base a rehabilitation procedure on awareness and intention in patients with a deficit in consciousness. How can a sustained overt orienting to the left be obtained from individuals whose pathology is exactly to remain unable to attend to the left? These techniques have produced significant results but are clearly exposed to several limitations. For example, rehabilitated patients may typically produce nearly-perfect performance on classical tests performed during a testing session and then hit the door when walking out of the room, i.e. their voluntary monitoring of attention is restricted to a specific context and does not apply as soon as more automatic control is required. Harvey and Milner [17] have also shown that the training of visual scanning in unilateral neglect may improve line bisection, which requires a sustained voluntary orienting of attention, but not other tasks. To act on higher level cognition in such a way as to bypass the impaired conscious awareness and intention, one should, at least in principle, find another entry route to space representation systems. The transformational theory proposed by Jeannerod and Biguer [18] proposed that it is the transformation of the sensory input into the motor output is impaired in unilateral neglect. Such theoretical position thus proposes an alternative view to the classical top-down approach to neglect rehabilitation; peripheral sensory stimulations can be viewed as bottom-up tools to act on higher-level spatial cognition.

Different manifestations of neglect may be alleviated by sensory stimulation (vestibular, optokinetic, transcutaneous electrical, transcutaneous mechanical vibration, auditory). The first improvement of visual neglect was reported by Silberpfenning [19] after a caloric vestibular stimulation applied to 2 patients who suffered from a cerebral tumor. The stimulation led to a reduction of head and eye deviation and reduction of neglect dyslexia. These preliminary cases were confirmed later by Rubens [12] who showed in 18 right brain-damaged patients with neglect that a left cold caloric stimulation might improve left visual neglect although a right stimulation might worsen the deficit. These exciting results had shown that a cognitive deficit related to damage of the right hemisphere might be positively influenced by a physiological stimulation (for a review of the effects of vestibular stimulation, (see [15]). They were followed by numerous studies assessing the effects of other kinds of stimulation and the nature of the improved symptoms. Following the study by Rubens [12] and the striking results obtained with caloric stimulation, many studies have replicated this result and extended its conclusions. Then several other types of sensory stimulation were proposed and tested in patients. Many manifestations of neglect have been shown to be alleviated by sensory stimulation (vestibular, optokinetic, transcutaneous electrical, transcutaneous mechanical vibration, auditory) (detailed reviews in [13-15]). The improvement has been mainly reported for extra personal neglect (classical neuropsychological testing), but many other aspects have been investigated: personal neglect, sensory and motor deficits of left hemi-body associated to neglect or extinction. Even productive manifestations of unilateral neglect, such as anosognosia or somatoparaphrenia, may be also reduced through sensory stimulation [20]. A transient improvement of force has been also observed in right brain damaged patients with unilateral neglect and hemiparesis [21]. Positive effects were also reported on postural instability in right brain damaged patients with neglect [10]. The second characteristic of effects reported through stimulation is their abrupt onset. In all studies the improvement was observed during, or immediately following the stimulation. Unfortunately however, a single application of these techniques produces positive effects lasting for only up to about 10-15 minutes, and then they vanish within minutes.

Another bottom-up approach that may produce more sustained effects is to produce sensory adaptations, which leaves traces in the brain after the sensory manipulation is stopped. One interesting aspect of sensori-motor relationships is that they are highly susceptible to adaptive processes. Simple reaching behaviour can be adapted to dramatic changes of the relationship between the body and its environment. One very interesting correspondence between prism adaptation and spatial neglect is that prism adaptation can produce a shift in manual straight-ahead pointing in a direction opposite to the visual shift, just as has been described in some patients with spatial neglect [22]. If a normal individual is exposed to right deviating prisms, he will exhibit a leftward deviation of his straight-ahead pointing, and the opposite is true for left-deviating prisms. When neglect patients were adapted to rightward visual shift produced by wedge prisms their straight ahead pointing was also shifted to the left [23, 24]. Further investigations have shown that many aspects of unilateral neglect could be improved by a short session of prism adaptation (see Fig 1, reviews: [9, 15]. Classical neuropsychological tests of visuo-motor [23-26] or visuo-verbal [24, 25] neglect were improved for at least 24h following a single adaptation session.

If one considers that both strategic and adaptive components of prism adaptation have been described (e.g. [27]), then one may question the idea that prism adaptation can be regarded as a pure bottom-up approach. However an interesting observation allows reinforcing this idea: none of the neglect patients we have examined so far (even when tested up to 28 years after the stroke) could notice anything special with the prism goggles (Rossetti et al. in preparation). This observation strikingly contrasts with the obvious surprise expressed by healthy subjects and by partners of neglect patients and thus strongly supports the idea that, unlike in normal, prism adaptation in unilateral patients can be considered as involving only sensori-motor adaptation processes.

3. Generalisation of Therapeutic Gains

It would seem logical that the effects of prism adaptation should be restricted to, or best for, visuo-motor tasks, because they have more common features with the visuo-manual adaptation procedure. In the original study, we observed that the best improvement was observed for the Schenkenberg bisection test (6/6 patients markedly improved) whereas the weakest improvement was obtained for text reading (2/6 patients markedly improved). Therefore many other tests of neglect were investigated (review: [9, 15]).

Rode et al. [28, 29] explored the effect of prism adaptation on visual imagery and found a clear-cut improvement in two patients who could initially not evoke cities on the western half of an internally generated map of France. This result strongly suggested that the after-effects of visuo-manual adaptation can no longer be considered to be restricted to visual and motor parameters [8]. Farnè et al. [25] compared visuo-motor tasks (including line and bell cancellation tests, and two sub-tests taken from the Behavioural Inattention Test (B.I.T) battery, namely letter cancellation and line bisection.) with visuo-verbal tasks (the visual scanning test, also taken from the B.I.T., requiring a verbal description of the objects depicted on a coloured picture; an object-naming task with 30 Snodgrass pictures of familiar objects intermingled with geometric shapes as distracters; word and non-word reading), in six patients. They observed that the two groups of tasks followed a strictly parallel improvement, which lasted for at least 24 hours.

The fact that other sensory modalities can be improved (haptic circle centring [26], dichotic listening [30], haptic object recognition [31] tactile extinction [32]) and that several non-manual tasks (postural control, wheel-chair driving, imagery, verbal reports) were also improved demonstrate that the effects of prism adaptation are not restricted to visuo-manual parameters as they are known to be in normal subjects. Recently Berberovic and Mattingley [33] have shown that even a non-spatial and non-manual aspect of neglect could be improved, namely temporal order judgement. These results strongly suggest that adaptation to wedge prisms somehow affects the very core of unilateral neglect.

Recently, Zorzi et al. [34] described a new feature of unilateral neglect. They used a simple test of mental number bisection, where patients where requested to report where they saw the middle between two numbers (e.g., between 1 and 19). They found that unlike control patients neglect patients exhibited a bias towards larger number, as if their mental representation of the mental number line was distorted as is classically described for line bisection. We have therefore investigated the effects of prism adaptation on this mental task without explicit spatial component. In patients who bisected lines to the left and numbers to the right, we found that prism adaptation strongly improved the bisecting bias [35].

More specifically in the motor domain, the visuo-manual plasticity triggered by the visual shift produced effects on non-trained tasks (most of the tests used differed from the pointing task performed during prism exposure) as well as on non-manual tasks (postural control [36], wheel-chair driving [9]). In addition the intentional component of neglect, assessed by a task where the patients had to reach for a central ball and then to throw it to the left or right side, can also be altered by adaptation [9]. In an experimental condition aimed at investigating the effect of prism on the intentional control of action, patients where asked to grasp a centrally presented ball and then to through it to the left or to the right. The kinematic analysis of the reach-to-grasp movement shows that neglect patients are overall slowed down when the secondary movement is directed to the left [37, 38]. After a short prism adaptation session, this asymmetry was modified for several movement parameters (reaction time, movement time, peak velocity, time to peak velocity) (see [9]). The pattern of result observed immediately after prism adaptation even showed the reverse pattern: reach movements were slower when the ball had to be sent to the right. This result demonstrates that the intentional control of action can be modified by prism adaptation.

Still at a motor level, there are several qualitative observations that prism adaptation can improve the motor behaviour of patients in everyday life [26]. Quantification of everyday life activity does also show that patients can be improved up to one year [39].

One of the crucial questions raised by the observation of a strong and sustained improvement of hemispatial neglect by a single short adaptation session is whether this plastic effect is restricted to the acute phase of the deficit. In our original study patients were tested between 3 weeks and 14 months post-stroke [24]. We have now collected data on a group of patients who were exposed to the adaptation procedure between five and twenty eight years post-stroke and amazingly found the same amount of improvement.

Retention over time is the second main important characteristics of a rehabilitation method. The effects of a single session last much longer for prism adaptation (at least 2h) than for other sensory stimulations (about 15 minutes). A group of patients showed a maintained improvement 24h after the training session [25], whereas individual cases may exhibit long lasting amelioration of neglect (e.g. [23, 26]).

Repetitive stimulation by sensory stimulation has been tested only recently. Frassinetti et al. [40] (2002) reported that a group of patients who benefited from two prism adaptation sessions daily over 2 weeks (a total of

10 sessions) exhibited an improvement over 5 weeks after the end of the treatment. Schindler et al. [41, 42] also explored the effects of repetitive neck vibration and found a sustained improvement following an intensive daily programme. Obviously such studies should be undertaken to determine the optimal training frequency and duration as well as the optimal combination of techniques that can used routinely for rehabilitation.

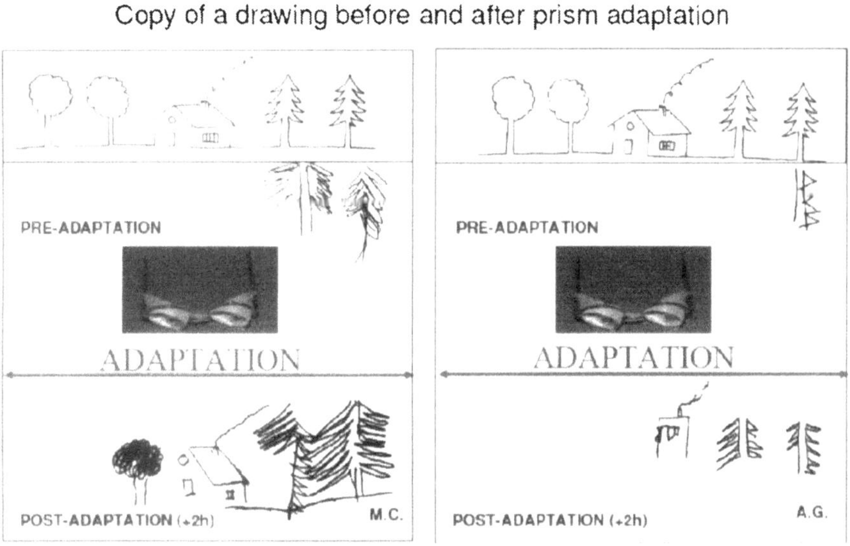

Fig.1. copy of Gainotti's drawing [43] before and after prism adaptation in two neglect patients (M.C. an A.G.). Before prism adaptation, only the right part of the drawing was copied whereas after prism adaptation more items on the left were copied.

References

1. Heilman KM, Bowers D, Valenstein E, et al (1987) Hemispace and hemispatial neglect. In: Neurophysiological and Neuropsychological aspects of Spatial Neglect. M Jeannerod (Ed), Noth-Holland: Elservier Science Publishers B.V. pp 115-150
2. Bisiach E, and Vallar G (1988) Hemineglect in humans. In: Handbook of physiology. F Boller, J Grafman (Eds), Elsevier Science Publisher. pp 195-222
3. Vallar G (1998) Spatial hemineglect in humans. Trends Cog Scie; 2: 87-97
4. Halligan PW, Marshall JC, Wade DT (1989) Visuospatial neglect: Underlying factors and test sensitivity. Lancet 14; 2(8668): 908-911

5. Denes G, Semenza C, Stoppa E, et al. (1982) Unilateral spatial neglect and recovery from hemiplegia. a follow-up study. Brain 105: 543-552
6. Fullerton KJ, McSherry D, and Stout RW (1986) Albert's test: a neglected of perceptual neglect. Lancet 22; 1: 430-432
7. Bergego, C, Bradat-Diehl, P, Taillefer, C et al (1994) "Evaluation et rééducation de l'apraxie d'utilisation des objets," in l'apraxie, D, Gall, and G Aubin, eds., SOLAL, Marseille, pp 214-223
8. Rossetti Y, Rode G, Pisella L, et al (1999) Sensorimotor plasticity and cognition: prism adaptation can affect various levels of space representation. In: Studies in Perception and Action. Grealy M. & Thomson J.A. (Eds.), Lawrence Erlbaum Associates pp. 265-269
9. Rode G, Pisella L, Rossetti Y, et al (2003) Bottom-up transfer of sensory-motor plasticity to recovery of spatial cognition: visuomotor adaptation and spatial neglect. Prog Brain Res142: 273-87
10. Rode G, Tiliket C., Honoré J., et al. (1998) Postural asymmetry is reduced by caloric stimulation in left hemiparetic patients. Scan. J. Rehab. Med. 30: 9-14
11. Diller, L., and Weinberg, J. (1997) Hemi-inattention in rehabilitation: the evolution of a rational remediation program. Adv. Neurol. 18: 63-82.
12. Rubens AB (1985) Caloric stimulation and unilateral visual neglect. Neurology 35: 1019-1024
13. Vallar G, Guariglia C, and Rusconi ML(1997) Modulation of the neglect syndrome by sensory stimulation. In Parietal lobe contribution to orientation in 3D space. Their P and Karnath HO (Eds), Springer Verlag, Heildeberg pp 555-579
14. Kerkhoff G. (2000) Neurovisual rehabilitation: recent developments and future directions. J Neurol Neurosurg Psychiatry 68(6): 691-706
15. Rossetti Y and Rode G (2002) Reducing spatiall neglect by visual and other sensory manipulations: noncognitive (physiological) routes to the rehabilitation of a cognitive disorder. In The cognitive and neural bases of spatial neglect Karnath HO, Milner AD, Vallar G (Eds), Oxford University press pp375-396
16. Bisiach E, Vallar G, Perani D, et al (1986) Unawareness of disease following lesions of the right hemisphere : anosognosia for hemiplegia and anosognosia for hemianopsia. Neuropsychologia 24: 471-482
17. Harvey, M, Milner, AD (1999) Residual perceptual distortion in 'recovered' hemispatial neglect. Neuropsychologia 37: 745-50
18. Jeannerod M and Biguer B (1987) The directional coding of reaching movements. A visuo-motor conception of spatial neglect. In: Neurophysiological and neuropsychological aspects of spatial neglect. Jeannerod M. (ed), North-Holland, Amsterdam. pp. 87-113.
19. Silberpfennig J (1941) Contributions to the problem of eye movements: disturbances of ocular movements with pseudo hemianopsia in frontal tumors. Confin Neurol, 4: 1-13
20. Rode, G, Charles, N, Perenin, MT, et al (1992) Partial remission of hemiplegia and somatoparaphrenia through vestibular stimulation in a case of unilateral neglect. Cortex 28: 203-208

21. Rode G, Perenin MT, Honoré J, et al. (1998a) Improvement of the motor deficit of neglect patients through vestibular stimulation: evidence for a motor neglect component. Cortex 3: 253-26
22. Jeannerod M and Rossetti Y (1993) Visuomotor coordination as a dissociable visual function: experimental and clinical evidence. In: Baillère's Clinical Neurology. Kennard C (Ed). London: Baillère Tindall. 439-460
23. Pisella L, Rode G, Farne A, et al (2002) Dissociated long lasting improvements of straight-ahead pointing and line bissection tasks in two hamineglect patients. Neuropsychologia 40(3): 327-334
24. Rossetti Y, Rode G, Pisella L, et al (1998) Prism adaptation to a rightward optical deviation rehabilitates left hemispatial neglect Nature 395: 166-169
25. Farné A, Rossetti Y, Toniolo S et al (2002) Ameliorating neglect with prism adaptation: visuo-manual and visuo-verbal measures. Neuropsychologia 40(7): 718 29
26. McIntosh RM, Rossetti Y, Milner AD (2002) Prism adaptation improves chronic visual and haptic neglect. Cortex 38: 309-320.
27. Redding, GM and Wallace B (1997) Adaptive spatial alignment. Mahwah, N.J., Lawrence Erlbaum Associates
28. Rode G, Rossetti Y, Boisson D (2001) Prism adaptation improves representational neglect. Neuropsychologia 39(11): 1250-4
29. Rode G, Rossetti Y, Li L, et al (1998) The effect of prism adaptation on neglect for visual imagery. Behav Neurol 11: 251-258.
30. Coutois-Jacquin S, Rossetti Y, Rode G, et al (2001) Effect of prism adaptation on auditory extinction : an attentional effect Presented at International Symposium on Neural Control of space coding and action production, Lyon, 22-24 March 2001
31. Toutounji N, Michel C, Luauté J, et al (2001) Prism adaptation improves haptic object recognition in hemispatial neglect. Société de Neuropsychologie de Langue Française, Paris, 2001.
32. Maravita A, McNeil J, Malhotra P, et al. (2003) Prism adaptation can improve contralesional tactile perception in neglect. Neurology 60(11): 1829-31.
33. Berberovic N, and Mattingley, JB (2003) Effects of prismatic adaptation on judgements of spatial extent in peripersonal and extrapersonal space. Neuropsychologia 41(4): 493-503
34. Zorzi M, Priftis K, Umilta C (2002) Brain damage: neglect disrupts the mental number line. Nature 417(6885):138-9.
35. Rossetti Y, Jacquin-Courtois S, Rode G, et al (2004) Does action make the link between number and space representation? Visuo-manual adaptation improves number bisection in unilateral neglect. Psychol Sci; 15(6): 426-30
36. Tilikete C, Rode G, Rossetti Y, et al (2001) Prism adaptation to rightward optical deviation improves postural imbalance in left hemiparetic patients. Cur Biol, 11: 524-528
37. Jacquin S. (2002) Influence de la négligence spatiale unilatérale sur le contrôle intentionnel de l'action. Unpublished MD dissertation, N° 2002-205, Lyon

38. Courtois-Jacquin S, Revol P, Rode G, et al (2005) Unilateral neglect affects execution of a sequence ending intention and movement to the left. Poster presented at the Motor Contor Conference, Sofia, 22-25 September
39. Parache M, Delecroix H, Berlinson S, et al (2002) Amélioration à long terme des activités de la vie quotidienne chez un patient négligent après adaptation prismatique. Poster presented at: Journées de Neurologie de Langue Française, Tours, 2-6 avril 2002. abstract : Revue Neurologique 158, 3, suppl 1, poster N26, pp 2S132)
40. Frassinetti F, Angeli V, Meneghello F, et al (2002) Long-lasting amelioration of visuospatial neglect by prism adaptation. Brain 125(3): 608-23
41. Schindler I, Kerkhoff G, Karnath H.O, et al (2002). Neck muscle vibration induces lasting recovery in spatial neglect. J Neurol, Neurosurg, and Psychiat, 73: 412-419
42. Schindler I, KerkhoffG, Karnath HO, et al (1999) Neck muscle vibration and visual exploration training – a cross-over study on neglect rehabilitation. Acta Neurobiologiae Experimentalis, 59: 69
43. Gainotti G, Messerli, P and Tissot, R (1972) Quantitative analysis of unilateral spatial neglect in relation to laterality of cerebral lesion. J Neurol, Neurosurg, and Psychiat 35(4): 545-550

Somatosensory Processing in the Postcentral, Intraparietal and Parietal Opercular Cortical Regions

Yoshiaki Iwamura

Department of Sensory Science, Kawasaki University of Medical Welfare, Okayama, Japan

Chapter Overview. Single neuronal recordings in the first somatosensory cortex of the monkey have shown the presence of systematic hierarchical processing along the rostrocaudal axis of the postcentral gyrus. Neuronal activity to code types of hand manipulation or features of tactile objects were found in the caudal part of the gyrus, indicating that the postcentral hand region is organized to process tactile and kinesthetic information for active touch. The hierarchical scheme led us to find neurons representing bilateral sides of the body in the postcentral-intraparietal region. Certain bilateral neurons could contribute to inter-manual transfer of tactile information. Other bilateral neurons appeared to represent bilateral body surfaces or limb positions. Still other bilateral neurons could represent the peripersonal space since they responded to the visual stimuli presented in the space over the somatosensory receptive field. We further studied neuronal receptive field properties in the parietal opercular region and confirmed that in the second somatosensory cortex (SII) neurons with larger and converging receptive fields covering remote and discontinuous body parts were common and consequently SII was organized much less somatotopically. Functional roles of the postcentral, intraparietal and SII regions in the somatosensory processing were discussed in the light of these single neuronal receptive field studies and recent brain imaging studies in humans.

Key Words. hierarchical somatosensory processing, SI, intraparietal region, SII, bilateral integration, cross-modal integration

1. Introduction

It has been established that somatosensory information is processed in serial and parallel manner in rather wide cortical regions, the postcentral, the intraparietal and the parietal opercular regions. I review what information is processed in each area, how these areas are interrelated and functionally differentiated, based on our own single neuronal recording studies in wake monkeys and on more recent human brain imaging studies.

2. Hierarchical Processing in the Postcentral Gyrus

2.1 Monkey studies

In monkey SI, size and complexity of receptive fields (RFs) increase with caudal progression from area 1 [1, 2] (Fig.1). Representations of different digits are somatotopic in area 3, overlap to some extent in area 1 and more extensively in area 2 [3] (Fig.2). Furthermore, neurons responding to the proximal skin or joint stimulation were found in the caudal part of the digit region [2]. Thus, information coming from digits, the wrist, the forearm and the elbow converge in the caudal part of the digit region. The increase in the complexity of RF properties was attributed to the convergence of multiple inputs onto single neurons via cortico-cortical connections and some additional thalamic projections to the caudal parts.

Complex types of neuronal responses were found in areas 1 and 2. There were unique neurons with selectivity to specific features of tactile objects such as shapes or texture in the caudal part of the gyrus [4,5]. Some neurons did not respond, or responded poorly to passive stimuli, but were activated better or solely by the monkey's active hand movements including reaching [6]. These results indicated that the postcentral hand region is organized to process tactile and kinesthetic information for active touch. In fact the monkey's manipulation performance was disturbed when small amount of muscimol was injected into the hand region

[7]. Presence of neurons responding to the proximal forelimb stimulation in the digit region is justified because they are necessary to be integrated with those from the hand in active hand use, as the proximal and distal forelimb structures move together and are stimulated together. Increase in the RF complexity toward the caudal part of the gyrus was seen also in the proximal arm /trunk region [8] or leg/foot region [9].

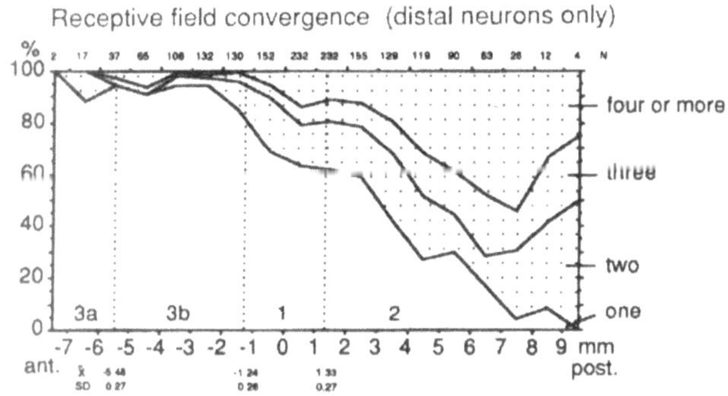

Fig.1. Increase in the RF convergence toward the caudal part of the postcentral gyrus (adapted from [2], with permission).

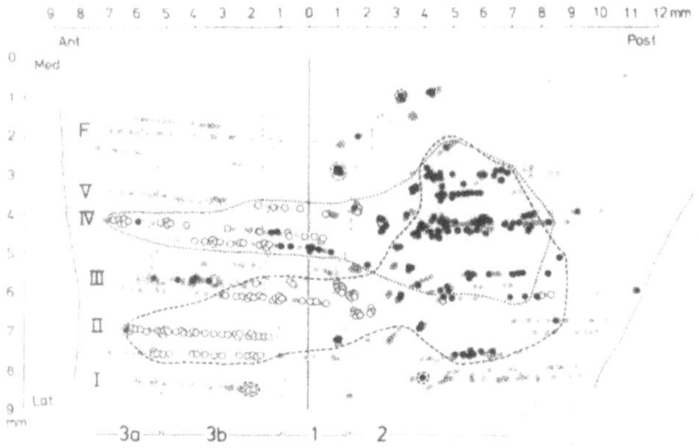

Fig.2. Convergence of digit representation in the caudal part of the postcentral region. The abscissa: the anterior-posterior; the ordinate: the medio-lateral direction. Large open circles, units with RFs on either 2nd or 4th digit; solid circles, units whose RFs involve two or three digits including either the 2nd or 4th digit; stars, units whose RFs involve three or more digits including both the 2nd and 4th digits; small open circles, all other units recorded in the two hemispheres of one monkey (adapted from [3], with permission).

Response latency of neurons to vibration stimulus was longer in area 2 than in area 3 or 1 [10]. Dipoles to generate P7, P10, N10, P12, and P18 in somatosensory evoked potentials (SEP) to median nerve stimulation in anesthetized monkeys were located in the thalamus, area 3, areas 1 and 2 and area 5 respectively [11].

2.2 Human studies

Digit representations examined by fMRI were somatotopic in area 3b but in areas 1 and 2 there were multiple foci and a considerable overlap of different digit representation [12, 13]. The absence of the digit somatotopy in area 2 was pointed out [14]. Digit maps were larger in area 1 than area 3b [15].

Systematic increase in the latency of evoked magnetic potentials along the rostrocaudal axis of the postcentral gyrus was reported recently [16].

In PET studies, moving stimuli over the volar surface of fingers activated more strongly area 1 and possibly area 2 than area 3b [17]. Area 2 was activated with preference for surface curvature changes and shape stimuli while areas 3b or 1 was activated by velocity and roughness of brushing. The anterior part of the supramarginal gyrus and the cortex lining the intraparietal gyrus were activated by shape discrimination, but not by other mechanical stimuli [18].

There is a case report with a laminar necrosis limited to the caudal edge of the lateral portion of the left postcentral gyrus: the patient could not discriminate the shape with his right hand. Otherwise, tests for light touch, pain, thermal sense, position sense and vibration sense showed no abnormalities [19].

3. Bilateral and Bimodal Integration in the Postcentral and Intraparietal Region

3.1 Bilateral integration in the monkey

Earlier studies indicated that the bilateral activity in SI was limited to the body midline including the dorsal or ventral trunk, occiput, perioral face, or oral cavity [20]. Our recent studies in macaque monkeys have demonstrated that the bilateral representation of the body is not restricted to the body midline in the postcentral somatosensory cortex [21, 22] (Fig.3).

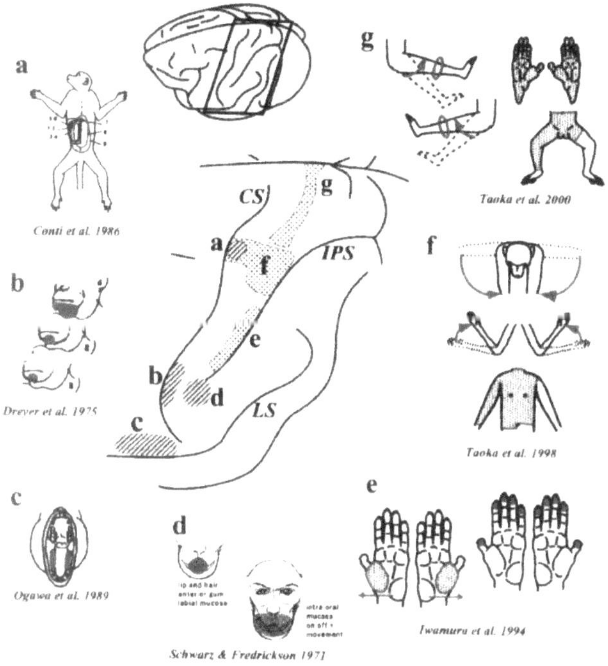

Fig.3. Summary of bilateral RF neurons in the postcentral gyrus of the macaque monkey. In the middle, dorsolateral views of the cerebral cortex and the postcentral gyrus of a monkey are shown with sites of recording of bilateral RF neurons (a-g). a, the trunk region (Conti et al 1986); b, the face/lips/jaw region (Dreyer et al 1975); c, the intraoral region (Ogawa et al 1989); d, the perioral face/lips region (Schwarz & Fredrickson 1971); e, the finger/hand region (Iwamura et al 1994); f, the trunk/ shoulder/arm region (Taoka et al 1998); g, the leg/foot region (Taoka et al 2000). CS:central sulcus; IPS:intraparietal sulcus;LS:lateral sulcus (adapted from [22], with permission).

We found a substantial number of neurons with bilateral RFs on the hand digits clustered in the caudalmost part of SI (areas 2, 5 and 7) in waking macaque monkeys [23]. Among neurons with RFs on digits, the multiple digit skin (and nail) types were the majority, and in general, the size and orientation of bilateral RFs were symmetrical. The preferred direction of directionally selective neurons as plotted by stimulation of one hand was also symmetrical with the other hand. In many cases, the spatial configuration of the RFs in the two hands was the same, but the intensity of the response was not necessarily the same, suggesting the presence of a hierarchy among these bilateral neurons.

After ablating completely or injecting ibotenic acid into the opposite hemisphere, no bilateral RFs were found in the postcentral gyrus. The results indicated the dependency of the ipsilateral response on the callosal route.

Bilateral neurons were also found in the more medial part of the postcentral gyrus, for the upperarm and the trunk [8]. We further investigated the leg and foot region, including the medial surface of the hemisphere [9]. Neurons with bilateral RFs existed in the hindlimb region also. They were much sparser in the foot region than in the hand region, especially those with RFs on multiple toes. It may reflect functional differences between the foot and the hand.

The bilateral RFs we found in the caudal part of the postcentral gyrus were the most complex type among all neurons found in this gyrus, and were considered to be at the highest level of the hierarchical chain in this gyrus. We interpret these results as suggesting that in the digit region, the bilaterality, possibly conferred by the callosal connections, is postponed until the very end of the hierarchical processing in this gyrus.

3.2 Bilateral integration in humans

Ipsilateral information may project to the postcentral somatosensory cortex in humans. Allison et al [24] recorded somatosensory evoked potentials (SEP) subdurally from patients and found a late component with onset latency of 40-50 msec driven by ipsilateral median nerve stimulation. They suggested from the latency that the ipsilateral potentials are generated not in area 3b, but rather in other regions of sensorimotor cortex including areas 4, 1, 2 and 7. More recently, recording of somatosensory evoked magnetic field (SEF) by stimulating the median nerve revealed the presence of ipsilateral components [25-27]. fMRI studies also demonstrated the ipsilateral activation of the primary somatosensory cortex during various types of natural tactile stimulation of the finger tips or palm [28-32]. Activation of the ipsilateral cortex is absent in callosotomized patients after tactile stimulation of ipsilateral hand [33].

In fMRI study, area 2 of the right hemisphere was activated in common by kinesthetic stimulation of either side of the body suggesting the presence of bilateral integration in this cortical region and also a right hemispheric dominance for perception of limb movements [34].

3.3 Bimodal integration in the monkey

Some neurons in area 7b were activated by both passive stroke of the trunk or arm skin (RFs were generally large) and visual stimuli moving toward or over the somatic RF [35]. The presence of similar bimodal neurons with RF on the face was shown in the deepest part of the intraparietal sulcus [36]. This cortical region was designated as VIP, a transitional zone from the postcentral to the parietal association region. These bimodal neurons were considered as representing awareness of the body surface or peripersonal space immediately surrounding the body.

We found that the caudalmost part of the postcentral hand and arm region contains similar bimodal neurons [37]. Among a substantial number of visually driven neurons, about a half were bimodal, with somatosensory RFs either on the face, trunk, arms, or hand, or combinations of these body parts, on the contralateral or ipsilateral side, or on the bilateral sides of the body. For these bimodal neurons, effective visual stimulus was to show an object to the animal within a space over or near the somatosensory RF of the neuron. Thus the visual RF was anchored to the somatosensory RFs.

We also found bimodal neurons of the proprioceptive types [38](Fig.4). They were intermingled with somatosensory neurons that responded to the movement of bilateral arms or shoulders or with those responding to stimulation of bilateral arms or trunk skin [8]. The bimodal neurons found in the dorsal bank of the intraparietal sulcus distributed from about 4mm deep from the corner of the sulcus to the bottom. The region where the majority of the bimodal neurons were found corresponded to the forearm projection zone in area 3b.

Bimodal neurons were classified by their somatosensory RF, into at least four types, 1) the trunk and arm type, 2) the arm proprioceptive type, 3) the hand type, and 4) the face type. The RFs of the hand type sometimes had a discontinuous counterpart on the face or on the trunk. These bimodal neurons were found in crude somatotopic manner. The trunk and arm type and the arm proprioceptive type most medially, the hand type next, and the hand type with discontinuous RFs more laterally, and finally, the face type most laterally. The lateral-most region for the bimodal neurons was just caudal to the thumb projection zone in area 3b. Interestingly, there was a tendency for bimodal neurons to cluster together in the digit region separated from the clusters of neurons with bilateral digit RF [23].

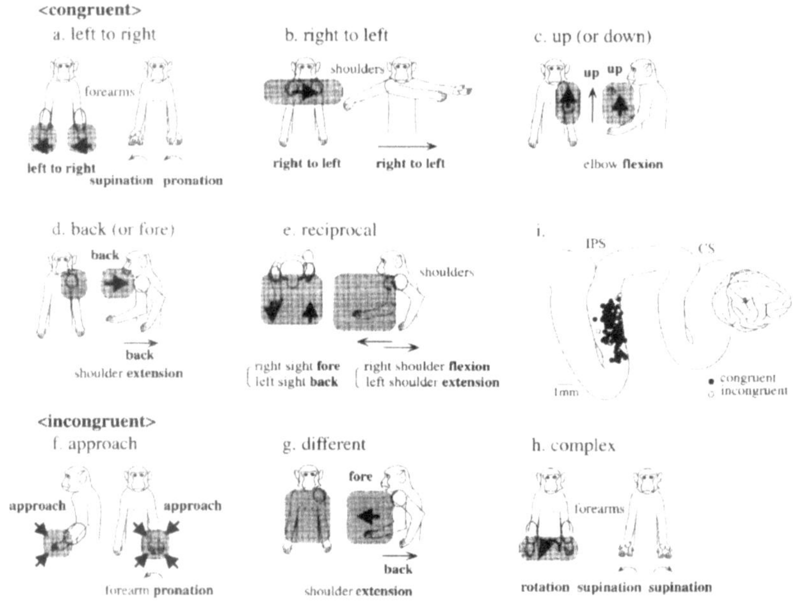

Fig.4. Receptive fields of bimodal neurons responding to forearm joint displacement and visual stimuli moving in particular directions. Visual receptive field in each neuron is shown as a shaded area (adapted from [38], with permission).

3.4 Cross-modal interactions in humans

Cross-modal interactions between vision and touch have been investigated in human mainly from the neuropsychological standpoint. A strong modulating effect of vision on touch was found when visual stimulus was presented near the hand [39,40]. The finding was explained by referring to the activity of bimodal neurons in premotor and parietal cortex of monkey [41]. Prolonged active use of tools can modify visual-tactile spatial integration [42]. The data are in line with a change in the body schema by tool use in monkeys [37]. Cross-modal visual-tactile extinction was investigated in right-hemisphere damaged patients [43].

Vision improves the spatial resolution of touch [44], and vision in fact modulates somatosensory cortical processing measured by event-related potentials [45].

Some human studies put the intraparietal cortical area in more direct focus, in the context of bimanual coordination [46], phantom limb phenomenon [47], motor processing [48] and selective attention to the space [49, 50].

4. The Second Somatosensory Cortex in Parietal Opercular Region

4.1 Monkey studies

It is still controversial in determining the extent of the second somatosensory cortex (SII) of macaque monkeys. There are conflicting reports of the number of areas that comprise this cortical region, either single or two. Several microelectrode-recording studies tried to delineate SII, as it possesses a somatotopic representation of the complete contralateral body surface [51-54]. However, the strict somatotopy was seen only when the mapping was performed under the anesthesia. Also some of mapping studies under the anesthesia ignored the ipsilateral inputs. In experiments done in unanesthetized preparation, SII neurons tend to have larger and more complex RFs including bilateral ones. Complexity of neuronal RF properties of SII may be attributed to the fact that SII is higher than SI as somatosensory inputs reach SII both directly from the thalamus and from the postcentral cortical regions. The ipsilateral information has been considered to come through the callosal connections both from the postcentral and the parietal opercular region of the opposite hemisphere.

Our own study performed in wake monkeys have demonstrated that, 1) in 44 % of isolated SII neurons RF properties were not identified by simple mechanical stimulation, 2) among the identified neurons 48 % were skin and 49 % were deep submodality, 3) 64 % of the identified neurons had bilateral RFs, 4) 74 % of the identified neurons had RFs on either the forelimb, hindlimb, trunk or head (single part type), 5) among the single part type, forelimb was the majority (66 %), 6) 26 % of the identified neurons had RFs on more than two body parts (combined type), 7) the combined type could be divided into the limb type and trunk type, 8) 75 % of the limb type had RFs confined within the distal portion of the limb, 9) 75 % of the trunk type had RFs extending to the limbs. Thus, in SII the integration goes beyond single limb or a body part. This was never seen in the postcentral region.

When the body representation map is drawn in SII, different body parts overlaps extensively. The head and forearm region were large and overlapped extensively with each other and there were two trunk regions.

The notion that SII is higher than SI in hierarchy was proposed based on their anatomical relationships. Each cytoarchitectonic subdivision of SI sends projections to SII, and SII projects back to SI to its superficial layers [56, 57]. Our results of single neuronal recording described above thus confirm such notion.

4.2 Human studies

Studies in human with non-invasive techniques indicated that the projection of the body to SII is bilateral [58], that the somatotopy is less fine-grained than SI [59,60]. In PET studies different part of SII was activated depending on the task, discrimination of either the roughness or shape. The posterior portion of SII received noxious information bilaterally where the hand and foot were not somatotopically represented [61]. These results suggest that the organization of SII in human may be as complex as that in monkeys.

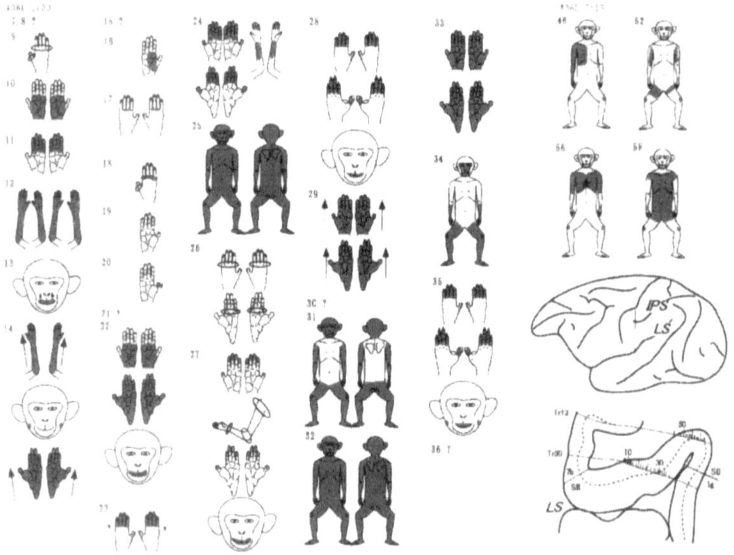

Fig. 5. Receptive fields of SII neurons in a monkey, collected along two penetration tracks as illustrated. Note great variation of receptive fields as classified in the text (adapted from [55], with permission).

5. Summary and Conclusions

Results of two types of studies, single neuronal recording performed in waking monkeys and brain imaging in human, were consistent in supporting a hierarchical scheme in somatosensory processing along the anterior-posterior axis of the postcentral gyrus. Both monkey and human studies have demonstrated the presence of the bilateral and visual-tactile bimodal integration in the intraparietal region, one of key stations for perception of the peripersonal space and for formation of the body image. I also presented evidence that the SII is the higher stage in the hierarchical stream and that it is unique for the multi-limb integration while the integration in the postcentral region remains within single limbs.

References

1. Iwamura Y. Hierarchical somatosensory processing. Curr Opin Neurobiol 1998; 8:522-528.
2. Iwamura Y, Tanaka M, Sakamoto M, Hikosaka O. Rostrocaudal gradients in neuronal receptive field complexity in the finger region of alert monkey's postcentral gyrus. Exp Brain Res 1993; 92:360-368.
3. Iwamura Y, Tanaka M, Hikosaka O. Overlapping representation of fingers in the somatosensory cortex (area 2) of the conscious monkey. Brain Res 1980; 197: 516-520.
4. Iwamura Y, Tanaka M. Postcentral neurons in hand region of area 2: their possible role in the form discrimination of tactile objects. Brain Res 1978; 150: 662-666.
5. Iwamura Y, Tanaka M, Hikosaka O, Sakamoto M. Postcentral neurons of alert monkeys activated by the contact of the hand with objects other than the monkeys own body. Neurosci Lett 1995; 186:127-130.
6. Iwamura Y, Tanaka M. Representation of reaching and grasping in the monkey postcentral gyrus. Neurosci Lett 1996; 214:147-150.
7. Iwamura Y, Tanaka M. Organization of the first somatosensory cortex for manipulation of objects: an analysis of behavioral changes induced by muscimol injection into identified cortical loci of awake monkeys. In: Franzen O, Westman J. Information processing in the somatosensory system. Wenner-Gren International symposium series, 57, New York, Stockton, 1991, pp.371-380.
8. Taoka M, Toda T, Iwamura Y. Representation of the midline trunk, bilateral arms and shoulders in the monkey postcentral somatosensory cortex. Exp Brain Res 1998; 132: 315-322.

9. Taoka M, Toda T, Iriki A, Tanaka M, Iwamura Y. Bilateral receptive field neurons in the hindlimb region of the postcentral somatosensory cortex in awake macaque monkeys. Exp Brain Res 2000; 134:139-146.
10. Lebedev MA, Nelson RJ. High-frequency vibratory sensitive neurons in monkey primary somatosensory cortex: entrained and nonentrained responses to vibration during the performance of vibratory-cued hand movements. Exp Brain Res 1996; 111:313-325.
11. Hayashi N, Nishijo H, Ono T, Endo S, Tabuchi E. Generators of somatosensory evoked potentials investigated by dipole tracing in the monkey. Neuroscience 1995; 68:323-338.
12. Kurth R, et al, fMRI shows multiple somatotopic digit representations in human primary somatosensory cortex. Neuroreport 2000; 11:1487-1491.
13. Deuchert M, et al, Event-related fMRI of the somatosensory system using electrical finger stimulation. Neuroreport 2002; 13:365-369
14. Blankenburg F, Ruben J, Meyer R, Schwiemann J, Villringer A. Evidence for a rostral-to-caudal somatotopic organization in human primary somatosensory cortex with mirror-reversal in areas 3b and 1. Cereb Cortex 2003; 13: 987-993.
15. Overduin SA, Servos P. Distributed digit somatotopy in primary somatosensory cortex. Neuroimage 2004; 23: 462-472.
16. Inui K, Wang X, Tamura Y, Kaneoke Y, Kakigi R. Serial processing in the human somatosensory system. Cereb Cortex 2004; 14: 851-857.
17. Bodegard A, Geyer S, Naito E, Zilles K, Roland PE. Somatosensory areas in man activated by moving stimuli: cytoarchitectonic mapping and PET. Neuroreport 2000; 11:187-191
18. Bodegard A, Geyer S, Grefkes C, Zilles K, Roland PE Hierarchical processing of tactile shape in the human brain. Neuron 2001; 31:317-328
19. Takeda K, Shozawa Y, Sonoo M, Shimizu T. The rostrocaudal gradient for somatosensory perception in the human postcentral gyrus. J Neurol Neurosurg Psychiat 2000; 69:692-693.
20. Manzoni T, Barbaresi P, Conti F, Fabri M. The callosal connections of the primary somatosensory cortex and the neural bases of midline fusion. Exp Brain Res 1989; 76: 251-266.
21. Iwamura Y. Bilateral receptive field neurons and callosal connections in the somatosensory cortex. Philos Trans R Soc Lond B Biol Sci 2000; 355:267-273.
22. Iwamura Y, Taoka M, Iriki A. Bilateral activity and callosal connections in the somatosensory cortex. Neuroscientist 2001; 7:419-429.
23. Iwamura Y, Iriki A, Tanaka M. Bilateral hand representation in the postcentral somatosensory cortex. Nature 1994; 369: 554-556.
24. Allison T, et al, Human cortical potentials evoked by stimulation of the median nerve. II. Cytoarchitectonic areas generating short-latency activity. J Neurophysiol 1989; 62: 694-710.
25. Korvenoja A, et al, Activation of ipsilateral primary sensorimotor cortex by median nerve stimulation. Neuroreport 1995; 6:2589-2593.

26. Schnitzler A, Salmelin R, Salenius S, Jousmaki V, Hari R. Tactile information from the human hand reaches the ipsilateral primary somatosensory cortex. Neurosci Lett 1995; 200:25-28.
27. Hoshiyama M, Kakigi R, Koyama S, Watanabe S, Shimojo M. Activity in posterior parietal cortex following somatosensory stimulation in man: magnetoencephalographic study using spatio-temporal source analysis. Brain Topogr 1997; 10: 23-30.
28. Boecker H, et al, High-resolution functional magnetic resonance imaging of cortical activation during tactile exploration. Human Brain Mapping 1995; 3:236-244.
29. Lin W, Kuppusamy K, Haacke EM, Burton H. Functional MRI in human somatosensory cortex activated by touching textured surfaces. J MRI 1996; 6:565-572
30. Polonara G, Fabri M, Manzoni T, Salvolini U. Localization of the first and second somatosensory areas in the human cerebral cortex with functional MR imaging. Am J Neuroradiol 1999; 20:199-205.
31. Hansson T, Brismar T. Tactile stimulation of the hand causes bilateral cortical activation: a functional magnetic resonance study in humans. Neurosci Lett 1999; 271:29-32.
32. Hamalainen H, Hiltunen J, Titievskaja I. fMRI activations of SI and SII cortices during tactile stimulation depend on attention. Neuroreport 2000; 11:1673-1676.
33. Fabri M, et al, Role of the corpus callosum in the somatosensory activation of the ipsilateral cerebral cortex: an fMRI study of callosotomized patients. Eur J Neurosci 1999; 11:3983-3994.
34. Naito E, et al, Dominance of the right hemisphere and role of area 2 in human kinesthesia. J Neurophysiol. 2005; 93:1020-1034.
35. Leinonen L, Hyvarinen J, Nyman G, Linnankoski I. Functional properties of neurons in lateral part of associative area 7 in awake monkeys. Exp Brain Res 1979; 34:299-320.
36. Duhamel J-R, Colby CL, Goldberg ME. Congruent representations of visual and somatosensory space in single neurons of monkey ventral intra-parietal cortex (area VIP). In: Paillard J, editor. Brain and space. Oxford: Oxford Univ, 1991. p.223-236.
37. Iriki A, Tanaka M, Iwamura Y. Coding of modified body schema during tool use by macaque postcentral neurons. Neuroreport 1996; 7:2325-2330.
38. Tanaka M, et al, Intraparietal bimodal neurones delineating extrinsic space through intrinsic actions. Psychologia 2004; 47:63-78.
39. Ladavas E, Zeloni G, Farne A. Visual peripersonal space centred on the face in humans.Brain 1998; 121:2317-2326.
40. Ladavas E, Farne A, Zeloni G, di Pellegrino G. Seeing or not seeing where your hands are. Exp Brain Res 2000; 131:458-467.
41. Ladavas E, Farne A. Visuo-tactile representation of near-the-body space. J Physiol Paris. 2004; 98:161-170

42. Maravita A, Spence C, Kennett S, Driver J. Tool-use changes multimodal spatial interactions between vision and touch in normal humans. Cognition 2002; 83:B25-34.
43. Farne A, Ladavas E. Dynamic size-change of hand peripersonal space following tool use. Neuroreport 2000; 11:1645-1649.
44. Kennett S, Taylor-Clarke M, Haggard P. Noninformative vision improves the spatial resolution of touch in humans. Curr Biol 2001;11:1188-1191.
45. Taylor-Clarke M, Kennett S, Haggard P. Vision modulates somatosensory cortical processing. Curr Biol 2002; 12:233-236.
46. Serrien DJ, Nirkko AC, Lovblad KO, Wiesendanger M. Damage to the parietal lobe impairs bimanual coordination. Neuroreport 2001; 12:2721-2724
47. Miyauchi S. Personal communication.
48. Macaluso E, Frith CD, Driver J. Directing attention to locations and to sensory modalities: multiple levels of selective processing revealed with PET. Cereb Cortex 2002; 12:357-368
49. Bremmer F, et al, Polymodal motion processing in posterior parietal and premotor cortex: a human fMRI study strongly implies equivalencies between humans and monkeys. Neuron 2001; 29:287-96.
50. Imaruoka T, Yanagida T, Miyauchi S. Attentional set for external information activates the right intraparietal area. Brain Res Cogn Brain Res. 2003; 16:199-209.
51. Whitsel BL, Petrucelli LM, Werner G. Symmetry and connectivity in the map of the body surface in somatosensory area II of primates. J Neurophysiol 1969; 32: 170-183.
52. Robinson CJ, Burton H. Somatotopographic organization in the second somatosensory area of M. fascicularis. J Comp Neurol 1980; 192: 43-67
53. Burton H, Fabri M, Alloway K. Cortical areas within the lateral sulcus connected to cutaneous representations in areas 3b and 1: a revised interpretation of the second somatosensory area in macaque monkeys. J Comp Neurol 1995; 355:539-562.
54. Krubitzer L, Clarey J, Tweedale R, Elston G, Calford M. A redefinition of somatosensory areas in the lateral sulcus of macaque monkeys. J Neurosci 1995; 15: 3821-3839.
55. Iwamura Y, Taoka M, Toda T. The second somatosensory cortex. In: Tanji J, Yoshizawa S (eds), The higher brain function (In Japanese) Asakura, Tokyo 2001, pp.3-14.
56. Burton H, Fabri M, Alloway K. Cortical areas within the lateral sulcus connected to cutaneous representations in areas 3b and 1: a revised interpretation of the second somatosensory area in macaque monkeys. J Comp Neurol 1995; 355:539-562.
57. Cauller LJ, Clancy B, Connors BW. Backward cortical projections to primary somatosensory cortex in rats extend long horizontal axons in layer I. J Comp Neurol 1998; 390:297-310.
58. Simoes C, Alary F, Forss N, Hari R. Left-hemisphere-dominant SII activation after bilateral median nerve stimulation Neuroimage. 2002; 15: 686-690.

59. Ruben J, et al, Somatotopic organization of human secondary somatosensory cortex. Cereb Cortex 2001; 11: 463-473.
60. Bingel U, et al, Somatotopic organization of human somatosensory cortices for pain: a single trial fMRI study. Neuroimage. 2004; 23: 224-232.
61. Ferretti A, et al, Functional topography of the secondary somatosensory cortex for nonpainful and painful stimuli: an fMRI study. Neuroimage. 2003; 20:1625-38.

Predicting Motor Intention

Mark Hallett and Ou Bai

Human Motor Control Section, NINDS, NIH, Bethesda, MD, USA

Chapter Overview. People have the perception that prior to making a voluntary movement, there is an intention to move. This subjective impression has been measured objectively using Libet's clock and is on average about 300 ms prior to EMG onset. The EEG shows activity prior to movement, and by the voltage measurement, there is a rising negativity called the Bereitschaftspotential beginning about 1.5 s prior to EMG onset. This indicates that the brain mechanisms for generating a voluntary movement begin prior to the subjective intention, that is, unconsciously. If it were possible to detect relevant EEG signals with single events in real time, then it would be possible to identify that movement is being prepared prior to the subjective experience. Work in our laboratory has been making progress in accomplishing this aim. Such a signal could also be used to drive a brain-computer interface.

Key Words. voluntary movement, EEG, Bereitschaftspotential, event-related desynchronization, intention

1. Introduction

It is commonly believed that we choose to make our movements. This is the idea behind the notion of "voluntary," and is what is meant by the concept of "free will." Thinking about the matter carefully, however, leads to the conclusion that this is a perception, a so-called qualia of consciousness. The perception is very clear, but it is not certain how it arises. It may reflect reality, that is, there is a force in the brain that chooses the next movement to make. In this circumstance, the perception is of this force. On the other hand, it could be deceptive, just an interpretation by the brain of unconscious brain events. As we do not understand consciousness and perception, it is rather difficult to discuss this. On the other hand, it is possible to get data on mechanisms of movement generation and also objective measurements of subjective impressions. There is some evidence that the brain begins to initiate movement prior to the subjective impression that a movement is being willed.

2. The Movement Related Cortical Potential (MRCP)

The MRCP is the EEG potential that can be recorded from the scalp in relation to a voluntary movement. It has a number of components [1, 2]. An early negativity preceding movement has two phases, an initial, slowly rising phase lasting from about 1500 ms to about 400 ms before movement, the Bereitschaftspotential or BP (also called the readiness potential in translation of the German), and a later, more rapidly rising phase lasting from about 400 ms to approximately the time of movement onset, the negative slope or NS'. (These two components could also be called the BP1 and the BP2.) The NS' peaks about 90 ms before the onset of EMG activity and is followed often by a brief decline in the negativity, called the premotor positivity. The next component is the motor potential or MP which begins before movement, peaks after movement onset and produces the highest negativity in the recording. The topography of the BP is generalized with a vertex maximum. With NS' the negativity begins to shift to the central region contralateral to the hand that is moving. There continues to be some controversy as to the structures that generate the MRCP. The main contributors to the BP appear to be the premotor cortex and the supplementary motor area, both bilaterally [3]. With the appearance of NS' the activity of the contralateral motor cortical region, including the primary motor cortex, predominates.

While the MRCP clearly is indicative of movement preparatory processes in the brain, its exact meaning is not clear. In particular, a relatively normal looking MRCP precedes unconscious movements as well as conscious movements. This was studied by looking at the brain events preceding unrecognized movements made by subjects at rest or engaged in

ceding unrecognized movements made by subjects at rest or engaged in a mental task [4].

3. Subjective timing of motor intention

The experiment that showed how this might be done was reported by Libet et al. in 1983 [5]. Subjects sat in front of a clock with a rapidly moving spot and were told to move at will. Subsequently, they were asked to say what time it was (where the spot was) when they had the first subjective experience of intending to act (this time was called W). They also were asked to specify the time of awareness of actually moving (this time was called M). There were two types of voluntary movements, one type was thoughtfully initiated and a second type was "spontaneous and capricious." As a control for the ability to successfully subjectively time events, subjects were also stimulated at random times with a skin stimulus and they were asked to time this event (called S). At the same time, EEG was being recorded and MRCPs were assessed to determine timing of activity of the brain. With thoughtful, preplanned movements, the BP begins about 1050 ms prior to EMG onset (the type I of Libet), and with spontaneous movements, the BP begins about 575 ms prior to movement (the type II of Libet) [6]. The type II may consist mainly of the NS' component.

Subjects were reasonably accurate in determining the time of S indicating that this method of timing of subjective experience was acceptable. W occurred about 300 ms prior to EMG onset, and M occurred about 90 ms prior to EMG onset. The onset of the BP type I occurred about 800 ms prior to W, and the onset of the BP type II occurred about 350 ms prior to W. The authors concluded "that cerebral initiation of a spontaneous, freely voluntary act can begin unconsciously, that is, before there is any (at least recallable) subjective awareness that a 'decision' to act has already been initiated cerebrally" [5].

4. Event-related desynchronization (ERD) with movement

The EEG can be analyzed in other ways than just the voltage or slow wave activity, described as the MRCP. The EEG can be broken down into its frequency components, classically described as delta, 0 to 4 Hz, theta, 4 to

8 Hz, alpha 8 to 12 Hz, beta, 12 to 30 Hz, and gamma, higher than 30 Hz. Gamma has low power and has not been studied as often. Over the motor cortex, the alpha activity sometimes has a characteristic arc shape and is called the mu activity. While the slow wave activity is thought to reflect the summed post-synaptic potentials, the oscillatory activity appears to reflect reverberating circuits. When a circuit has high oscillatory activity, it is likely largely idling. It is a very old observation that the alpha power drops with movement; this is the called "blocking of the mu-activity." Power actually drops prior to movement onset in both alpha and beta bands, and this has been called event-related desynchronization or ERD [7-9]. The term desynchronization implies that the activity in that frequency band is no longer synchronized and that this is the reason for the power drop. After movement, the power in alpha and beta rebound to greater than before movement, and this is called event-related synchronization or ERS.

5. Can we work in real time?

If the MRCP and ERD begin up to 1 s prior to the subjective recognition of the intention to move, then it might be possible to predict movement intention. That is, it might be possible to identify the "intention of the brain" before an individual recognizes the intention. There is a major problem with this, however. The MRCP and ERD changes are ordinarily recognized only in averages, typically of about 100 trials. To be predictive, we would have to be able to work with single trials and in real time. This is clearly difficult. The brain makes a good deal of EEG activity, and there are many artifacts in EEG recordings; thus, extracting what we want is a challenge.

Should we be successful in this endeavor, in addition to being of physiological interest, the real time signal would be appropriate to drive a brain computer interface.

The first step is to produce very high quality (averaged) MRCPs and ERDs, both in time and scalp distribution. This should give information that can be exploited for single trial analysis. One observation, based on the literature and long experience, is that ERD is a much more reliable signal than MRCP. Some normal individuals have very low amplitude MRCP. Hence, ERD is a better choice for this work.

In our high-resolution EEG study subjects were asked to perform voluntary finger tapping using either hand [10]. The timing of movement and which hand to move were freely chosen. Bilateral ERD in the beta band was observed over sensorimotor cortex before both left and right hand movements. From 2 s to 1 s before movement onset, ERD was only observed on left hemisphere before right hand movement, while ERD was present over both hemispheres before left hand movements. These results indicate that we might be able to predict movement onset and, as well, the laterality of human voluntary movement intention as early as 1 s before movement. However, our study also showed that the amplitude of early ERD is small, causing likely analytical difficulty for single-trial prediction.

6. The Brain-Computer Interface to Virtual Reality (BCI2VR) Project

We started a project "Brain-Computer Interface to Virtual Reality" (BCI2VR) to develop an intelligent system to enable processing and decoding EEG activity in real-time. The system integrates hardware interfacing, software programming and high-speed computing technologies. Testing results on a two-CPU PC workstation shows that delay by mathematical computation is in a reasonable range of several ms. The kernel of the BCI2VR system is the computational algorithms for signal processing and pattern recognition. The computational procedure consists of four steps: data preprocessing; time/frequency feature extraction; spatial feature extraction; and feature classification. We have incorporated multiple methods in order to optimize prediction/decode accuracy. Data preprocessing includes: signal reference methods (Laplacian, common reference); infinite impulse response/finite impulse response (IIR/FIR) frequency filters for noise reduction. Time/frequency feature extraction includes: filter methods; FFT-based classic spectrum estimation methods; auto-regressive model-based spectrum estimation methods; time-frequency representation; continuous/discrete wavelet transform. Spatial feature extraction includes: principal component analysis/independent component analysis (PCA/ICA) rotation/deduction; common spatial patterns (CSP) representation; discrete cosine transform/discrete wavelet transform (DCT/DWT); hidden Markov process model. Feature classification includes: linear distance classifier; Baysian probability classifier; artificial neural network (ANN) classifier; supporting vector machine (SVM). Further, BCI2VR reserves plug-in function to incorporate add-on computational methods. BCI2VR also provides output interface for controlling external devices. We have so far completed the frame work for BCI2VR, and are still working on the details of system components.

We have conducted real-time prediction of human movement intention using the current BCI2VR system with a few normal subjects so far. The experiment consists of two sessions: (1) data calibration for construction of prediction model; (2) real-time prediction of movement intention with/without feedback of prediction information. The subjects were asked to make decision (at their own pace) of which hand to move after a "start" cue was given. They then executed the movement as soon as they made their decision. EMG signals were recorded to monitor movement tasks performed. According to empirical knowledge of EEG features during voluntary movement, we reduced the number of electrodes from 122 to 29. The 29 electrodes were placed on sensorimotor area, supplementary motor area and pre-motor area bilaterally. We employed following computational strategy: Laplacian reference, FFT estimation of power spectrum with logarithm representation, CSP spatial filter and Baysian probability classification. We employed the Fisher Criterion to measure mean distances among classes. The Fisher Criterion is considered as a good indicator of classification accuracy as it is judged from both mean distance and sample variance among different classes. The Fisher assessment spans all possible time and frequency windows, and the best timing and frequency components were automatically judged for making prediction of movement intention.

We have studied several subjects extensively so far. Classification was made upon EEG spatiotemporal feature vector, and probability was calculated using pre-determined model. We determined that a good prediction time to be at 0.8 s before movement onset. The real-time prediction of human voluntary movement was achieved with an accuracy as high as 80%.

References

1. L. Deecke, "Electrophysiological correlates of movement initiation," *Rev Neurol (Paris)*, vol. 146, pp. 612-9, 1990.
2. M. Jahanshahi and M. Hallett, "The Bereitschaftspotential: Movement Related Cortical Potentials." New York: Kluver Academic/Plenum Publishers, 2003, pp. 334.

3. K. Toma, T. Matsuoka, I. Immisch, T. Mima, D. Waldvogel, B. Koshy, T. Hanakawa, H. Shill, and M. Hallett, "Generators of movement-related cortical potentials: fMRI-constrained EEG dipole source analysis," *Neuroimage*, vol. 17, pp. 161-73, 2002.
4. I. Keller and H. Heckhausen, "Readiness potentials preceding spontaneous motor acts: voluntary vs. involuntary control," *Electroencephalogr Clin Neurophysiol*, vol. 76, pp. 351-61, 1990.
5. B. Libet, C. A. Gleason, E. W. Wright, and D. K. Pearl, "Time of conscious intention to act in relation to onset of cerebral activity (readiness-potential). The unconscious initiation of a freely voluntary act," *Brain*, vol. 106, pp. 623-42, 1983.
6. B. Libet, E. W. Wright, Jr., and C. A. Gleason, "Readiness-potentials preceding unrestricted 'spontaneous' vs. pre- planned voluntary acts," *Electroencephalogr Clin Neurophysiol*, vol. 54, pp. 322-35, 1982.
7. G. Pfurtscheller, A. Stancak, Jr., and G. Edlinger, "On the existence of different types of central beta rhythms below 30 Hz," *Electroencephalogr Clin Neurophysiol*, vol. 102, pp. 316-25, 1997.
8. G. Pfurtscheller and A. Berghold, "Patterns of cortical activation during planning of voluntary movement," *Electroencephalogr Clin Neurophysiol*, vol. 72, pp. 250-8, 1989.
9. L. Leocani, C. Toro, P. Manganotti, P. Zhuang, and M. Hallett, "Event-related coherence and event-related desynchronization/synchronization in the 10 Hz and 20 Hz EEG during self-paced movements," *Electroencephalography and Clinical Neurophysiology*, vol. 104, pp. 199-206, 1997.
10. O. Bai, Z. Mari, S. Vorbach, and M. Hallett, "Asymmetric Spatiotemporal Patterns of Event-Related Desynchronization Preceding Voluntary Sequential Finger Movements: A High-Resolution EEG Study," *Clinical Neurophysiology*, in press, 2005.

Cortical Field Potentials and Cognitive Functions

Hisae Gemba

Department of Physiology, Kansai Medical University, Moriguchi, Japan

Chapter Overview. Cortical field potentials were recorded by surface and depth electrodes implanted in various cerebral cortices in monkeys on self-paced or visuo-initiated hand movements, or go/no-go discriminative hand movements with auditory or visual stimuli, and then analyzed. Readiness potentials appeared in the premotor, motor, somatosensory and posterior parietal cortices. The rostral bank of the left arcuate sulcus and the dorsal bank of the principal sulcus were found to be related to recognition learning and decision-making, respectively.

Key Words. Cortical field potential, decision-making, monkey, readiness potential.

1. Introduction

Voluntary movements may be divided into self-paced and stimulus-initiated movements. Self-paced movements are initiated by internal information related to instinct or homeostasis, and stimulus-initiated movements by external information received by a sensory organ. Central nervous mechanisms concerning the two types of movement are considered to differ from each other. In order to find out the functional neuronal circuits on the movements, simultaneous recording of field potentials on the surface and at a 2.0-3.0 mm depth in a particular cortex is a useful method, because it enables us to predict which kind of synaptic input will be activated [1,2,3]. Cortical field potentials are mainly due to excitatory postsynaptic potentials generated in apical dendrites of pyramidal cells in the cerebral cortex. Synaptic inputs into the superficial part of the apical dendrite of the pyramidal cell are due to superficial thalamo-cortical projections, while synaptic inputs into the deeper part of the apical dendrite of the cell are due to deep thalamo-cortical and cortico-cortical projections.

In the present paper, monkeys were trained to lift a lever at self-pace (self-paced hand movements) or in response to visual stimulus (visuo-initiated hand movements). Monkeys were also trained for go/no-go discriminative hand movements with auditory or visual stimuli. Field potentials were recorded by electrodes implanted on the surface and at a 2.0-3.0 mm depth in various cortical areas in a monkey on the movements. The potentials were analyzed in connection with movement onset to find out cortices producing readiness potentials for self-paced hand movements [4,5]. On the other hand, the potentials were analyzed in connection with stimulus start and behavioral observations, in visuo-initiated hand movements and go/no-go discriminative hand movements with auditory or visual stimuli.

2. Methods

Japanese monkeys (Macaca fuscata; 3.0-8.5 kg) were used in the present study. Experimental procedures were approved by the Animal Care Committee of the Kansai Medical University. Attention was also paid to the ethical principles adopted by the Physiological Society of Japan: "Guiding principles for the care and use of animals in the field of physiological sciences".

2.1 Field potential recording on self-paced movements

Monkeys were trained for self-paced hand movements. Silver needle electrodes (0.25 mm in diameter; insulated except at their pointed tips) were used as cortical surface and depth electrodes, being implanted in the cortical loci in the monkey under anesthesia as shown in the upper part of Fig. 1, II. Several five-electrode arrays were implanted in the posterior parietal cortex (Fig. 1, II, lower part). The remainder of the method for recording field potentials in various cortical areas was the same as reported in previous papers [4,5]. A mechanogram (MECH) during hand movements was recorded by means of a transducer attached to the lever. The right cerebellar hemispherectomy was performed in two monkeys after full recording of cortical field potentials. Influences of the lesion upon cortical field potentials were investigated together with behavioral observations.

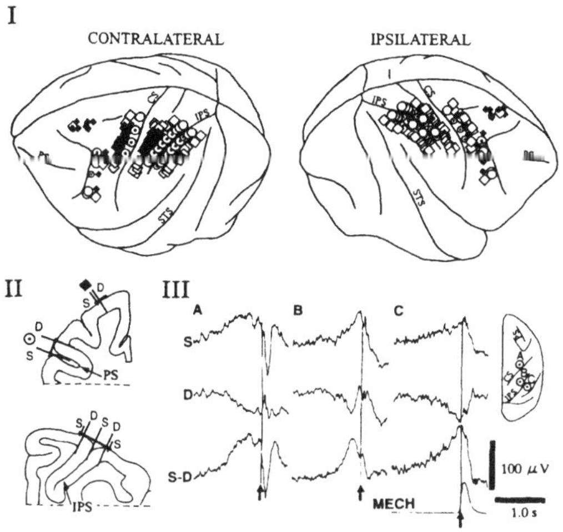

Fig.1. I: Distribution of readiness potential in the hemispher contralateral (CONTRALATERAL) and ipsilateral (IPSILATERAL) to a moving hand. Closed diamonds and open circles with a dot show sites of the readiness potential, while open diamonds and circles sites of no readiness potential. II: Top figure: electrodes in the dorsolateral cortex (closed diamond) and in the dorsal bank of the PS (principal sulcus) (open circle with a dot). S, surface electrode; D, depth electrode. III: Specimens associated with self-paced movements of left hand. Data are from the right motor (A) and posterior parietal (B-C) cortices. S, surface potential; D, depth poitential; S-D, surface minus depth potential; MECH, mechanogram; 100 times averaging by movement onset pulse (vertical line and upward arrow); 100 μV for cortical potentials (negativity upward); 1.0 s for all records. (Source: see Ref. [8])

2.2 Field potential recording on visuo-initiated movements

Cortical field potentials were recorded in various cortical areas in a naive monkey learning to lift a lever in response to a visual stimulus (green with 0.5-1.0 s duration) appropriately and quickly (visuo-initiated hand movements), using the above-mentioned electrodes. The stimuli were given to the monkey at random intervals (5.0-10.0 s). The monkey was rewarded for hand movements performed within the stimulus duration. The remainder of the method for recording field potentials in various cortical areas was the same as reported in previous papers [5,6,7]. The histogram of reaction times (RT) from the stimulus onset to the movement onset was also recorded.

2.3 Field potential recording on GO/NOGO task

Monkeys were trained for go/no-go discriminative hand movements with auditory or visual stimuli. Tones of 1000 Hz and 2000 Hz with a duration of 500 ms were used as go and no-go auditory stimuli respectively, while green and red colors as go and no-go visual stimuli respectively. The go and no-go stimuli were given to the monkey in random order and at random intervals (5.0-10.0 s). When the monkey lifted the lever by hand within 500 ms after the onset of the go stimulus, it was rewarded, whereas it was not rewarded after the no-go stimulus, regardless of whether it lifted or did not lift the lever.

After electrophysiological investigations over several months, the monkeys were sacrificed under deep pentobarbital anesthesia and the sites of recording electrodes were verified morphologically, and histological studies on the cerebellum were also performed

3. Results and Discussion

3.1 Motor execution

Typical specimens of cortical field potentials associated with self-paced hand movements in a monkey are shown in Fig. 1, III. A surface-negative, depth-positive (s-N, d-P) potential was found to start about 1.0 s before movement in the forelimb motor (A) and posterior parietal (B, C) cortices contralateral to the moving hand. The premovement potential was considered to be readiness potential for hand movement, being responses due to superficial thalamo-cortical projections [1,2,4]. Readiness potential was marked in the upper part of the ventral premotor cortex (PMCv), the forelimb area of the motor (MC) and somatosensory (SSC) cortices, and in the anterior and posterior banks of the intraparietal sulcus (IPS) on the side contralateral to the moving hand. None or a small potential was recorded medially in both banks of the IPS (Fig. 1, I, CONTRALATERAL). Marked potential was recorded in the dorsal part of the PMCv on the side ipsilateral to the moving hand, while none or a small potential was recorded in the MC, SSC and posterior parietal cortex (PPC) in the ipsilateral hemisphere (Fig. 1, I, IPSILATERAL). A small potential was recorded in the dorsal premotor cortex (PMCd) on both sides (Fig. 1, I, CONTRALATERAL and IPSILATERAL) [8,9]. Hemicerebellectomy on the right side almost completely eliminated the readiness potential in the forelimb motor cortex contralateral to the lesion (not shown) [10].

On the other hand, two previous papers using electrical stimulation and morphological methods suggested that the readiness potential of the PMCv, facial part of MC, SSC and PPC originated mainly in the area X, VPLo, LP and pulvinar nuclei in the thalamus respectively [11,12].The present study indicated that the readiness potential in the MC originated in the cerebello-thalamo-cortical responses.

In the previous study aiming at how preparative activities in the PPC for hand movements activated the forelimb muscle, electromyographic activities elicited in the muscle by electric stimulation of the MMC and PPC were recorded. Conduction time of nerve impulses was also measured between the PPC – PMCv, PMCv – MC, and PPC – MC respectively, using electric stimulation. It was found that excitatory activities in the PPC could be transmitted to the forelimb muscle without passing through the PMCv and MC [9]. It is possible that many cortices producing readiness potential – motor areas in broad sense – could be simultaneously activated to initiate self-paced movements through parallel processing, as seen in Fig. 2.

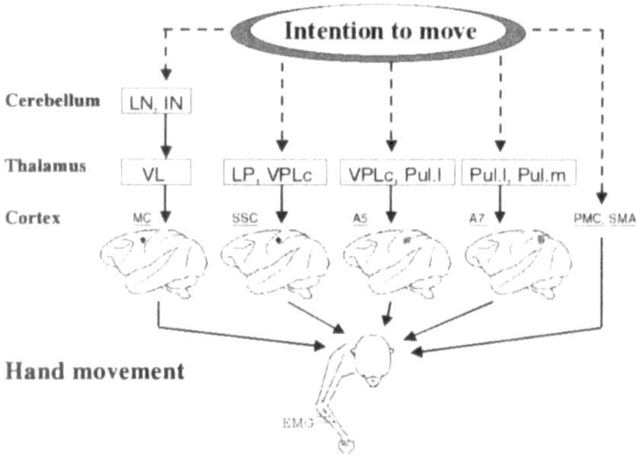

Fig.2. A model of the process from internal stimulus (Intention to move) to initiation of hand movements (Hand movement) via the cerebellum (Cerebellum), thalamus (Thalamus) and cerebral cortex (Cortex). Unknown neural connections are indicated by broken lines with arrows; highly probable neural connections as solid lines with arrows. EMG, electromyogram. (Source: see Ref. [9])

3.2 Motor learning

Changes in cortical field potentials during the learning processes of visuo-initiated movements of right hand in a single monkey are shown in Fig. 3. Data were from the prefrontal (A, B, C), premotor (D), forelimb motor (E), prestriate (F) and striate (G) cortices in the left hemisphere. The data at stage I were recorded on the first training day, when the monkey lifted the lever regardless of stimuli, as seen in RT (I, RT). At stage II, recorded on the 21st training day, an upward (s-N, d-P) potential appeared in the prefrontal (C) and prestriate (F) areas, although the monkey still lifted the lever independent of stimuli (II, C, F and RT). As the upward potential in C and F became bigger along with further training, the monkey came to respond to stimuli with the movements (III, C, F and RT). The early learning period stages, I to III, were called recognition learning [13,14]. With training repeated further, the movement gradually became quicker and more skillful. In 6-8 weeks after the start of the training, monkeys almost accomplished the movement (stage IV). At stage IV, recorded on the 59th day, more marked upward (s-N, d-P) potential was recorded in the motor cortex (E) with shorter and less variable reaction times (RT) than at stage III (III and IV, E and RT). The late learning stages, III to IV, were called skill learning [14,15].

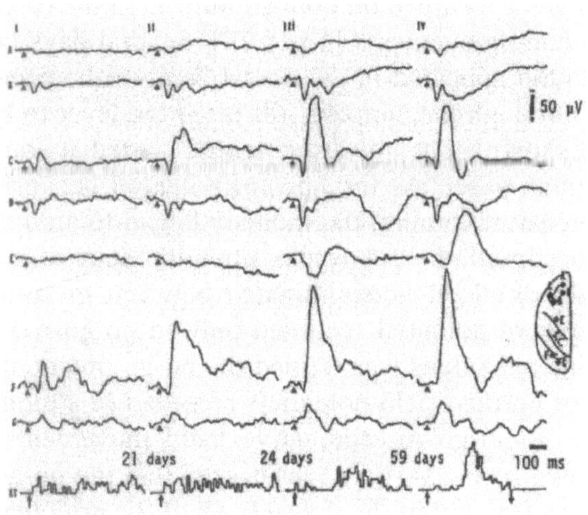

Fig. 3. Changes of cortical field potentials (S-D) in learning processes of visuo-initiated hand movements in same monkey. 100 times averaging by stimulus onset pulse (triangle and upward arrow); downward arrow, end time of stimulus; RT, reaction time histogram. Columns I-IV: Learning stages I-IV. Data are from the prefrontal (A-C), premotor (D), motor (E), prestriate (F) and striate (G) cortices, contralateral to a moving hand. (Source: see Ref. [13])

The previous study on changes in cortical field potentials in learning processes of reaction time hand movements with complex tone indicated that the rostral bank of the left arcuate sulcus played a significant role for a monkey to associate a stimulus with movement – recognition learning – in audio-initiated movements, as in visuo-initiated movements [16]. It was shown by the resection studies of the cerebellum that the upward (s-N, d-P) potential in the motor cortex originated in the cerebello-thalamo-cortical responses, and that skill learning was due to cerebro-cerebellar interactions, in both the visuo- and audio-initiated hand movements [10,14,17].

3.3 Decision-making

Cortical field potentials were recorded and analyzed while monkeys learned go/no-go discriminative hand movements with auditory stimuli. Tones of 1000 Hz and 2000 Hz were used as go and no-go stimuli. On the first training day, the monkey responded to both go and no-go stimuli, and field potentials were recorded on both stimuli, as in the case of simple-task, audio-initiated hand movements [14,16,17]. Several days later, an upward (s-N, d-P) potential appeared in the dorsal bank of the principal sulcus on no-go stimuli, although the monkey still lifted the lever to both go and no-go stimuli (not shown). The upward potential started at about 100 ms after the no-go stimulus onset. As the upward potential became more marked, with further intensive training, the monkey began to stop the hand movement on no-go stimuli. In 3-4 weeks after the start of the training, the monkey almost completely discriminated between go and no-go stimuli [14,18]. The upward potential recorded only in no-go trials in the dorsal bank of the principal sulcus was termed the no-go potential [18,19]. Typical specimens of cortical field potentials recorded in a monkey, which accomplished the go/no-go discriminative hand movements with auditory stimuli, are shown in Fig. 4, I, A.S. It is seen that the no-go potential was recorded in the rostral part of the dorsal bank of the principal sulcus (A).

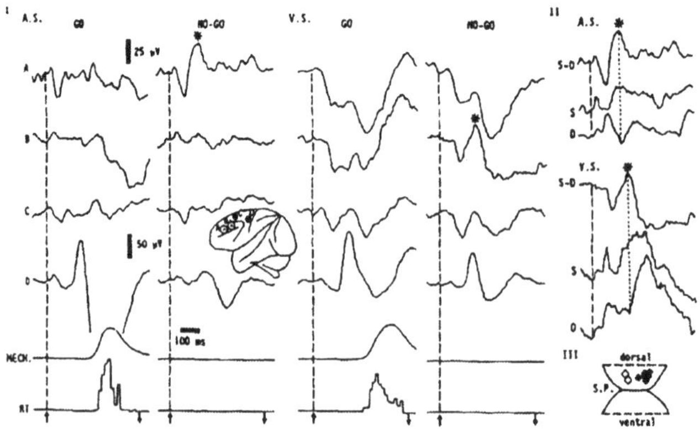

Fig. 4. I: Cortical field potentials (S-D) on go/no-go discriminative movements with auditory (A.S.) and visual (V.S.) stimuli in same monkey. Data are from the rostral (A) and caudal (B) parts of the dorsal bank of the principal sulcus (S.P.), and the premotor (C) and forelimb motor (D) cortices. 100

times averaging by respective onset (upward arrow and broken line) pulses of go and no-go stimuli; 25 μV for cortical potentials in A-C; 50μV for cortical potentials in D. Note that no-go potentials (asterisks) were recorded in A and B, respectively on auditory and visual stimuli. II: No-go potentials in A and B in I are shown with the cortical surface (S) and depth (D) potentials. III: Distribution of no-go potential in the dorsal bank of the S.P., which is schematically unfolded. Open diamond, no-go potential on auditory stimuli; filled diamond, no-go potential on visual stimuli. (Source: see Ref. [18])

Monkeys were also trained for go/no-go discriminative hand movements with visual stimuli. Green and red colors were used as go and no-go visual stimuli, and cortical field potentials were recorded in various cortical areas. Typical specimens of cortical field potentials recorded in the same monkey as in A.S., which also accomplished the go/no-go discriminative hand movements with visual stimuli, are shown in Fig. 4, I, V.S. It is seen that the no-go potential was recorded in the caudal part of the dorsal bank of the principal sulcus (B). The no-go potential was assumed to be related to the decision not to move, as in a previous experiment using electric stimulation in the monkey [20].

No-go potentials for visual stimuli were recorded in the caudal part of the dorsal bank of the principal sulcus, as seen in Fig. 4, III, and in the rostroventral corner in the prefrontal cortex on both sides (not shown) [20]. No-go potentials for auditory stimuli were recorded in the rostral part of the dorsal bank of the principal sulcus (Fig. 4, III). No-go potentials were also recorded in humans [21].

4. Conclusion

The present paper showed at least in part neural connections in the process from internal stimulus to initiation of hand movements. Intention to move a hand at self-pace could be expressed as nerve impulses possibly in the limbic system, and then, via unknown neural connections, the nerve impulses could be conveyed to the cerebellum and thalamus. The nerve impulses are transmitted to the PMC, SSC and PPC by the superficial thalamo-cortical projections, and to the MC by the cerebello-thalamo-cortical projections. These cortices could be simultaneously activated to initiate forelimb muscle activities for hand movements, possibly through parallel processing. It was also shown that external information was transmitted into the prefrontal cortex from a sensory organ through the sensory cortex, and that the specific response to the information – i.e. to move or not to move, was judged and decided on in the prefrontal cortex. The external information could be conveyed to the MC, SSC and SMA (supplementary motor area) by cerebro-cerebellar interactions or other neural connections after it is processed in the prefrontal and premotor cortices. The MC, SSC and SMA could be then activated to initiate visuo- or audio-initiated hand movements.

Acknowledgment

The author thanks Ms. K. Nakao and Dr. R. Matsuzaki for their technical help.

References

1. K. Sasaki, "Cerebro-cerebellar interconnections in cats and monkeys," in Cerebro-cerebellar Interactions, J. Massion and K. Sasaki, Eds. Amsterdam: Elsevier, 1979, pp. 105-124.
2. K. Sasaki, H. Gemba and S. Hashimoto, "Premovement slow cortical potentials on self-paced hand movements and thalamocortical and corticocortical responses in the monkey," Exp. Neurol., vol. 72, pp. 41-50, 1981.
3. K. Sasaki and H. Gemba, "Action of the cerebello-thalamo-cortical projection upon visually initiated reaction-time hand movements in the monkey," Stereotac. Func. Neurosurg., vol. 60, pp. 104-120, 1993.
4. S. Hashimoto, H. Gemba and K. Sasaki, "Analysis of slow cortical potentials preceding self-paced hand movement in the monkey," Exp. Neurol., vol. 65, pp. 218-229, 1979.
5. H. Gemba and K. Sasaki , "Distribution of potentials preceding visually initiated and self-paced hand movements in monkeys," Brain Res., vol. 306, pp. 207-214, 1984.
6. H. Gemba, S. Hashimoto and K. Sasaki , "Cortical field potentials preceding visually initiated hand movements in monkeys," Exp. Brain Res., vol. 42, pp. 435-441, 1981.
7. H. Gemba, K. Sasaki and J. Ito, "Different cortical potentials preceding self-paced and visually initiated hand movements and their reciprocal transition in the same monkey, " Brain Res., vol. 306, pp. 207-214, 1982.

8. Reprinted from Neuroscience Letters, Vol 357, Gemba et al, Preparative activities in posterior parietal cortex for self-paced movement in monkeys, pp 68-72, Copyright (2004), with permission from Elsevier.
9. Reprinted from Current Trends in Neurology, Vol 1, Gemba et al, The process from internal stimulus to movement initiation, pp 109-123, Copyright (2005), with permission from Research Trends.
10. K. Sasaki and H. Gemba, "Cortical field potentials preceding self-paced and visually initiated hand movements in one and the same monkey and influences of cerebellar hemispherectomy upon the potentials," Neurosci. Lett., vol. 25, pp. 287-292, 1981.
11. S. Kyuhou, R. Matsuzaki and H. Gemba, "Cerebro-cerebellar projections onto the ventral part of the frontal cortex of the macaque monkey," Neurosci. Lett., vol. 230, pp. 101-104, 1997.
12. R. Matsuzaki, S. Kyuhou, K. Matsuura-Nakao and H. Gemba, "Thalamo-cortical projections to the posterior parietal cortex in the monkey," Neurosci. Lett., vol. 355, pp. 113-116, 2004.
13. K. Sasaki and H. Gemba, "Development and change of cortical field potentials during learning processes of visually initiated hand movements in the monkey," Exp. Brain Res., vol. 48, pp. 459-473, 1982.
14. H. Gemba, "Motor programming for hand and vocalizing movements," in Principles of Frontal Lobe Function, DT. Stuss and RT. Knight, Eds. New York: Oxford University Press, 2002, pp. 127-148.
15. K. Sasaki and H. Gemba, "Learning of fast and stable hand movement and cerebro-cerebellar interactions in the monkey," Brain Res., vol. 277, pp. 41-46, 1983.
16. H. Gemba, N. Miki and K. Sasaki, "Field potential change in the prefrontal cortex of the left hemisphere during learning processes of reaction time hand movement with complex tone in the monkey," Neurosci. Lett., vol. 190, pp. 93-96, 1995.
17. H. Gemba and K. Sasaki, "Cortical field potentials associated with audio-initiated hand movements in the monkey," Exp. Brain Res., vol. 65, pp. 649-657, 1987.
18. Reprinted from Brain Research, H. Gemba and K. Sasaki, Vol 537, Potential related to no-go reaction in go/no-go hand movement with discrimination between tone stimuli of different frequencies in the monkey, pp 340-344, Copyright (1990), with permission from Elsevier.
19. K. Sasaki and H. Gemba, "Electrical activity in the prefrontal cortex specific to no-go reaction of conditioned hand movement with colour discrimination in the monkey," Exp. Brain Res., vol. 64, pp. 603-606, 1986.
20. K. Sasaki, H. Gemba and T. Tsuzimoto, "Suppression of visually initiated hand movement by stimulation of the prefrontal cortex," Brain Res., vol. 495, pp. 100-107, 1989.

21. H. Gemba and K. Sasaki, "Potentials related to no-go reaction in go/no-go hand movement task with color discrimination in human," Neurosci. Lett., vol. 101, pp. 263-268, 1989.

FMRI Studies on Passively Listening to Words, Nonsense Words, and Separated KANAs of Japanese

Chang Cai[1], Takanori Kochiyama[2], Tokuhiko Satake[3], Kunihiico Osaka[3], and Jinglong Wu[2]

[1] Graduate School of Engineering, Kagawa University, Takamatsu, Japan
[2]Faculty of Engineering, Kagawa University, Takamatsu, Japan
[3]Osaka Neurosurgery Hospital, Takamatsu, Japan

Chapter Overview. Japanese differs from English because it uses KANAs as basic pronunciation units, which also can be used to build words directly via certain combinations. Therefore, Japanese better facilitates the separation of underlying cognitive components in spoken language processing, e.g., phonemes, combinations of phonemes, and words with meanings. In this study, we measured the brain activities of subjects by using functional magnetic resonance imaging (fMRI) as they were passively listening to a series of words (word condition), combinations of phonemes without meanings (nonsense word condition), and temporally separated KANAs (separated KANA condition). The three conditions all activated the bilateral superior temporal gyrus, while the detailed location and volume of activated areas varied across the three kinds of stimuli. Listening to the words and separated KANAs activated more areas than did listening to the nonsense words, perhaps owing to different familiarity levels; areas activated in the separated KANA condition were slightly posterior to those activated in the word and nonsense word conditions. Compared with the nonsense word and separated KANA conditions, the word condition also activated part of the right primary motor cortex and part of the left premotor cortex, which may be related to articulatory-gestural representations of word access. There were no obvious areas activated in the anterior STG and STS, which have been related to semantic representations of words.

Key Words. fMRI, superior temporal area, Japanese, speech perception

1. Introduction

Recent studies of the anatomical and functional organization of the non-human primate auditory cortical system point to multiple, hierarchically organized processing pathways that involve the temporal, parietal, and frontal cortices [1, 2, 6]. Based on these studies, the auditory cortex in primates comprises several cortical fields organized into core (primary), belt (secondary), and parabelt regions. Anatomical and electrophysiological studies indicate that adjacent regions are interconnected and that information proceeds from core to belt to parabelt, and to more distal areas as processing demands become more complex.

Functional neuroimaging studies also provide evidence that human speech perception might be based on multiple, hierarchical processing pathways and that different kinds of representations could be preferentially treated in different streams. In a comprehensive synthesis of neuroimaging literature, Binder and colleagues outlined a hierarchical model of the processing of nonspeech sounds to speech [3]. According to this model, 1) dorsal areas bilaterally surrounding Heschl's gyrus were activated more by FM tones than by noise, suggesting a role in processing simple temporally encoded auditory information; 2) regions centered bilaterally in the superior temporal sulcus were activated more by speech stimuli than by FM tones; and 3) based on the combined results of several studies, the left ventral temporal and temporoparietal regions were likely involved in processing lexical-semantic or syntactic information associated with words. Although some studies favored the third point of the above model, other studies [8-10] did not find any difference with respect to the word and non-word conditions [11].

By combining the findings of several studies from neuroanatomy and functional imaging, Scott and his colleagues hypothesized a two-pathway model of speech perception in which the anterior pathway from the primary auditory core might be important for mapping acoustic-phonetic cues onto lexical representations, whereas the posterior pathway from the core might process articulatory-gestural representations of speech acts [4, 7]. This analysis might resolve the longstanding debate between accounts of speech perception that favor acoustic feature processes and those that emphasize gestural motor processing.

In this study, we aimed primarily to distinguish the brain activations with respect to the phonological perception of Japanese words and nonsense words while subjects were passively listening. Comparing the reversed words or pseudowords used by Binder [3] to Japanese nonsense words and words, the Japanese words have common acoustic complexity and potential variability, which is important for differentiating between the word and nonsense word conditions in an fMRI experiment. Considering the independency of KANAs, we also compared the difference between the word and the nonsense word conditions with the separate KANA conditions, which are similar to the former two conditions, in order to find clues about basic pronunciation unit recognition in brain.

2. Methods

2.1 Stimuli

We used three kinds of stimuli in our experiments: Japanese words, Japanese nonsense words, and separated KANAs. We selected 60 commonly used words from [15] and 60 selected nonsense words, produced by Matlab (Mathworks Inc., Sherborn, MA, USA), all of which consisted of four randomly sequenced KANAs. We used 71 different individual KANAs, which are the pronunciation units of Japanese, as the separated KANAs. Examples of the words, nonsense words, and separated KANAs are listed in Table 1.

2.2 Subjects

Seven normal volunteers participated in this study (all males, 21-23 years old). All subjects were right-handed, native Japanese speakers. After full explanation, all subjects gave a written informed consent for all studies according to a protocol approved by the institutional research review board.

2.3 Tasks

In this study, we used a block design to reveal the areas of brain activity in response to the three conditions: word, nonsense word, and separated KANAs. The duration of each session was 150 s, divided into five 30-s blocks: OFF, ON, OFF, ON, and OFF. Two sessions were presented to each subject for each condition, but using different sound materials. Therefore, every subject was presented with six sessions, the sequence of which was random. During the entire scanning process, the subjects were asked to pay attention to what was presented to them and to listen.

Table 1. Examples of three kinds of stimuli: words, nonsense words, and separated KANAs.

Category	Examples	Meaning or Explanation
Words	えんぴつ	pencil
	せきたん	coal
	せんせい	teacher
None sense words	まめうぎ	Nonsense stimuli, in form of words, but no meaning. Words and nonsense words are presented to subjects once every 2 seconds.
	のろざい	
	みきよへ	
Separated KANAs	ち、ず、づ、た、ゆ、ら、よ、ご、さ	Separated KANAs were presented to subjects one by one, every 0.5 s.

2.4 MRI acquisition

Image scanning was performed on a 1.5T scanning system (EchoSpeed 1.5T, GE) using a standard frequency head coil for signal transmission and reception. The functional images consisted of 20 consecutive slices, covering the whole cerebral cortex. The T2*-weighted gradient echo-planner imaging sequence was used with following parameters: TR/TE=6000/50 msec; FA=90°; matrix size = 64×64; voxel size = 4×4×5.5mm. Before the acquisition of functional images, a T2-weighted anatomical image was obtained in the same plane as the functional image using a spin echo sequence (voxel size=1×1×5.5).

2.5 Image analysis

Image and statistical analyses were performed with the Statistical Parametric Mapping package SPM2 (http://fil.ion.ucl.uk/spm) implemented in Matlab (Mathworks Inc., Sherborn, MA, USA).The pro-processing included religning, co-registering, normalizing, and smoothing.

Within the framework of a fixed effect analysis, we searched for voxels significantly activated in all subjects in response to our conditions. The task-related neural area activated during each condition was modeled with a box-car function, convoluted with a canonical hemodynamic response function.

3. Results

All stimuli produced extensive activation of the superior temporal region bilaterally, while the activated area and intensity in the left hemisphere were greater than those in the right hemisphere across the three kinds of stimuli, as illustrated in Fig. 1. Moreover, activated regions in the right temporal lobe were slightly shifted anteriorly and laterally relative to the activated regions in the left temporal. In all of the conditions, the activation along Heschl's gyrus (HG) and the planum temporale (PT) extended further medially in the left hemisphere than in the right, while the activation of the lateral super temporal gyrus (STG) and the planum polare (PP) was more prominent in the right hemisphere. The detailed results of activation were shown in Fig. 1 and Table 2.

Table 2. Locations of activation peaks and clusters for the three kinds of stimuli: words, nonsense words, and separated KANAs (p<0.001, uncorrected).

			Voxel - leve	Cordinates		
	Anatomical location	Cluster size	$Z_{uncorrected}$	x	y	z
Words	middle STS	1268	7.72	62	-10	-2
	middle STS		6.63	66	-14	10
	posterior STS		4.18	58	-32	10
	MTG	1264	6.24	-64	-24	2
	STG		6.2	-56	-8	-2
	precentral gyrus	52	4.1	54	4	40
	central sulcus	39	3.7	-50	-4	22
Nonsense	STG	363	5.03	64	-16	2
Words	MTG			56	2	-8
	STG	468	4.97	-56	-10	-2
	MTG		4.76	-64	-26	4
Separate	middle STS	1374	7.56	62	-12	0
KANAs	STG		7.24	66	-18	4
	MTG	1562	7.07	-62	-24	4
	STS		6.4	-52	-28	8
	middle STS		5.48	-60	-12	0

Coordinates are in standard sterotaxic space 18.. Voxel size: 2.0 x 2.0 x 2.0 mm. STG, superior temporal gyrus; STS, superior temporal sulcus; MTG, middle temporal gyrus.

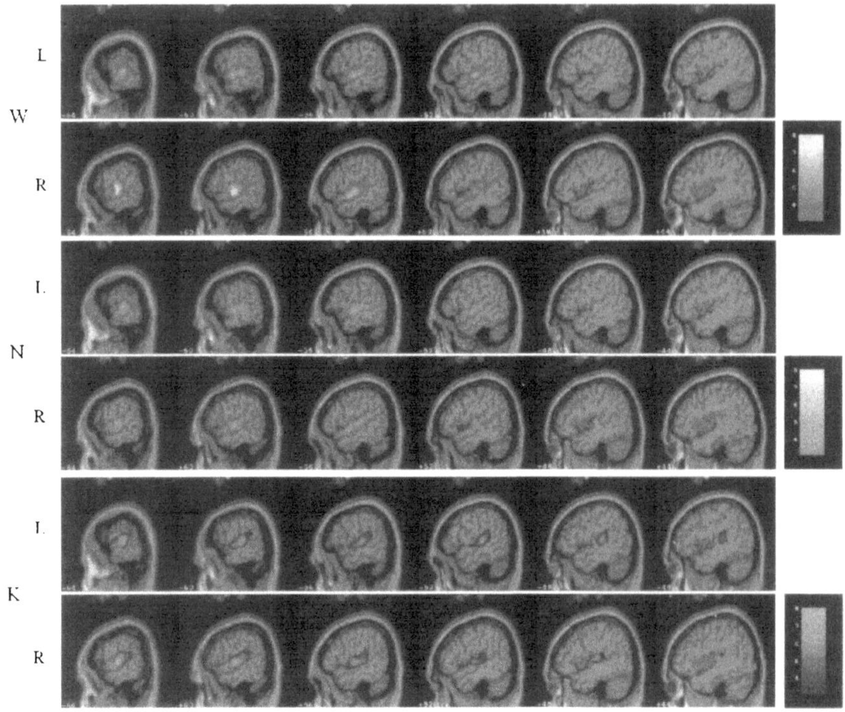

Fig. 1. The areas of the brain activated in the three conditions: words, nonsense words, and separated KANAs (P< 0.001, uncorrected). In the first two rows, the red areas are the areas activated in the word condition. In the third and fourth rows, the green areas are the areas activated in the nonsense word condition. In the last two rows, the blue areas are activated in the separated KANA condition.

4. Discussion

The responses of the superior temporal cortex in the human brain to Japanese words, nonsense words, and separated KANAs were obtained using blood oxygenation level-dependent (BOLD) fMRI. The dorsal temporal surfaces of the HG and PT were activated by all stimuli, but there was some variability across the three kinds of stimuli with respect to the volumes and location. In the region of the superior temporal cortex, the words and separated KANAs activated almost the same volumes; however, the areas activated by the separated KANAs were slightly posterior to those activated by the words and nonsense words. The volume activated by the nonsense words was smaller than that activated by the words and separated

KANAs, perhaps owing to the different familiarities of the stimuli. Compared with the nonsense word and separated KANA conditions, the word condition activated areas in addition to the superior temporal cortex; these included the primary motor cortex (BA 4) in the right hemisphere and the ventral pre-motor cortex (BA 6) in the left hemisphere. No areas in the anterior STG and STS, which are related to the semantic representations of words, were obviously activated [7].

Superior temporal activation occurred bilaterally for all three kinds of stimuli, as in previous studies [3-6]. As all stimuli bilaterally activated the HG and PT, these areas would not be language specific. The right temporal areas of activation were shifted slightly more anterolaterally relative to the homologous activations in the left hemisphere, which was also found in the studies of Binder and Penhune [3, 16, 17]. Right auditory cortex activation included more of the anterolateral STG and PP than did left auditory activation. This arrangement may compensate for the fact that the horizontal portion of the PT is generally larger in the left hemisphere [3].

In the present study, we did not find any hierarchical sound processing trends such as those found in animal studies [1, 6] and human functional imaging studies [3, 5], perhaps owing to the simplicity of the task (passively listening) and the similarity of the sound stimuli (words, nonsense words, and separated KANAs). In addition to the activated areas in the primary auditory cortex, such as the HG and PT, some posterior and medial areas were also activated, especially in the word and separated KANA conditions. The posterior activation favors the hypothesis by Scott [7], in which the posterior stream of speech perception might be related to articulatory-gestural representations. The posterior activations also favor the results of some studies on lesions, such as aphasia conditions [13], in which the posterior left temporal lobe damage could hinder the patients' ability for repetition while preserving speech comprehension. Moreover, in the present study, we did not find any obvious activations in the anterior superior temporal gyrus and sulcus, which are important for representing and accessing the content meaning of utterances, based on animal studies [23], lesion studies in humans with respect to semantic dementia [14, 24], and functional imaging studies [3]. As a result of the passive listening to words, nonsense words, and separated KANAs, and the fast broadcasting speed, it may not have been possible for subjects to access the meaning of the words, let alone the nonsense words and separated KANAs.

In this study, words and separated KANAs activated more areas than did nonsense words in both hemispheres (Table 2 and Fig. 1). Considering almost the same speed and spectural characteristics among these three kinds of stimuli, we postulated that the differences in the volumes activated by words, separated KANAs, and nonsense words were mainly related to the differences in the familiarity levels. Japanese speakers are very familiar with KANAs, which are the pronunciation units of the language; the words used in the experiment were familiar words often used in ordinary life, as illustrated in Table 1. The nonsense words, however, sounded like novel Japanese words to the subjects; hence, the subjects were not familiar with them. Given that all of the tasks involved passive listening and that the words and separated KANAs activated nearly the same volumes, the differences in the volumes activated by words, separated KANAs, and nonsense words would not be related to meaning access but rather to the familiarity of the stimuli.

The word condition activated the primary motor cortex (BA 4) in the right hemisphere and the ventral pre-motor cortex (BA 06) in the left hemisphere, in addition to the superior temporal cortex, as shown in Fig. 1 and Table 2. In another study [19], the right primary motor cortex was also activated when subjects were doing semantic matching tests. However, in our study, it was not necessary for the subjects to access the meanings of most of the words because the subjects were required to only passively listen to the stimuli (including words), which were broadcast at a rapid (one word or nonsense word every 2 s). Thus, the activation of the right primary motor cortex may be related to semantic control, which could trigger the access of semantic information/representations, especially the access of word articulation representations, from auditory or visual input [7, 19]. In animal studies, the ventral pre-motor cortex contains neurons that discharge not only when a monkey grasps or manipulates objects but also when it observes similar actions performed by others. Such neurons are called "mirror neurons." Furthermore, neurons in this part of the brain appear to respond similarly to both mouth actions and manual gestures and to respond to the sounds produced by those manual gestures [7, 21]. Recent studies in humans have revealed that speech perception, either watching speaking mouths or listening to speech without visual input, enhanced the excitability of the motor cortex in the left hemisphere relative to nonlinguistic baseline perceptual tasks [7, 22]. Taken together with the animal studies and previous human imaging studies, these findings indicate that the ventral pre-motor area in the left hemisphere is related to articulatory-gestural representations of words, which could be accessed during word perception.

5. Conclusions

The results of our study support the model of Scott [7], in which the articulatory-gestural representations may be stored and perceived in the posterior stream. The activation of the right primary cortex and left pre-motor cortex by words may be related to the control of semantic information access and articulatory-gestural representations of learned words, respectively. Owing to the fast broadcasting speed of the words in our study, the passive listening task may not have produced obvious semantic access. We found that the words and separated KANAs activated more areas than did the nonsense words, perhaps because they are more familiar.

References

1. J. P. Rauschecker et al., "Processing of complex sounds in the macaque nonprimary auditory cortex," Science, Vol. 268, pp. 111-114, 1995.
2. J. P. Rauschecker and B. Tian, "Mechanisms and streams for processing of 'what' and 'where' in auditory cortex," Proc. Natl. Acad. Sci. U. S. A., pp. 11800-11806, 2000.
3. J. R. Binder, J. A. Frost and T. A. Hammeke et al. "Human temporal lobe activation by speech and nonspeech sounds," Cerebral Cortex, Vol. 10, pp. 512-528, 2000.
4. S. K. Scott, C. C. Blank, and S. Rosen, et al. "Identification of a pathway for intelligible speech in the left temporal lobe," Brain, Vol. 123, pp. 2400-2406, 2000.
5. M. H. Davis and I. S. Johnsrude. "Hierarchical processing in spoken language comprehension," Journal of Neuroscience, Vol. 23, no. 8, pp. 3423-3431, 2003.6.
6. J. Kaas and T. Hackett, "Subdivisions of auditory cortex and processing streams in primates," Proc. Natl. Acad. Sci. U.S A., pp. 11793-11799, 2000.
7. S. K. Scott and I.S. Johnsrude, "The neuroanatomical and functional organization of speech perception," Trends in Neurosciences, Vol. 26, no. 2, February 2003.
8. J-F. Démonet, C. Price et al., "Differential activation of right and left posterior sylvian regions by semantic and phonological tasks: a positron emission tomography study in normal human subjects," Neuroscience Letters, Vol. 182, pp. 25-28, 1994.
9. D. Perani, S. Dehaene et al., "Brain processing of native and foreign languages," Neuroreport, Vol. 7, pp. 2439-2444, 1996.
10. C. J. Price, R. J. S. Wise et al., "Hearing and saying: The functional neuro-

anatomy of auditory word processing," Brain, Vol. 119, pp. 919-931, 1996.
11. S. Hirano, Y. Naito et al., "Cortical activation by monaural speech sound stimulation demonstrated by positron emission tomography," Exp Brain Res, Vol. 113, pp. 75-80, 1997.
12. S. K. Scott and R. J. S. Wise, "The functional neuroanatomy of prelexical processing in speech perception," Cognition, Vol. 92, pp. 13-45, 2004.
13. H. Axer, A. G. von Keyserlingk et al., "Supra-and infrasylvian conduction aphasia," Brain and language, Vol. 76, pp. 317-331, 2001.
14. J. R. Hodges, K. Patterson et al., "Semantic dementia: progressive fluent aphasia via temporal lobe atrophy," Brain, Vol. 115, 1783-1806, 1992.
15. T. Ogawa, "52 kategori ni zokusuru go no shutsugen dohyo (Frequency association norms for 52 categories)," Jinbun Ronkyu, Vol. 22, 1-61, 1972.
16. V. R. Penhune, R. J. Zatorre, J. D. MacDonald et al., "Interhemispheric anatomical differences in human primary auditory cortex: probabilistic mapping and volume measurement from magnetic resonance scans," Cerebral Cortex, Vol. 6, 661-672, 1996.
17. J. T. Devlin, J. Raley, E. Tunbridge et al., "Functional asymmetry for auditory processing in human primary auditory cortex," The Journal of Neuroscience, Vol. 23, no. 37, pp. 11516-11522, 2003.
18. J. Talairach and P. Tournoux, "Co-planar stereotaxic atlas of the human brain. Thieme," Thieme, New York, 1988.
19. B. Reginald Adams and P. Janata, "A comparison of neural circuits underling auditory and visual object categorization," NeuroImage, Vol. 16, pp. 361-377, 2002.
20. G. Rizzolatti and M. A. Arbib, "Language within our grasp," Trends Neurosciences, Vol. 21, pp. 188-194, 1998.
21. Kohler, E. et al. "Auditory mirror neurons in the ventral premotor cortex of the monkey," Soc. Neurosci., Abstr., Vol. 27, pp. 1299, 2001.
22. K. E. Watkins* , A. P. Strafella, and T. Paus, "Seeing and hearing speech excites the motor system involved in speech production," Neuropsychologia, Vol. 41, pp. 989-994, 2003.
23. T. A. Hackett, I. Stepniewska et al., "Subdivisions of auditory cortex and ipsilateral cortical connections of parabelt auditory cortex in macaque monkeys," J. Comp. Neurol., Vol. 394, pp. 475-495, 1998.
24. J. R. Hodges, J. Spatt et al., "'What' and 'how': Evidence for the dissociation of object knowledge and mechanical problem-solving in the human brain," Proc. Natl. Acad, Sci. U.S.A., Vol. 96, pp. 9444-9448, 1999.

Interpretation of Increases in Deoxy-Hb during Functional Activation

Yoko Hoshi[1], Shunji Kohri[1], and Norio Kobayashi[2]

[1]Department of Integrated Neuroscience, Tokyo Institute of Psychiatry
[2]Department of Pediatrics School of Medicine, Hokkaido University

Chapter Overview. Increases in deoxy-Hb in activated brain areas are often explained by a lower increase in regional cerebral blood flow compared with an increase in regional cerebral metabolic rate of oxygen. However, our two studies, in which by using a near-infrared spectroscopy (NIRS) instrument, we examined activity-dependent changes in the Hb oxygenation state in the somatosensory cortex of the rat brain during contrateral tibial nerve stimulation and in the occipital regions of human neonates during photic stimulation, provide other possible explanations.

Key Words. Near intra-red Spectroscopy, and Hemodynamics.

1. Introduction

Negative BOLD signals in activated brain areas are often explained by increases in deoxy-Hb attributed to uncoupling between regional cerebral blood flow (rCBF) and regional cerebral metabolic rate of oxygen ($rCMRO_2$). For example, "initial dip" is generally accounted for by transient venous hypoxia, since an increase in rCBF lags that in $rCMRO_2$ [1, 2]. This is supported by observations by optical imaging using intrinsic signals, in which an early (< 3 sec) increase in deoxy-Hb is followed by its delayed decrease [3, 4]. Another example is negative BOLD signals during visual activation in infants [5, 6]. Near-infrared spectroscopy (NIRS) also detected increases in deoxy-Hb with increases in oxy-Hb and t-Hb during photic stimulation in infants [7]. This increase in deoxy-Hb is explained by a lower increase in rCBF compared with an increase in $rCMRO_2$, which is related to brain development: rapid synaptic formation occurs in infants, accompanying increased metabolism. However, our two studies suggest that increases in deoxy-Hb during functional brain activation do not necessarily mean relatively hypoxic conditions.

2. Study 1

2.1 Methods

Seven male Wister rats, weighing 180-250 g, were anesthetized by intraperitoneal injection of pentbarbital. They were tracheotomized, and a femoral artery and vein were cannulated. For the measurement of somatosensory evoked potential (SEP), skin and muscle overlying the calvaria were reflected. A needle electrode was inserted into the left parietal bone 1 mm posterior to the bregma and 3.5 mm lateral to the sagital suture, and a reference electrode was inserted into the nasal bone. Rats were paralyzed with an intravenous injection of panchronium bromide and mechanically ventilated with room air. The tidal volume and respiratory rate were adjusted to give an arterial pCO_2 value of 37-42 mmHg.

SEPs were obtained following transcutaneous stimulation of the right posterior tibial nerve with rectangular pulses of 2 mA intensity and 0.5 ms duration at a frequency of 5 Hz. Three hundred or 200 responses were averaged and displayed on a signal averager (Synax ER 1100; NEC, Tokyo, Japan).

A portable NIRS apparatus was build. In this, NIR light from a halogen lamp passed through three interference filters (wavelengths; 700, 730, and 750nm) and illuminated the rat's head. Measurement was performed by two methods in each rat. In one, light was illuminated through a light guide with a diameter of 2 mm onto the left parietal bone at which the recording needle electrode for SEP was inserted, and reflected light was measured through a light guide with a diameter of 1mm. In the other, light was illuminated on the palate and transmitted light was measured on the parietal bone at which the recording needle was inserted.

NIRS measurements were performed simultaneously while measuring SEP. When an atypical SEP or no response was observed, data were excluded from analysis. Changes in relative concentrations of oxy-Hb, deoxy-Hb, and t-Hb were calculated from absorbance changes at the three above-mentioned wavelengths every 500 ms (reflected light) or 1000 ms (transmitted light).

2.2 Results

2.2.1 Measurements of reflected light (measurements of the somatosensory cortex)

Two patterns of changes in the Hb oxygenation state were observed during activation of the somatosensory cortex: an increase in oxy-Hb and a reciprocal decrease in deoxy-Hb with no significant change in t-Hb (pattern 1, Fig. 1A), and increases in oxy-Hb, deoxy-Hb, and t-Hb (pattern 2, Fig. 1B). Changes in NIRS parameters were reproducible in each rat. Pattern 2 was observed in only 1 rat. These changes occurred within 1 second after the start of stimulation and lasted for about 20 seconds after the cessation of stimulation. In 2 rats, early increases in deoxy-Hb and t-Hb with no change in oxy-Hb, which occurred within 500 ms after the start of stimulation, were observed (Fig. 1C). About 5 seconds after the start of stimulation, oxy-Hb started to increase and deoxy-Hb started to decrease.

2.2.2 Measurements of transmitted light (measurements of the somatosensory cortex and the deeper structures of the brain)

Two patterns of changes in the Hb oxygenation state were also observed in measurements of transmitted light: an increase in deoxy-Hb and a reciprocal decrease in oxy-Hb with no significant change in t-Hb (pattern 1, Fig. 2A), and decreases in oxy-Hb, deoxy-Hb, and t-Hb (pattern 2, Fig. 2B). The Hb oxygenation state returned to the original level about 20 seconds or more after the cessation of stimulation. In one rat that showed the pattern 2, deoxy-Hb first did not change and then increased (Fig. 2C).

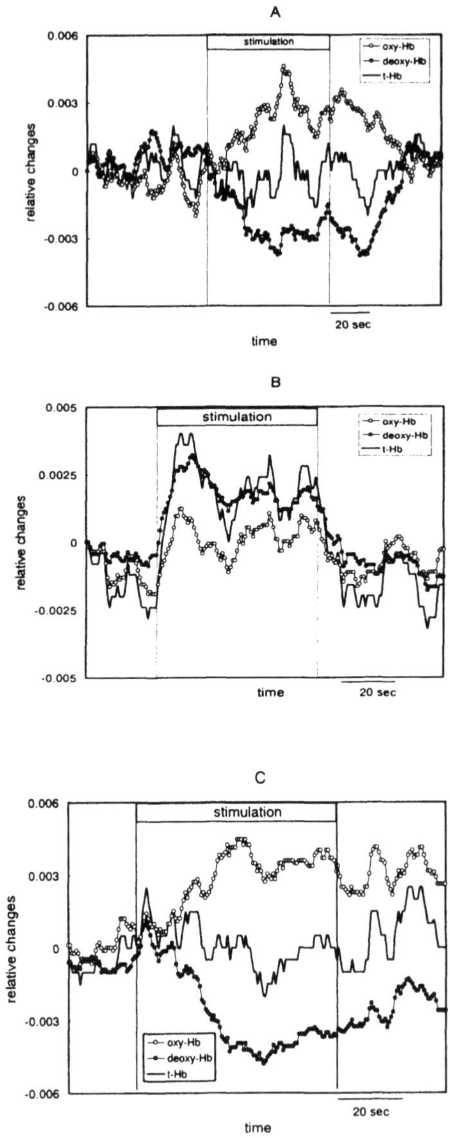

Fig. 1. Changes in oxy-Hb, deoxy-Hb, and t-Hb in the rat somatosensory cortex during contrateral tibial nerve stimulation.

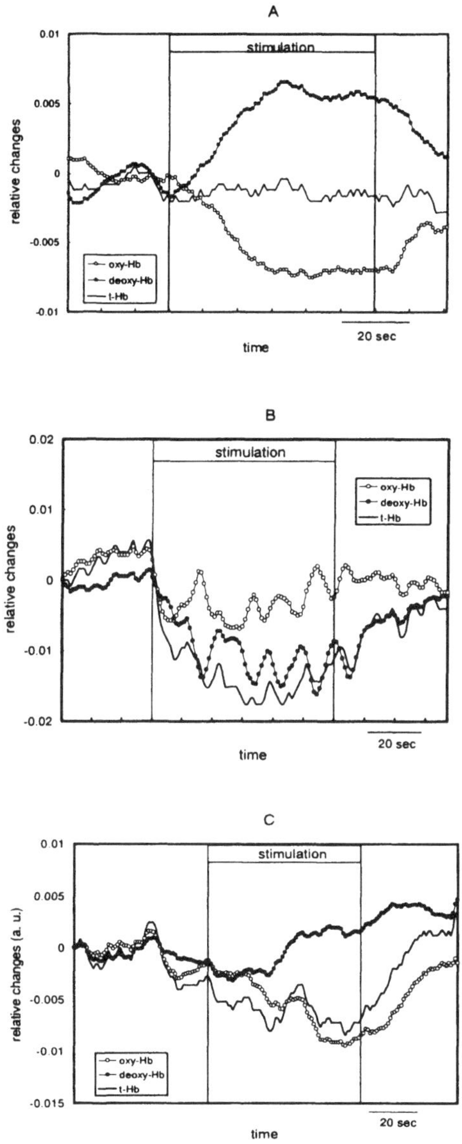

Fig. 2. Changes in oxy-Hb, deoxy-Hb, and t-Hb in the rat somatosensory cortex and the deeper structures of the brain during contrateral tibial nerve stimulation

2.3 Discussion

The NIR signals observed here were attributed to changes in blood oxygenation and the diameter of blood vessels (blood volume). Because of the dual wavelength analysis, the contribution of changes in scattering was eliminated from the present NIRS signals. Increases in oxy-Hb and t-Hb with a decrease in deoxy-Hb are typically observed in activated areas. This pattern, which was not observed in the present study, results from combined effects of the dilation of arterioles and a decrease in deoxy-Hb in venous blood due to overcompensation of blood flow [8]. When changes in rCBF is small, an increase in oxy-Hb and a reciprocal decrease in deoxy-Hb without a change in t-Hb are observed [9]. When an increase in rCBF causes venous dilation, increases in all the three NIRS parameters are observed. It is, thus, supposed that the measurements of reflected light in this study detected changes occurring rather in a venous area in the activated cortex. As can be seen in Fig. 1C, early increases in deoxy-Hb and t-Hb with no change in oxy-Hb are consistent with the observations from microscopic approaches [3, 4], although the duration of increases in deoxy-Hb in the present study was longer than previously reported. This discrepancy might be related to the difference in stimulation.

In the measurement of transmitted light, signal changes occurring in the somatosensory cortex were reversed by overlapping signals attributed to the deeper structures in the brain. Pattern 1 observed in measurements of transmitted light (Fig. 2A) meant that blood flow velocity decreased with no significant change in blood volume. Pattern 2 (Fig. 2B) meant that rCBF decreased accompanying decreases in local blood volume. It is, thus, suggested that while the somatosensory cortex was activated, other brain regions were deactivated or blood flow in these regions were stolen to the somatosensory cortex.

Under the present experimental conditions, it could not be exactly determined to what depth from the brain surface detected light penetrated in reflectance mode. However, it is conceivable that the detected light passed through not only the somatosensory cortex but also brain regions in which rCBF was decreased during stimulation. When both activated and rCBF-decreased regions are illuminated, NIRS signals represent the summation of changes in oxy-Hb and deoxy-Hb in the two regions. This provides a different explanation for the early increases in deoxy-Hb and t-Hb with no change in oxy-Hb (Fig. 1C) from that of Maloneck et al., who concluded that in the very early phase of activation, local capillary blood volume increases, but arterioles have not yet dilated and the blood flow has not yet increased [3]. That is, in the early phase, the magnitude of increases in deoxy-Hb in the deeper brain structures might be larger than that in the activated area, while that of the change in oxy-Hb might be the same in both areas. This could also explain the early increases in deoxy-Hb observed in microscopic approaches, although penetration depth of the visible light is shallower compared with the NIR light.

3. Study 2

3.1 Methods

Seven healthy neonates, 4 or 5 days old (gestation = 37 - 41 weeks) were studied. The prenatal and postnatal courses were uncomplicated in all of these neonates. None had any abnormalities likely to affect brain function. Before each study, informed parental consent was obtained. The measurements were performed during spontaneous sleep after feeding.

The neonates were spine position on a cot under dim light. Using a 6-channel NIRS monitoring system (OMM-100, Shimadzu, Kyoto, Japan), we measured three areas in the occipital region simultaneously (Fig. 3). This system consists of three semiconductor laser diodes (wavelengths = 780, 805, and 830nm) as light sources. The laser pulses are directed by a moving mirror to one of six light guides that lead to corresponding source positions on the subject every 400 ms. Three of the six channels were used in this study, and sampling time was 2 seconds. Visual stimulation was performed with a stroboscopic white light flashing at 10 Hz projected on the neonate's eyelids through a cut filter of the NIR region for 30 seconds. In this study, an increase in oxy-Hb was used as an indicator of an activity-dependent increase in rCBF [9]. When an increase in oxy-Hb was not observed in any channel, the position of the probe was moved higher up or laterally (or both) on the order of millimeters. When an increase in oxy-Hb was observed, the measurement was repeated for 2 minutes after the cessation of the previous stimulation to confirm reproducibility.

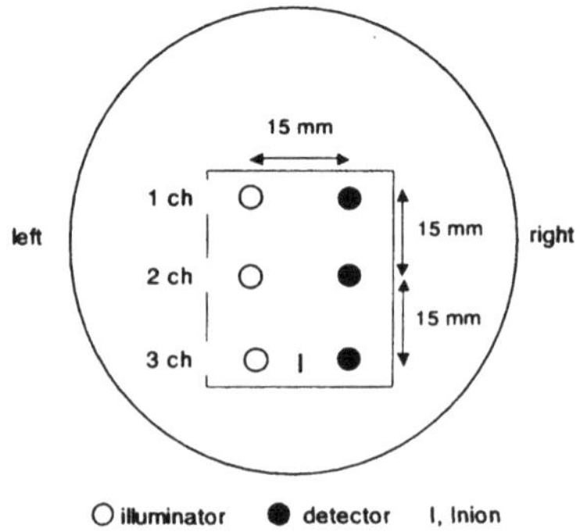

Fig. 3. Positions of light guides

3.2 Result

Figure 4 demonstrates the NIRS data for one subject who showed an increase in oxy-Hb in channel 3. In this neonate, deoxy-Hb and t-Hb also increased in channel 3. In contrast, all the three NIRS parameters decreased in channel 2, and no change was observed in channel 1. An increase in oxy-Hb was observed only in one channel in all the subjects. The second stimulation caused the same changes observed during the first stimulation, except for the change in direction of deoxy-Hb in the activated area. The change in deoxy-Hb varied with each subject and with each stimulation even in the same subject. For example, one subject had an increase in deoxy-Hb during the first stimulation, yet the second stimulation caused a slight decrease. No change in deoxy-Hb in either of the two measurements in three subjects. Decreases in oxy-Hb, t-Hb, and deoxy-Hb near the activated area, typical examples of which are seen in channel 2 of Figure 4, were observed in five of the seven subjects.

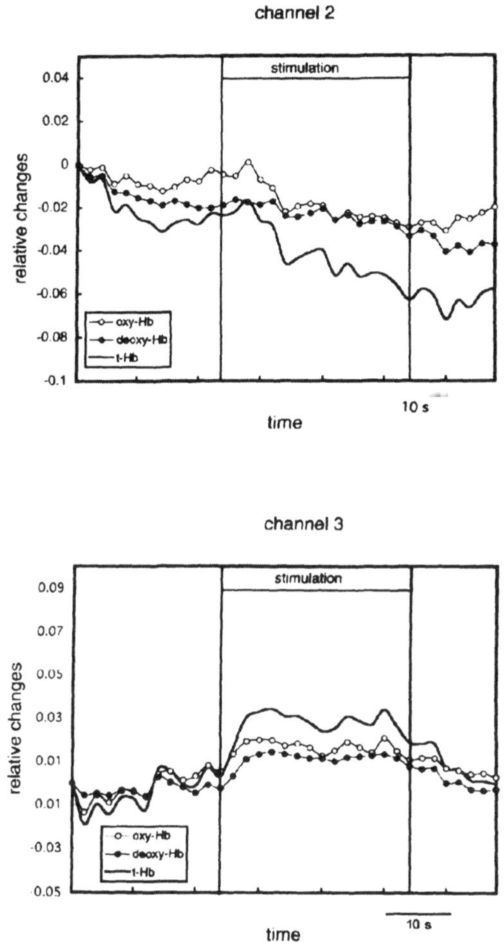

Fig. 4. Example traces of changes in oxy-Hb, deoxy-Hb, and t-Hb in channels 2 and 3 during photic stimulation. In channel 1, no significant changes were observed in this subject. From reference [10] , with permission.

3.3 Discussion

The present study demonstrated that deoxy-Hb increases in the visual cortex of the neonatal brain. This is in accordance with previous NIRS and fMRI studies on infants [5-7]. Widely accepted explanation for this decrease in deoxy-Hb is that the degree of increases in $rCMRO_2$ exceeds that of rCMR in infants, especially in infants older than 8 weeks, which corresponds with a period of rapid formation of synapses [6]. However, this cannot explain our observation that the change in the direction of deoxy-Hb varied not only with each subject but also with each measurement even in the same. The direction of the changes in deoxy-Hb observed in NIRS vary with the venous blood oxygenation and changes in the diameter of venules [9]. Thus, another possible explanation is that the increase in deoxy-Hb observed in neonates is related to the venous dilation caused by increases in rCBF.

4. Conclusion

To interpret increases in deoxy-Hb during functional activation, heterogeneity of brain activity and cerebral blood volume, as well as venous hypoxia, should be taken into account.

References

1. Ernst T, Hennig J (1994) Observation of a fast response in functional MR. Magn Reson Med 32: 146-149.
2. Menon R S, Ogawa S, Hu X, et al (1995) BOLD based functional MRI at 4 Tesla includes a capillary bed contribution: echo-planar imaging correlates with previous optical imaging using intrinsic signals. Magn Reson Med, 33: 453-459.
3. Malomeck D, Dirnagl D, Liduer U, et al (1997) Vascular imprints of neuronal activity: Relationships between the dynamics of cortical blood flow, oxygenation, and volume change following sensory stimulation. Proc Natl Acad Sci USA, 94: 14826-14831.
4. Nemoto M, Nomura Y, Sato C, et al (1999) Analysis of optical signals evoked by peripheral nerve stimulation in rat somatosensory cortex: Dynamic changes in hemoglobin concentration and oxygenation. J Cereb Blood Flow Metab, 19: 246-259.
5. Born P, Leth H, Miranda MJ, et al (1998) Visual activation in infants and young children studied by functional magnetic resonance imaging. Pediatr Res, 44: 578-583.
6. Yamada H, Sadato N, Konishi Y, et al (1997) A rapid brain metabolic change in infants detected by fMRI. Neuroreport, 8: 3775-3778.
7. Meek JH, Firbank M, Elwell CE, et al (1998) Regional hemodynamic response to visual stimulation in wake infants. Pediatr Res, 43: 840-843.
8. Fox PT, Raichle ME (1986) Focal Physiological uncoupling of cerebral blood flow and oxidative metabolism during somatosensory stimulation in human subjects. Proc Natl Acad Sci USA, 83: 1140-1144.
9. Hoshi Y, Kobayashi N, Tamura M (2001) Interpretation of near-infrared spectroscopy signals: a study with a newly developed perfused rat brain model. J Appl Physiol, 90: 1657-1662.
10. Hoshi Y, Kohri S, Matsumoto Y, et al (2000) Hemodynamic responses to photic stimulation in neonates. Pediatr Neurol 23: 323-327.

Probing the Plasticity of the Brain with TMS

John Christopher Rothwell

Sobell Department of Motor Neuroscience and Movement Disorders, Institute of Neurology, Queen Square, London WC1N 3BG, UK

Chapter Overview. Transcranial magnetic stimulation (TMS) was originally developed as a convenient way of delivering an electrical stimulating pulse across the resistive barrier of the skull and scalp into the brain. The currents induced in the brain are brief, and similar in magnitude and time course to those produced by a conventional peripheral nerve stimulator. The advantage is that the stimulation is almost painless and can readily be applied in conscious human subject, allowing us for the first time to directly manipulate brain activity. It turns out that if stimuli are given repetitively, it is possible to induce effects on the brain that outlast the period of stimulation for minutes or even hours and days. It is these effects that may provide a window to probe mechanisms of neural plasticity in the intact human brain.

Key Word. TMS(Trans-Cranial Magnetic Stimulation)

1. Principles of TMS

The magnetic stimulator consists of a coil of wire connected to a large electrical capacitance. When the capacitance is discharged through the coil a very large current flows. This reaches a peak value of several thousand amps within about 200μs and then decays to zero. The current produces a magnetic field oriented perpendicular to the coil which can reach values of up to 3 Tesla (Barker, 1991). The precise stimulating characteristics depend upon the model of stimulator used. For example, the Magstim 200 stimulator produces a magnetic field which rises to peak within about 150μs and then decays slowly to zero over the next millisecond. Such a rapidly changing magnetic field induces electrical eddy currents in any conductive structures nearby. Because the skull presents a low impedance to magnetic fields of this frequency, eddy currents are produced in the brain, and it is these which stimulate neural tissue. The induced current is proportional to the rate of change of magnetic field, with the result that the peak current is achieved within about 75μs falls to zero by 150μs, and then is followed by a small reverse current lasting for the next 500-1000μs.

Currents induced on the scalp by magnetic stimulation are much less than those produced by transcranial electric stimulation so that the sensation produced by magnetic stimulation is very slight. The magnetic field falls off rapidly with distance from the coil: with a typical 12cm diameter round coil the strength falls by half at a distance of 4-5cm from the coil surface. Since the cerebral cortex can be 1 to 2cm from the surface of the scalp, and since the central sulcus itself can be 2cm deep in man, this means that stimulation is severely attenuated at deep sites such as basal ganglia or thalamus. In addition, since the resistance of white matter is greater than the grey matter, currents induced in the sub-cortical tissues are likely to be small in comparison with those induced in surface layers of cerebral cortex. Indeed, experiments with transcranial magnetic stimulation of motor cortex in monkey suggest that even at the highest intensities of stimulation activation of cortico-spinal fibres does not spread below the grey matter (Edgley, Eyre et al., 1992).

2. TMS Induced Plasticity

Single pulses of TMS are thought to activate the axons of neurones in the cortex and set up local synpatic activity at the site of stimulation. In the motor cortex, it has been shown that a single stimulus can lead to repetitive activity in neural circuits that lasts several milliseconds before being cut short by a longer lasting period of inhibition. Thereafter cortical excitability recovers to baseline levels.

However, repeated pulses of TMS have now been demonstrated to lead to after-effects on cortical excitability that outlast the period of stimulation for several minutes or hours. The exact nature of these after effects is unknown but they are thought by many authors to involve changes in the strength of synaptic connections that are analogous to long term potentiation and depression (LTP and LTD respectively) in experiments on reduced animal preparations. If the after effects of repeated TMS do involve plasticity of synaptic connections then it is possible that they could interact with and perhaps be used to investigate physiological processes that are thought to depend on synaptic changes such as learning and memory.

There are presently two main methods for inducing long lasting effects: repetitive TMS (rTMS) and paired associative stimulation (PAS). In both cases, the majority of the experiments have been conducted on the motor cortex. This is because single pulse stimulation here evokes recordable EMG responses in contralateral muscles (MEPs), the size of which can be used as an indicator of the excitability of the motor cortical output. It is thought, however, that similar principles apply to other areas of the neocortex.

rTMS simply consists of giving repeated TMS pulses at 1-50 Hz (Siebner and Rothwell, 2003). Several hundred stimuli usually are given with the intensity of each single pulse set to be around the threshold for evoking an EMG response in resting muscle (resting motor threshold, RMT). In principle, rTMS is similar to the conditioning protocols that are often used in animal experiments to produce long term synaptic changes such as LTP and LTD in hippocampal circuitry.

PAS involves repeated pairing of a TMS pulse with a carefully timed electrical stimulus to a peripheral nerve (Stefan, Kunesch et al., 2000). In a typical experiment, the electrical stimulus would be given to the median nerve at the wrist, and then 25ms later, a TMS pulse would be applied over the contralateral motor cortical hand area. Both stimuli would then be repeated 100 times or so every 4-10s. For about 30min after the end of this conditioning paradigm, the excitability of the motor cortical projection to hand muscles innervated by the median nerve (e.g. the abductor pollicis brevis) is increased. Interestingly, if the timing between the stimuli is reduced to 10ms, then the after effect is reversed, and cortical excitability is reduced. PAS is thought to be analogous to spike timing dependent plasticity in animal experiments. The sensory input from the median nerve and the activity produced by the TMS pulse are thought to converge on a common neurone. Depending on the temporal relation between the events, the TMS input is either strengthened or suppressed. This form of TMS induced plasticity is sensitive to drugs that interfere with the function of NMDA receptors and hence probably involves LTP/LTD like processes in the cortex.

One important feature of all these protocols is that the basal state of the cortex at the time of conditioning can change the long term effects that occur. For example, the suppression of motor cortex excitability that follows 1 Hz rTMS applied at rest is abolished if the stimulation is performed while subjects contract their hand muscles (Touge, Gerschlager et al., 2001); conversely the effect of 5 Hz rTMS on muscles in the forearm is reversed depending on whether the subject contracts the flexor or extensor muscles during the period of stimulation (Fujiwara and Rothwell, 2004). This is likely to be because the response to each single TMS pulse depends on the excitability of the cortex at the time the stimulus is applied. If different sets of connections are activated by each pulse of a repeated train, then this will change the pattern of long term effects that are observed.

As pointed out by many authors, the processes of neural plasticity have to be carefully controlled (Turrigiano and Nelson, 2000). Plasticity that is too easy to induce might lead to formation of unwanted associations, whereas plasticity that is too difficult to produce leads to problems in learning anything new. This is apparent if we consider increasing the excitability of neural connections with LTP-like processes (Bienenstock, Cooper et al., 1982). The positive-feedback nature of LTP carries the risk of triggering an uncontrolled increase in synaptic effectiveness which becomes potentially destabilizing, and overpowers all other inputs in the system. Evidence suggests that this can be prevented by making the amount of LTP dependent on the level of activity in the postsynaptic neuron: the greater the ongoing activity, the less effective are processes leading to LTP whilst processes leading to LTD are enhanced. Conversely, the lower the activity of the postsynaptic neurones, the more effective are processes that lead to LTP. This is known as "homeostatic" plasticity and is formalized in the model originally described by Bienenstock, Cooper and Munro (BCM model).

3. Relationship between Brain Plasticity and Behavioural Learning

Some of the best evidence for a connection between synaptic plasticity and learning comes from a series of experiments by Donoghue and colleagues in rat motor cortex showing that LTP and LTD at intracortical synapses is influenced by behavioural learning paradigms (Rioult-Pedotti, Friedman et al., 2000). Using trains of electrical stimuli to activate horizontal pathways in slices of motor cortex, they found that the cortex of rats trained in a forelimb task showed less LTP than slices from untrained animals; conversely LTD was enhanced. The effects were consistent with the BCM theory of bidirectional synaptic plasticity in which high prior levels of synaptic activity reduce the sensitivity of process leading to LTP and increase the sensitivity of LTD.

Ziemann et al (Ziemann, Iliac et al., 2004) recently performed an analogous experiment in human motor cortex. Their test of synaptic plasticity was the PAS model of paired associative learning. As in the previous animal experiments, Ziemann et al showed that a prior period of training in a hand motor task reduced the amount of LTP-like plasticity that could be evoked in the PAS paradigm whereas it increased the amount of LTD-like plasticity. The conclusion was that these indirect measures of cortical synaptic plasticity in the human motor cortex could be used to investigate, and perhaps even to influence, processes of behavioural learning. To date there have been no equivalent studies on rTMS induced plasticity and behavioural learning, but there seems no reason to suppose that they would behave in a fundamentally different way to the results obtained with PAS.

4. A New Method for Inducing Rapid Plasticity in the Human Motor Cortex with rTMS

Although rTMS is commonly used to induce plastic changes in the excitability of the cortex, the results are often weak, highly variable from one individual to another (Maeda et al., 2000), and rarely last longer than half an hour. In addition, stimulation protocols can often last quite some time (e.g. 25min if 1500 stimuli are given at 1 Hz).

There are several possible reasons for the previous disappointing results of rTMS: first, even in animal experiments, LTP/LTD is difficult to demonstrate in the cortex of awake and freely moving animals without the use of extended or repeated sessions of stimulation (Froc, Chapman et al., 2000). Second, concerns over safety have limited many humans studies to relatively low frequencies of stimulation (usually <10 Hz) (Wassermann, 1998) whereas animal studies often use much higher frequencies such as the "theta burst" paradigm (3-5 pulses at 100 Hz repeated at 5 Hz). Third, TMS in humans is relatively non-focal, and therefore cannot be used to target spatially specific neural connections. In most instances, this means that rTMS will activate a mixture of systems that potentially could have interacting effects that make the final outcome difficult to predict.

Huang et al (Huang, Edwards et al., 2005) recently introduced a rapid method of employing rTMS that more closely replicates common paradigms used in animal experiments. The TMS pulses are applied in high frequency bursts of 3 pulses at 50 Hz, repeated five times per second. These are "theta burst" paradigms, so called because the theta rhythm in EEG has a frequency of 5 Hz. Bursts that are applied intermittently (2s on, 8s off, repeated 20 times; 600 total TMS pulses) cause facilitation whereas continuous theta bursts for 40s (a total of 600 pulses) lead to suppression. The authors speculated suggest that in humans TBS produces a mixture of facilitatory and inhibitory effects on synaptic transmission, with facilitation building up faster than inhibition. If we assume that both facilitation and inhibition saturate at some level, then it is possible to explain the main features of the results as long as we allow inhibition to dominate in the long run. Thus, a short, intermittent protocol such as iTBS would favour rapid build up of facilitation. In contrast, a longer lasting continuous protocol such as cTBS would initially produce facilitation, but eventually this would saturate and inhibitory effects which build up slower, but saturate at a higher level would dominate.

References

1. Barker AT (1991) An introduction to the basic principles of magnetic nerve stimulation. J.Clin.Neurophysiol. 8: 26-37
2. Bienenstock EL, Cooper LN, Munro PW (1982) Theory for the development of neuron selectivity: orientation specificity and binocular interaction in visual cortex. J.Neurosci. 2: 32-48
3. Edgley SA, Eyre JA, Lemon RN, Miller S (1992) Direct and indirect activation of corticospinal neurons by electrical and magnetic stimulation in the anesthetized macaque monkey. Journal Of Physiology-London 446: 224
4. Froc DJ, Chapman CA, Trepel C, Racine RJ (2000) Long-term depression and depotentiation in the sensorimotor cortex of the freely moving rat. J.Neurosci 20: 438-445
5. Fujiwara T, Rothwell JC (2004) The after effects of motor cortex rTMS depend on the state of contraction when rTMS is applied. Clin.Neurophysiol. 115: 1514-1518
6. Huang YZ, Edwards MJ, Rounis E, Bhatia KP, Rothwell JC (2005) Theta burst stimulation of the human motor cortex. Neuron 45: 1-6
7. Rioult-Pedotti MS, Friedman D, Donoghue JP (2000) Learning-induced LTP in neocortex. Science 290: 533-536
8. Siebner HR, Rothwell J (2003) Transcranial magnetic stimulation: new insights into representational cortical plasticity. Exp.Brain Res. 148: 1-16

9. Stefan K, Kunesch E, Cohen LG, Benecke R, Classen J (2000) Induction of plasticity in the human motor cortex by paired associative stimulation. Brain 123 Pt 3: 572-584

10. Touge T, Gerschlager W, Brown P, Rothwell JC (2001) Are the after-effects of low-frequency rTMS on motor cortex excitability due to changes in the efficacy of cortical synapses? Clin.Neurophysiol. 112: 2138-2145

11. Turrigiano GG, Nelson SB (2000) Hebb and homeostasis in neuronal plasticity. Curr.Opin.Neurobiol. 10: 358-364

12. Wassermann EM (1998) Risk and safety of repetitive transcranial magnetic stimulation: report and suggested guidelines from the International Workshop on the Safety of Repetitive Transcranial Magnetic Stimulation, June 5-7, 1996. Electroencephalogr.Clin.Neurophysiol. 108: 1-16

13. Ziemann U, Iliac TV, Pauli C, Meintzschel F, Ruge D (2004) Learning modifies subsequent induction of long-term potentiation-like and long-term depression-like plasticity in human motor cortex. J.Neurosci. 24: 1666-1672

Time Series of Awake Background EEG Generated by a Model Reflecting the EEG Report

Shigeto Nishida[1], Masatoshi Nakamura[2], Akio Ikeda[3], Takashi Nagamine[4], and Hiroshi Shibasaki[5]

[1]Department of Computer and Communication Engineering, Fukuoka Institute of Technology, Fukuoka, Japan
[2]Department of Advanced Systems Control Engineering, Saga University, Saga, Japan
[3]Department of Neurology, Kyoto University Graduate School of Medicine, Kyoto, Japan
[4]Human Brain Research Center, Kyoto University Graduate School of Medicine, Kyoto, Japan
[5]National Institute of Neurological Disorders and Stroke, National Institute of Health, Washington, USA

Chapter Overview. Awake background electroencephalogram (EEG) is interpreted by electroencephalographer (EEGer), and the results of EEG interpretation are described in EEG reports. Desired EEG reports should include enough information so that the time series of the raw EEG can be imagined from them. Then, it is important to know how much information about the characteristics of EEG is included in the report. In this study, a method for constructing an EEG model, which reflects the EEG report is proposed. First, EEG parameters for EEG interpretation items are determined based on EEGer's report, then the model parameters are calculated from the determined EEG parameters. The proposed models were constructed from the EEG reports of 7 subjects. The time series of the constructed models were visually evaluated by the EEGer, and satisfactory results were obtained. The information included in the EEG report can be represented through the time series of the model constructed by the proposed method. The proposed method can be utilized for the training of EEG report making.

Key Words. Awake background EEG, Model construction, EEG report, EEG interpretation

1. Introduction

Awake background electroencephalogram (EEG) is recorded from the scalp of a subject. As the EEG reflects at least a part of the functional status of the subject's brain, the electroencephalographer (EEGer) could interpret EEG to diagnose brain diseases. Through the visual inspection of EEG, the EEGer makes the EEG report, in which a final judgment and its details (the characteristics of EEG) are described. Automatic integrative interpretation method for the awake background EEG, consisting of quantitative EEG interpretation part and EEG report making part, has been developed by the authors, and satisfactory results were obtained [1,2]. Desired EEG reports should include enough contents so that the time series of the raw EEG can be imagined from them. Therefore, it is important to know how much information about the characteristics of EEG is included in the report. Many researchers have proposed automatic methods for judging the degree of abnormality of EEG from the time series of raw EEG, for example, extraction of features of EEG waveform [3], judgment of sleep stage [4], detection of spike [5]. However, there is no method for reproducing the time series of raw EEG from the results of EEG interpretation.

In this study, a method for generating the time series reflecting the EEG report by use of an EEG model is proposed. In this method, the parameters of the EEG model were appropriately determined based on the EEG report, and the model output was generated as the time series that reflects the EEG report. By reproducing the time series from the EEG report, in which the summary of the characteristics of raw EEG is described, it is possible to know how much information is included in the EEG report. The proposed method was tested with EEG data of 7 subjects, and the generated time series was visually evaluated by EEGer.

2. Methods

2.1 EEG Model

The EEG model, which was developed in our previous study [6] was slightly improved, and represented as

$$x^{p}(n\Delta t) = \sum_{k=1}^{4} a_k^{p}(n\Delta t)\sin(2\pi\, m_k n\Delta t + \theta_k^{p}(n\Delta t)) \tag{1}$$

$$a_k^{p}((n+1)\Delta t) = \gamma_k a_k^{p}(n\Delta t) + \xi_k^{p}(n\Delta t)$$

where $x^p(n\Delta t)$ is the model output including 4 different oscillations, which correspond to the δ, θ, α and β waves respectively; m_k the mean frequency; γ_k coefficients of the first order Markov process; $\xi_k^p(n\Delta t)$ the independent random numbers of Gaussian distribution with zero mean and variance of $(\sigma_k^p)^2$; p the scalp region number (p=1,2, ⋯, 16), and Δt the signal sampling interval (Δt=0.01 s). The common mean frequencies m_k were used at all scalp regions. The region number p is removed and wave number k =1,2,3,4 is represented as $k=\delta,\theta,\alpha,\beta$ to simplify the representation of symbol.

2.2 Determination of EEG Parameters Reflecting EEG Report

In this study, the EEG model reflecting the EEG report was constructed in two steps. In the first step, the values of the EEG parameters were determined based on the EEG report. In the second step, the EEG model parameters m_k, γ_k, σ_k ($k=\delta,\theta,\alpha,\beta$) were calculated from the determined EEG parameters.

A final judgment in four-grade (Normal, Mildly abnormal, Moderately abnormal, Markedly abnormal) and its details (the characteristics of EEG) are described in the EEG report. The described details were categorized into 12 EEG interpretation items as shown in Table 1, and judgment of each item was corresponding to the four-grade [1]. The EEG parameters represent the feature of each item quantitatively. The dominant rhythm was represented by using the α wave model component ($k=\alpha$). The values of EEG parameters were determined according to the judgment of each item based on the EEG report as shown in Table 1. The details of determination of the EEG parameters for each item are described as follows.

1) Existence of dominant rhythm

   When the dominant rhythm exists, the values of EEG parameters for the dominant rhythm are determined according to the following items 2)-8). If the dominant rhythm is judged as absence, the values of EEG parameters are determined to y=2.6, M_α=10Hz, P_α=8μV in all scalp regions, and the following items 2)-8) are skipped.

2) Organization of dominant rhythm

   The value of organization parameter y is determined as given in Table 1.

3) Asymmetry of organization

   The value of organization parameter on more abnormal side (right(r) or left(l)) is increased according to the degree of abnormality.

4) Frequency of dominant rhythm

   The value of frequency M_α is determined as given in Table 1.

5) Asymmetry of frequency

Table 1. Determination of values of EEG parameters

EEG interpretation item	EEG parameters	Normal	Mildly abnormal	Moderately abnormal	Markedly abnormal
Dominant rhythm					
Existence		Yes			No[a]
Organization	y	0	0.6	1.6	2.6
Asymmetry	$\mid y^{(l)} - y^{(r)} \mid$	0	0.5	0.8	1.5
Frequency	M_α	10	8.5	7.0	6.0
Asymmetry	$\mid M_\alpha^{(l)} - M_\alpha^{(r)} \mid$	0	0.7	1.5	2.0
Amplitude	P_α	50[b]	110	140	-
Asymmetry	$\mid P_\alpha^{(l)} - P_\alpha^{(r)} \mid / P_\alpha$	0	0.55	0.7	0.9
Extension	Ratio of P_α	see	Table 2		
Beta wave					
Amplitude	P_β	10	70	110	-
Asymmetry	$\mid P_\beta^{(l)} - P_\beta^{(r)} \mid / P_\beta$	0	0.55	0.7	0.9
Theta wave					
Duration	R_θ	(P_θ=10)	0.04	0.2	0.5
Delta wave					
Duration	R_δ	(P_δ=17)	-	0.3	0.5

[a] y=2.6, M_α=10Hz, P_α=8 μV, γ_α=0.93

[b] When organization is markedly abnormal, P_α is set to 30 μV

The value of frequency on more abnormal side (right or left) is decreased according to the degree of abnormality.

6) Amplitude of dominant rhythm

The values of amplitude P_α at O_1 and O_2 are determined as given in Table 1. The amplitudes at other scalp regions are determined in the item 8). When organization is markedly abnormal, P_α is set to 30.

7) Asymmetry of amplitude

The value of amplitude on more abnormal side (right or left) is decreased according to the degree of abnormality.

8) Extension of dominant rhythm

The amplitudes of dominant rhythm at regions except O_1, O_2 are determined according to the degree of abnormality of extension as given in Table 2. The values in Table 2 show the ratio of the amplitudes at occipital region (O_1, O_2) and those at other regions. The amplitudes are independently determined in each hemisphere.

9) Amplitude of β wave

The values of amplitude P_β at all scalp regions are determined as shown in Table 1.

Table 2. Extension of dominant rhythm (amplitude ratio)

	O_1,O_2	P_3,P_4	C_3,C_4	F_3,F_4	F_{p1},F_{p2}	T_5,T_6	T_3,T_4	F_7,F_8
Normal	1.0	0.9	0.5	0.3	0.0	0.6	0.3	0.0
Mildly abnormal	1.0	0.9	0.6	0.5	0.0	0.6	0.3	0.0
Moderately abnormal	1.0	0.9	0.8	0.8	0.3	0.6	0.6	0.5
Markedly abnormal	1.0	0.9	0.8	0.8	0.6	0.6	0.6	0.6

10) Asymmetry of amplitude of β wave
The value of amplitude on more abnormal side (right or left) is decreased according to the degree of abnormality.
11) Duration of θ wave
The value of the ratio $R_\theta(=S_\theta/S_T)$ of amount(S_θ) is determined according to the degree of abnormality at abnormal scalp regions. The value of amplitude P_θ is set to 10 μV at normal scalp regions, where θ wave does not appear.
12) Duration of δ wave
The value of the ratio $R_\delta(=S_\delta/S_T)$ of amount(S_δ) is determined according to the degree of abnormality at abnormal scalp regions. The value of amplitude P_δ is set to 17 μV at normal scalp regions, where δ wave does not appear.

Amount S_θ and S_δ in item 11) and 12) are power of θ and δ waves respectively, and S_T is the summation ($S_\delta+S_\theta+S_\alpha+S_\beta$). Table 3 shows the relationship between the terminology about the appearance of slow wave (δ and θ waves) in the EEG report, and the scalp regions regarded as abnormal in item 11) and 12).

2.3 Calculation of Model Parameters from EEG Parameters

The model parameters are calculated from the determined EEG parameters as follows.

a) Mean frequency m_k
The mean frequency of dominant rhythm m_α is the same as M_α, and those of other waves are set to m_δ=2Hz, m_θ=6Hz, and m_β=19Hz.

b) Coefficient γ_k
In dominant rhythm, γ_α is calculated as

$$\gamma_\alpha = 2-\cos(2\pi\Delta t F_\alpha)-\sqrt{\{2-\cos(2\pi\Delta t F_\alpha)\}^2-1} \tag{2}$$

Table 3. Relationship between the terminology about the appearance of slow wave in the EEG report, and the scalp regions regarded as abnormal

Terminology	Scalp regions
diffusely	all regions
more on the left hemisphere	F_{p1},F_3,C_3,P_3,O_1,F_7,T_3,T_5
anteriorly	F_{p1},F_3,C_3,F_7,T_3,F_{p2},F_4,C_4,F_8,T_4
posteriorly	C_3,P_3,O_1,C_4,P_4,O_2
on the left anterior quadrant	F_{p1},F_3,C_3,F_7,T_3
on the left posterior quadrant	C_3,P_3,O_1
on the left frontal region	F_{p1},F_3
on the left posterior region	P_3,O_1
on the left temporal region	F_7,T_3,T_5
on the left	F_{p1},F_3,C_3,P_3,O_1,F_7,T_3,T_5

where F_α is half width of the theoretical power spectrum of the α wave model component, and calculated as

$$F_\alpha = \{y - 0.49 + 0.013(P_\alpha)^2/16 - 4.82\times10^{-5}(P_\alpha)^4/16^2 + 0.41(P_\alpha)^2/(16S_T) - 3.12R_\delta\}/0.58 \quad (3)$$

$$S_T = \{(P_\alpha)^2/16 + (P_\beta)^2/16\}/(1 - R_\delta - R_\theta) \quad (4)$$

The γ_α is calculated by (2) for occipital scalp regions (O_1, O_2), and the calculated value is used for other scalp regions. The lowest limit of γ_α is set to 0.98. If the organization of dominant rhythm is judged markedly abnormal, the γ_α is set to 0.93. For slow waves, the parameters are set to γ_δ=0.98 and γ_θ=0.98 if the slow wave is judged rhythmic. Otherwise, the parameters are set to γ_δ=0.95 and γ_θ=0.9.

c) Standard deviation σ_k

The parameter σ_k is calculated as

$$\sigma_k = P_k\sqrt{1-(\gamma_k)^2}/2, \qquad (k = \delta, \theta, \alpha, \beta). \quad (5)$$

In slow waves, if P_k is not obtained and R_k is given, P_k is calculated from R_k as

$$P_k = 4\sqrt{R_k S_T}, \qquad (k = \delta, \theta). \quad (6)$$

3. Results

The data of 7 subjects (4 moderately abnormal, and 3 markedly abnormal subjects, 16 - 81 years) were used to show the validity of the proposed method. Here, the results for 1 subject are shown in detail. Figure 1 shows the results for Subject 1, whose EEG was judged moderately abnormal by the EEGer, who is one of the authors, (a) raw EEG recorded from 16 scalp regions, (d) EEG report described by the EEGer, (b) generated time series of the EEG model constructed from the EEG report by use of the proposed method. For comparison, (c) shows time series of the EEG model constructed from the raw EEG time series. In (c), the model parameters were determined by minimizing the square sum of the difference between the power spectrum of the raw EEG and the theoretical power spectrum of the EEG model [6]. The time series of the EEG model (b) were evaluated by the visual inspection of the EEGer, and were judged that the characteristics

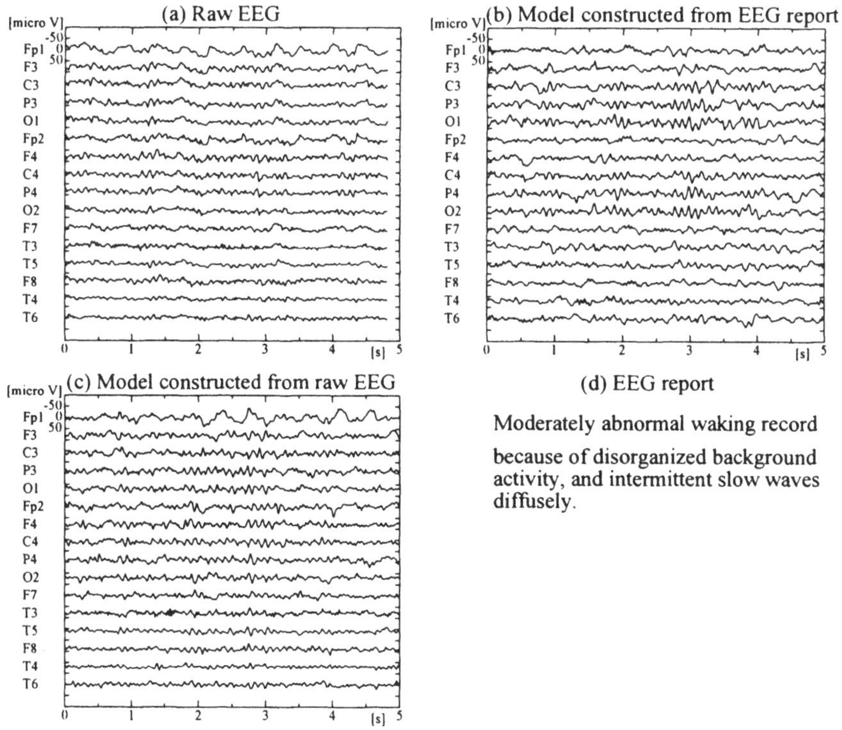

Fig. 1. Results for Subject 1

of the model time series were in agreement with the contents described in the EEG report, though the time series of the model was slightly different from the raw EEG. Figure 1 (a) and (b) show that the features of the time series generated from the EEG model (b) are similar to those of the raw EEG (a).

The time series of EEG model constructed from the raw EEG in Figure 1 (c) included more information about the raw EEG than the model constructed from the EEG report. Then, the features of the time series of this model (c) are in close agreement with those of the raw EEG (a), for example the amplitude at F_{p1}, P_4, O_2, T_4 and T_6.

In this study, the proposed method was applied to the data of 7 subjects including the above subject. The EEGer evaluated the model time series by visual inspection, and judged that the characteristics of the model time series reflected the EEG report for all subjects.

4. Discussion

4.1 Evaluation of the EEG Model

The background EEG consists of 4 rhythms (δ, θ, α and β waves). The EEG model used in this study can produce time series similar to the rhythm of the EEG, since each EEG rhythm is represented by the sinusoidal wave with a constant frequency m_k and Markov process amplitudes $a_k(n\Delta t)$. The rhythmic waveform of the EEG can be represented by adjusting the coefficient γ_k of the Markov process. As γ_k approaches unity, the model approaches sinusoidal wave. Consequently, the time series approaches rhythmic waveform. By varying the amplitude of sinusoidal wave by the Markov process, the shape of the power spectrum of the model can be approximated to that of the raw EEG in frequency domain. Then, the Markov process amplitude model can represent the features of the background EEG in both time and frequency domains very well.

4.2 Determination of EEG Parameters

The constructed model time series were visually inspected by the EEGer, and confirmed that they reflect the information in the EEG report for all subjects. The reason why those satisfactory results were obtained, was that the EEG parameters were appropriately determined from the EEG report and the model parameters were calculated from the EEG parameters. In our previous study, the automatic EEG interpretation method was developed [1], in which, the EEG parameters were obtained from the power spectrum of the raw EEG, and the EEG report was automatically constructed by grading the abnormality in each item based on the values of the EEG parameters. In the current study, the automatic EEG interpretation method was utilized inversely for determining the EEG parameters. The EEG parameters were obtained from the model parameters, and the determination of the EEG parameters was made at the median (standard values) of the abnormality in the automatic EEG interpretation method [1]. And, the model parameters were calculated from the determined EEG parameters. Therefore, the satisfactory results were obtained.

4.3 Differences between the Generated Model Time Series and the Raw EEG

The matching between the constructed model and the raw EEG time series is not perfect as the EEG report contains only abnormal items, which are minimum information for EEG interpretation. Namely, these differences between the model time series and the raw EEG were caused by the information, which were not described in the EEG report. The time series of the model constructed from the raw EEG time series were in close agreement with those of the raw EEG as this model was constructed by using more information than those in the EEG report. If all information about the characteristics of the raw EEG are described in the EEG report, the time series of the constructed model would be even more close to the raw EEG.

Various time series are imagined from the EEG report for the normal or mildly abnormal EEGs, because the range of characteristics of EEG judged as normal in each EEG interpretation item is wide. Then, the results for the moderately and markedly abnormal EEGs, whose EEG reports had many items about abnormality, were shown in this paper.

4.4 Signification of the Proposed Method

Desired EEG reports should include enough information so that the time series of the raw EEG can be imagined from them. The proposed method reveals the ability of reproducing EEG time series from the EEG report. Therefore, the information included in the EEG report can be represented through the time series of the model, which was constructed by the proposed method. Furthermore, the proposed method can be utilized for the training of EEG report making.

5. Conclusion

A method for constructing an EEG model, which reflects the EEG report was proposed. The EEG parameters were appropriately determined from the EEG report by using the automatic EEG interpretation method inversely, which was proposed in our previous study. And, the model parameters were calculated from the determined EEG parameters. The proposed models were constructed from the EEG reports of 7 subjects. The constructed model time series were visually inspected by the EEGer and confirmed that they reflect the information in the EEG report for all subjects. By using the proposed method, it is possible to know how much information about the EEG characteristics is included in the EEG report through the time series of the constructed model.

References

1. Nakamura M, Shibasaki H, Imajoh K, et al (1992) Automatic EEG interpretation : a new computer-assisted system for the automatic integrative interpretation of awake background EEG. Electroenceph. clin. Neurophysiol. 82: 423-431
2. Nakamura M, Sugi T, Ikeda A, et al (1996) Clinical application of automatic integrative interpretation of awake background EEG: quantitative interpretation, report making, and detection of artifacts and reduced vigilance level. Electroenceph. clin. Neurophysiol. 98: 103-112
3. Fujimori B, Yokota T, Ishibashi Y, et al (1958) Analysis of the electroencephalogram of children by histogram method. Electroenceph. clin. Neurophysiol. 10: 241-252
4. Smith JR, Karacan I (1971) EEG sleep stage scoring by an automatic hybrid system. Electroenceph. clin. Neurophysiol. 31: 231
5. Gotman J, Gloor P (1976) Automatic recognition and quantification of interictal epileptic activity in the human scalp EEG. Electroenceph. clin. Neurophysiol. 41: 513-529
6. Nishida S, Nakamura M, Shibasaki H (1986) An EEG model expressed by the sinusoidal waves with the Markov process amplitude (in Japanese). Japanese Journal of Medical Electronics and Biological Engineering 24: 8-15

Event-related Changes in the Spontaneous Brain Activity during 3D Perception from Random-dot Motion

Sunao Iwaki[1,2], Giorgio Bonmassar[2], and John W. Belliveau[2]

[1]Institute of Human Science and Biomedical Engineering, National Institute of Advanced Industrial Science and Technology (AIST), Osaka, Japan
[2]Athinoila A. Martinos Center for Biomedical Imaging, Massachusetts General Hospital, Boston, USA

Chapter Overview. Perception of three-dimensional structure from visual motion requires the visual motion to be integrated spatially as well as to be recognized as a 3-D object. Psychophysical studies have been made to investigate how the visual system extracts the three-dimensional structure of objects from the two-dimensional motion of random dots (3-D structure-from-motion: 3D-SFM), though the neural dynamics underlying the reconstruction of a 3-D structure from optic flow is not fully understood. We used neuromagnetic measurements to detect the dynamic brain responses to 3-D structure perception from visual motion in human. Three different visual stimuli were used; (A) 3-D structure-from-motion (SFM): dots moved as if they belonged to a spherical surface rotating horizontally , (B) In-plane rotation (IPR): dots moved as if they belonged to a circle rotating in-plane, or (C) Random motion (RM): dots moved with random direction and random displacement. The results of event-related time-frequency analysis showed significant suppression of alpha band (7-13 Hz) activity in the right occipito-temporal region, and the suppression of beta band (15-25 Hz) activity in the bilateral superior occipital region in the 'SFM' condition compared to the 'IPR' condition. Spatiotemporal imaging of the brain activity using adaptive beamformer method showed that these areas were also involved in the generation of the event-related MEG components observed during the perception of 3-D objects between the latencies of 140 and 220 ms after the onset of random-dot motion. These results suggest that the perception of moving 3-D object from 2-D random-dot motion is accomplished by the activation both in the ventral and dorsal visual pathways.

Key Words. Magnetoencephalography, Random-dot Motion, 3D structure perception from motion, event-related desynchronization, and adaptive beamformer.

1. Introduction

Perception of three-dimensional structure from visual motion requires the visual motion to be integrated spatially as well as the recognition of the object shape. Many psychophysical studies have been made to investigate how the visual system extracts the three-dimensional structure of objects from the two-dimensional motion of random dots (structure-from-motion: SFM) [1-4]. Electrophysiological studies of non-human primates suggested the involvement of middle-temporal (MT) and medial-superior-temporal (MST) neurons in the perception of 3-D structure from the optical flow [5,6]. Recent neuroimaging studies using functional magnetic resonance imaging (fMRI) techniques suggested the involvement of the parieto-occipital junction (POJ), the superior-occipital gyrus (SOG), and the ventral occipito-temporal junction (OTJ) in the perception of 3-D structure from motion [7,8] though the neural dynamics underlying the reconstruction of a 3-D structure from optic flow is not fully understood. In this study, we used neuromagnetic measurement (magnetoencephalography: MEG) to detect the dynamic changes in the spontaneous brain activities related to 3-D structure perception from visual random-dot motion in human. We also use linearly-constrained minimum-variance (LCMV) beamformer to reconstruct the spatiotemporal distributions of the brain activity during the perception.

2. Methods

2.1 Subjects and visual stimuli

Seven normal right-handed subjects (22-50 years) with no history of neurological disorders were participated in the study as volunteers. Written informed consent was obtained from each subject prior to the experiment complying with the human research policies of internal review board of the National Institute of Advanced Industrial Science and Technology.

The visual stimuli consisted of 100 random dots, which started to move 500 ms after the presentation. In this experiment, three different stimuli were used; (A) 3 D structure-from-motion (SFM) condition: dots moved as if they belonged to a spherical surface with a radius of 7.5 degree in visual angle rotating horizontally, (B) In-plane rotation (IPR) condition: dots moved as if they belonged to a circle rotating in-plane, or (C) Random motion (RM) condition: dots moved with random direction and random displacement (Fig. 1).

2.2 Data acquisition

Neuromagnetic signals were measured during subjects viewing visual stimuli with a whole-cortex-type 122-channel DC-SQUID system (Neuromag122^{TM}, Elekta Neuromag Ltd., Helsinki, Finland) [9]. The system has 122 planar gradiometer s at 61 locations distributed over the scalp [10]. The coordinate system (head coordinate system) used in this study was determined from the left and right preauricular points and the nasion, with the x-axis connecting the two preauricular points (positive direction to the right), the y-axis running perpendicular to the x-axis and passing through the nasion (positive direction to the nasion), and the z-axis running perpendicular to the x-y plane passing through the origin. The sensor positions with respect to the head coordinate system were determined by measuring the magnetic fields generated by three marker coils located on the scalp [11], whose locations in relation to the head coordinate system were determined before the experiment using a three dimensional digitizer

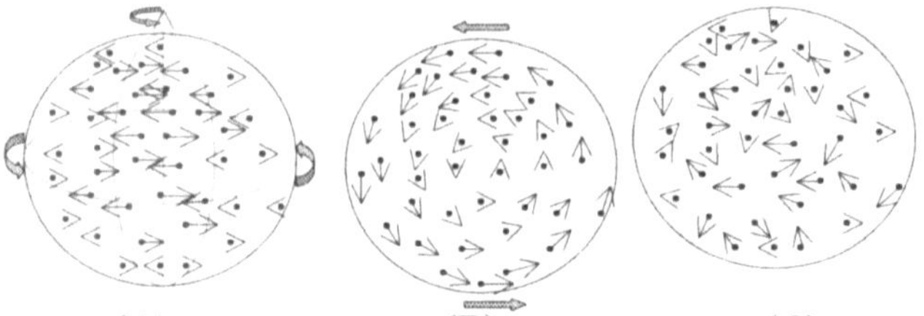

(A) SFM (B) IPR (C) RM

Fig. 1. Visual stimuli used in this study. (A) Structure from motion (SFM) codition: dots moved as if they belonged to a spherical surface, (B) In-plane rotation (IPR) condition: dots moved as if they belonged to a circle rotating in-plane, and (C) Random motion (RM) condition: dots moved with random direction and random displacement.

(Polhemus, Inc., Colchester, VT, U.S.A.). The stimulus-related epochs of 2000 ms, including a 1000 ms pre-stimulus baseline, were recorded with a pass-band of 0.03 - 100 Hz and a sampling rate of 550 Hz. Electrooculogram (EOG) were simultaneously recorded. The epochs with the EOG change exceeding 150 μV, and/or an MEG signal change exceeding 2000 fT/cm were considered to be contaminated by either eye blink artifacts or external magnetic noise and discarded from further analysis.

2.3 Time-frequency analysis of the spontaneous activities

The event-related time-frequency wavelet decomposition (using Morlet's wavelet basis) [12,13] of the recorded MEG raw signals between 5 and 60 Hz was used to assess event-related changes in spontaneous oscillatory brain activity, and the time-frequency energy of each single trial was averaged separately for each stimulus condition in the time-frequency domain. Time-frequency representations of the recorded MEG data are obtained by using the wavelet transformation. The MEG signals are convolved by complex Morlet wavelet basis $w(t, f_0)$ [14] having a Gaussian shape both in the time domain (with standard deviation (SD) of σ_t) and in the frequency domain (with SD of σ_f) at its central frequency f_0,

$$w(t, f_0) = A \cdot \exp(-t^2 / 2\sigma_t^2) \cdot \exp(2i\pi f_0 t) \tag{1}$$

With

$$\sigma_f = \frac{1}{2\pi\sigma_t} \tag{2}$$

Wavelets are normalized so that their total energy is 1 with the normalization factor A,

$$A = (\sigma_t \sqrt{\pi})^{-1/2} \tag{3}$$

A wavelet family is characterized by the 'width' of the wavelet,

$$m = f_0 / \sigma_f \tag{4}$$

which should be practically chosen greater than 5 [15]. In this study, we used the wavelet family defined by $m = 6$ with f_0 ranging from 5 to 60 Hz.

The time-varying energy of the MEG signal, $E(t,f_0)$, in a frequency band around f_0 is defined as the squared norm of the result of the convolution of a complex wavelet $w(t,f_0)$ with the signal $s(t)$,

$$E(t, f_0) = \left| w(t, f_0) \times s(t) \right|^2 \tag{5}$$

This method provides a tool to evaluate event-related changes of time-varying energy of the MEG signal in each frequency band.

2.4 Reconstruction of the spatiotemporal brain activity

Linearly-constrained minimum-variance (LCMV) beamformer [16] was used to reconstruct brain activity from MEG measurements, which allows us to compare event-related changes in spontaneous activities and the spatiotemporal brain activities evoked by the same events. The LCMV beamformer was implemented as a weighted sum of the data recorded at the MEG sensor array. These weights were determined so as to minimize the filter output power subject to a linear constraint, which forces the filter to pass a specific distribution of the measured signal generated from a brain electrical source at a specified location, while the minimization of power attenuates activity originating at other locations [16]. Raw MEG data extracted from 600 ms post-stimulus periods were used to estimate signal covariance matrix, and those extracted from 300 ms pre-stimulus periods were used to estimate noise covariance matrix for the beamformer synthesis. Dynamic Statistical Parametric Maps (dSPMs) [17] were calculated and superimposed on the individual reconstructed brain surface.

3. Results

3.1 Results of the individual time-frequency analysis

Figure 2 shows the results of the individual time-frequency analysis in the typical subject perceiving the 3-D structure from motion. Event-related suppression of alpha-band (8 - 12 Hz) spontaneous activity in the bilateral occipital, occipito-temporal and parieto-occipital regions in response to the start of motion was dominant in all subjects in all conditions.

To assess the differences in event-related changes of spontaneous brain activity between the stimulus condition with 3-D object perception (SFM) from coherent random-dot motion and that without 3-D perception (IPR), difference in event-related time-frequency energy between SFM and IPR conditions was tested. The results of inter-subject paired t-test, which was performed to detect the statistically significant suppression of the time-frequency energy during SFM condition (TF(SFM)) compared to the IPR condition (TF(SFM) < TF(IPR)), were shown in Fig. 3. Significant suppression of alpha band activity (8 - 13 Hz) in the SFM condition compared to the IPR condition was detected in the parietal and the right occipito-temporal regions. Suppression of beta band (15 - 25 Hz) activity was also observed in the bilateral parietal and superior occipital regions.

3.2 Reconstruction of the spatiotemporal brain activity during 3-D object perception

Figure 4 shows the event-related neural activities estimated by LCMV beamformer in the typical subject while the subject perceiving the 3-D rotating sphere from 2-D random-dot motion. Activations in the right occipito-temporal (Brodmann's area (BA) 18,19), bilateral parietal (BA7), parieto-occipital (BA19), parieto-temporal (BA39), and posterior inferior temporal (BA19,37) regions were observed in the latency range between 140 ms and 220 ms after the onset of random-dot motion. Estimated activities in the parietal and parieto-temporal regions were less pronunced in the IPR condition compared to the SFM condition, and these activities were not observed in RM condition.

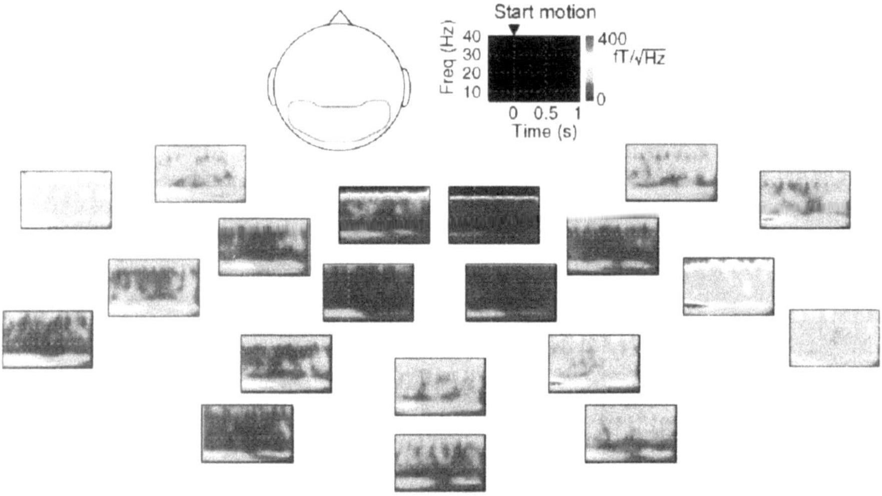

Fig. 2. The results of the time-frequency analysis in typical subjects during the perception of 3-D objects from 2-D random-dot motion. Event-related suppression of alpha-band (8 - 12 Hz) spontaneous activity in the bilateral occipital, occipito-temporal and parieto-occipital regions in response to the start of motion was dominant in all conditions throughout the subjects.

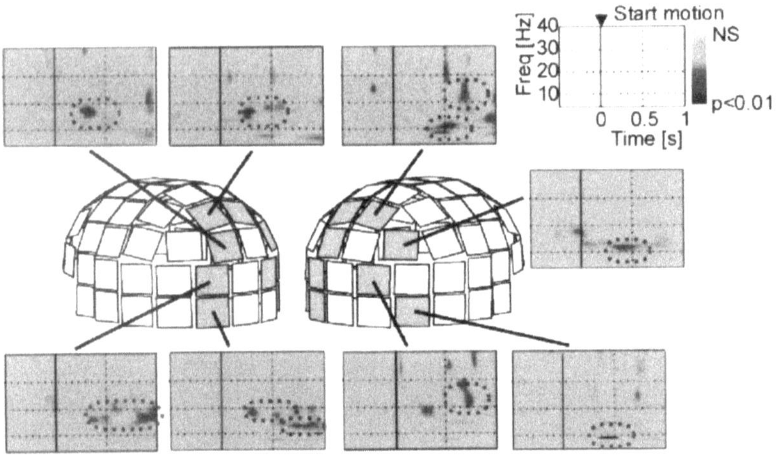

Fig. 3. The results of the inter-subject paired t-test to detect statistically significant suppression of the time-frequency energy during SFM condition compared to the IPR condition (TF(SFM) < TF(IPR)). Significant suppression of alpha band activity (8-13 Hz) in the SFM condition compared to the IPR condition was detected in the right occipito-temporal region. Suppression of beta band (15-25 Hz) activity was also observed in the bilateral superior occipital regions.

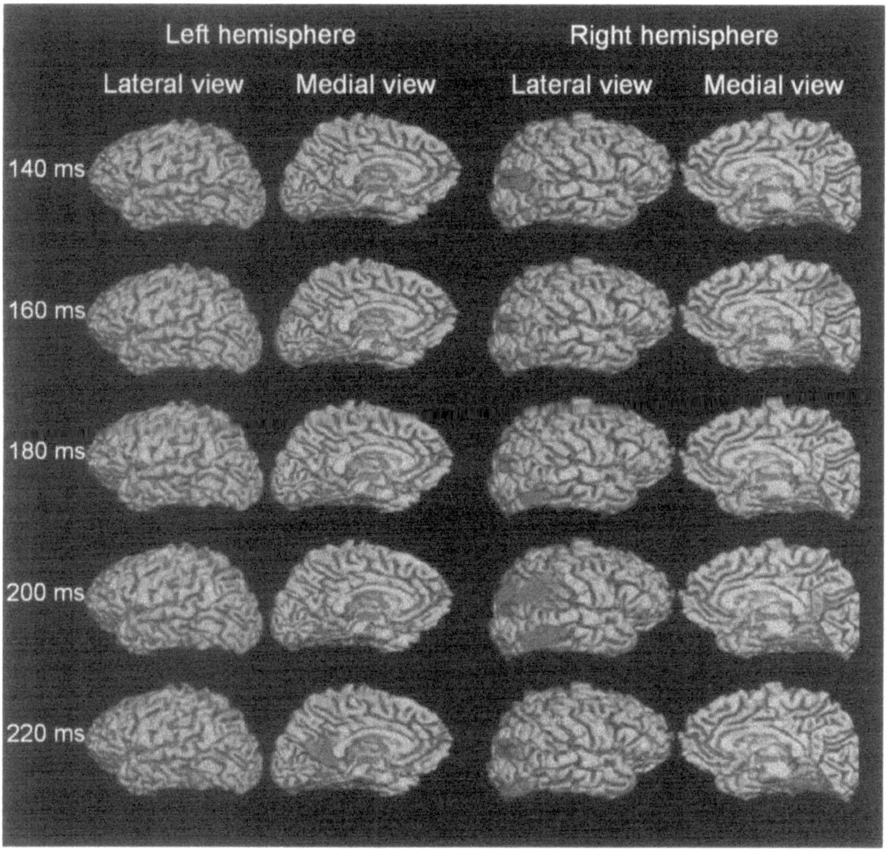

Fig. 4. Event-related neural activities estimated by LCMV beamformer in a typical subject during the perception of 3-D objects from 2-D random-dot motion. Activations in the right occipito-temporal and parieto-occipital regions as well as the bilateral superior parietal regions were observed in the latency range between 140 ms and 220 ms after the onset of random-dot motion.

4. Discussion

In this study, both (a) the event-related suppression of alpha- and beta-band spontaneous brain activities assessed by Wavelet time-frequency analysis, and (b) the event-related neural activities estimated by using beamformer approach, were detected in the occipito-temporal, bilateral parietal, and bilateral parieto-occipital regions. The suppression of spontaneous activities started about 300 ms after the onset of random-dot motion which was used as a cue for 3-D shape perception, while the event-related neural activities were observed between the latencies of 140 and 220 ms.

Current results of the spatiotemporal imaging of the event-related brain activation suggest that these regions are involved in the perception of 3-D structure from motion, which are in agreement with the results from the previous studies of 3-D structure perception from motion using fMRI [7, 8] in terms of the location of the activation, and adding the further insight into the temporal characteristics of the neural activities in these regions. The results were also consistent with the hypothesis that the local suppression of spontaneous brain activities, especially in the alpha-band, in the occipital, the occipito-temporal, and the parieto-occipital areas indicates the engagement of these regions to the relevant sensory functions [18]. Transient decrease of beta-band activities were reported in the temporal, the parieto-temporal, and the sensorimotor regions during the auditory [19] and the tactile stimuli [20], suggesting that the local transient suppression of beta-band activities might be associated with the event-related neural activities in these regions during the task performance.

Equations (2) and (4) define the trade off between the temporal resolution and the frequency resolution in the time-frequency analysis used in this study. Unlike short-time Fourier transformation (STFT), the wavelet transformation analyzes the signals at different frequencies with different resolution. Wavelet duration ($2\sigma_t$) becomes 95.5 ms at the signal frequency of 20 Hz. Although the temporal resolution of the wavelet analysis used in this study is not sufficient to detect the changes of the on-going spontaneous brain activity in millisecond precision, the results shown here indicated that the method was capable of observing slow changes in alpha and beta band activities.

The change in the activation in the occipito-temporal motion-sensitive area in conjunction with the perception of global motion is also reported in the previous fMRI studies [21-23], and the intra-parietal region were reported to be activated during the mental imagery processing [24,25], which suggests that the perception of moving 3-D object from 2-D random-dot motion includes both perception of global motion and 3-D mental image processing, that are accomplished by the cooperative activation in the ventral and dorsal visual pathways.

Acknowledgment

The authors would like to thank Drs. Mitsuo Tonoike and Masahiko Yamaguchi for detailed advices in recording human MEG signals and the development of visual stimulus presentation system. This study was supported in part by the Japan Society for the Promotion of Science (JSPS), National Instituted of Advanced Industrial Science and Technology (AIST), Nissan Science Foundation, and National Institute of Health (NIH).

References

1. Wallach H, O'Connell DN (1953) The kinetic depth effect. J Exp Psychol 45:205-217
2. Baunstein ML (1962) Depth perception in rotating dot patterns: effects of nu merosity and perspective. J Exp Psychol 64:415-420
3. Rogers BJ, Graham M (1979) Motion parallax as an independent cue for depth perception. Perception 8:125-134
4. Caudek C, Proffitt DR (1993) Depth perception in motion parallax and stereokinesis. J Exp Psychol: Hum Percept Perform 19:32-47
5. Bradley DC, Chang GC, Andersen RA (1998) Encoding of three-dimensional structure-from-motion by primate area MT neurons. Nature 392:714-717
6. Sugihara H, Murakami I, Shenoy KY, et al (2000) Response of MSTd eurons to simulated 3D orientation of rotating planes. J Neurophysiol 87:273-285
7. Orban GA, Sunaert S, Todd JT, et al (1999) Human cortical regions involved in extracting depth from motion. Neuron 24:929-940
8. Paradis AL, Cornilleau-Peres V, Droulez J, et al (2000) Visual perception of motion and 3-D structure from motion: an fMRI study. Cereb Cortex 10:772-783
9. Hämäläinen M, Hari R, Ilmoniemi RJ, et al (1993) Magnetoencephalography – theory, instrumentationm and applications to noninvasive studies of the working human brain. Rev Mod Phys 65:413-497
10. Ahonen A, Hämäläinen M, Kajola M, et al (1993) 122-channel SQUID instrument for investigating the magnetic signals from the human brain. Physica Scripta T49:198-205
11. Ahlfors S, Ilmoniemi RJ (1989) Magnetometer position indicator for multi channel MEG. In: Advances in Biomagnetism, Williamson SJ et al (eds) Plenum Press, New York, pp. 673-676
12. Tallon-Baudry C, Bertrand O, Delpuech C, et al (1996) Stimulus specificity of phase-locked and non-phase-locked 40 Hz visual responses in human. J Neurosci 16:4240-4249
13. Jensen O, Gelfand J, Kounios J, et al (2002) Oscillations in the alpha band (9-12 Hz) increase with memory load during retention in a short-term memory task. Cereb Cortex 12:877-882
14. Kronland-Martinet R, Morlet J, Grossmann A (1987) Analysis of sound patterns through wavelet transforms. Int J Pattern Recognit Artificial Intelli-

gence 1:273-302
15. Grossmann A, Kronland-Martinet R, Morlet J (1989) Reading and understanding continuous wavelets transforms. In: Wavelets, Time-Frequency Methods and Phase Space, Combes JM, et al (eds) Springer, New York, pp. 2-20
16. Van Veen BD, van Drongelen W, Yuchtman M, et al (1997) Localization of brain electrical activity via linearly constrained minimum variance spatial filtering. IEEE Trans Biomed Eng 44:867-880
17. Dale AM,, Liu AK, Fischl BR, et al (2000) Dynamic statistical parametric mapping: combining fMRI and MEG for high-resolution imaging of cortical activity. Neuron 26:55-67
18. Williamson SJ, Kaufman L, Lu ZL, et al (1997) Study of human occipital alpha rhythm: the alphon hypothesis and alpha suppression. Int J Psychophysiol, 26:63-76
19. Cacace AT, McFarland DJ (2003) Spectral dynamics of electroencephalographic activity during auditory information processing. Hear Res. 176: 25-41
20. Cheyne D, Gaetz W, Garnero L, et al (2003) Neuromagnetic imaging of cortical oscillations accompanying tactile stimulation. Brain Res Cogn Brain Res 17:599-611
21. Peuskens H, Claeys KG, Todd JT, et al (2004) Attention to 3-D shape, 3-D motion, and texture in 3-D structure from motion displays. J Cogn Neurosci 16:665-682
22. Vaina LM, Belliveau JW, des Roziers EB, et al (1998) Neural systems underlying learning and representation of global motion. Proc Natl Acad Sci 95: 12657-12662
23. Castelo-Branco M, Formisano E, Backes W, et al (2002) Activity patterns in human motion-sensitive areas depend on the interpretation of global motion. Proc Natl Acad Sci. 99:13914-13919
24. Iwaki S, Ueno S, Imada T, et al (1999) Dynamic cortical activation in mental image processing revealed by biomagnetic measurement. Neuroreport 10: 1793-1797
25. Jordan K, Heinze HJ, Lutz K, et al (2001) Cortical activations during the mental rotation of different visual objects. Neuroimage 13:143-152

Effects of Electric or Magnetic Brain Stimulation on Cortical and Subcortical Neural Functions in Rats

Tetsuo Touge[1], Daniel Gonzalez[2], Takanori Miki[3], Chiharu Hiramine[1], and Hiroaki Takeuchi[4]

[1]Health Sciences, School of Nursing, Faculty of Medicine, Kagawa University, Kagawa, Japan
[2]Third department of Internal Medicine, Faculty of Medicine, Kagawa University, Kagawa, Japan
[3]Department of Anatomy and Neurobiology, Faculty of Medicine, Kagawa University, Kagawa, Japan

Chapter Overview. We studied the effects of repetitive electrical (rES) or transcranial magnetic stimulation (rTMS) on somatosensory evoked potentials (SEPs) and neural structures in rats. rES was performed with a frequency of 1 or 0.2 Hz for 3 hours and 20 minutes using two stainless screws inserted into the skull. The right somatosensory areas were magnetically stimulated with a frequency of 1 Hz for 2 hours and 20 minutes using a small eight-figure coil. SEPs elicited by stimulating the left hind limb was recorded before and after rES or rTMS. Following scarifying rats, immunohistochemical analysis of rat brain were performed. As a result, amplitudes of cortical components of SEPs were significantly reduced after 1 Hz and 0.2 Hz rES compared with those in the control. rTMS was also similar effects to those for rES on cortical SEP components. Immunohistochemical analysis demonstrated c-Fos immunoreactive neurons in the wide spread cortical and subcortical areas of the hemisphere stimulated by rES. rTMS induced c-Fos immunoreactivity (IR) of cortical neurons in relatively localized areas.

The present results support that low frequency rES and rTMS have transient inhibitory effects on cortical neurons. We presume that c-Fos IR is a good maker for rapidly reactive neurons stimulated by rES or rTMS in rats.

Key Words. Trans Cranical Magnetic Stimulation, Subcortical Neural Functions, and Cortical Neural Functions

1. Introduction

Recent studies showed that repetitive magnetic or electric stimulation of the brain transiently modulated cortical or subcortical neuron excitabilities, which was shown by recordings of motor evoked potentials (MEPs) or cortical brain potentials, and by measuring regional cerebral blood flow or regional cerebral metabolic rates using single photon emission tomography (SPECT) or positron emission tomography (PET)[1,2,3]. However, the mechanism of modulating effects of electric or magnetic brain stimulation on neural functions was still unknown. To evaluate the effects of repetitive electrical stimulation (rES) and transcranial magnetic brain stimulation (rTMS) on cortical or subcortical neuron excitabilities in rats, we studied changes of somatosensory evoked potentials (SEPs) before and after rES or rTMS. In addition, immunohistochemical analysis of rat brain was performed to elucidate the cellular mechanism associated with effects of rES and rTMS on cortical or subcortical neural function.

2. Subjects and Methods

This study was approved by the ethics committee of our institution.

Sprague-Dawly rats were employed to evaluate the effects of rES or rTMS on cortical neuron activities. Rats were anesthetized by urethane (1.3 g/kg) solved by saline (1.3g/2 ml), and were mounted on a stereotactic apparatus. Two bar holes of 1 mm diameter were punctuated using an electric drill; a bar hole located at a point 2 mm right and 1 mm caudal to the bregma which distributed at the primary sensorimotor area and another bar hole located at 3 mm right and 2 mm caudal to the bregma (Fig. 1). A pair of stainless screws of 1 mm diameter were inserted into the bar holes. A pair of stainless steel needles was attached at the right hind leg to stimulate the sciatic nerve with a frequency of 0.5 Hz using a rectangular pulse of 200 μs duration. The stimulus intensity was fixed to induce the minimum twitches of the right footpau. To record SEPs, electric signals were induced with a bandpass between 10 Hz and 3 KHz and a sensitivity of 10 μV/division via the stainless steel screws described above, and were averaged 20 times.

rES was performed using stainless steel screws inserted into the scalp described above. The one positioned at the primary sensorimotor area was set as negative for cathodal stimulation. Stimulus intensity was just below ones which produce minimum twitch of scalp muscles. Stimulation with frequencies of 1 Hz or 0.2 Hz was continued for 3 hours and 20 minutes with breaks. SEPs were recorded before stimulation and every 1 hour after the onset of stimulation except for the first 20 minutes.

rTMS was performed using a single magnetic stimulator with a small double-cone coil. The double-cone coil of 44 mm diameter with an angle of 150° was set as the center of coil was positioned at the primary sensorimotor area. rTMS was delivered with a stimulus frequency of 0.2 Hz for 2 hours and 20 minutes, with breaks of 5 minutes after each 20 minutes to cool down the coil. Median nerve SEPs were recorded at the same time points as those for rES.

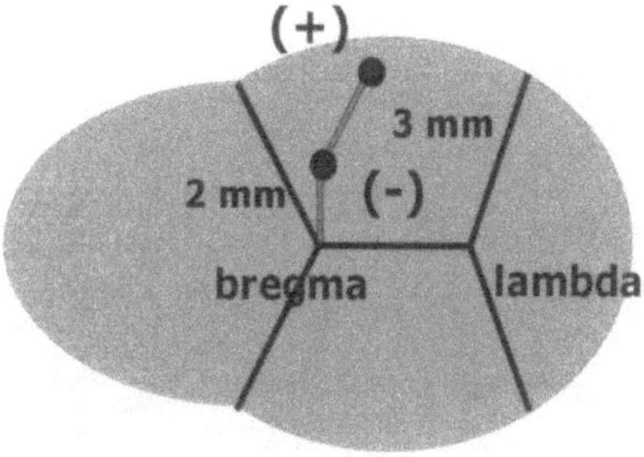

Fig. 1. 20 minutes to cool down the coil. Median nerve SEPs were recorded at the same time points as those for rES.

Each peak of SEPs was labeled as Fig. 2. Peak-to-baseline and peak-to-peak amplitudes and peak latencies of SEPs were measured and were statistically anlyzed. After recording SEPs, rats were sacrificed for immunohistochemical analysis of the brain. Nissl and Haematoxyline-eosin stainings were performed after the brains were embedded in paraffin and cut coronally. For immunohistochemical procedure, the brains were removed and cryoprotected in 30% (w/v) sucrose at 4°C. Serial 40 μm-thick frozen sections were cut coronally and processed for immunohistochemistory using c-Fos and glial fibrillary acidic protein (GFAP) antibodies.

3. Results

3.1 Effects of rES on SEPs

The amplitudes of a cortical component of SEPs were linearly increased according to time after the onset of rES. Compared to control, the amplitudes of N1 were significantly inhibited from 20 minutes after the onset of 1 Hz rES ($P<0.001$) or 1 hour after 0.2 Hz rES ($P<0.0252$). After 1 Hz rES, the amplitudes of P2, P1-N1, P2-N2 and N2-P3 were significantly inhibited, and the amplitudes of P1-N1 and N1-P2 were also significantly attenuated after 0.2Hz rES. The inhibition of N1-P2 amplitudes after rES was significantly larger in 1 Hz rES than in 0.2 Hz rES. The latencies of SEP components did not significantly change.

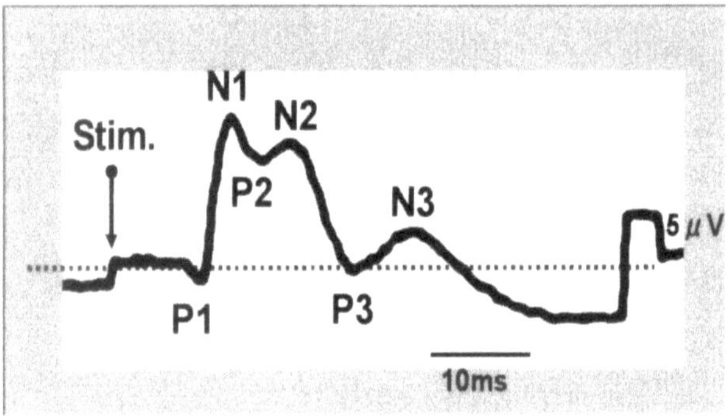

Fig. 2. A typical SEP waveform. Each peak is labelled in order of appearance.

3.2 Effects of rTMS on SEPs

The N1 amplitudes were significantly inhibited at 20 minutes and following time points after the onset of rTMS ($P<0.05$). However, the N1 amplitudes increased after the end of rTMS. The amplitudes and latencies of other SEP components did not significantly change.

3.3 Immunohistochemical changes of cortical or subcortical neurons by rES or rTMS

Nissl and H-E staining showed no microscopical changes of cortical and subcortical neurons after rES or rTMS. Immunohistochemical analysis demonstrated c-Fos immunoreactive (IR) neurons predominantly in the wide spread cortical areas of the hemisphere stimulated by rES. Subcortical neurons including the putamen also showed c-Fos IR. Neurons in the contralateral hemisphere did not markedly present c-Fos IR. rTMS induced c-Fos IR of cortical neurons relatively localized in the stimulated cortical areas over the motor cortex. rES and rTMS did not induce GFAP IR in rat brain.

4. Discussion

The present study showed that low frequency rES with 1 or 0.2 Hz significantly inhibited the cortical components of SEPs, the N1 and P1-N1 amplitudes. 0.2 Hz rTMS also significantly inhibited the N1 amplitudes compared to the control. The inhibitory effects of rES or rTMS on SEPs transiently disappeared. Previous studies in humans showed that low frequency rTMS applied to the motor cortex reduced cortico-spinal neuron excitabilities for about 30 minutes after the stimulation [1, 2]. A feasible mechanism for this inhibitory effect is that rTMS inhibited synaptic transmission of electrical currents following rTMS initially activates cortical interneurons. The amplitudes of cortical components, N1 or P1-N1 appear to represent activities of ascending thalamocortical fibers. Since antidromic currents of the thalamocortical fibers was not transmitted to presynaptic sites, we presume that rES or rTMS affected neural activities of the somatosensory cortex themselves. However, it is not deniable that rES or rTMS affects synaptic transmission of electrical currents from the thalamocortical fibers to cortical somatosensory neurons.

Immediate early gene c-Fos is a rapid indicator for activated neurons by various sensory or mental stimuli. The present study demonstrated that rES induced c-Fos IR in the wide spread cortical areas of the hemisphere stimulated, and subcortical neurons including the basal ganglia showed c-Fos IR. These findings suggest that electrical currents were transmitted over the cortex via cortico-cortical or cortico-subcortical fiber connections. Since c-Fos IR in the contralateral hemisphere was not remarkable, spreading of electrical currents in the subdural space did not play an important role in induction of c-Fos immunoreactive neurons. On the other hand, the present result that rTMS induced c-Fos IR neurons localized in the stimulated motor area are coincident with the results of previous reports [4, 5]. Although rTMS may generate weaker electrical currents in the brain than rES, the stimulating method including coil setting and stimulus parameters may play a crucial role in effects of rTMS on neural functions. The fact that rES or rTMS did not induce GFAP-IR neurons supports an aspect in a previous report that rES or rTMS did not activate glial astrocytes [6].

Finally, we conclude that physiological and immunohistochemical changes of cortical neurons by rES or rTMS well corresponded.

Acknowledgment

This study was supported by a Grant-in-Aid for Science Research (C) from the Ministry of Education, Science, Sports and Culture of Japan.

References

1. Touge T, Gerschlager W, Brown P, Rothwell JC (2001) Are the after-effects of low-frequency rTMS on motor cortex excitability due to changes in the efficacy of cortical synapses? Clin Neurophysiol 112: 2138-2145
2. Ikeguchi M, Touge T, Kaji R, et al (2005) Durable effect of low-frequency repetitive transcranial magnetic stimulation for modulating cortico-spinal neuron excitability. In: Tsuji S, Tobimatsu S, Kakigi R, Uozumi T, Akamatsu N (eds) Excerpta Medica International Congress Series 1278, pp272-275
3. Paus T, Jech R, Thompson CJ, et al (1998) Dose-dependent reduction of cerebral blood flow during rapid-rate transcranial magnetic stimulation of the human sensorimotor cortex. J Neurophysiol 79: 1102-1107
4. Ji RR, Schlaepfer TE, Aizenman CD, et al (1998) Repetitive transcranial magnetic stimulation activates specific regions in rat brain. Proc Natl Acad Sci 95: 15635-15640
5. Doi W, Sato D, Fukuzako H, Takigawa M (2001) c-Fos expression in rat brain after repetitive transcranial magnetic stimulation. NeuroReport 12: 1307-1310
6. Hausmann A, Marksteiner J, Hinterhuber H, Humpel C (2001) Magnetic stimulation induces neuronal c-fos via tetrodotoxin-sensitive sodium channels in organotypic cortex brain slices of the rat. Neurosci Lett 310: 105-108

Event-related Potentials and Stimulus Repetition in Explicit and Implicit Memory Tasks

Hisao Tachibana

Department of General Internal Medicine, Hyogo College of Medicine, Nishinomiya, Japan

Chapter Overview. The Event-related potentials (ERPs) elicited by the second presentation of an item are more positive than the ERPs to the first presentation (the ERP 'repetition effect'). The repetition effect is related to both implicit and explicit aspects of memory. This paper deals with the ERP repetition effect in normal aging and patients with Parkinson's disease during implicit and explicit memory tasks.

The results may indicate that implicit memory is relatively preserved in normal aging, but declines in patients with Parkinson's disease. On the other hand, explicit memory declines in normal aging, and declines even more in patients with Parkinson's disease.

Key Words. Explicit memory, Implicit memory, N400, Normal aging, and Parkinson's disease.

1. Introduction

Two kinds of memory are labeled explicit and implicit. Explicit memory is tested directly by requiring explicit recollection of past experience, while implicit memory is tested indirectly by means of repetitive priming, in which overt recollection of previous episodes is not required.

The event-related potentials (ERPs) provide complementary information to that inferred from behavioral measures during the test phases (whether implicit or explicit) of memory experiences [1]. Some studies [2] have investigated ERPs evoked by 'old' and 'new' words during direct tests of recognition memory. Others [2,3] have employed indirect memory tests in which word (and nonword) repetition was incidental to the task. Consistent findings on direct recognition testing and during repetition priming in an indirect task are that ERPs elicited by the second presentation of an item are more positive than the ERPs to the first presentation (the ERP 'repetition effect' or 'old / new effect') [2, 4]. Thus, the repetition effect is related to both implicit and explicit aspects of memory. In the present study, we evaluated whether memory tested using ERP was affected by normal aging and also by Parkinson's disease (PD) [5, 6].

2. Methods

2.1 Lexical decision task (implicit memory task)

2.1.1 Subjects

Study participants included 17 normal young subjects (age 24.4±1.8 years), 14 normal elderly subjects (age 62.7±5.5 years), and 21 patients with PD (age 63.8±8.6 years). MMSE (Mini-Mental State Examination) scores were 29.9±0.2 for young subjects, 29.2±1.1 for elderly subjects, and 28.8±1.7 for PD patients. There was no difference in MMSE score between elderly subjects and PD patients.

2.1.2 ERP measurement

The stimuli consisted of words and pronounceable nonwords, each three letters in length. Words were Japanese nouns of three syllables and moderate familiarity. Nonwords were speech sounds without meaning in Japanese. Stimulus duration was 381±69msec and the interstimulus interval was 2s (onset to onset). The ERPs were recorded as subjects listened to word or pronounceable nonword lists. Words were repeated at intervals of 0 (lag 0; 2sec), 5 (lag 5; 12sec) and 11-77 (lag 11-77; 120-240sec) intervening items. The subjects were required to push a button in response to occasional nonwords. An EEG was obtained using silver-silver chloride electrodes placed at scalp sites Fz, Cz and Pz and referred to the left earlobe. An EOG was attached below the left inferior orbital margin and on the left outer canthus. The EEG and EOG activity were amplified using a 0.05-to 30-Hz filter.

2.1.3 Results

The grand average ERP waveforms for each group are illustrated in Fig.1. The mean N400 amplitude was smaller in patients with PD than in either normal group, but there was no difference between the two normal groups. In the young subjects, new words elicited a clear N400, which was maximal at Fz and Cz and peaked at approximately 550ms. In the elderly subjects, the N400 amplitude between 300 and 800msec after stimulus onset in Fz site was smaller than those in Cz and Pz sites. In both groups N400 was attenuated for repeated words over all electrodes, and the attenuation was pronounced for immediate (lag0) than for both delayed repetition (lag5 and lag11-77). In PD patients, attenuation was noted only for immediate repetition.

2.2 Recognition memory task (explicit memory task)

2.2.1 Subjects

Seventeen normal young subjects (age, 25.7±3.7 years), 17 normal elderly subjects (age 63.6±10.1 years) and 17 patients with PD (age, 65.8±8.2 years) were studied. All normal young subjects scored 30 points on the MMSE. The mean MMSE in elderly subjects and PD patients was 28.6±1.8 and 27.4±2.4, respectively. No difference was found between elderly subjects and patients with PD.

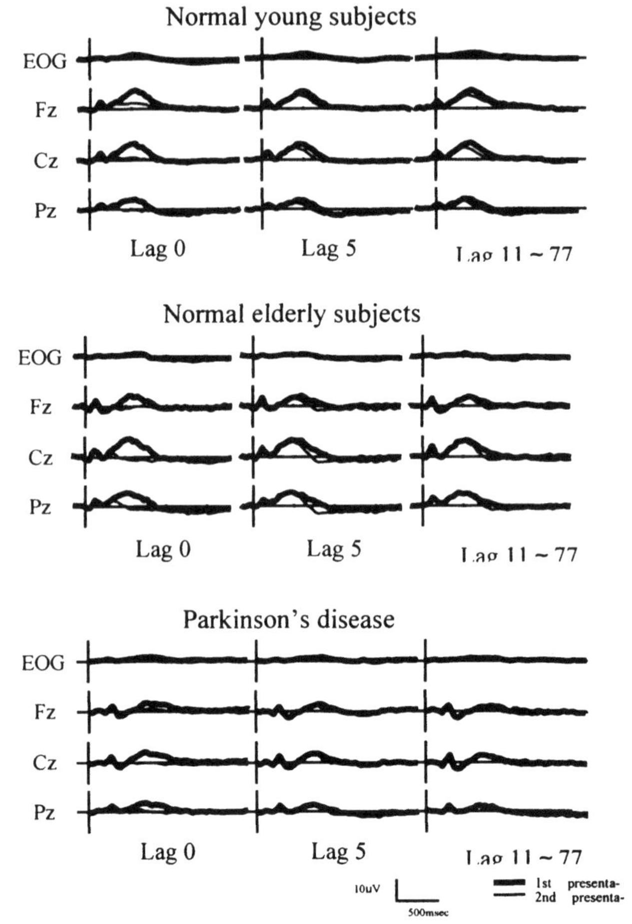

Fig. 1. Grand average ERP waveforms elicited by words at first presentation and second presentation for lag 0, lag 5 and lag 11-77 in normal subjects and patients with PD during implicit memory task. In the normal N400 was attenuated for repeated words, and attenuation was more prominent for immediate than for both delayed repetitions. In patients with PD, the N400 attenuation was identified only for immediate repetition.

2.2.2 ERP measurement

Stimuli consisted of Japanese nouns; each three letters in length and each of three syllables and moderate familiarity. Stimulus duration was

381±69ms, and the mean interstimulus interval was 3s (onset to onset). The ERPs were recorded as subjects listened to word lists. Words were repeated at intervals of 0 (lag0; 3sec), 1 (lag1; 6sec) and 6 (lag6; 21sec) intervening items. The subjects were required to push a button with the right thumb upon first presentation of a word and with the left thumb upon repeat presentation. EPR recordings were essentially the same as those of lexical decision task.

2.2.3 Results

Subject differences were seen in new/old response accuracy at lags 1 and 6 between the two normal groups. Response accuracy in PD patients was significantly decreased compared to the elderly group for both the first presentation and lag 6 repetition (Table 1). The grand average ERP waveforms for the normal young and elderly groups and the PD group at each electrode are shown in Fig.2. In normal young subjects, new words elicited a clear N400 that was maximal at the Fz or Cz site, and peaking at approximately 470 ms after stimulus onset. The ERPs for repeated words (lag0) began to diverge from the ERPs for new words at approximately 200-250ms, and the effects persisted to 750-800ms. Repeated words at lags 1 and 6 also showed a N400 attenuation, although the magnitude of attenuation was less prominent. In the normal elderly subjects, new words elicited a clear N400 that peaked at approximately 500 ms. Repeated words at lags 0 and 1 showed an attenuation of N400, while no attenuation was found for lag 6. No significant difference was seen in N400 between young and elderly normal groups, although N400 peak latency and amplitude in the elderly group were somewhat later and smaller. In patients with PD, a clear N400 peaking at about 500 ms was also elicited for new words. The N400 attenuation for repeated words was evident for immediate repetition (lag 0) and lag 1 repetition, as seen also in the elderly control group. There was a significant difference in N400 amplitude for the first presentation of words between normal elderly subjects and PD patients.

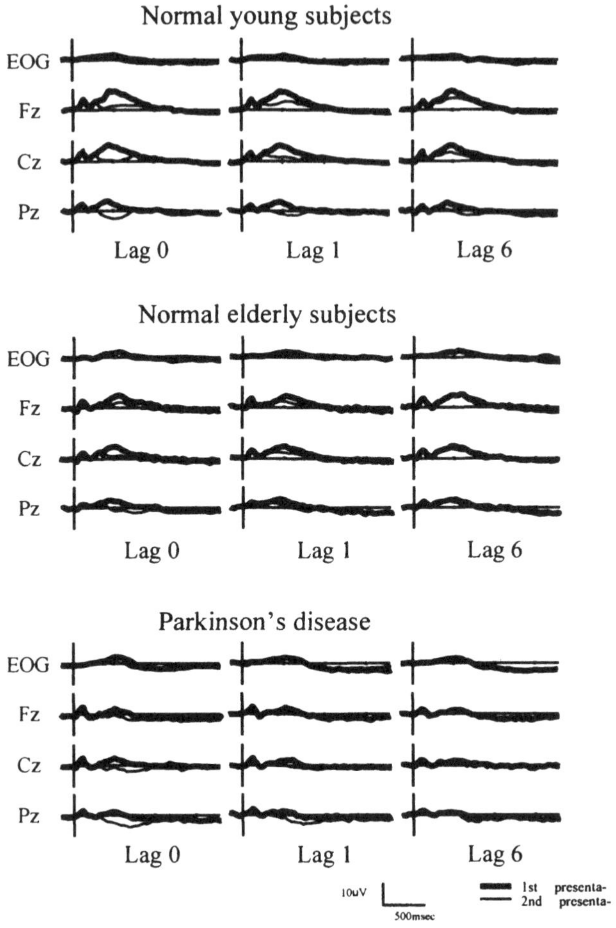

Fig. 2. Grand average ERP waveforms elicited by words at first presentation and second presentation for lag 0, lag 1 and lag 6 in normal subjects and patients with PD during explicit memory task. In the normal young subjects, the N400 is attenuated for repeated words. Attenuation is more pronounced for lags 0 and 1 than for lag 6. In the normal elderly subjects, and patients with PD, the N400 is attenuated for repeated words at lags 0 and 1, but no attenuation is evident at lag 6.

Table 1. Response accuracy (%) for new and repeated words

	Normal young subjects	Normal elderly subjects	Parkinson's disease
New words	98.2±2.2	98.3±1.9[1)]	88.8±16.5[1)]
Lag 0	99.6±0.8	99.0±1.3	93.2±14.7
Lag 1	99.9±0.4[2)]	97.2±4.5[2)]	92.0±15.6
Lag 6	88.6±7.4[3)]	76.9±14.3[3),4)]	64.8±18.7[4)]

1) $p<0.001$ 2) $p<0.005$ 3) $p<0.02$ 4) $p<0.05$ by Mann-Whitney U-test

3. Discussion

This implicit memory task demonstrated a decrease in the N400 amplitude in patients with PD compared to those in both control groups. Although the nature of the process that is underlying the N400 component has not been fully clarified, the most likely hypothesis, at present, views the 400 component as reflecting a context integration process [7]. By this account, stimuli such as words lead to the generation of an internal representation that provides the context within which an additional stimulus can be integrated. On the basis of this hypothesis, our results suggest that the generation of an internal representation within episodic memory is impaired in patients with PD and that this impairment leads to reduced N400 amplitude in the present task.

The present findings in the control groups agree with those of Hamberger and Friedman [8] in failing to show a reduction in the magnitude of ERP repetition effects in the normal elderly. Repetition effects of lag 0 remained normal in PD patients, but those at lag 5 and lag 11-77 were reduced. If episodic theories of repetition effects are correct, the reduced repetition effect in PD patients could be explained by an impairment of episodic memory [7]. Several authors [9], however, have demonstrated a dissociation between ERP and behavioral data and have suggested that ERP repetition effects are independent of the process responsible for explicit memory. In addition, in the present study, the repetition was incidental to the task, although explicit strategies might influence behavioral and ERP measures. Therefore, as suggested by Friedman et al. [3], an alternative account of reduced N400 repetition effects in PD patients is that they reflect processes subserving implicit memory. These processes may have been intact in the normal elderly subjects, as shown in the previous neuropsychological tests.

The ERP repetition effects were apparent for all lags in the young subjects during explicit memory task. Although normal elderly subjects showed ERP repetition effects for short lags such as 0 and 1, the region of the waveform occupied by the long-lag repetition effect showed no reliable effects. Immediate lag 0 repetition stimuli presented at the 3 s produced rapid, accurate responses and the N400 attenuation in young and elderly normal subjects. In the elderly subjects, auditory recognition memory was associated with impaired performance and loss of the N400 attenuation after an interval of 21 s (lag 6). Thus, while the rapid memory process in the 3-s range appears to be intact, there is selective impairment of sustained memory processes at longer retention intervals which appears to occur in normal aging.

The PD patients showed a lower N400 amplitude for the first presentation of words than elderly control subjects, suggesting that the context integration process per se is impaired in patients with PD. Our N400 data suggest that PD patients fail to generate a full memory representation of the first presentation of words. Moreover, this impoverished memory representation can explain the disproportional decline in performance of PD patients with a second presentation of words at lag 6. The present ERP findings support previous reports that PD patients exhibit deficits in encoding information [10]. In conclusion, both normal aging and PD are associated with impairment of recognition memory. Recognition memory deficits in PD may result, at least partly from impaired integration of items with their priming context. In contrast, the context integration process per se is relatively preserved in normal aging.

Present two experiments may indicate that implicit memory is relatively preserved in normal aging, but declines in patients with PD. Explicit memory, however, declines in normal aging, and declines even more in patients with PD.

References

1. Rugg MD, Mark RE, Glchrist J, Roberrts RC (1997) ERP repetition effects in indirect and direct tasks: effects of age and interitem lag. Psychophysiology 34:572-586
2. Rugg M, Furda J, Lorist M, (1988) The effect of task on the modulation of event-related potentials by word repetition. Psychophysiology 25:55-63
3. Friedman D, Hamberger M, Stern Y, Marer K (1992) Event-related potentials (ERPs) during repetition priming in Alzheimer's patients and young and older controls. J. Clin. Exp. Neuropsychol. 14:448-462
4. Friedman D, Hamberger M, Ritter W (1993) Event-related potentials as indicators of repetition priming in young and older adults: am;plitude, duration, and scalp distribution. Psychol. Aging 8:120-125

5. Tachibana H, Miyata Y, Takeda M, Sugita M, Okita T (1999) Event-related potentials reveal memory deficits in Parkinson's disease. Cogn. Brain Res. 8:165-172
6. Minamoto H, Tachibana H, Sugita M, Okita T (2001) Recognition memory in normal aging and Parkinson's disease: behavioral and electrophysiologic measures. Cogn. Brain Res. 11:23-32
7. Rugg MD, Doyle MC (1994) Event-related potentials and stimulus repetition in direct and indirect test of memory. Cognitive Electrophysiology. Birkhauser, Boston, pp124-148
8. Hamberger M, Friedman D (1992) Event-related potential correlates of repetition priming and stimulus classification in young, middle-aged, and older adults. J. Gerontol. (Psychol Res.) 47:395-405
9. Rugg MD, Pearl S, Walker P, Roberts RC, Holdstock JS (1994) Word repetition effects on event-related potentials in healthy young and old subjects, and in patients with Alzheimer-type dementia. Neuropsychologia 32:381-398
10. Cooper JA, Sagar HJ (1993) Encoding deficits in untreated Parkinson's disease. Cortex 29:251-265

Multichannel Surface EMGs to Assess Function of Spinal Anterior Horn Cells

Katsuya Ogata[1], Tomomi Kurokawa-Kuroda[1], Yoshinobu Goto[2], and Shozo Tobimatsu[1]

[1]Department of Clinical Neurophysiology, Neurological Institute, Faculty of Medicine, Graduate School of Medical Sciences, Kyushu University
[2]Department of Occupational Therapy, Faculty of Rehabilitation, International University of Health and Welfare.

Chapter Overview. Peristimulus time histograms (PSTHs) have been constructed from needle electromyographic (nEMG) recordings, and used to evaluate the function of spinal anterior horn cells to inputs from the corticospinal tract or Ia afferents. However, nEMG is a painful and uncomfortable technique, and to overcome these difficulties, we have recorded surface EMGs (sEMGs) from an 8 electrode array with interelectrode distances of 5 mm. With this array, we were able to record motor unit potentials (MUPs) from which the PSTHs can be constructed. The electrode array was aligned with the long axis of the right abductor pollicis brevis while transcranial magnetic stimulation (TMS) was directed to the motor area of the right hand. With a careful control of voluntary muscular contraction, a single MUP could be isolated, and PSTHs could be constructed. Our results show that the properties of these PSTHs were very similar to those constructed from nEMGs. We conclude that multichannel sEMGs can be used to construct PSTHs non-invasively, which then enables us to explore dysfunctions of the pyramidal tract such as in cases of amyotrophic lateral sclerosis.

Key Words. Motor Unit Potential (MUP), Multichannel surface EMG, Needle EMG, and Peristimulus Time Histogram (PSTH).

1. Introduction

Motor evoked potentials (MEPs) elicited by transcranial magnetic stimulation (TMS) are widely used to evaluate the function of the corticospinal tract (CST) because abnormal MEPs indicate functional alterations of the CST and/or anterior horn cells. Peristimulus time histograms (PSTHs) are used to analyze the timing of motor unit potentials (MUPs) elicited by external stimuli. The stimuli can be given to either the peripheral nerve from Ia afferents [1], or to the central pathway through the CST [2] by TMS or by electrical stimulation. PSTHs represent the excitatory postsynaptic potentials (EPSPs) of the anterior horn cells, and abnormal findings have been reported in disorders such as amyotrophic lateral sclerosis (ALS), [3, 4], multiple sclerosis, and cerebral infarcts [5, 6].

Needle electromyography (nEMG) is conventionally used to record single MUPs. However, nEMG is painful and uncomfortable, which is a stress for the subjects. Recording surface EMGs (sEMGs), on the other hand, is painless and can be used to assess involuntary movements or compound muscle action potentials (CMAPs) while measuring MEPs. sEMGs are the summated potential changes from the entire muscle, and are thus difficult to use to examine single MUP because it is influenced by volume conduction.

We have developed a multiple electrode array with small inter-electrode distances to increase the spatial resolution. This then allowed the examiner to select an optimum recording pair that included a single MUP from the seven channels.

We have constructed PSTHs of the small hand muscle with this 8 electrode array from the MUPs elicited by TMS. The results indicated that this non-invasive technique can be used to assess the function of anterior horn cells in patients.

2. Subjects and Methods

Eleven normal volunteers (men:women = 4:7; ages: range 21-35 years; mean ± SD 27.8 ± 5.0 years) were studied. All subjects were right-handed. They were fully informed about the nature of the experiment and a written informed consent was obtained. The experimental procedures were approved by Ethics Committee of Kyushu University and conformed to the tenets of the Declaration of Helsinki.

Subjects were comfortably seated on an armchair, and the multichannel electrode array was positioned on their right abductor pollicis brevis

(APB) along the long axis of the muscle fiber. The electrode array was a custom-made array of 8 surface electrodes with inter-electrode distances of 5 mm (Fig. 1). Each electrode was 1.5 mm in diameter and was made of silver. Bipolar recordings were made between each pair of adjacent electrodes. The signals were amplified with a bandpass of 50 to 3000 Hz and displayed on a CRT monitor (Neuropack 8, Nihonkoden, Japan).

The subjects were asked to observe the 7 MUP signals displayed on the monitor screen, and to flex the APB muscle weakly. The subjects learned to maintain a constant contraction by watching the monitor screen and listening to an audio feedback system. Under these conditions, a single MUP could be discriminated on at least one channel.

The TMS was delivered by a magnetic stimulator, Magstim 200 (Magstim, UK), using a circular coil of 9 cm in diameter. The center of the coil was placed over the Cz, and the current direction in the coil was selected to flow counterclockwise so that the edge of the coil would stimulate the left motor cortex effectively. The stimulus intensity was selected to be the lowest intensity (threshold) which would evoke CMAPs of more than 100 μV in at least 5 of 10 trials with a mild voluntary contraction of the target muscle. TMS was delivered at about 0.2 Hz and 150 stimuli were given to each subject.

Epochs of -250 to +250 msec from the beginning of the TMS were recorded and digitized at 10 kHz. The data were stored on a personal computer and analyzed off-line by a signal processing program (Multiscope PSTH, Medical try-system, Japan). The slice level to detect a single MUP peak was determined arbitrarily after playing back each recording. We rejected sweeps in which CMAPs were evoked. We also excluded records

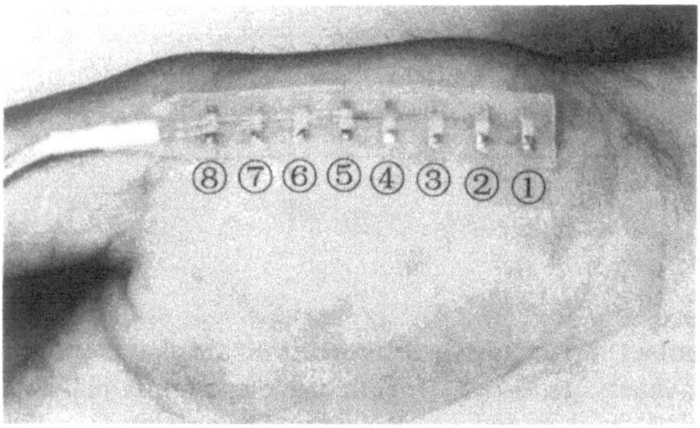

Fig. 1. Photograph of the multichannel surface electrodes placed on the right abductor pollicis brevis. The numbers indicate the position of each electrode from proximal to distal site.

containing artifacts or polyphasic MUPs.

The histogram was constructed from the MUP counts in 1 msec bins relative to the onset of the stimulus. The range of consecutive bins with counts exceeding 2 standard deviations (SDs) above the mean of background firing rate was determined around 20 ms after the stimulus, was defined as the primary peak, and its onset latency and duration were measured. The means and SDs of the background counts were also calculated from those between –220 and –20 ms. The MUPs in the primary peak were counted, and the background-firing rate was subtracted from the count to give the extra counts. The extra counts were normalized for the mean stimulus number of 100 and for the mean interspike interval of 100 msec.

nEMGs and multielectrode sEMGs were recorded simultaneously in one subject. The PSTHs constructed from the nEMGs and from the sEMGs were compared.

3. Results

One or more motor units were detected in the multichannel sEMGs (Fig. 2). The identification of an individual MUP depended greatly on the degree of voluntary contraction. After a short practice time, a single MUP

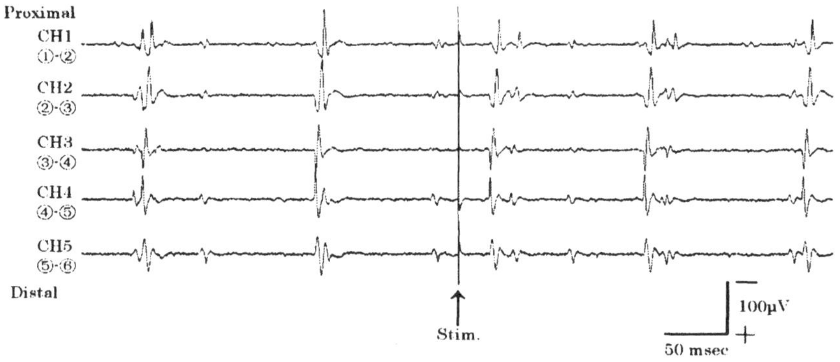

Fig. 2. Multichannel surface EMGs recorded from the abductor pollicis brevis. Five of the 7 channels are shown. Upper channels are proximal and lower are distal electrodes. The numbers indicate the pair of electrodes as in Figure 1 for each channel. An arrow indicates the onset of stimulation. Note the difference of amplitudes in the different channels but the relatively stable firing pattern. A phase reversal is noted between channels 3 and 4, which indicates that the motor point is located there.

could be evoked by all of the subjects with a contraction intensity of about 5% or less. The amplitudes of the MUPs were different in the different channels; however, their shapes were relatively constant in each channel with constant contraction. Thus, we were able to choose the best channel from which to construct the PSTHs.

A primary peak and silent period were always observed in the PSTHs, and secondary peaks were also seen in several subjects (Fig. 3). Because multichannel sEMGs are bipolar recordings, the site of phase reversal was identified, which was supposed to indicate the point of origin of the MUP. Thus, we corrected the onset latency by the difference of onset latency between the motor point and the one used for the MUP detection. After the correction, the onset latency was calculated to be 20.5 ± 1.1msec (mean +

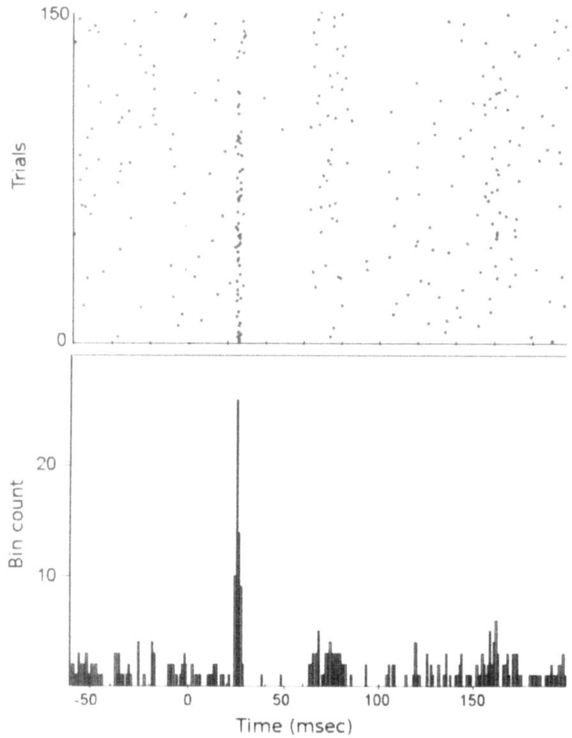

Fig. 3. PSTH constructed by surface EMG.
The upper panel shows the raster view of MUP firings while the lower represents the PSTH. A primary peak (arrow) and silent period (SP) can be seen.

SD) while the duration of the primary peak was 3.4 ± 0.7 msec. The extra count of the primary peak was 27.7 ± 15.3. The duration of silent period was determined by subtracting the end of the primary peak to the onset of the second peak, which was 60.2 ± 28.6 ms.

The simultaneous recordings of the sEMG and nEMG in one of the subjects showed that the same motor units were recorded in both recordings (Fig. 4) although the duration of motor unit potentials in the sEMG was longer than that in the nEMGs. The parameters of the PSTH constructed from both recordings were approximately the same. The onset latency was 19.0 ms for nEMG and 18.7 for sEMG. The duration of primary peak was 3 ms for both methods. The silent period was 46 ms for the two methods.

4. Discussion

With our 8 surface electrode array, we were able to record sEMGs of one MUP non-invasively from which a PSTHs could be constructed. Because the waveforms of the MUPs were not dependent on the method, the use of

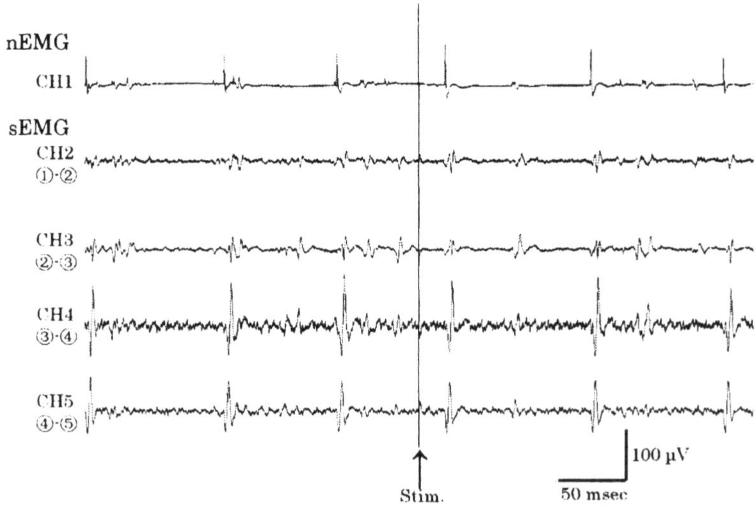

Fig. 4. Simultaneous recording of needle EMG (nEMG) and surface EMG (sEMG)
The top trace corresponds to a MUP of nEMG while the remaining traces are derived from sEMG. With appropriate muscle contraction level, sEMG can detect the same MUPs as observed in nEMG.

nEMGs need not be used as long as the MUP firing was recorded. Multichannel electrodes have been used to estimate muscle fiber conduction velocity (for review, see Zwartz and Stegeman [7]). We were able to obtain better signals not only by the small interelectrode distances that minimized the effect of volume conduction but we could also select the optimum recording channel from the different channels.

The waveform of the sEMG was stable and reproducible for long periods under constant contraction with its fixed position on the targeted muscle. Thus, sEMG has the advantage that it does not require careful stabilization of a needle electrode where slight movements can result in large changes in the waveform. However, the MUPs recorded with sEMG are small and greatly influenced by the intensity of contraction. If the subject changes the intensity of voluntary contraction, the MUPs are easily contaminated by other MUPs because the effect of volume conduction cannot be reduced as much as with nEMG. Therefore, it was important for us to ask the subjects to maintain a fine control of voluntary contraction. Once the subjects learned how to maintain the muscle contraction, they performed very well and kept the contraction level of 5% or less during the recording.

We were able to obtain the primary peak, silent period, and secondary peak in the PSTH. Each component had properties similar to those previously reported for nEMG [3, 8-12]. In addition, when we compared the PSTHs of nEMG with that of sEMGs, the parameters of PSTH were not significantly different. Therefore, our PSTH technique can be used to estimate the dysfunction of anterior horn cell in such disorders as ALS.

Klein et al [13] reported the usefulness of a 126-channel high-density sEMG (10 columns, 13 rows) to detect MUPs and to construct PSTHs. With his 2 dimensional electrode array, several different MUPs were detected simultaneously representing the two-dimensional distribution of each MUP. However, his electrode was too large and difficult to apply to small hand muscles. In contrast, our electrode is narrow and small so that it is possible to record single MUPs of the APB muscle. Moreover, this method does not require complex computations although we need to be certain that the subjects maintain a constant degree of voluntary contraction.

In conclusion, our results indicate that our 8 surface multi-electrode array can isolate individual MUPs from which PSTHs can be constructed to assess the functional state of CST. Additional studies are needed to determine the value of this new method to test its use in neurological disorders.

References

1. Ashby P, Labelle K (1977) Effects of extensor and flexor group I afferent volleys on the excitability of individual soleus motoneurones in man. J Neurol Neurosurg Psychiatry 40:910-919.
2. Day BL, Rothwell JC, Thompson PD et al (1987) Motor cortex stimulation in intact man. 2. Multiple descending volleys. Brain 110 (Pt 5) :1191-1209.
3. Kohara N, Kaji R, Kojima Y et al (1996) Abnormal excitability of the corticospinal pathway in patients with amyotrophic lateral sclerosis: a single motor unit study using transcranial magnetic stimulation. Electroencephalogr Clin Neurophysiol 101:32-41.
4. Nakajima M, Eisen A, McCarthy R et al (1996) Reduced corticomotoneuronal excitatory postsynaptic potentials (EPSPs) with normal Ia afferent EPSPs in amyotrophic lateral sclerosis. Neurology 47:1555-1561.
5. Boniface SJ, Mills KR, Schubert M (1991) Responses of single spinal motoneurons to magnetic brain stimulation in healthy subjects and patients with multiple sclerosis. Brain 114 (Pt 1B):643-662.
6. Boniface SJ, Schubert M, Mills KR (1994) Suppression and long latency excitation of single spinal motoneurons by transcranial magnetic stimulation in health, multiple sclerosis, and stroke. Muscle Nerve 17:642-646.
7. Zwarts MJ, Stegeman DF (2003) Multichannel surface EMG: basic aspects and clinical utility. Muscle Nerve 28:1-17.
8. Mills KR (1995) Motor neuron disease. Studies of the corticospinal excitation of single motor neurons by magnetic brain stimulation. Brain 118 (Pt 4):971-982.
9. Eisen A, Entezari-Taher M, Stewart H (1996) Cortical projections to spinal motoneurons: changes with aging and amyotrophic lateral sclerosis. Neurology 46:1396-1404.
10. Nakajima M, Eisen A, Stewart H (1997) Diverse abnormalities of corticomotoneuronal projections in individual patients with amyotrophic lateral sclerosis. Electroencephalogr Clin Neurophysiol 105:451-457.
11. Eisen A, Nakajima M, Weber M (1998) Corticomotorneuronal hyperexcitability in amyotrophic lateral sclerosis. J Neurol Sci 160 Suppl 1:S64-S68.
12. Weber M, Eisen A, Nakajima M (2000) Corticomotoneuronal activity in ALS: changes in the peristimulus time histogram over time. Clin Neurophysiol 111:169-177.
13. Kleine BU, Blok JH, Oostenveld R et al (2000) Magnetic stimulation-induced modulations of motor unit firings extracted from multi-channel surface EMG. Muscle Nerve 23:1005-1015.

Cortical Processing of Sound in Patients with Sensorineural Hearing Loss

Yasushi Naito

Department of Otolaryngology, Kobe City General Hospital, Kobe, Japan

Key Words. Magnetoencephalogram (MEG), Positron Emission Tomography (PET), Auditory cortex, Hearing loss, and Cochlear implant.

1. Introduction

Cortical processing of sound in patients with mild to moderate sensorineural hearing loss, which is common and experienced by most people with age, has rarely been reported. In this article, our three previous studies are reviewed to investigate how disorders in the peripheral auditory organ influence the brain activation pattern, which may provide further information about the strategy used by the brain to process and recognize incomplete sensory signals delivered to it.

2. Effects of Acoustic Structures of Sound Stimuli on Auditory Evoked Magnetic Fields (AEFs)

We measured AEFs by presenting white noise and speech-related sounds in normal subjects and in patients with mild inner-ear hearing loss and compared them with those in normal hearing subjects [1]. The purpose of this study was to investigate the effects of acoustic structures of sound stimuli on AEFs and their alteration induced by inner-ear hearing loss.

Subjects: Eight normal subjects and seven patients with mild sensorineural hearing loss were examined. The hearing levels of normal subjects were within 20 dB at all frequency tested. The mean hearing levels of the right and left ear among the patients were 29.4 dB and 33.7 dB respectively.

Sound stimulation (Figure 1): Sound stimuli applied were: (1) white noise, (2) 170 Hz pure tone (the fundamental frequency (F0) of the vowel /a/), (3) synthetic complex sound (F0123) composed of F0 and three formants of the vowel /a/, and (4) the vowel /a/. The intensity of the sound stimuli was adjusted to 80 dB SPL. All patients reported sensation of distortion or abnormal quality for the present sound stimuli.

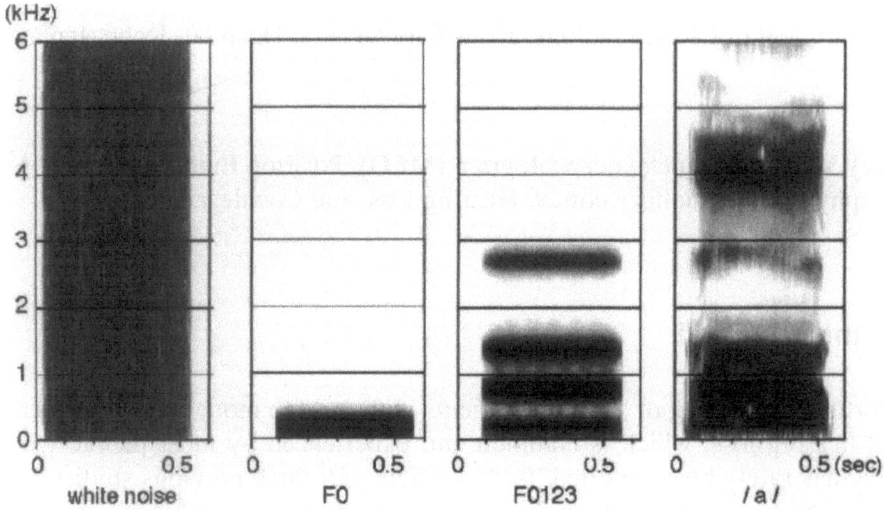

Fig. 1. Sound spectrograms of sound stimuli

MEG recording: AEFs were recorded with a whole-head magnetometer, which has 122 first-order planar gradiometers (Neuromag-122) (Figure 2).
N100m latency: In normal subjects, the N100m latency was significantly longer for F0 than for F0123 and /a/ stimulation in the left hemisphere. In the right hemisphere, the N100m latency for F0 was significantly longer than for the other stimuli, and the N100m latency for /a/ stimulation was shorter than that for noise stimulation. The N100m latencies in the right hemisphere with the left ear stimulation were significantly shorter than those in the left hemisphere with the right ear stimulation.

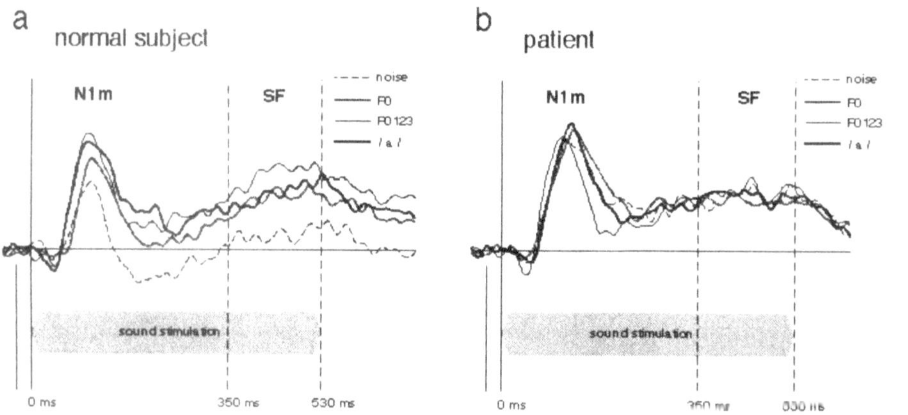

Fig. 2. AEF waveforms of a normal subject and a patient with sensorineural hearing loss

Representative AEF waveforms at the channel, which gave the largest N100m response, are shown (recorded in the right hemisphere with left ear stimulation). In the normal subject (a), the response waveforms varied according to the stimulus, and the response to noise was smaller than those to other stimuli both in N1m and SF latency ranges. In contrast, the wave form variation was less prominent in the patient with sensorineural hearing loss (b)(patient 7). Calibration: 100 fT/cm.

In the patient group, the N100m latencies in the right hemisphere with the left ear stimulation did not significantly differ from those in the left hemisphere with the right ear stimulation, indicating that the interhemispheric difference in N100m latency found in normal subjects disappeared in the patient group.

N100m dipole moment (Figure 3): In normal subjects, the dipole moments of N100m for F0123 and /a/ stimuli were significantly larger than that for noise stimulus in the left hemisphere, and the N100m moment for F0123 was significantly larger than that for noise in the right hemisphere. The dipole moments in the right hemisphere with the left ear stimulation were significantly larger than those in the left hemisphere with the right ear stimulation.

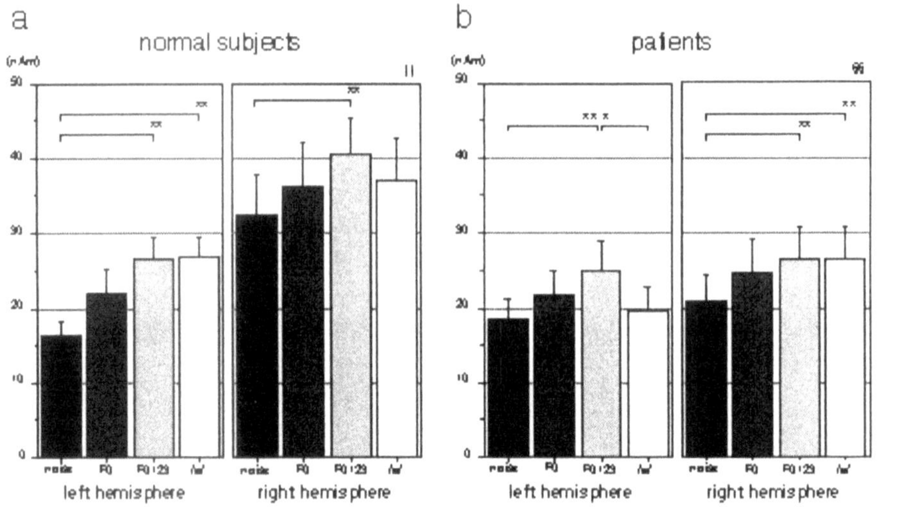

Fig. 3. N100m dipole moments in normal subjects and in patients with sensorineural hearing loss

a: In normal subjects, the dipole moments of N100m for F0123 and /a/ stimuli were significantly larger than that for noise stimulus in the left hemisphere, and N1m moment for F0123 was significantly larger than that for noise in the right hemisphere. The dipole moments in the right hemisphere with the left ear stimulation were significantly larger than those in the left hemisphere with the right ear stimulation (ll, P<0.01).

b: In patients with sensorineural hearing loss, the dipole moment of N100m for F0123 stimulus was significantly larger than that for noise stimulus in both hemispheres. N100m dipole moment for /a/ stimulus was significantly smaller than that for F0123 stimulus in the left hemisphere, and was significantly larger than that for noise stimulus in the right hemisphere. The dipole moments in the patient group were significantly smaller than those in normal subjects in the right hemisphere (§§, P<0.01).

(*: P<0.05, **: P<0.01; contrast comparison)

In patients with hearing loss, the dipole moment of N100m for F0123 stimulus was significantly larger than that for noise stimulus in both hemispheres. N100m dipole moment for /a/ stimulus was significantly smaller than that for F0123 stimulus in the left hemisphere, and in the right hemisphere, it was significantly larger than that for noise stimulus. No significant interhemispheric difference of the N100m dipole moment was noted in the patient group. The dipole moments in the patient group were significantly smaller than those in normal subjects in the right hemisphere, while the difference was not significant in the left hemisphere.

Comments: In normal subjects, N100m dipole moments evoked by a noise were smaller than those evoked by F0123 in both hemispheres. Such differences in AEF between noise and other stimuli, however, were less prominent in the patient group. The right hemispheric dominance observed in the present normal subjects was lost in patients for N100m. The N100m latencies in the right hemisphere in the current patient group were significantly longer than in normal subjects, and the N100m moments in the right hemisphere were significantly smaller than in normal subjects, indicating weaker AEF responses in the right hemisphere in the patient group. In contrast, the differences in these parameters between normal subjects and patients were not significant in the left hemisphere. Reduced AEF responses in the right hemisphere and subsequent disappearance of the right hemispheric dominance in the patient group might be related to an impairment in the processing of tone quality and loudness in inner-ear hearing loss.

3. AEF in Patients with Loudness Recruitment Phenomenon

We measured AEFs in patients who had mild to moderate inner-ear hearing loss with loudness recruitment, and compared them with those of normal-hearing subjects to elucidate the cortical correlates of abnormal loudness sensation [2].

Subjects: Eight patients with inner-ear hearing loss were studied. They had bilateral sensorineural hearing impairment, and their hearing thresholds at 1 kHz ranged from 30 to 50 dB HL (hearing level). All patients exhibited loudness recruitment in both ears. AEFs were measured in each ear. The control group consisted of 14 healthy adults with normal audiograms. Stimulus and measurement: A 1 kHz pure tone was presented monaurally and the intensity was set at 40, 50, 60 or 70 dB HL. Recordings were performed in a magnetically shielded room with the subjects seated under a helmet shaped dewar. AEF was recorded with a 122-channel whole-head DCSQUID magnetometer.

Peak latency of N100m: The latency to ipsilateral stimuli was significantly longer than that to contralateral stimuli. The latency in patients decreased greater than that in healthy subjects as a function of stimulus sound intensity.

Moment of ECD for N100m: The moment of ECD for N100m increased as a function of sound intensity in both groups. Overall, the moment was larger to contralateral than to ipsilateral stimuli and larger in patients than in healthy subjects. The moment of ECD for N100m in patients was larger than that in healthy subjects at 60 dB and 70 dB. The ratio of the ECD moment for N100m at 60 and 70 dB to that at 50 dB was 1.91 and 2.51 in patients, and 1.22 and 1.33 in healthy subjects.

Source location for N100m: All the source areas were located over the superior temporal auditory cortices in each hemisphere. No systematic change was observed as a function of stimulus intensity in either the healthy or the patient group.

Comments: In a previous study, which used electroencephalography (EEG), patients with sensorineural hearing loss exhibited larger amplitude of responses at the threshold level, with greater increase as a function of stimulus intensity than in healthy hearing subjects [3], a finding consistent with our results. However, neither source location nor interhemispheric difference in evoked responses was evaluated due to limitations of the EEG method.

Loudness recruitment is a symptom commonly associated with inner-ear hearing impairment, but the neural mechanisms responsible for it are poorly understood. In our study, increase in cortical activation as a function of stimulus intensity was more prominent in patients. This larger increase in activation of auditory cortex in patients with inner-ear hearing loss is analogous to psychological characteristics of loudness recruitment, which include a rapid rise in loudness sensation with increase in stimulus intensity. This can be considered one of the cortical correlates of subjective symptoms accompanying inner-ear hearing loss.

4. Cortical Activation by Speech in Patients with Moderate Hearing Loss

We examined cortical activation by speech in patients with moderate inner ear hearing loss to investigate the influence of inner ear hearing loss on auditory networks in the brain using positron emission tomography (PET) [4].

Subjects: Five right-handed Japanese subjects were involved in the present study. All subjects were diagnosed as having inner ear hearing loss, and average pure-tone threshold was 55.6 - 75.3 dB HL in the right ear, and 53.7 - 73.6dBHL in the left ear, and there was no significant difference between them.

Sound stimulation paradigm: A prescan examination was performed for each subject to distinguish words that were perceived well from those that were perceived poorly. The word list consisted of 180 Japanese words used in daily life, such as tsukue (desk) and inu (dog). They were presented monaurally in each ear using an earphone at a rate of 1 word every 3 s. According to the result of the pre-scan examination, well-perceived words and poorly-perceived words were picked from the word list for each ear for each subject, and lists of well-perceived words and poorly perceived words specific to each ear of each subject were used for this study. Regional cerebral blood flow measurement and data analysis: For each stimulation condition, 15-O labeled water was injected into the right cubital vein and the rCBF was measured for 60 s. Each subject's brain was scanned using a General Electric Advance tomograph (GE Medical Systems, Milwaukee, Wis., USA). The data acquired were analyzed with the Statistical Parametric Mapping (SPM96, Wellcome Department of Cognitive Neurology, London, UK) software.

Brain activation: For right ear stimulation, well perceived word stimulation induced rCBF increase in the bilateral superior and middle temporal gyri, left inferior frontal gyrus (Broca's area) and its right hemisphere homologue, and left angular gyrus. Poorly perceived word stimulation induced rCBF increase in the left superior and middle temporal gyri, Broca's area and its right hemisphere homologue, and the right superior temporal gyrus. There was no significant activation in the right middle temporal gyrus and in the left angular gyrus. During left ear stimulation, well-perceived words induced rCBF increase in the bilateral superior and middle temporal gyri, Broca's area, and its right hemisphere homologue. The left angular gyrus exhibited weak activation with a Z value of 3.14. Poorly perceived word stimulation induced rCBF increase in the right superior and middle temporal gyri, Broca's area and its right hemisphere homologue. There was no significant activation in the left temporal lobe and in the left angular gyrus.

Comments: During listening to speech, speech sounds undergo acoustic-phonetic and lexico-semantic processing and are perceived as meaningful words. Patients with inner ear hearing loss have difficulty in identifying phonemes and in perceiving words correctly, suggesting that insufficient activation of ipsilateral temporal lobe in our study may be related to the failure of acoustic-phonetic or lexico-semantic processing. However, inactivation of the ipsilateral auditory cortex may not to be due to the failure of meaning comprehension, since hearing reverse-played speech, which has no meaning, is reported to activate the temporal lobes bilaterally. Thus, inactivation of the ipsilateral auditory cortex in our study may reflect an alteration of acoustic-phonetic processing caused by distorted information from the damaged cochlea. In the present results, activation of the temporal lobe was symmetrical irrespective of the ear stimulated, while that of the angular gyrus was restricted to the left hemisphere. Left hemisphere dominance for language processing is widely accepted and many functional activation studies have shown left hemisphere lateralization in cortical activation during auditory word comprehension. However, some reports point to a bilateral temporal speech processing system, at least at the prelexical level, by investigating the level of language processing in detail. It is reported that temporal lobe activation was seen bilaterally in acoustic and phonetic processing, while left lateralized activation in the temporoparietal regions was recognized during processing lexical semantic and/or syntactic information [5]. These reports indicate that the bilateral temporal lobe is related to the prelexical processing of speech sound, which is consistent with our results.

5. Cortical Processing of Speech in Profoundly Deaf Cochlear Implant Users

To investigate the cortical activities during listening to noise and speech in cochlear implant users, we compared cerebral blood flow in postlingually deafened cochlear implant (CI) users with that in normal hearing subjects using positron emission tomography. Twelve normal hearing volunteers and twelve postlingually deafened CI users were included in the present study. All normal volunteers were male, with an average age of 29.3 ± 3.8 (mean ± SD) years. Of the twelve patients with postlingual deafness, six patients had the implant in the right ear, while the remaining six patients had the implant in the left ear. The average age of the patients was 41.8 ± 11.8 years. Duration of deafness ranged from 0.5 to 14 (mean: 3.6) years, and the length of CI use from 1 to 96 (mean: 21.6) months. Patients' vowel perception test performance (percentage of correct answers) was 81.1 ± 20.5 % (score for chance performance on the test was 20 %), and consonant (consonant-vowel syllable) perception test performance was 33.0 ± 17.9 % (score for chance performance on the test was 7.7 %). The performances of normal subjects for these tests were all 100 %. While noise activation in CI users did not significantly differ from that in normal subjects, hearing speech activated more cortical areas in CI users than in normal subjects. Comparison of speech activation in these two groups revealed higher activation in CI users not only in the temporal cortices but also in Broca's area and its right hemisphere homologue, the supplementary motor area and the anterior cingulate gyrus. In postlingually deafened subjects, hearing of speech coded by CI may be accompanied by increased activation of both the temporal and frontal cortices.

6. Conclusions

We observed excessive activity in the auditory cortex at high stimulus intensities in patients with inner-ear hearing loss. AEF responses in patients were less dependent on the acoustic structure of the stimuli than in normal subjects. Although the N100m latency was shorter and its dipole moments were larger in the right hemisphere in normal subjects, such interhemispheric differences were not identified in patients. In patients with inner ear hearing loss, insufficient activation of the temporal lobe ipsilateral to the ear stimulated might correlate with less accurate word comprehension. Injury to the peripheral auditory organs modifies information processing in the auditory cortex, which may correlate with various subjective symptoms that accompany inner-ear hearing loss.

References

1. Naito Y, Fujiki N, Nagamine T, et al. Auditory evoked magnetic fields in patients with hearing disorders. Recent advances in biomagnetism (ed by Yoshimoto T, Kotani M, Kuriki S, Karibe H, Nakasato N). pp521-524, Tohoku University Press, Sendai, 1999.
2. Morita T, Naito Y, Nagamine T, et al. Enhanced activation of the auditory cortex in patients with inner-ear hearing impairment: An MEG study. Clin Neurophysiol 2003; 114: 851-859.
3. Cody DT, Griffing T, Taylor WF. Assessment of the newer tests of auditory function. Ann Otol Rhinol Laryngol 1968; 77: 686-705.
4. Tateya I, Naito Y, Hirano S, et al. Inner ear hearing loss modulates ipsilateral temporal lobe activation by monaural speech stimuli. NeuroReport 2003; 14: 763-767.
5. Zahn R, Huber W, Drews E, et al. Hemispheric lateralization at different levels of human auditory word processing: a functional magnetic resonance imaging study. Neurosci Lett 2000; 30: 287: 195-198.
6. Naito Y, Tateya I, Fujiki N, et al. Increased cortical activation during hearing of speech in cochlear implant users. Hear Res 2000; 143:139-146.

Efficacy of the Levodopa on Frontal Lobe Dysfunction in Patients with de Novo Parkinson's Disease; A Study Using the Event-related Potential

Koichi Hirata[1], Yuka Watanabe [1], Akinori Hozumi [1], Hideaki Tanaka [1], Mio Arai[1], Yoshiaki Kaji[1], Masako Saito[2], and Kayoko Iwata[2]

[1]Department of Neurology
[2]Institute for Medical Science Dokkyo University School of Medicine, Tochigi, Japan

Chapter Overview. We investigated the efficacy of the levodopa on cognitive function in patients with de novo Parkinson's disease (PD) using the event related potential (ERP) measure and neuropsychological test batteries. We studied 20 PD patients who had never received anti Parkinson medications. After administration of levodopa, the patients revealed significant increase of the achieved categories and decrease of preservation errors in the WSCT, shortening of time in Trail Making Test although MMSE score showed no significant difference. The P3 peak attenuated although there were no differences in peak latency or on scalp topography. These findings suggest that levodopa affect to the neural circuit connect frontal cortex with the striatum and normalize its function, and it causes decrease of P3 peak reflecting appropriate resource allocation. In addition, P3 might be more appropriate indicator rather than psychological test

Key Words. Parkinson's disease, and Event-related potential.

1. Introduction

Parkinson's disease (PD) is characterized by the loss of dopaminergic neurons projecting from the substantia nigra to the striatum. This results in altered function of cortical - striatal - pallidal-thalamic cortical circuits that mediate normal movement, resulting in bradykinesia and rigidity. It has been well known that cognitive dysfunction affects patients with idiopathic PD, even though they do not have clinical dementia. In 1974, Albert et al. reported that patients with subcortical disease, including PD, show cognitive disorders that differ from Alzheimer's disease [1]. There are various reports about the cognitive disorders of Parkinson's patients. PD patients are impaired on a variety of cognitive tasks that depend on frontal lobe function and/or basal ganglia. Studies using various psychological batteries demonstrated a wide variety of mental deficits in patients with PD such as impairment of tests of planning, Attention set-shifting, skill leaning, habit learning [2]. P3 is affected by the dopaminergic and cholinergic systems, both of which are impaired in PD. P3 has been shown to be abnormal in PD with cognitive decline or dementia, but investigations of P3 in patients with PD without clinical dementia (non-demented PD) remains controversial [3]. A surprising result reported by Green et al. [4] was that the P3 amplitude elicited by target was increased in non-medicated and mild PD. A similar result was reported later by Tanaka et al. [5]. The latter compared PD with differences in conditions such as duration of illness, with or without dementia, the P3 amplitude increased only in non-demented PD but not in demented patients.

Concerning the P3 latency, most studies showed consistent results in that PD did not differ from normal controls. Few studies, however, revealed the P3 with delayed latency in non-demented PD. We investigated the influence of Levodopa (L-dopa) on cognitive function in de novo PD patients using the event related potential P3 measurement and neuropsychological test batteries. The purpose of the present study is not only to evaluate the effect of L-dopa administration changes but also the effects in difference of severity and duration of illness using ERP P3 as an objective tool in mild non-demented de novo PD patients.

2. Methods

2.1 Subjects

We studied 20 patients with PD who had never received anti Parkinson medications. We excluded subjects with clinical dementia and an MMSE score of less than 24. None of the subjects had evidence of other brain damage as confirmed by computed tomography or magnetic resonance imaging. The subjects were classified to the subgroups based on duration and severity of illness. The severity of illness was determined by Hoehn and Yahr. All subjects were fully informed as to the nature and purpose of the study and gave their consent to the study. Table.1 shows the clinical background of all subjects.

Table1. the clinical background of subjects

De novo PD patients	20
Gender (M/F)	8 /12
Age (mean ±SD)	68.5 ± 8.92 years
Duration of illness (median)	1.475 years
<1 year	11/20
>1 year	9/20
Severity of illness (Hohen&Yahr)	
I	5/20
II	7/20
III	8/20

2.2 Neuropsychological testing

We used neuropsychological tests such as the Mini-Mental State Examination (MMSE), New modified Wisconsin card sorting test and Trail Making Test and Stroop Test. Wisconsin Card Sorting test was used to evaluate the frontal lobe function as it is one of the few tests to detect a clear deficit specific to patients with frontal lobe dysfunction. The achieved categories classification score and the perseverant errors reported by Nelson were used to evaluate the results of this test.

2.3 Stimuli and experimental conditions

Auditory oddball P3 paradigm as given for the subjects. Two tones auditory oddball paradigm was used to elicit P3. Multichannel ERPs were averaged with 21 channels using the international 10/20 system for electrode location. A topography mapping system (Brain Atlas) was used for recording and averaging. Levodopa with benserazide was administrated. Initial dose was 100 mg/day; it was then increased up to 200 mg/day per week. Finally, patient was given Levodopa with benserazide 300 mg/day. Before and after administration of the drugs we measured P3 and ran the psychological tests. For the ERP's components, global field power (GFP) which is reference-independent measures was determined, and P3 GFP peak, peak la-tency and topography were assessed [6].

3. Results

3.1 Neuropsychological examination

MMSE score showed no significant difference before and after administration of L-dopa. On the other hand, Wisconsin Card Sorting test score showed significant increase of the achieved categories and decrease of perseveration errors after the administration especially in patients with short duration (short group) and slight severity of illness (Hoehn and Yahr I). Stroop test showed no significant difference before and after administration of L-dopa, but Trail Making test showed significant improvement especially in the patients whose duration of illness are long.

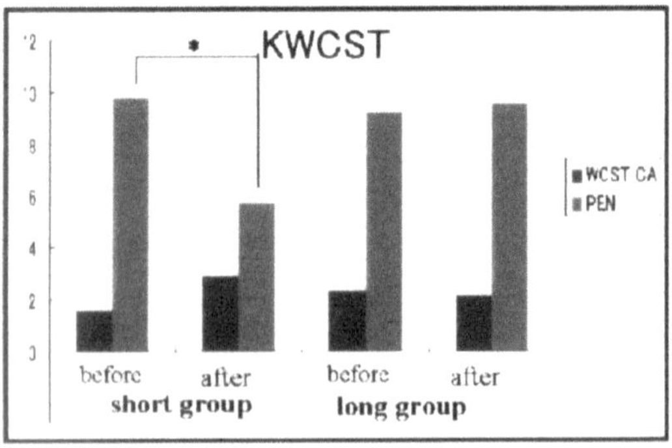

Fig.1. Change inWCST after administration of L-dopa

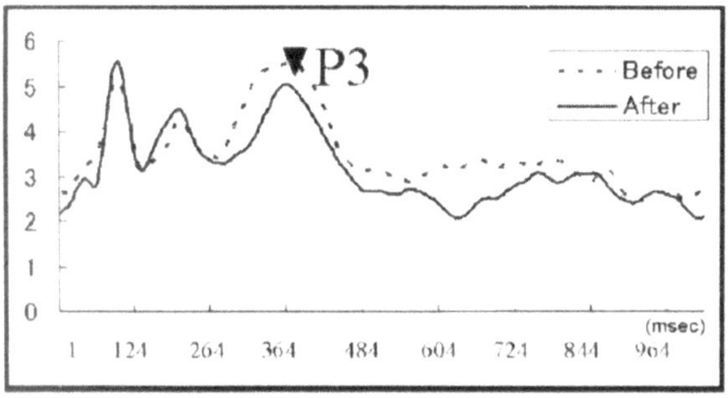

Fig.2. Change in the grand average P3 GFP waveform after administration of L-dopa

However, there is no difference of P3 latency and location in topography. P3 GFP peak decreasing is significant especially in patients with long duration (long group) and sever illness (Hoehn and Yahr III) in contrast to the results of neuro-psychological examinionan.

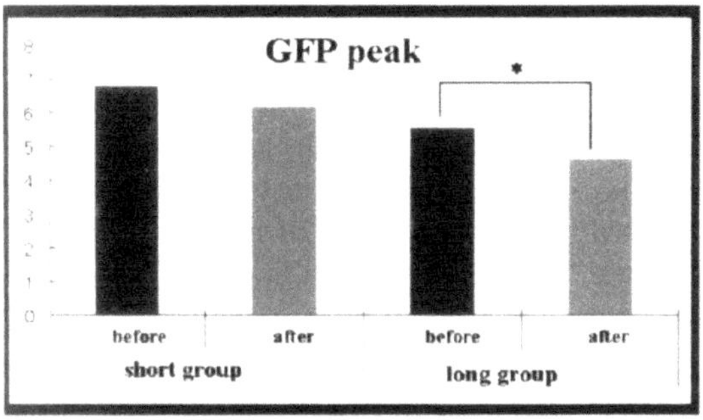

Fig.3. Change in P3 GFP in difference of duration of illness.

4. Discussion

Many studies using various psychological batteries demonstrated an attention set-shifting, skill leaning, habit learning based on frontal lobe and basal ganglial dysfunction in patients with PD [1, 2, 7]. We investigated the influence of L-dopa on cognitive function in de novo PD patients using the event-related potential P3 measurement and neuropsychological test batteries. As a result we have demonstrated that P3 GFP peak decreased significantly after administration of L-dopa, and that its ERP results may have close relationship with the results of psychological tests, especially the Wisconsin Card Sorting test. Arnsten et al. [8] administered Dopamine (DA) D1 receptor compounds in monkeys for the effects on the working memory functions of the prefrontal cortex. The partial agonist which is D1 agonist improved the aged or young reserpine-treated D1 depleted monkeys, but did not improve young control animals.

Therefore they have concluded the importance of DA D1 mechanisms in cognitive function, and provide functional evidence of DA system degeneration. Based on their reports, our result suggests that the L-dopa improves the function of dopaminergic neural circuit between frontal lobe cortex and basal ganglia in PD.

P3 amplitude is related to how the stimulus delivers its task-relevant information, and that it is proportional to the amount of attention resources in terms of processing capability that is employed in a given task. P3 amplitude changes have been interpreted reflecting neural activity related to the basic aspects of cognition. P3 amplitude would be altered relative to the attention processing resources to be allocated in a task. P3 amplitude variation may provide valuable information about the intensity and the timing of energy allocation to various sub-processes of information processing. Green et al. [4] hypothesized that enlarged P3 amplitude reflects abnormality in use of attention resources to compensate for frontal dysfunction in PD. We have mentioned in the other study, P3 GFP was significantly increased in the early stage of non-demented PD patients. These results are confirmed as well as the findings of Green et al. [4] and Tanaka et al. [8]. Therefore we think that L-dopa administration decreases P3 GFP, which reflects an improvement of attention processing resources in de novo PD patients.

The purpose of the present study is not only to evaluate the effect of L-dopa administration changes but also the effects in difference of severity and duration of illness using ERP P3 as an objective tool in mild non-demented de novo PD pa-tients. Although the amelioration of Wisconsin Card Sorting test score was more significant in patients with short duration and slight severity of illness, P3 GFP improvement seen only in the severe and long duration group after the administration of the drug. These results suggest that P3 GFP change linearly correlates with duration and severity of illness. The reason of discrepancy of 2 measurements may be that the psychological test can be affected learning effects. Therefore we think that P3 GFP more objectively indicate ameliorating effect of L-Dopa rather than the psychological test.

Acknowledgment

This research was supported in part by a grant- in-aid from the smoking research foundation and Japanese ministary (K.H).

References

1. Lees AJ: Parkinson's disease and dementia. Lancet, vol 1, 43-44. 1985.
2. Lees A.J. and Smith E. Cognitive deficits in the early stage of Parkinson's disease. Brain, vol106, 257-270, 1983
3. Hozumi A, Hirata K, Tanaka H, Yamazaki K : Perseveration for novel stimuli in parkinson's disease: An evaluation based on event-related potentials topography. Movement Disord, vol 15, 835-842, 2000.
4. Green J, Woodard JL, Sirockman BE, Zakers GO, Maier CL, GreenRC, Watts RL: Event-related potential P3 change in mild Parkinson's disease. Mov Disord, vol 11, 32-42, 1996.
5. Tanaka H., Koenig T., Pascual-Marqui R.D., Hirata K., Kochi K. and Lehmann D.Event-related potential and EEG measures in Parkinson's disease without and with dementia Dement Geriatr Cogn Disord, vol 11, 39-45, 2000.
6. Kanazawa A, Mizuno Y, Narabayashi H: Executive function in Parkinson's disease. Rinsho Shinkeigaku, vol 41, 167-172 (Japanese), 2001.
7. Lehmann D. and Skrandies W. Reference-free identification of components of checkerboard evoked multichannel potentials field. Electroencephalogr Clin Neurophysiol, vol 48, 609-621, 1980.
8. Flowers KA, Robertson C: The effect of Parkinson's disease on the ability to maintain a mental set. J Neurol Neurosug Psychiatry, vol 48, 517-529, 1985.
9. Arnsten AF, Cai JX, Murphy BL, Goldman-Rakic PS: Dopamin D1 recepter mechanisms in the cognitive performance of young adult and aged monkeys. Psychopharmacol (Berl) vol 116, 143-151, 1994.

Part 6
Complex Bioinformatics

A Wireless Integrated Immunosensor

Tomohiro Ishikawa[1], Turgut-Sefket Aytur[2], and Bernhard E. Boser[2]

[1]Sharp Corp., Chiba, Japan
[2]Berkeley Sensor and Actuator Center, University of California at Berkeley, California, United States

Chapter Overview. A 1.5mm x 1.5mm bioassay chip has been fabricated using a 0.25μm RF-CMOS process. The chip is dipped into antibody solution and antibodies coat its surface. The chip provides a 32-by-32 Hall-sensor array designed to detect magnetic beads immobilized onto the chip surface by immuno-chemistry. The signal frequency is up-converted by the Hall sensor to suppress the flicker noise, and the current direction on the Hall sensor is rotated in channel plane to cancel the offset in the magnetic field. The chip requires no additional packaging: it is added directly into the sample solution to perform immuno-assay. A wireless link through the on-chip coil provides the power as well as the clock to the chip. The measurement data are also retrieved through the link.

Key Words. Immunosensor, and Biosensor.

1. Introduction

Despite significant advances in medical-evaluation and treatment technologies, infectious diseases still remain as a major common cause of death. In addition, nowadays we have potential risks of being ex-posed to various hazardous substances in our every-day-living environment. In order to provide broad- and large-scale screenings for daily-use diagnoses, immuno-sensors must satisfy two essential requirements: low cost and ease of use.

For immunological assays, Enzyme-Linked Immuno-Sorbent Assay

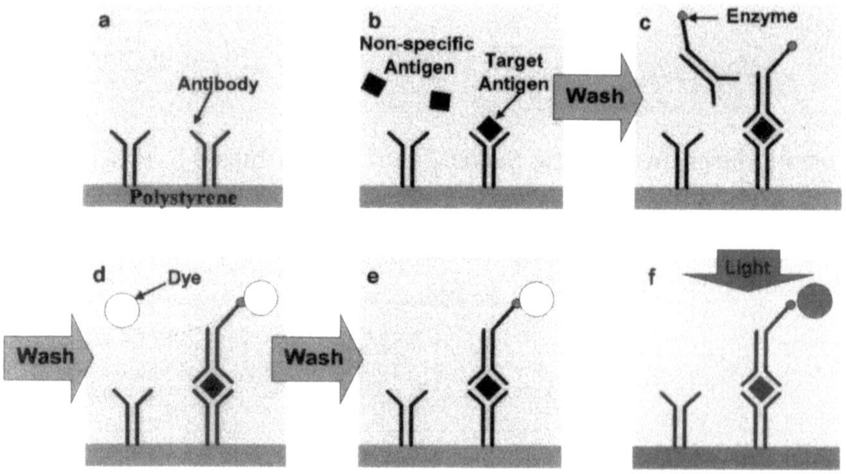

Fig. 1. Conventional technique, ELISA

Antibodies are attached to the *Polystyrene* substrate in advance (**a**). Antibody catches only *target antigen* specifically (**b**). The site that caught the antigen immobilizes *enzyme*-linked antibody (**c**). A fluorescent *dye* is added (**d**). The Enzyme acts on dye and produces color (**e** and **f**).

(ELISA) is considered as the benchmark. However, as Fig. 1 shows, the ELISA protocol is complicated and it requires the 'washing' process between each step. In addition, obtaining a quantitative analysis requires expensive external equipment. This makes field and at-home diagnostics difficult. Furthermore, ELISA is poorly suited for parallel assays.

Earlier work has shown that magnetic labels can simplify the imuno-assay protocol and allow parallel assays [1]. However, these devices rely on sensing technology that is expensive and difficult to manufacture. Recently, micro-scale on-chip RF-ID applications have been proposed and demonstrated [2]. The RF-ID chip is activated by external RF signals, and it responds with its ID. This technology has not been exploited in bioassay applications yet. In this report, the device employs the on-chip RF-ID concept and is fabricated using a conventional, standard-CMOS-IC process. It is low cost and is easy to use, making it suitable for large-scale screenings for daily-use diagnoses.

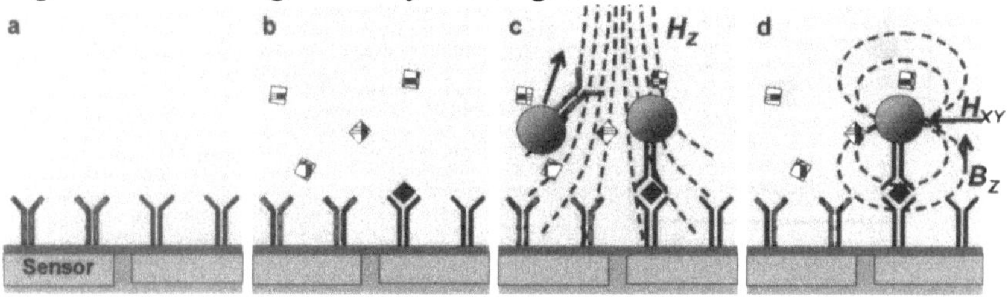

Fig. 2. 'Magnetic wash' and sensing

Antibodies are attached onto the Hall *Sensor* in advance (**a**). Antibody catches only target antigen specifically (**b**). Antibody-antigen coupling immobilize the magnetic bead and an uneven magnetic field pulls off unbound beads (**c**). In a horizontal parallel magnetic field, bead induces local vertical component of flux, B_z (**d**).

2. Magnetic Labeling

With the magnetic label, magnetic beads are used instead of dye-colored beads (used in ELISA). A bio-assay sample and the beads are added to the sample solution, as in the ELISA protocol (Fig. 2-a). Fig. 2-c shows uneven vertical magnetic field can pull off un-bound and weakly bound beads. Only strongly bound bead stay on the sensor surface and this process serves as magnetic 'washing' process.

In Fig. 2-d, a horizontal even magnetic field is applied. The hall sensor detects the vertical component of the magnetic flux, Bz, which is induced by the bead, and thereby the sensor detects the existence of the bead anchored on the sensor's surface [3].

3. Sensor Design

3.1 Sensor array

Fig. 3 shows the NMOS-Hall-sensor array on the chip. The sensor array consists of 32-by-32 NMOS Hall-effect sensing elements. The Row and

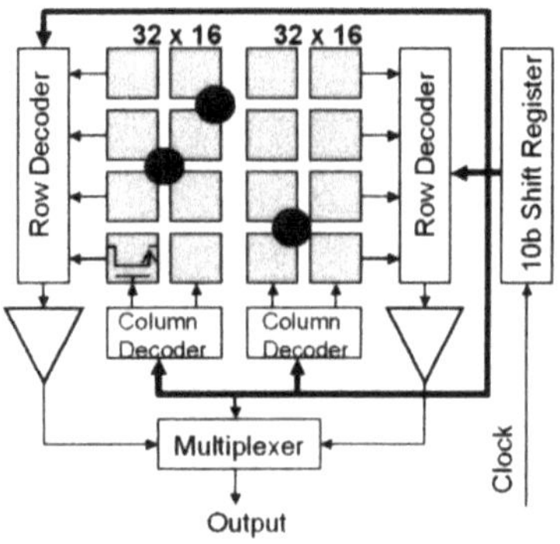

Fig. 3. A 32 x 32 NMOS Hall sensor array

Column decoders access each element sequentially. Each sensing element in the array consists of a paired Hall sensor with the currents flowing in the opposite directions. Each Hall sensor is drawn as 4-um x 4-um square four terminal device so that it is match to the local field from the magnetic bead.

3.2 Frequency conversion

Fig. 4 shows the schematic diagram of a sensor element. The magnetic field ($H_0\sin(\omega_m t)$) is generated by an external electromagnet. Because of the flicker noise and 1/f noise, CMOS devices deteriorate at lower frequencies. On the other hand, if we operate the sensor at higher frequency to avoid flicker and 1/f noises, the magnitude of current through the electromagnet decreases, and, as a result, the magnitude of the magnetic field also decreases. To solve this trade off, a radio-frequency modulation is applied to the gates of the NMOS transistors as:

$$V_{gs}(t) = V_{DC} + V_{AC}\sin\omega_e t \tag{1}$$

and

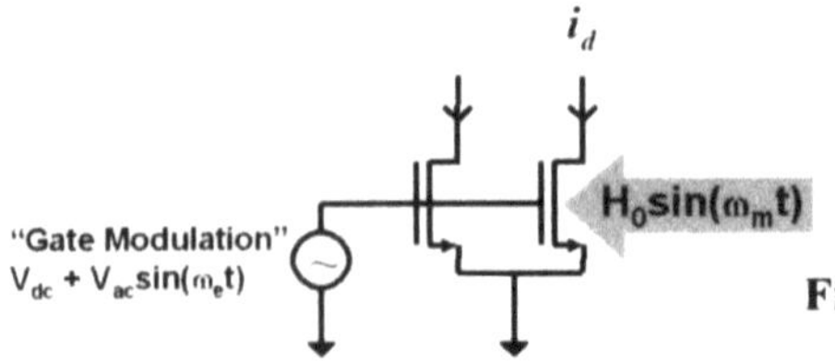

Fig. 4. Schematic of sensor element

$$H_{xy}(t) = H_0 \sin \omega_m t \tag{2}$$

Where ω_c and ω_m are the modulation frequencies for the NMOS gates (electrical components) and the electromagnet (magnetic component), respectively.
Then, the sensor's output current (id) contains the term:

$$i_d \propto \sin(\omega_e \pm \omega_m)t \tag{3}$$

This driving method clearly has a couple of advantages. The electrical-modulation frequency can be very high, and as a result, the signal from the sensor is amplified in the high-frequency region with less flicker noise. On the other hand, the magnetic-modulation frequency can be low enough to provide a sufficiently strong magnetic field. In addition, parasitic coupling of the magnetic field in the experimental setup is eliminated.

3.3 Current rotation

Since the sensor measures the vertical component of the magnetic flux, any miss alignment of the chip causes offset of the magnetic field and causes errors in the measurement. Fig. 5 shows the more detailed diagram of a paired, Hall-effect sensing element. Since the currents are flowing in opposite directions in the coupled pair of Hall sensor, the offset due to any misalignment is canceled, and the combined sensing element measures only the local magnetic field from the nearby bead. In addition, the connection to the Hall sensor sequentially switches and rotates the current direction on the sensor.

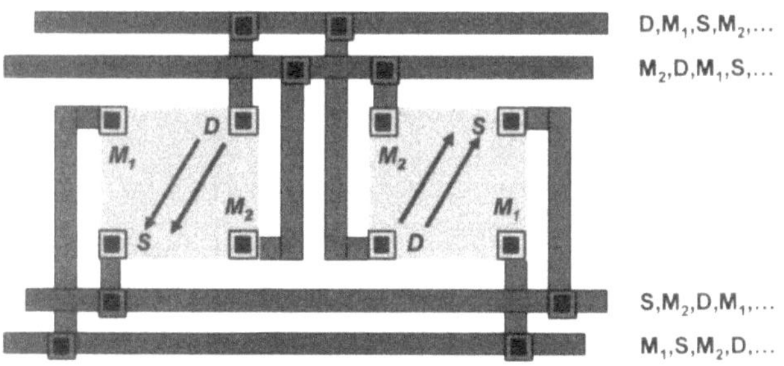

Fig. 5. Structure of a sensor element

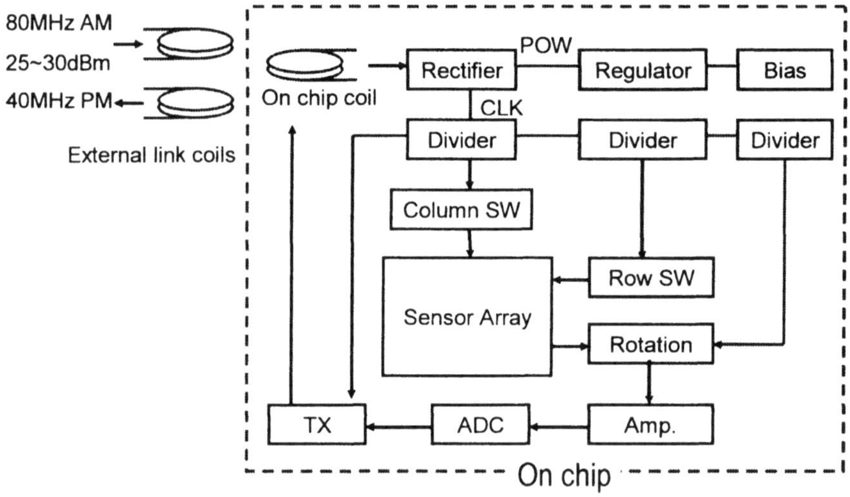

Fig. 6. Block diagram of the chip

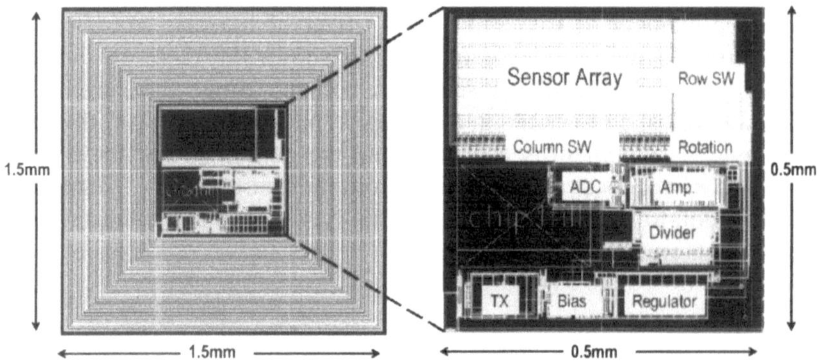

Fig. 7. Chip layout

In Fig. 5, on the left device, first, the right upper and the left bottom terminals serve as Drain and Source, respectively. Then, the left upper and the right bottom ones do. As a result the direction of the drain current is rotated in 90 degree. This rotation reduces the device mismatch and offset [4].

4. Chip Design

Figs. 6 and 7 show the block diagram and layout of the chip, respectively. The chip was fabricated using a 0.25-um RF-CMOS process. A 40-turn, 2.2-uH, on-chip coil is implemented using the top metal (copper) layer of the process. It has Q>2 and a self-resonance frequency of 80 MHz in the air.

The on-board link coil feeds on the 80MHz external carrier whose power is 30-dBm. The on-chip coil provides a 500uA output at 2V at a distance of 1mm. The output is rectified to DC voltage by zero-Vt NMOS devices connected as diodes. Then, the regulator provides 1.7V-DC power to the other functioning blocks on the chip. The rectifier circuit also extracts 8MHz AM signal, and it serves as a system clock on the chip.

During the measurement, the chip is placed in an external 2-kHz magnetic field. The row and the column switches (SW) scan the sensor array and the Ana-log-Digital-Converter (ADC) digitizes the signal from each sensor element. The serial output of ADC is sent back through on-chip coil with the 40-MHz Phase Shift Keying (PSK)-modulated carrier.

The completed chips and an onboard link coil are shown in Figure 8.

Fig. 8. The completed chips, 1.5mm x 1.5mm (left) and link coil (right)

5. Experimental Result

The chip was evaluated with a test circuit board. The board consists of AM transmitter and PSK receiver. AM transmitter provides 80MHz carrier in power of 30dBm with 8MHz amplitude modulation to the on-board link coil shown in Fig. 8.

A 2-kHz, 50kA/m sinusoidal magnetic field was applied by using an external electromagnet. The received signal passed through low-noise amplifier and high-Q 40MHz band-pass filters. Then the signal was digitized by an oscilloscope and analyzed using MATLAB.

Fig. 9 shows the decoded output signal. The signal contains 10 flaming bits for data synchronization and 22 data bits for combined 2ch of 11bits data.

6. Conclusion

A magnetically-labeled immunoassay chip has been fabricated and demonstrated. The chip was fabricated in a standard RF-CMOS process and

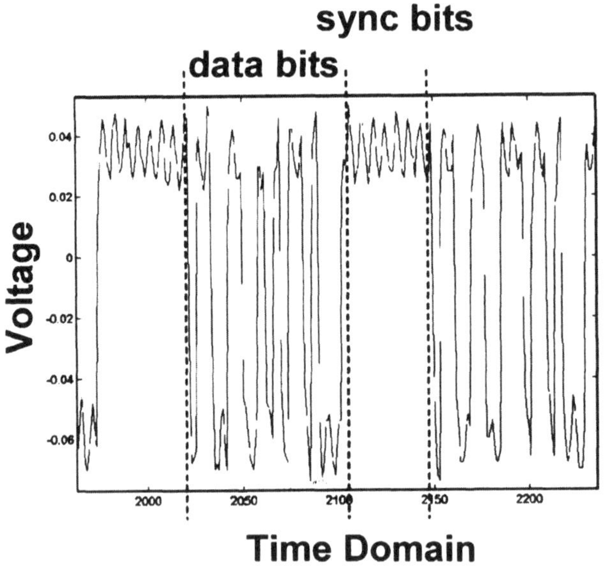

Fig. 9. Digitized time-domain output

requires no additional packaging. Our protocol is much simpler than the present standard-assay ELISA. The chip can be directly put into bioassay samples, and it passes in-formation through the wireless link circuit. The power and the clock for the chip were supplied through the wireless link established by using on-chip coil, and the measurement result was also sent back through the same on-chip coil but at a different frequency.

The bioassay proposed in this work will enable low-cost, large-scale screening, daily-diagnosis applications in the future.

Acknowledgment

The authors wish to thank the Acumen Fund for their support of this project, and the School of Public Health at the University of California, Berkeley for their cooperation.

References

1. Baselt, D, Lee G U, Natesan M, et al., "A Biosensor Based on Magnetoresistance Technology", Biosensors and Bioelectronics, Vol. 13, pp.731-739, 1998.
2. Usami M, Sato A, Sameshima K, et al., "Powder LSI: An Ultra Small RF Identification Chip for Individual Recognition Applications", Proc. IEEE ISSCC, San Francisco, CA, Jan. 2003.
3. Aytur T S, Ishikawa T, Boser B E, et al., "An Immunoassay Platform Based on CMOS Hall Sensors", Digest Solid-State Sensors and Actuators Workshop, Hilton Head, SC, June 2001.
4. Taranow S, et al., "Method for compensation of nonequipontential voltage in the Hall voltage and means for its realization", German Patent Appl. 2333080, 1973.

Dynamics of Cortical Neurons and Spike Timing Variability

Takashi Tateno

Department of Mechanical Science and Bioengineering, Osaka University, Japan

Chapter Overview. Neurons and dynamical models of spike generation display two different classes of threshold behavior, with steady current stimulation: type 1 and type 2. The dynamics at threshold could have profound effects on the encoding of input as spikes, the sensitivity of spike generation to noisy input, and the coherence of population firing. To comprehend the consequences of spike timing variability during more natural fluctuating input, we compared the postsynaptic firing variability of type-1 regular-spiking pyramidal cells and type-2 fast-spiking interneurons, applying a conductance injection method to rat acute slice preparations *in vitro*. Using two statistics, the reliability of spike occurrence and the spike jitter, we studied the sensitivity of firing variability to the rate of independent, shunting inhibition and to the correlation in time of synthesized synaptic inputs. The results indicate that differences between the two cell types are consistent with a role of regular-spiking neurons as rate-coding integrators, and a role of fast-spiking neurons as resonators controlling the coherence of synchronous firing.

Key Words. Action potential, Conductance injection, Cortical cell, and neural coding.

1. Introduction

Recent studies have revealed that cortical neurons in the mammalian recorded *in vivo* exhibit highly variable responses to the repeated presentation of the same stimulus [1]. *In vivo* recordings of conductance in cortical pyramidal cells indicate that cortical networks are highly active and induce a large fluctuating background conductance [2]. Background presynaptic activity is thus an important source of unreliability and variability in spike timing of cortical cells *in vivo*. In contrast, *in vitro* recordings show that spike generation in cortical neurons can be precise and reproducible to repeated injections of a fast fluctuating identical current [3] and conductance [4]. Several modeling and experimental approaches have suggested that the high variability of firing *in vivo* could have two main causes: synchrony or correlation of inputs in time [5] and independent inhibition [6].

In a study of the responses of axons isolated from *Carcinus maenas* to various intensities of rectangular current stimuli, Hodgkin found that some axons could show a continuous transition from zero frequency to arbitrarily low frequencies of firing, whereas others show an abrupt onset of repetitive firing at a nonzero firing frequency [7]. These types of threshold dynamics are recently referred to as "type 1" and "type 2," respectively. Recently, moreover, there has been accumulating experimental evidence that with steady current or conductance stimulation, regular-spiking (RS) pyramidal neurons and fast-spiking (FS) inhibitory interneurons in mouse and rat somatosensory cortical slices respectively demonstrate these two types of the threshold behaviors and firing properties [8, 9].

To shed more light on differences in postsynaptic encoding by the two types of threshold dynamics and to understand the significance of threshold type during more natural fluctuating input, we used the conductance injection technique [10] to compare the input-output relationships and spike-timing variability of RS pyramidal cells and FS interneurons in layer 2/3 rat somatosensory cortical slices. For the purpose, trains of unitary conductance transients, electrically mimicking simultaneous fast excitatory (AMPA-type) and inhibitory ($GABA_A$-type) inputs were injected [4]. This study is an attempt to elucidate how the interaction of inhibitory and excitatory inputs determine linear and nonlinear gain control of firing rate, and modulate spike-timing reliability in RS and FS cells in the cortical network. We discuss the possible functional roles of the two types in the cortical network, in the context of concerted activity driven by fluctuating transient synaptic inputs.

2. Materials and Methods

We recorded from layers 2/3 pyramidal neurons and FS cells in slices of somatosensory cortex from Wister rats aged 15–21 days. For details of the preparation, solutions, and whole-cell patch-clamp recording technique, see Ref. [9]. For conductance injection (dynamic clamp) stimulation [10], the opening of a population of receptor channels at synapses is modeled by an excitatory (AMPA) receptor synaptic conductance and an inhibitory ($GABA_A$) receptor synaptic conductance [4]. Depending on the changing membrane voltage V(t), an injected current is described by $I(t)=g_E(t)[E_E-V(t)]+g_I(t)[E_I-V(t)]$, where E_E and E_I are the reversal potentials for the AMPA- and $GABA_A$-type conductances, respectively. For each cell, E_I was set at the membrane resting potential and E_E =0 mV. For two types of receptor conductances, the waveform of a single synaptic input was specified by the linear combination of two exponentially decaying functions and the time constant values were the same as used in [4]. Once parameters of unitary synaptic input are fixed, the time series of the unitary input completely determine the time course of conductance stimuli. As reported before [4], we used two point-process models: (i) Poisson and (ii) doubly stochastic process models. Poisson stimulus trains were constructed by summing unitary events, e.g. AMPA- or $GABA_A$-type, at intervals denoted by a random variable t_i, with the probability density: $p(t_i)=\lambda\exp(-\lambda t_i)$. The intensity of stimulation was varied by changing the Poisson rate λ in excitatory and inhibitory inputs. The ratio of λ's between excitatory and inhibitory conductance inputs was independently changed to investigate the effect of their balance on spike timing. To simulate correlated or synchronous firing of synaptic events, in addition, we used a doubly stochastic process model which mimics bursty presynaptic events. We use another parameter S as a synchrony index to indicate correlated input in time; the higher S, the more correlated in time. In more detail, see Ref. [4]. For reproducibility of spike times, two measures, termed "jitter" and "reliability", were examined and they are similar to those used in [3,4].

3. Results

In response to sustained current injection, RS cells showed a strong frequency adaptation. Their frequency vs. current intensity (*f-I*) relation is continuous at threshold. In contrast, FS cells showed relatively little frequency adaptation with a longer decay time. They were able to fire at high frequencies with relatively little spike frequency adaptation, which occurred mainly over the first few spikes. FS cells also showed an abrupt onset of repetitive firing with increasing current - there is a clear discontinuity at a critical frequency f_c in their *f-I* relationship at threshold. FS neurons are unable to support maintained regular firing below the critical frequency, which varied between 10 and 30 Hz for different cells. Above f_c, the instantaneous firing rate increased monotonically with current strength. To examine the effects of inhibitory input on the postsynaptic firing rate and interspike-interval variability of both RS and FS cells, inhibitory $GABA_A$-type and excitatory AMPA-type Poisson conductance stimuli were synthesized and simultaneously injected into cells. For RS cells, average firing frequency vs. AMPA-type Poisson rate (*f-R*) relationships for three rates (e.g., 1, 2, and 3k Hz) of $GABA_A$-type Poisson conductance were monotonically increasing and strongly nonlinear, and inhibitory input shifted *f-R* curves rightward. In addition, inhibitory inputs reduced the nonlinearity of the curves because the inputs released action-potential depolarization block for higher levels of excitation, further enhancing firing frequency. In RS cells, the average coefficient of variation (CV) of interspike intervals (ISIs) at the three levels of inhibitory input and in the control condition was around 0.45 and was not easy to distinguish for the three levels of inhibition. Although, for FS cells, inhibitory inputs shifted *f-R* curves rightward as for RS cells, *f-R* curves of FS cells were more weakly nonlinear than in RS cells. Moreover, unlike in RS cells, *f-R* curves for several inhibitory levels were parallel with each other and did not cross at higher rates excitatory Poisson input, in the range of parameters investigated.

It is well-established that values of CV of ISIs (CV(ISI)) *in vivo* are typically greater than one [2, 5]. In contrast, for both RS and FS cells, CV(ISI) of *in vitro* firing during excitation by Poisson AMP-type input is much lower than those of *in vivo* firing. However, inhibitory synaptic events could add more variance to the input so that spike-time variability was increased. The increased variability can be seen more clearly in FS cells with higher levels of inhibition. Therefore, we investigated it in more detail. To address the relationship with input correlation in time (synchrony, *S*), we used a doubly-stochastic process input, with low or high synchrony.

In RS cells, inhibition caused a number of spikes to drop out and the firing rate decreased as the level of inhibition increased for both lower and higher synchronous modes of simultaneous excitatory and inhibitory inputs. The initial firing rate in the control was selected at 13-20 spikes/s in this experiment. Increasing the ratio of inhibition to excitation reduced firing frequency in an approximately linear fashion, at both low and high synchrony. Inhibition caused a decrease of CV(ISI) at high synchrony, whereas it does an increase of CV(ISI) at low synchrony. We also found that reliability increased, and jitter decreased, as the proportion of inhibition increased, at both levels of synchrony. Furthermore, jitter decreased for both the lower and higher synchronous mode as the inhibition was increased. The same tendency was observed in all 8 cells. These data suggest that shunting inhibition is unlikely to account for the high firing variability observed *in vivo*, as reported in Ref. [4].

In FS cells, inhibition similarly caused a number of spikes to drop out and the firing rate decreased as the proportion of inhibition increased, at high and low levels of *S*. In the experiment, the initial firing rate in control was selected at 15-30 spikes/s. Increasing the ratio of inhibition reduced firing frequency in an approximately linear fashion for both levels of *S*. Increased inhibition caused a decrease of CV(ISI) at high synchrony, but an increase of CV(ISI) at low synchrony. For FS interneurons, in contrast with RS cells, reliability showed a clear maximum as a function of the proportion of inhibition, as illustrated in both lower and higher synchrony, while spike timing jitter simultaneously went through a clear minimum. Hence, this evidence suggests that for FS neurons, there is an optimum level of shunting inhibition for achieving the most precise spike time encoding. This result was observed over a range of different degree of synchrony, or correlation.

Because firing of cortical neurons is often correlated with population activity in the network, we examined how clustering of conductance transients in time affects the measures of firing variability and the temporal fine structure of responses. In RS cells, at a specific level of inhibition (e.g., λ_I / λ_E=0.6) increasing synchrony caused decreased firing rate in an approximately linear fashion. As expected, synchrony induced increasing CV(ISI). Both reliability and jitter, increased with increasing the level of synchrony. For FS cells, in contrast, at a specific level of inhibition increasing synchrony caused increasing firing rate in an approximately linear fashion. Similarly to RS cells, synchrony caused increasing CV(ISI). Reliability increased with synchrony, while spike jitter did not change drastically at the parameter range.

The characteristic difference of responses between RS and FS cells can be expected to emerge when they are stimulated by inputs close to their spike thresholds, owing to the difference of their threshold dynamics. To understand the behavior of two cell types, we used a 1-s decaying synaptic conductance input composed of synchronous transient unitary events [4]. Raster plots we obtained in the experiment show that spike times become markedly less precise with time during the response, owing to intrinsic noise in the spike-generating mechanism of the neuron. Several characteristic stages can be seen in the evolution of the spike time jitter. For mean spike times below about 200 ms, jitter is very low. It then shows a rapid stage of increase until the end of each stimulus. However, the final stage was different between RS and FS cells. For the last several events, in FS cells, jitter declined because the number of spikes fell and action potentials to a weak stimulus close to the spike threshold dropped. For RS cells, in contrast, although jitter declined as well, it still remained above a certain level (e.g., jitter > 1.5 ms) in the final stage.

4. Discussion

Variability of spike-time responses comes from two factors: the variability of the presynaptic input and the unreliability of the postsynaptic neuron. The latter is clearly separated by measuring ensembles of responses to identical input. In *in vitro* preparations, stimuli with fast fluctuations resembling synaptic activity *in vivo* are well-known to produce more reliable responses than steady input. To compare with previous work [4], we used several statistical measures of spike variability averaged over hundreds of spikes arising from different trajectories of membrane potentials in different stimulation conditions.

In this study, both RS and FS cells showed much less variability of spiking in response to steady Poisson AMPA excitation than *in vivo*, where CV(ISI) exceeds 1. Adding $GABA_A$-type Poisson conductance in the excitation caused increase of CV(ISI) slightly in both cell types, but the values were still below 1. In addition, it has recently become clear that a cluster of spikes or a burst in vivo is driven by concerted activity in multiple nearby cells. Near-coincident activation of multiple synaptic inputs in a network induces synchronous firing owing to leaky-integration behavior in neurons and their network. Therefore, we introduced correlation in time or synchrony into excitatory and inhibitory conductance stimuli, using a nonstationary Poisson process as well as the previous study [4] to compare with the result. Spike-time reliability increased in both RS and FS cells with increasing synchrony. This result indicates that variability is partly controlled by synchrony in RS and FS cells. Thus, spike clusters or bursts are more reliable codes in both cell types than are single spikes.

In cortical neurons neither independent Poisson trains of IPSCs nor GABA-type conductance input was able to reproduce *in vivo* variability, although CV(ISI) increased to some extent. Our results indicate that in both RS and FS cells increasing the level of inhibition caused opposite effects on CV(ISI), depending on the level of synchrony; increasing the ratio of inhibition to excitation increased CV(ISI) at low synchrony, but decreased it at high synchrony. Furthermore, for FS neurons, there appears to be an optimum level of shunting inhibition for achieving the most precise spike time encoding. This was observed over a range of different degrees of synchrony, or correlation, in the excitatory and inhibitory input.

Because pyramidal RS cells and FS interneurons have different intrinsic membrane properties and their sensitivity to external inputs, they are likely to play dissimilar functional roles in the cortical network. In the neocortex, pyramidal RS neurons appear to collect widely distributed inputs delivered in different layers and make excitatory synapses to other neurons. Owing to strong adaptation and type-1 threshold dynamics, for weak inputs or the late stage of transient ones coherent firing of RS cells will fail easily. Thus, as rate-code integrators, RS cells could encode different tempos in a continuous fashion and may promote high spike-time variability in information transmission. RS cells might thus be suitable for rate cording and the transmission of transient information rather than temporal coding and maintaining of coherent rhythms. In contrast, inhibitory interneurons of the same type (e.g., FS or low-threshold spiking cells) are strongly interconnected by both electrical and chemical synapses, and are believed to promote synchronous firing as temporal-code resonators. FS interneurons also have type-2 threshold dynamics and little spike-adaptation. Furthermore, because they have subthreshold oscillations, the phase of rhythmic firing among FS cells can be maintained even if input transiently falls below the subthreshold level. Although, because of discontinuous *f-I* relationship, FS cells have a harder onset of spike initiation and are consequently less reliable than RS cells, FS cells have little jitter in the late stage of transient input. Thus, threshold dynamics exert powerful and subtle effects on spike-timing variability.

References

1. Tolhurst DJ, Movshon JA, Dean AF (1983) The statistical reliability of signals in single neurons in cat and monkey visual cortex. Vision Res 23:775-785
2. Destexhe A, Rudolph M, Pare D (2003) The high-conductance state of neocortical neurons in vivo. Nature Rev Neurosci 4:739-751
3. Mainen ZF, Sejnowski TJ (1995) Reliability of spike timing in neocortical neurons. Science 268:1503-1506
4. Harsch A, Robinson HPC (2000) Postsynaptic variability of firing in rat cortical neurons: the roles of input synchronization and synaptic NMDA receptor conductance. J Neurosci 20:6181-6192

5. Softky WR, Koch C (1993) The highly irregular firing of cortical cells is inconsistent with temporal integration of random EPSPs. J Neurosci 13:334-350
6. Shadlen MN, Newsome WT (1998) The variable discharge of cortical neurons: implications for connectivity, computation, and information coding. J Neurosci 15:3870-3896
7. Hodgkin AL (1948) The local electric changes associated with repetitive action in a non-medullated axon. J Physiol 107:165-181
8. Kawaguchi Y (1995) Physiological subgroups of nonpyramidal cells with specific morphological characteristics in layer II/III of rat frontal cortex. J Neurosci 15:2638-2655
9. Tateno T, Harsch A, Robinson HPC (2004) Threshold firing frequency-current relationships of neurons in rat somatosensory cortex: type 1 and type 2 dynamics. J Neurophysiol 92:2283-2299
10. Robinson HPC, Kawai N (1993) Injection of digitally synthesized synaptic conductance transients to measure the integrative properties of neurons. J Neurosci Methods 49:157-16

Can Newtonian Mechanics Aid in the Development of Brain Science?: A Challenge to Bernstein's Degrees-of-Freedom Problem

Suguru Arimoto[1,2,3] and Masahiro Sekimoto[1]

[1] Department of Robotics, Ritsumeikan University
[2] The 21st Century COE Program, Ritsumeikan University
[3] BMC Research Center, RIKEN

Chapter Overview. A natural control scheme for resolving the ill-posedness of inverse kinematics of multi-joint reaching movements under redundancy of degrees-of-freedom (DOF) is proposed, which need neither introduce any artificial performance index to determine the inverse kinematics uniquely nor calculate the pseudo-inverse of the Jacobian of task coordinates with respect to joint coordinates. The control signal is composed of a linear superposition of three terms 1) damping shaping, 2) task-space position-error feedback with a single stiffness parameter, and 3) gravity compensation by means of estimation of uncertainty in the potential energy. It is shown through computer simulation and experiments that the signal can generate smooth reaching motions provided that a set of damping factors and the stiffness parameter are selected in a synergistic way. A physical meaning of each term of the control signal is discussed from the viewpoints of Newtonian mechanics and neuro-physiology. It is claimed that generation of such neuro-motor signals can be interpreted naturally by hypothesizing existence of a virtual spring and damper setup in parallel that would draw the arm endpoint. It can be concluded from those illustrative examples that human can acquire skilled motions through repeated trials of movements toward achieving the principle of least action associated with each corresponding imposed task.

Key Words. Bernstein's problem, multi-joint reaching, redundant robots, inverse kinematics, ill-posedness of inverse

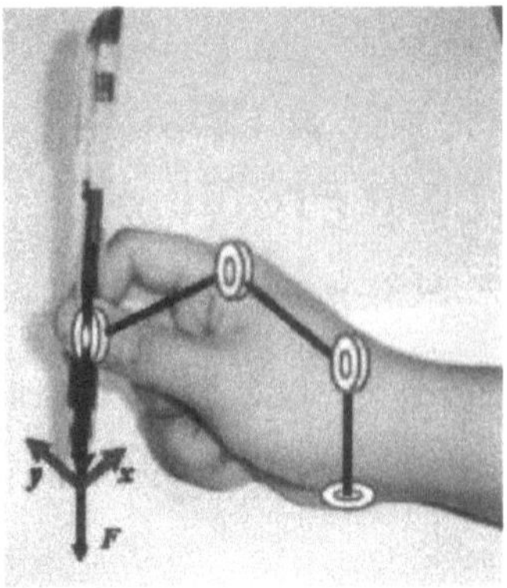

Fig. 1. A handwriting robot with surplus DOFs.

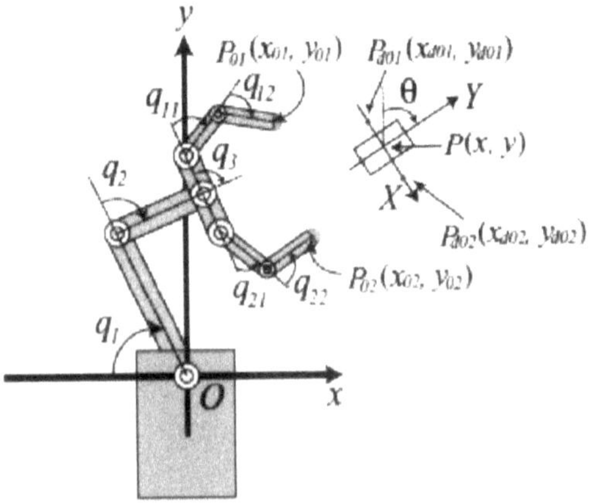

Fig. 2. Two dimensional (2-D) reaching by a whole arm model when double endpoint targets are specified.

Introduction

This paper aims at challenging Bernstein's Degrees-of-Freedom problem posed very early on (more than a half century ago) but not yet solved from the viewpoints of robotics, kinesiology, and neuro-physiology. As illustrated by examples as shown in Fig. 1 and Fig. 2, the problem is how to find control signals that are exerted at movable joints and lead to movement of the endpoint of the whole arm toward a given target point specified in xy -plane. In the case of a handwriting robot of Fig. 1, the objective task $\vec{x}_d$ is given in the task space $\vec{x} = (x, y)^{\mathrm{T}}$ and the joint coordinates $q = (q_1, \cdots, q_4)^{\mathrm{T}}$ are of four dimension. Therefore, there arises an infinite number of inverses q_d that realize $\vec{x}(q_d) = \vec{x}_d$ and thereby the problem of obtaining inverse kinematics from the task description space $X(=\Re^2)$ to the 4-dimensional joint space $\Re^4$ becomes ill-posed. Similarly to this case, a planar arm-hand robot with seven joints is also redundant in DOF if the objective task is to move its two endpoints P_{01} and P_{02} approach the specified points respectively and shape the hand to grasp a given object. In this case the number of physical variables for description of the task is four which is by three less than 7 DOFs. Hence, the problem of inverse kinematics from $\vec{x}_{d1}$ and $\vec{x}_{d2}$ to $q = (q_1, \cdots, q_7)^{\mathrm{T}} \in \Re^7$ becomes ill-posed.

A variety of ideas for solving such ill-posedness of inverse kinematics for redundant systems with excess DOFs has been proposed in the vast literature [1,2,3,4,5,6,7,8,9,10,11,12,13,14,15] not only in the area of robotics but also in physiology. According to the literature in robotics [2,3,4,5,6,7], if an endpoint trajectory $\vec{x}_d(t)$ is given by determining its velocity profile $\dot{\vec{x}}_d(t)$ then the velocity vector of joints can be calculated by the form

$$\dot{q}_d(t) = J^+(q(t))\dot{x}_d(t) + (I - J^+(q)J(q))\vec{v} \tag{1}$$

where $\vec{v}$ should be computed so as to optimize a certain performance index related to joint position variables (for example, manipulability index [4], obstacle avoidance [3], etc.). In (1), $J(q)$ stands for the Jacobian matrix of $\vec{x}$ in q, that is, $(\partial\vec{x}/\partial q^{\mathrm{T}})$, and denotes $J^+(q)$ the pseudo-inverse of $J(q)$. The original idea of use of the pseudo-inverse $J^+(q)$ is due to Whitney [2]. Once a desired joint velocity $\dot{q}_d(t)$ is planned in such a way, it is claimed that the computed torque method can be applied for determining the control input that must generate the whole motion of the robot. This is called "inverse kinematics approach." Another idea of direct generation of a control signal that is called "inverse dynamics approach" is based upon a form of control input

$$u = H(q)J_H^+(q)\{\ddot{\vec{x}}_d - \dot{J}_H(q)\dot{q}\} + h(q,\dot{q}) + (I - J^+(q)J(q))\vec{v} \tag{2}$$

where $\vec{v}$ is computed so as to optimize a certain performance index related to velocity or acceleration variables (for example, kinetic energy [7], torque [6], energy dissipation, etc.). In (2), $H(q)$ denotes the inertia matrix of the robot arm and $h(q,\dot{q})$ means compensation for the remaining nonlinear function including centrifugal and Coriolis forces and the gravity effect. In the physiological literatures [8,9,10,11,12,13,14,15], main concerns are focused on the question why human skilled multi-joint reaching movements exhibit typical characteristics that 1) endpoint trajectory becomes a quasi-straight line and less variable, 2) velocity profiles of the endpoint velocity becomes bell-shaped, though 3) joint trajectories are rather variable trials-by-trials [10,11]. Then, a variety of cost functions for derivation of such properties of point-to-point reaching movements has been proposed, among which a quadratic function of endpoint jerk (rate of acceleration) was the first [8] and successively a cost function based on joint torques was introduced [9] for planning not only an endpoint trajectory but also joint trajectories. However, in the physiological literature, there is a dearth of papers that attempted to directly deal with reaching movements with redundant joints, though the importance of Bernstein's DOF problem [1] has been widely known among physiologists.

Differently from the traditional approaches, a simpler control method for multi-joint reaching movements was proposed very recently by the authors [16,17,18,19] and shown to be effective even in the case of human or robotic arms with redundant DOFs. In those papers, only planar motions confined to a horizontal plane are treated and therefore the control signal is free from gravity with a simple form:

$$u = -C\dot{q} - J^{\mathrm{T}}(q)k\Delta\vec{x} \tag{3}$$

Notwithstanding this simpler form than (2), once damping factors $C = \mathrm{diag}(c_1,\cdots,c_n)$ and single stiffness parameter k chosen carefully, it generates smooth reaching movements by realizing a quasi-straight endpoint trajectory, bell-shaped velocity profiles, and double-peaked acceleration profiles typically seen in case of human skilled multi-joint movements [10,11]. However, all these treatments are restricted to planer motions as well as most of the previous literature on multi-joint reaching movements.

This paper extends our previous approach [16,17,18,19] to the following three types of human-like multi-joint reaching movements under redundancy in joint DOFs: 1) 2-D reaching with double endpoint targets (see Fig. 2), 2) 3-D Point-to-Point reaching under geometric constraint (a handwriting robot as an illustrative example, see Fig. 1), and 3) 3-D reaching under the effect of gravity. In all cases it is shown that dexterous reaching motions can be realized by using a simple control signal similar to (3) together with compensation for the gravity effect, provided that damping factors together with a single stiffness parameter k are carefully chosen so that viscous-like forces at joints work synergistically through exertion of spring-like forces evoked from the total potential energy. In conclusion, we claim that such a swift and smooth voluntary movement acquired through learning based on repeated practices is going to be compatible with the natural "principle of least action." Instead of the traditional EP-hypothesis (Equilibrium-Point hypothesis [12,14] or End-Point hypothesis [13]), we introduce "Virtual-Spring hypothesis," which explains more directly and explicitly generation of synergistic spring-like forces. It is also discussed in light of recent noteworthy research results on acto-myosin activities in relation to muscle contraction [20,21,22] that the term $-kJ^{\mathrm{T}}(q)\Delta\vec{x}$ in the control signal of (3) can be generated in a feedforward manner as observed first in action of the vestibulo-ocular reflex arc through the extra-ocular muscles to determine the posture and movement of the eyes [23,24]. It is also claimed that control signals for gravity compensation based on estimates of uncertain parameters in the potential energy must be adjusted through the spinal-reflex loop regulated from the cerebellum. Finally, an important role of coactivation of agonistic and antagonistic muscles in generation of large damping forces at a final stage of convergence to the target position is discussed from the Newtonian mechanics point of view. It is claimed that damping factors for shoulder and elbow joints can be reduced more considerably to the level of passive viscosity of muscles if an additional virtual damper is incorporated in parallel with the virtual spring. This leads reasonably to the Virtual Spring-Damper hypothesis as a more natural metaphor that can lead to emergence of movements of skilled multi-joint reaching.

Virtual-Spring Hypothesis

First we define the virtual-spring hypothesis for explaining a physical meaning of the term $-kJ^{\mathrm{T}}(q)\Delta\vec{x}$ in (3). In order to gain an in-depth insight into its principal role, we treat a simple 2-D reaching movement with excess DOFs as shown in Fig. 3, which has been extensively analyzed in our previous papers [16,17].

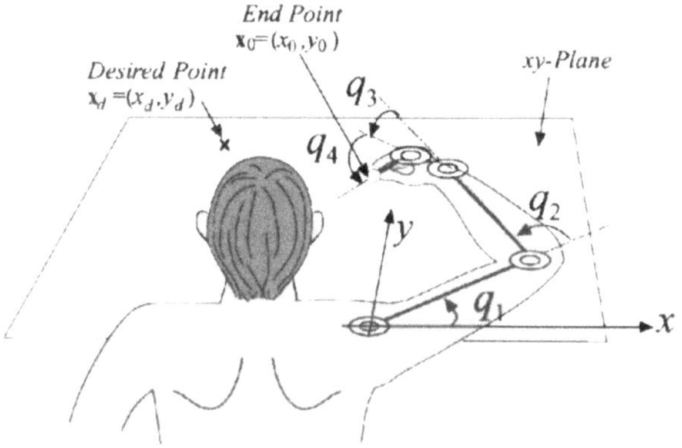

Fig. 3. "Reaching" by means of a surplus DOF system of hand-arm dynamics.

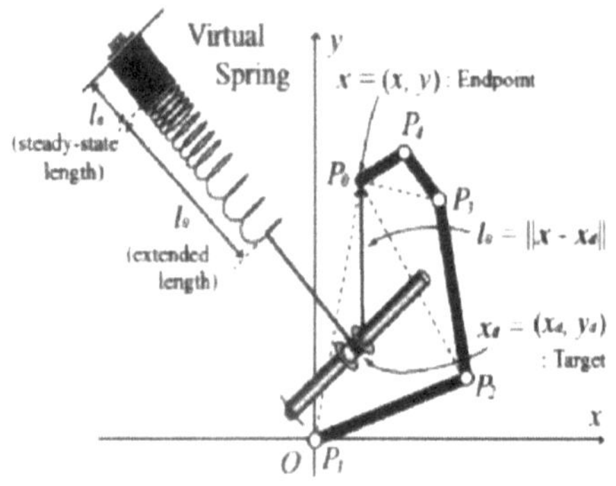

Fig. 4. Virtual-spring hypothesis.

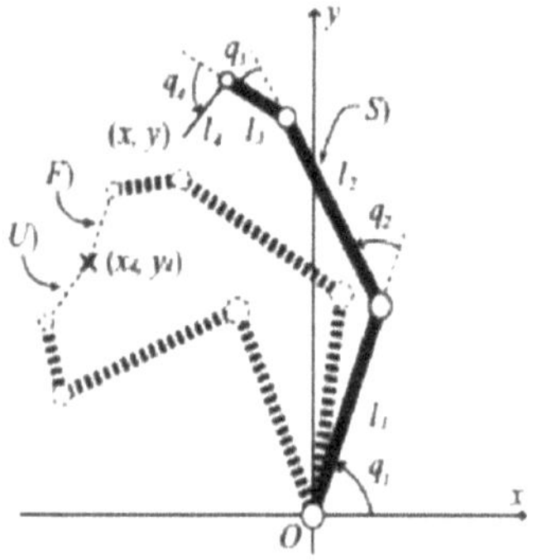

Fig. 5. S) Starting posture, F) Final posture, U) Unreasonable posture.

In the case of control of a robotic arm, the term $-kJ^{\mathrm{T}}(q)\Delta\vec{x}$ in (3) is called a task-space feedback based on measurements of the endpoint position $\vec{x}(t)$ by visual sensing. In the case of human reaching, however, exerted torques $-kJ^{\mathrm{T}}(q)\Delta\vec{x}$ must be generated from a group of muscles which are endowed with a total potential energy generated by excitation of neuro-motor signals from the CNS (central nervous system). This fact can be well interpreted by hypothesizing a virtual spring as shown in Fig. 4, drawing the end of the whole arm to the target with the force equivalent to the vector $k(\vec{x}_d - \vec{x}(t))$, which is equivalent to $-k\Delta\vec{x}$. In this hypothesized model, the i-th joint is exerted by the torque $-kJ_i^{\mathrm{T}}\Delta\vec{x}$ and also subject to the damping force $-c_i\dot{q}_i$, where

$$J(q) = (J_1(q), \cdots, J_4(q)), \quad J_i(q) = \frac{\partial \vec{x}}{\partial q_i} \tag{4}$$

It should be remarked that the magnitude and signature of each spring-like moment of force at each joint are not directly related to the magnitude of $\Delta\vec{x}$ but to the quantity of k times the inner product of two-dimensional vectors $J_i(q)$ and $\Delta\vec{x}$. Note that the magnitude of $-kJ_i^{\mathrm{T}}\Delta\vec{x}$ is equivalent to that of moment of force $\overrightarrow{P_iP_0} \times k\Delta\vec{x}$, where symbol "×" denotes the external product and $\overrightarrow{P_iP_0}$ denotes the position vector from the i-th joint to the endpoint (see Fig. 4).

Table 1. Initial conditions and the target for short-range reaching

Arm posture (deg)	: $q_1(0)=83.00$, $q_2(0)=45.00$, $q_3(0)=40.00$, $q_4(0)=74.00$
Endpoint position (m)	: $x(0) = -0.2734$, $y(0) = 0.4788$
Target position (m)	: $x_d = -0.3500$, $y_d = 0.3500$
Distance (m)	: $\|\Delta x(0)\| = 0.1173$

Table 2. Lengths of upper arm (l_1), forearm (l_2), palm (l_3), and index finger (l_4) together with corresponding masses (m_i: $i = 1,\cdots,4$) and inertia moments (I_{gi}; $i = 1,\cdots,4$). The data are taken from an average male adult

Regions	Upper arm ($i = 1$)	Forearm ($i = 2$)	Palm ($i = 3$)	Index finger ($i = 4$)
l_i : Length (m)	0.2800	0.2800	0.09500	0.09000
l_{gi} : Center of mass (m)	0.1400	0.1400	0.04750	0.04500
r_i : Cylinder radius (m)	0.04000	0.03500	N/A	0.009500
h_i : Cuboid height (m)	N/A	N/A	0.08500	N/A
d_i : Cuboid depth (m)	N/A	N/A	0.03000	N/A
m_i : Mass (kg)	1.407	1.078	0.2423	0.02552
I_{gi} : Inertia moment (kgm^2)	9.758×10^{-3}	7.370×10^{-3}	2.004×10^{-4}	1.780×10^{-5}

Actually, when damping factors are chosen in such a way that

$$c_1 = 1.89, c_2 = 1.21, c_3 = 0.29, c_4 = 0.070\,(\mathrm{Nms}) \tag{5}$$

then the endpoint trajectories for various stiffness parameter k starting from the posture S) shown in Fig. 5 together with Table 1 become as shown in Fig. 6. In this computer simulation, physical parameters of a human adult shown in Table 2 are used. In Figs. 7–9 we show transient responses of $x(t), y(t), q_2(t)$, and $q_4(t)$ together with velocity profiles $\dot{x}(t)$ and $\dot{y}(t)$. As seen from Fig. 5 the torque $-kJ_1^{\mathrm{T}}\Delta\vec{x}$ exerting to the shoulder joint is plus at $t=0$ and therefore $q_1(t)$ is increasing with t as well as $q_2(t)$ seen in Fig. 8(a), but the torque $-kJ_4^{\mathrm{T}}\Delta\vec{x}$ exerting to the finger joint is minus at $t=0$ and therefore $q_4(t)$ is decreasing with t as seen in Fig. 8(b).

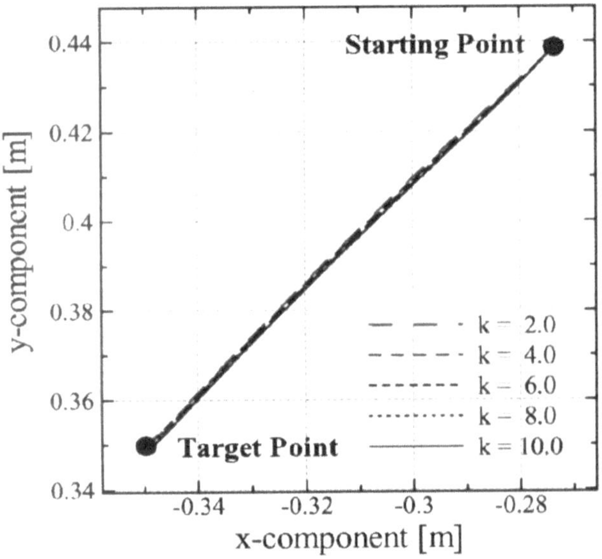

Fig. 6. Endpoint trajectories of reaching movements when damping factors of (5) is used.

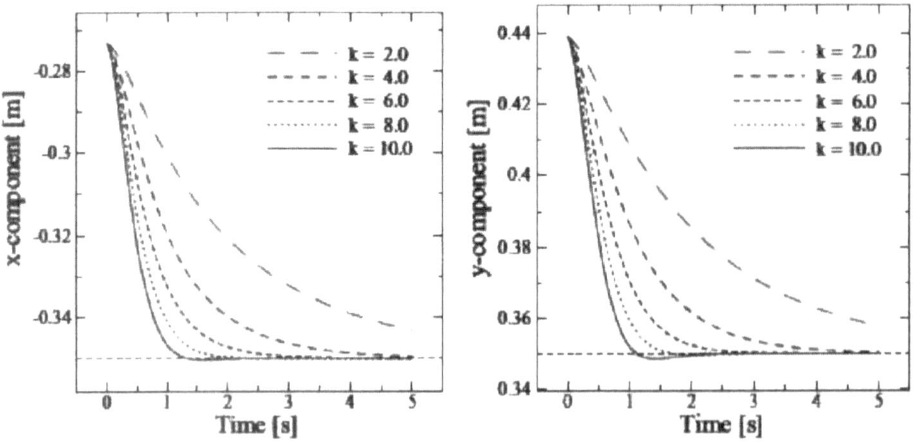

(a) $x(t)$ (b) $y(t)$

Fig. 7. Transient responses of x and y when damping factors of (5) are chosen.

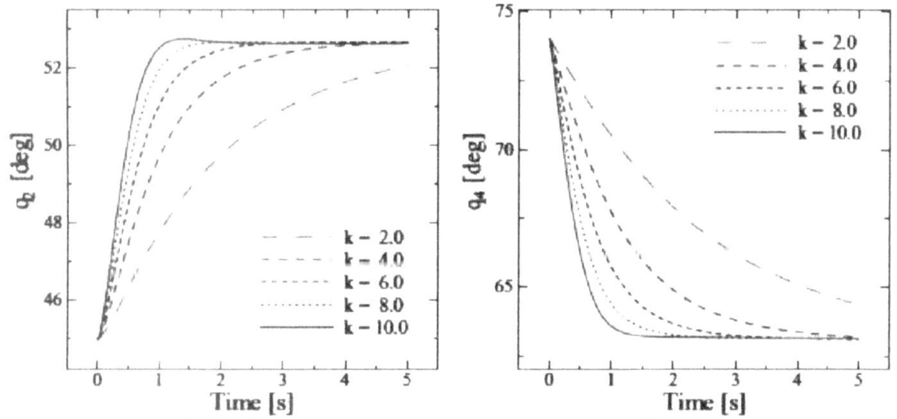

(a) $q_2(t)$ (b) $q_4(t)$

Fig. 8. Transient responses of joint angles q_2 and q_4 when damping factors of (5) are chosen.

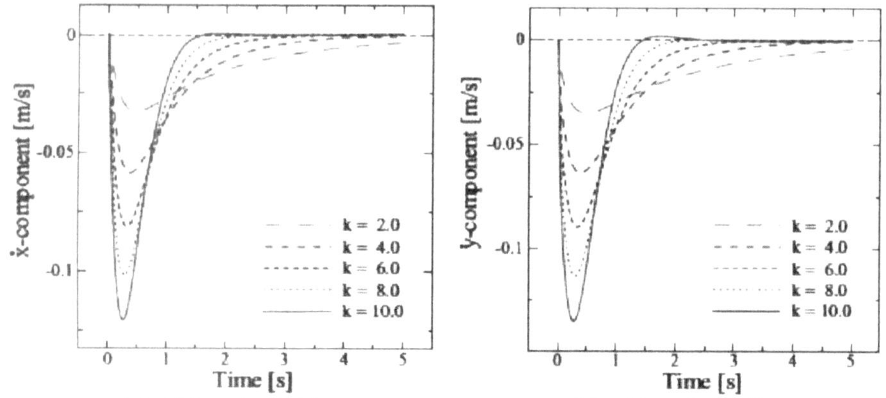

(a) $\dot{x}(t) = dx(t)/dt$ (b) $\dot{y}(t) = dy(t)/dt$

Fig. 9. Transient responses of velocities $\dot{x}$ and $\dot{y}$ when damping factors of (5) are chosen.

Basis on this virtual-spring hypothesis, we claim that a skilled reaching movement must emerge from synergistic generations of potential energies in a group of muscles involved in these four joints, whose total is equivalent to $(1/2)k\,\|\Delta\vec{x}(0)\|^2$, and synergistic formation of damping factors generating viscous-like forces in each subgroup of muscles corresponding to each joint. Variability of joint motions is mainly due to permissible fluctuations of damping effects, which are rather insensitive to the endpoint trajectory. In fact, it becomes a quasi-straight line as seen in Fig. 6 and less variable for changes of stiffness k as first observed in human multi-joint movements [10] and claimed in [11]. Each joint movement is fairly dependent on the posture being subject to exertion of each corresponding torque $-kJ_i^{\mathrm{T}}(q)\Delta\vec{x}$.

2-D Reaching with Double Endpoint Targets

This virtual-spring hypothesis can be also applied for movement of a preshaping form of 2-D whole arm model depicted in Fig. 2 which consists of 7 joints. 2 for the arm (shoulder and elbow), 1 for the wrist, and 2 and 2 for a pair of fingers. The task can be described by a pair of two given points $\vec{x}_{d1}$ and $\vec{x}_{d2}$ so that two finger endpoints $\vec{x}_i = (x_i, y_i)$, $i = 1, 2$, approach and converge asymptotically the desired targets $\vec{x}_{di}$ respectively. In this case, since there are 4 physical position variables that specify the objective task, 3 DOFs of the whole arm are redundant in comparison with the whole 7 DOFs. If we suppose two virtual springs that draw corresponding endpoints respectively as shown in Fig. 10, then the control signal is given in the form:

$$u = -C\dot{q} - kJ_1^{\mathrm{T}}(q)\Delta\vec{x}_1 - kJ_2^{\mathrm{T}}(q)\Delta\vec{x}_2 \tag{6}$$

where $J_i(q) = \partial\vec{x}_i/\partial q$ and $\Delta\vec{x}_i = \vec{x}_i - \vec{x}_{di}$ for $i = 1, 2$. Computer simulation results are shown in Figs. 11–12 when the object is placed considerably away from the initial position of the hand for Fig. 11 and relatively close to that for Fig. 12. In the former case (Fig. 11), the dotted lines show the final preshaping form for grasping when improved damping factors $c_1 = 0.15$ and $c_2 = 0.20$ for the shoulder joint ($i = 1$) and elbow joint ($i = 2$) respectively are chosen in comparison with $c_1 = 0.22$ and $c_2 = 0.10$ for the final posture with continuous lines. From com-

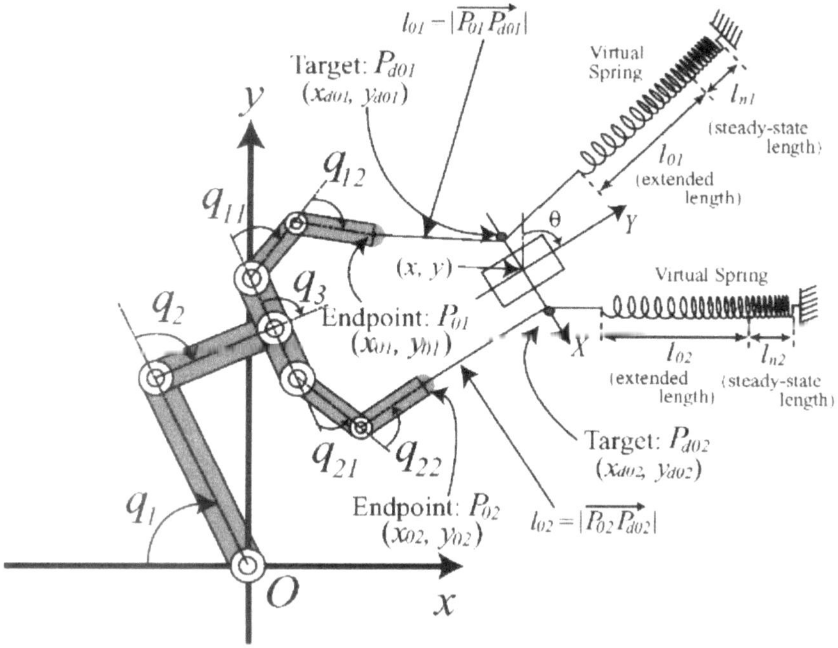

Fig. 10. Two virtual springs drawing corresponding endpoints toward targets P_{d0i} (i = 1, 2) respectively.

parison of these figures, in the former case (see Fig. 11) the damping for shoulder should be smaller to generate a large motion but in the latter case (see Fig. 12) the damping for elbow should be smaller, relatively to the standard damping factors $c_1 = 0.22$ and $c_2 = 0.10$ which are determined to be approximately equal to $\sqrt{H_{11}(q)}$ and $\sqrt{H_{22}(q)}$ where $H_{ii}(q)$ stands for the i-th diagonal entry of the inertia matrix $H(q)$ of the whole arm. In both cases, other damping coefficients are chosen and fixed in such a way that $c_i = 0.005$ for $i = 3,\cdots,7$ and k_i is set as $k_1 = k_2 = 8.0$.

Reaching under Geometric Constraint

Robot dynamics such as a handwriting robot as shown in Fig. 13 can be described by the Lagrange equation of motion with a holonomic constraint:

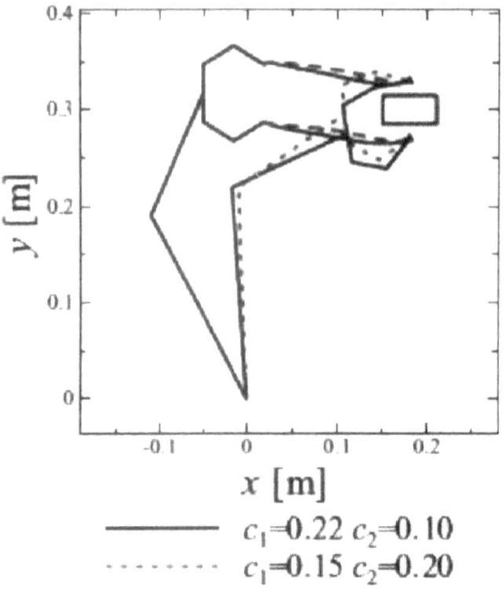

Fig. 11. Preshaping motion for grasping when the target object is far away from the initial position.

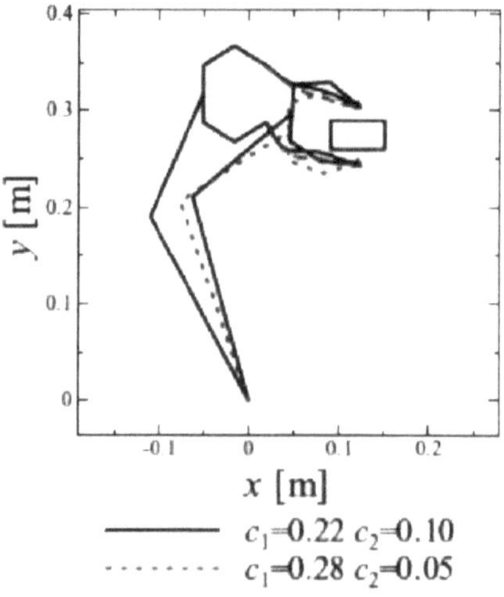

Fig. 12. Preshaping motion for grasping when the object is nearby from the initial position.

$$H(q)\ddot{q}+\left\{\frac{1}{2}\dot{H}(q)+S(q,\dot{q})\right\}\dot{q}+g(q)-\left(\frac{\partial\phi}{\partial q}\right)^{\mathrm{T}}\lambda=u \tag{7}$$

where $g(q)$ stands for the gravity term such that $g(q)=\partial P(q)/\partial q$ for the potential energy $P(q)$, and $\phi(q)$ is given by

$$\phi(q)=z(q)=l_1+l_2\mathrm{c}_2+l_3\mathrm{c}_{23}+l_4\mathrm{c}_{234} \tag{8}$$

and hence

$$\phi(q)=0 \tag{9}$$

expresses the holonomic constraint such that the nib of the pen (the last link) is constrained on the horizontal plane $z(q)(=\phi(q))=0$ (see Fig. 13). Hence, λ denotes a Lagrange multiplier corresponding to the holonomic constraint of (9) and therefore the term $(\partial\phi/\partial q)^{\mathrm{T}}\lambda$ expresses the torque vector composed of rotational moments at joints induced by the constraint force with the magnitude λ and the direction normal to the xy-plane.

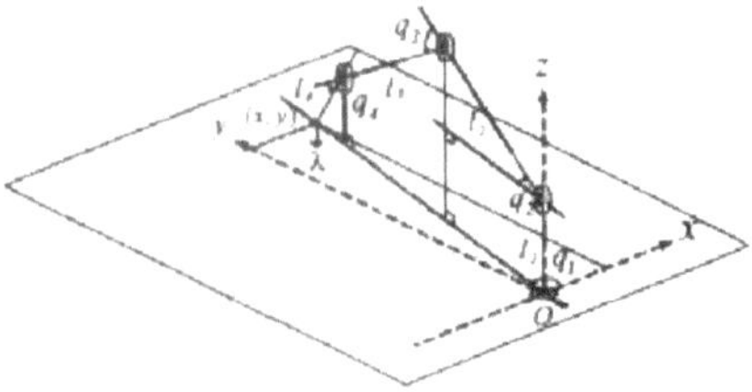

Fig. 13. A handwriting robot with four joints (four degrees of freedom).

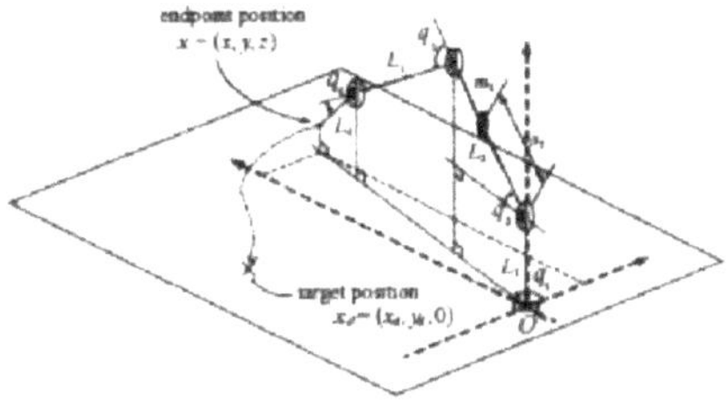

Fig. 14. 3D-reaching under the effect of gravity force.

Consider now a control scheme for this system such that, given a desired position $\vec{x}_d = (x_d, y_d)^{\mathrm{T}}$ and a desired contact force λ_d, the nib of the pen should move toward $\vec{x}_d$ in touch with the plane and at the same time the contact force λ converges to λ_d. At first it is assumed that the gravity term $g(q)$ is known or can be computed in real-time based on measurement data by optical encoders of joint actuators. Under this assumption, the following control signal is analyzed in details in our recent paper [17]:

$$u = g(q) - C\dot{q} - J_{\vec{x}}^{\mathrm{T}} k \Delta\vec{x} - \left(\frac{\partial \phi}{\partial q}\right)^{\mathrm{T}} \lambda_d \tag{10}$$

where $\vec{x} = (x, y)^{\mathrm{T}}$, $\Delta\vec{x} = \vec{x} - \vec{x}_d$, and $J_{\vec{x}} = \partial\vec{x}(q)/\partial q$ denotes the 2×4 Jacobian matrix of $\vec{x}$ in q. According to computer simulation and experimental results, this control signal works well even in the case of trajectory tracking when the desired target $\vec{x}_d$ is given as a time-varying signal $\vec{x}_d(t)$. The details, however, must be omitted in this paper due to shortage of given pages.

3-D Reaching under the Effect of Gravity

The arguments developed in Sects. 2–4 can be extended to cope with three-dimensional reaching movements under the effect of gravity force. Consider the handwriting robot with a pen whose endpoint is far above the xy-plane (see Fig. 14). The problem is to control the robot to move the nib of the pen to reach a given desired point $\vec{x}_d = (x_d, y_d, z_d = 0)^{\mathrm{T}}$ without attaching the xy-plane during its motion. In this case the motion equation is expressed as

$$H(q)\ddot{q} + \left\{\frac{1}{2}\dot{H}(q) + S(q,\dot{q})\right\}\dot{q} + g(q) = u \tag{11}$$

In this case, a control signal composed of (3) with an additional estimator $\hat{g}(q)$ is proposed in our recent separate paper [25], which is defined as

$$u = -C\dot{q} - kJ^{\mathrm{T}}(q)\Delta\vec{x} + Y(q)\hat{\Theta}(t) \tag{12}$$

where the last term corresponds to $\hat{g}(q)$ and

$$Y(q) = g\begin{bmatrix} 0 & 0 & 0 \\ c_2 & c_{23} & c_{234} \\ 0 & c_{23} & c_{234} \\ 0 & 0 & c_{234} \end{bmatrix} \tag{13}$$

$$\Theta = (m_2 s_2 + m_3 l_2 + m_4 l_2, m_3 s_3 + m_4 l_3, m_4 s_4)^{\mathrm{T}} \tag{14}$$

$$\hat{\Theta}(t) = \hat{\Theta}(0) - \int_0^t \eta^{-1} Y^{\mathrm{T}}(q(\tau)) s(\tau) \mathrm{d}\tau \tag{15}$$

$$s(t) = \dot{q}(t) + C^{-1} k J^{\mathrm{T}}(q(t)) \Delta\vec{x}(t) \tag{16}$$

where η denotes a positive constant. Substituting (12) into (11) yields

$$H\ddot{q} + \left(\frac{1}{2}\dot{H} + S\right)\dot{q} + C\dot{q} + kJ^{\mathrm{T}}\Delta\vec{x} - Y\Delta\Theta = 0 \tag{17}$$

where $\Delta\Theta = \hat{\Theta} - \Theta$.

The details on theoretical analysis of the closed-loop equation (16) together with computer simulation results will be presented in our separate paper [25]. In this paper we mention that there exists a non-negative definite scalar function V with positive parameters η, β, γ such that

$$\frac{\mathrm{d}}{\mathrm{d}t} V \leq -\gamma V \tag{18}$$

where V can be defined as

$$V = \frac{1}{2}\dot{q}^{\mathrm{T}}H(q)\dot{q} + k\|\Delta\vec{x}\|^2 + \frac{\eta(1+\beta)}{2}\|\Delta\Theta\|^2 + \dot{q}^{\mathrm{T}}H(q)C^{-1}\{kJ^{\mathrm{T}}\Delta\vec{x} - \beta Y\Delta\Theta\} \quad (19)$$

Convergence of V to zero as $t \to \infty$ implies that $\Delta\vec{x} \to 0$ and $\Delta\Theta \to 0$ as $t \to \infty$.

Discussions from the Physiological Viewpoint

From the physiological viewpoint, some types of human voluntary movements must be controlled in a feedforward manner as first observed in action of the vestibulo-ocular reflex arc through the extra-ocular muscles to determine the posture and movement of the eyes relative to the external world [23,24]. As already discussed in the previous paper [19], each component of the task-space feedback term $-kJ^{\mathrm{T}}(q)\Delta\vec{x}$ in the signal u of (3) can be regarded as a total of spring-like forces generated by muscle-contractions associated with each joint, which corresponds to each total amount of potential energies produced by a population of acto-myosin activities in sarcomere. A little more details of the discussions have been presented in a recent paper [19] on the basis of "Virtual-Spring" hypothesis as an enhancement of the well-known "Equilibrium-Point" hypothesis. Another important term denoted by $Y(q)\hat{\Theta}$ in u of (12) did not appear in the case of multi-joint movements in the horizontal plane. This term plays a crucial role in readjustment of joint torques to withstand changeable loads when our hand grasps a thing. Apparently this term $Y(q)\hat{\Theta}$ should quickly reflect the change of total gravity effect and therefore be constructed quickly through the spinal-reflex loop that may be also regulated from the cerebellar motor control system. As to the role of viscous-like torques $-c_i\dot{q}_i$ in synergistic convergence of the endpoint trajectory to the target there is a dearth of papers even in physiological journals. However, the magnitudes c_1 and c_2 in (5) must be too big to be compared with actual viscosity of contractile muscles. Therefore, we postulate that time-delayed coactivation of agonistic and antagonistic muscles may lead to generation of relatively large dampings. Nevertheless, the magnitudes c_1 and c_2 in (5) corresponding to passive damping factors for shoulder and elbow joints are too big to be compared with actual viscosity of contractile muscles [26,27].

To reduce the damping coefficients in control signals of (3), let us consider another form of control signal

$$u = -\zeta_0 C\dot{q} - J^{\mathrm{T}}(q)(\zeta\sqrt{k}\dot{\vec{x}} + k\Delta\vec{x})$$
$$= -\left\{\zeta_0 C + \zeta\sqrt{k}J^{\mathrm{T}}(q)J(q)\right\}\dot{q} - J^{\mathrm{T}}(q)k\Delta\vec{x} \tag{20}$$

where ζ_0 and ζ denote positive constants. Apparently, this control signal plays a role of a parallel pair of mechanical damper and spring that draws the endpoint of the whole arm to the target as shown in Fig. 15.

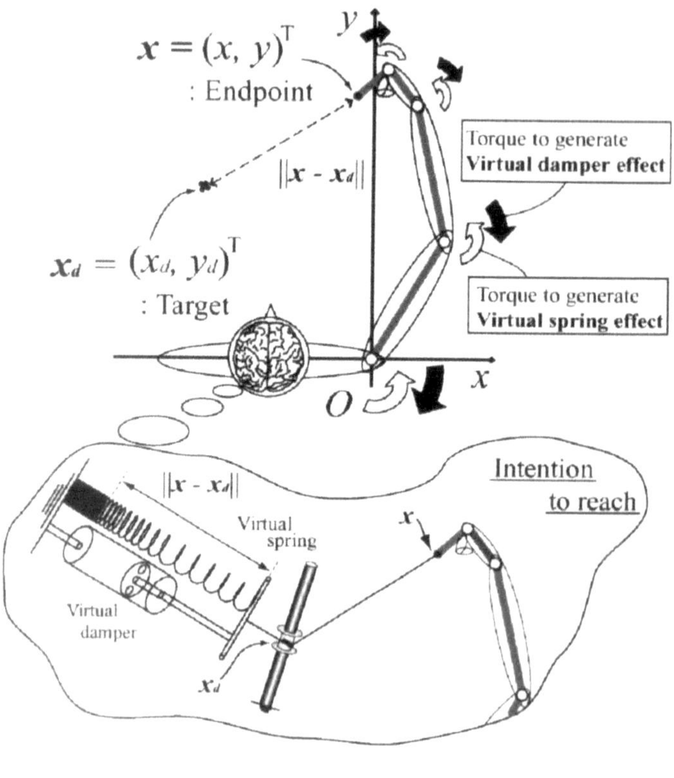

Fig. 15. Virtual spring-damper hypothesis.

Table 3. Initial conditions and the target for middle-range reaching

Arm posture (deg)	: $q_1(0)=59.00$, $q_2(0)=43.00$, $q_3(0)=25.00$, $q_4(0)=95.00$
Endpoint position (m)	: $x(0) = -0.03806$, $y(0) = 0.5295$
Target position (m)	: $x_d = -0.3500$, $y_d = 0.3500$
Distance (m)	: $\|\Delta x(0)\| = 0.3599$

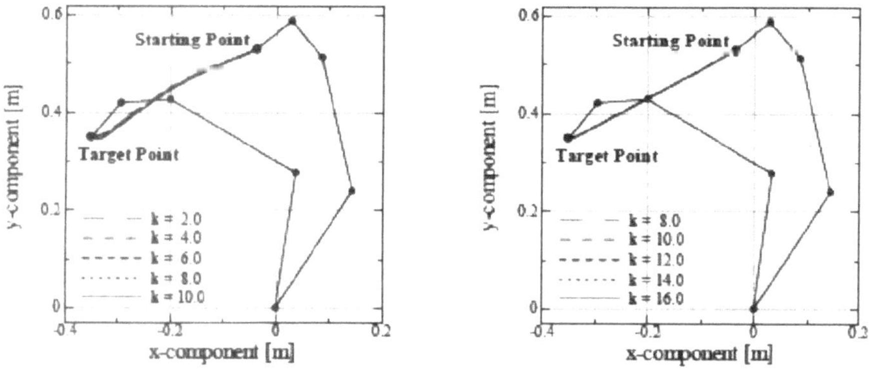

(a) Based upon the virtual spring hypothesis

(b) Based upon the virtual spring-damper hypothesis

Fig. 16. Endpoint trajectories in the case of planar middle-range reaching by the multi-joint arm with 4 DOFs.

To show how the control signal of (20) generates effectively smooth and natural multi-joint movements, we conducted numerical simulation of middle-range reaching by using the same human arm model and initial posture and target position as given in Tables 2 and 3. For comparison of the effectiveness of the control signal of (20) based upon the virtual spring-damper hypothesis with that of (3) based upon the virtual spring hypothesis, we also conducted numerical simulation based upon the virtual-spring hypothesis under the same condition. When damping factors of (5) are set in both control signals and ζ_0 and ζ in (20) are chosen as $\zeta_0 = 0.2$ and $\zeta = 1.5$ respectively, numerical simulation results are shown in Figs. 16–18. Each passive damping factor $\zeta_0 c_i$ in the case of virtual-spring hypothesis can be reduced to the range of 20% of c_i in the case of virtual spring-damper hypothesis. As seen from Fig. 16, endpoint

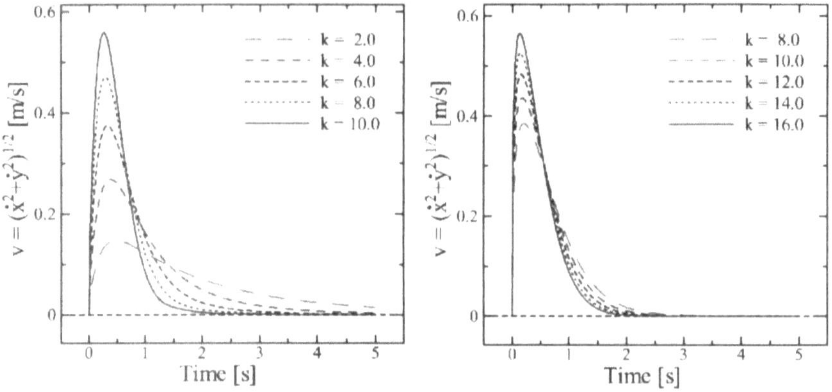

(a) Based upon the virtual spring hypothesis

(b) Based upon the virtual spring-damper hypothesis

Fig. 17. Endpoint velocity profiles in the case of planar middle-range reaching.

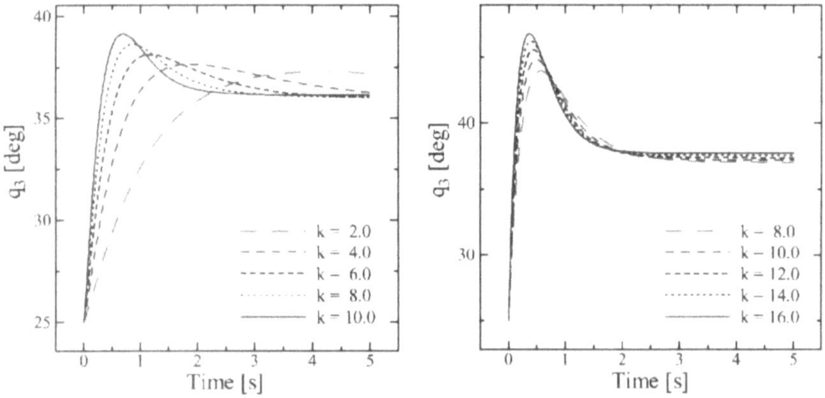

(a) Based upon the virtual spring hypothesis

(b) Based upon the virtual spring-damper hypothesis

Fig. 18. Transient responses of joint angle $q3$ in the case of planar middle-range reaching.

trajectories in the case of virtual-spring hypothesis are almost rectilinear but curve at a final convergent stage in the vicinity of the target point. This implies that the damping factors chosen synergistically in the case of short-range reaching are unfit in the middle-range reaching movements. However, endpoint trajectories in the case of virtual spring-damper hypothesis become rectilinear almost completely from the beginning to the final stage even in middle-range reaching. Moreover, it is interesting to note that profiles of the endpoint trajectory in xy-plane are almost invariant for changing the stiffness k in both cases. On the other hand, the endpoint velocity $v(=\sqrt{\dot{x}^2+\dot{y}^2})$ increases considerably according to the increase of k and profiles of each joint trajectory are rather variable according to change of k even in middle-range reaching movements (see Figs. 17–18). It can be seen from Figs. 17 and 18 that grades of variability of the endpoint velocity and each joint trajectory can be reduced considerably by hypothesizing a virtual damper in parallel with the spring. Thus, from simulation results of Figs. 16–18, it is shown that the control scheme based upon the virtual spring-damper hypothesis can lead to movements of skilled multi-joint reaching. A theoretical reason of the effectiveness of this virtual spring-damper hypothesis must be explored, but it will be presented in another paper because the argument becomes rather mathematically elaborative and more pages enough for the detailed argument are needed.

References

1. Bernstein NA (Latash ML translated from the Russian) (1996) On Dexterity and Its Development. Lawrence Erlbaum Associates, Inc
2. Whitney DE (1969) Resolved motion rate control of manipulators and human prostheses. IEEE Trans on Man-Machine Systems MMS-10:47–53
3. Nakamura Y (1991) Advanced Robotics: Redundancy and Optimization. Addison-Wesley, Reading, MA
4. Yoshikawa T (1984) Analysis and control of robot manipulators with redundancy. In: Brady M, Paul R (eds) Robotics Research : The 1st International Symposium, MIT Press, Cambridge, MA, pp 735–747
5. Lenarcic J (ed) (1991) Special Issue on Redundant Robots. Laboratory Robotics and Automation 6
6. Hollerbach JM, Suh KC (1987) Redundancy resolution of manipulators through torque optimization. IEEE J of Robo and Autom RA-3:308–316
7. Khatib O (1987) A unified approach for motion and force control of robot manipulators: the operational space formulation. IEEE J of Robotics and Automation RA-3:43–53
8. Flash T, Hogan N (1985) The coordination of arm movements: an experimentally confirmed mathematical model. J of Neuroscience 7:1688–1703
9. Uno Y, Kawato M, Suzuki R (1989) Formation and control of optimal trajectory in human multijoint arm movement. Biol Cybern 61:89–101
10. Morasso P (1981) Spatial control of arm movements. Experimental Brain Research 42:223–227
11. Latash ML (1998) Neurophysiological Basis of Movement. Human Kinetics Pub, NY
12. Feldman AG (1966) Functional tuning of the nervous system with control of movement on maintenance of a steady posture: III. Mechanographic analysis of execution by arm of the simplest motor tasks. Biophysics 11:766–775
13. Bizzi E, Polit A, Morasso P (1976) Mechanisms underlying achievement of final head position. J of Neurophysiology 39:435–444
14. Bizzi E, Accornero N, Chapple W, Hogan N (1984) Posture control and trajectory formation during arm movement. J of Neurosci 4:2738–2744
15. Hogan N (1984) An organizing principle for a class of voluntary movements. J of Neuroscience 4:2745–2754
16. Arimoto S, Bae J-H, Hashiguchi H, Ozawa R (2004) Natural resolution of ill-posedness of inverse kinematics for redundant robots under constraints. Communications in Information and Systems 4:1–28
17. Hashiguchi H, Arimoto S, Ozawa R (2004) Control of a handwriting robot with DOF redundancy based on feedback in task coordinates. J of Robotics and Mechatronics 16:381–387
18. Arimoto S, Sekimoto M, Hashiguchi H, Ozawa R (2005) Natural resolution of ill-posedness of inverse kinematics for redundant robots: A challenge to Bernstein's degrees-of-freedom problem. Advanced Robotics 19:401–434

19. Arimoto S, Sekimoto M, Ozawa R (2005) A challenge to Bernstein's degrees-of-freedom problem in both cases of human and robotic multi-joint movements. IEICE Trans on Fundamentals of Electronics, Communications, and Computer Sciences E88-A:2484–2495
20. Spudich JA (1994) How molecular motors work. Nature 372:515–518
21. Vale RD, Milligan RA (2000) The way things move: looking under the hood of molecular motor proteins. Science 288:88–95
22. Kitamura K, Tokunaga M, Iwane AH, Yanagida T (1999) A single myosin head moves along an action filament with regular steps of 5.3 nanometers. Nature 397:129–134
23. Ito M (1970) Neurophysiological aspects of the cerebellar motor control system. Intl J of Neurology 7:162–176
24. Ito M (1972) Neural design of the cerebellar motor control system. Brain Research 40:81–84
25. Arimoto S, Hashiguchi H, Sekimoto M (2005) Natural resolution of DOF redundancy in execution of robot tasks under the gravity: A challenge to Bernstein's problem and applications to handwriting robots. In: Proc of the 12th Intl Conf on Advanced Robotics. Seattle, pp 51–57
26. Tsuji T, Morasso PG, Goto K, Ito K (1995) Human hand impedance characteristics during maintained posture. Biol Cybern 72:475–485
27. Winters JM, Stark L (1987) Muscle models: what is gained and what is lost by varying model complexity. Biological Cybernetics 55:403–420

Novel Registration Method for DSA Images Based on Thin-plate Spline

Jian Yang[1], Songyuan Tang[1], Qin Li[2], Yue Liu[1], and Yongtian Wang[1]

[1] Department of Optoelectronic Engineering, Beijing Institute of Technology, P. R. China
[2] Department of Biomedical Engineering, Beijing Institute of Technology, P. R. China

Chapter Overview. A new algorithm based on thin-plate spline is proposed for the non-rigid registration of Digital Subtraction Angiography (DSA) images. The presented method is novel on three aspects. Firstly, the rigid matching of sub-image blocks divided from mask and live images is achieved by a template matching method based on mutual information (MI). Secondly, the control points are selected from the center points and four valid corner points extracted in the original and transformed mask images. With the five selected control points of each sub-image blocks, the information of translation and rotation of the affine transformation can be utilized effectively. Then by using thin-plate spline as the mapping function, the global elastic registration is achieved. Thirdly, a scatter point interpolation is performed by weighting inverse proportion to the distance of the neighbor points to obtain a smoothly matched image. Experimental results show that our method is robust and effective.

Key Words. Digital subtraction angiography, Mutual information, Thin-plate spline

1. Introduction

Digital Subtraction Angiography (DSA) is a powerful technique for the visualization of blood vessels. Crystal clear images of abdominal, renal, cerebral and peripheral blood vessels can be obtained by the highly sophisticated equipment. It represents the most common image modality applied in clinical practice to assist vessel diagnosis and therapy [1]. With this technique, a bolus of radiopaque contrast material is injected into blood vessels of interest through plastic catheter, a sequence of X-ray projection images is acquired during (live image) and prior (mask image) to the passage of the material in the blood vessels. By subtracting live image with mask image, background structures in the live images are largely removed. However, mask and live images are acquired at different time, the position of background tissue around the vessels often changed during the elapsed time due to the respiration and organ movement of the patient, if direct subtracting of these images, motion artifacts occur. Such artifacts may hamper proper interpretation of the images and even lead to misdiagnosis. In order to reduce artifacts, several techniques have been developed, such as TID (Time Interval Difference) subtraction, energy subtraction, respiratory-gated subtraction, hybrid subtraction and so on. But artifacts are the combination of several factors. More complicated techniques are needed in order to eliminate the artifacts completely.

Image registration technique is a useful method in correcting artifacts. Correspondences between pixels in mask and live image are calculated, and then a certain warping method is applied to mask image. By doing this, artifacts can be eliminated effectively in the subtracted image. During the past twenty years, enormous efforts have been put on this subject, great progresses have been achieved. Overview of the registration techniques of DSA images have been made by Erik H et al. [2] in 1999. However the non-rigid motion of the tissue inside human body is most complicated, there are global and local disparities between the mean gray-levels of the live and mask images. Simply shift or rotation of the misregistered image can not eliminate those artifacts. Up to now none techniques can deal with all kind of artifacts, the only widely used method for motion correction on commercial DSA devices is manual pixel shifting [3], which is time consuming and tedious.

For the technique of coronary angiography, due to the pulsate of the heart, there has large global and local non-rigid warping between mask and live images. Registration for such kind of angiography images is much more complicated. In this paper a new registration method is proposed for coronary angiography images. Firstly we suppose that the non-rigid motion in the projective image of the heart tissues can be decomposed of translation and rotation of many sub-image blocks. By the template matching method, rigid registration of sub-image blocks can be achieved. And then we make up pairs from the center points and four valid corner points extracted in original and transformed mask image, and these point pairs are selected as control points. After this, the global elastic registration can be achieved by using thin-plate spline. Then by the scattered data interpolation method, continuous and smooth matched image can be obtained. All these registration procedures are fully automatic.

This paper is organized as follows: the rigid registration algorithm of sub-image blocks is presented in section 2. The global elastic registration method based on thin-plate spline is presented in section 3. The scattered data interpolation method is derived in detail in section 4. Discussion and conclusion are found in section 5.

2. Rigid Registration Algorithm of Sub-image Blocks

Generally two classes of registration method are involved: area-based methods and feature-based methods [5]. If the salient feature in the images can be extracted easily and the number of common elements of the detected sets of features is sufficiently high, feature-based methods is proposed. But for cardiac DSA images, pixels' warping between mask and live images are very high, in some cases even as large as 80 pixels. It's very hard for the people to detect pixel sets with the same properties. In this paper, area-based method is applied for the registration of sub-image blocks.

Mutual information, basic concept of information theory, was first used in registration of medical images in 1995 by Frederik.M [4] et al... Efficient registration results had been obtained. It's a kind of statistical dependence between two random variables, i.e., the information that one variable carries about the other. It has been proved that the mutual information criterion allows for accurate, robust and completely automated registration of multi-modality medical images [5].

The two images that are to be registered can be taken as random variables A and B respectively, x and y donate the intensity values of a pair of corresponding voxels of the two images. Then marginal distributions $p(x)$, $p(y)$ and joint distributions $p(x,y)$ can be written as follows:

Where $p(x)$, $p(y)$, $p(x,y)$ can be obtained by normalization of the joint and marginal images intensity histograms of both images. Then Under Kullback-Leibler distance [6], mutual information $I(A,B)$ can be defined as:

Mutual information measure the amount of information that image B contains about A. So if the images are geometrically aligned, mutual information of corresponding voxel pairs of the images will reach its maximization. Then image registration techniques based on mutual information come down to find a best geometric transformation T_p, that made mutual information $I(T_p(A),B)$ of image A and B reach its maximization, which can be summarized by the following equations:

We assume that in our sub-image blocks there are only rigid transformation, i.e., translation (tx,ty) and rotation θ, then transformation model can be defined as follows:

Therefore rigid registration of sub-image blocks can be thought as an optimization problem. The Powell algorithm is a kind of multi-dimensional optimization method, which need not to calculate differential of parameters and is rapid for convergence [7]. So in this paper the Powell algorithm has been chosen to find the best transformation that make the mutual information of image pairs reach its maximization.

$$p(x) = \Pr\{X = x; x \in A\} \tag{1}$$

$$p(y) = \Pr\{Y = y; y \in B\} \tag{2}$$

$$p(x,y) = \Pr\{X = x, Y = y; x \in A, y \in B\} \tag{3}$$

$$I(A,B) = \sum_{x,y} p_{AB}(x,y) \log \frac{p_{AB}(x,y)}{p_A(x)p_B(y)} \tag{4}$$

$$T_p^{*} = \arg\max_{T_p}(A, T_p(B)) \tag{5}$$

$$T_P(x,y) = \begin{bmatrix} \cos(\theta) & \sin(\theta) & tx \\ -\sin(\theta) & \cos(\theta) & ty \\ 0 & 0 & 1 \end{bmatrix} \begin{bmatrix} x \\ y \\ 1 \end{bmatrix} \tag{6}$$

$$E_{TPS(f)} = \sum_{a=1}^{n} \|\mu_a - f(v_a)\|^2 + \lambda \iint \left[\left(\frac{\partial^2 f}{\partial x^2}\right)^2 + 2\left(\frac{\partial^2 f}{\partial x \partial y}\right)^2 + \left(\frac{\partial^2 f}{\partial y^2}\right)^2 \right] dxdy \tag{7}$$

$$f(v) = v \cdot d + \phi(v) \cdot w \tag{8}$$

$$E_{TPS}(d,w) = \|X - Vd - \phi w\|^2 + \lambda trace(w^T \Phi w) \tag{9}$$

$$F_i(x,y) = \frac{\sum_{k=1}^{u} \frac{f_{ik}(x,y)}{d_{ik}^2(x,y)}}{\sum_{k=1}^{u} \frac{1}{d_{ik}^2(x,y)}} \qquad i = 1,2,...,m \times n \text{ , } u = 1,2,...,m \times n \tag{10}$$

3. Thin-plate Spline Algorithm

One of the most important attribute of thin-plate spline is its ability to decompose a space transformation into a global affine transformation and a local non-affine warping component [8]. Under the restriction of corresponding points, matching matrixes and mapping parameters can all be achieved. The using of thin-plate spline interpolation as point-based elastic registration method for medical images was first proposed by Bookstein [9] in 1989.

Assume there are two sets of corresponding points U and V, represented as $\{u_a, a = 1,2,..n\}$ and $\{v_a, a = 1,2,...,n\}$ respectively. Then energy function of thin-plate spline can be defined as:

Where f is the mapping function between point sets v_a and u_a. $(1, v_{ax}, v_{ay})$ and $(1, u_{ax}, u_{ay})$ donate the homogeneous coordinate of v_a and u_a respectively. The first and the second term in above equation have different physical meanings. The first term is the approaching probability between point sets v_a and u_a. The second term, on the other hand, is a smoothness constraint. Different λ means different degree of warping. When λ is close to zero, corresponding points match exactly.

According to this energy function, there exists a minimizing function $f(v), v \subset R^2$ for any fixed λ, which can be represented as follows:

Where d is a 3×3 affine transformation matrix, w is a $n \times 3$ non-affine warping coefficient matrix, $\phi(v)$ is a $1 \times n$ vector decided by thin-plate spline kernel. For each point of v, there exists a $\phi_a(v)$, which can be defined as: $\phi_a(v) = c\|v - v_a\|^2 \log\|v - v_a\|$. Where c is a constant. When the solution of equation (8) is substituted into equation (7), we get:

where ϕ is $n \times n$ matrix formed from the $\phi(v_a)$. According to above equation affine matrix d and warping parameters w can be calculated.

By the rigid registration of all the pairs of sub-image blocks, we get a lot of corresponding point sets between mask and live images. These pairs of points can be selected as control points. Then by using thin-plate spline, the global elastic registration can be achieved. The results of elastic registration are mostly decided by the numbers of control points and the positions of points in the image spaces. The transformation parameters (or response) and computational time will increase if the numbers of control points increased. Also if control points were not selected properly, iterative search of parameters may get invalid results. In this paper, the center and four valid corner points extracted in original and transformed mask images are selected as control points, so when we divided mask and live images into $m \times n$ sub-image blocks, $5 \times m \times n$ control points can be achieved. Using the methods above, the information of translation and rotation of affine transformation can be utilized effectively.

4. Scattered Data Interpolation

Scattered data interpolation is a technique of constructing a continuous function of two, three, or more independent variables that exist in two dimensional plane, three dimensional or ND space [10]. The method of scattered data interpolation can be classified into two kinds: global methods and local methods. For the global methods, interpolated value is influenced by all of the data, but for local methods, the interpolated value is only influenced by the values at nearby points from the scattered point sets.

The transformation between mask and live image can be got by thin-plate spline. In succession, the corresponding pixel coordinates of the registered new mask image can be achieved by applying the transformation to every pixel of the mask image. However, the coordinates of the transformed pixels can not be integer exactly. If we only use the nearest point to form the new graphic, some of the pixels that have large scale of warping will disappear and some of the pixels will be utilized repeatedly. In order to resolve such kind of question, inverse distance weighted interpolation method is introduced. The basic assumption of this method is that the interpolated values should be influenced more by nearby points and less by the more distant points, which can be defined as followed:

where $d_{ik}(x,y)=\sqrt{(x-x_{ik})^2+(y-y_{ik})^2}$ donates the Euclidean distance between point (x,y) and point (x_{ik},y_{ik}). Through this interpolation method, continuous and smoothness images can be arrived.

Different interpolation methods have been applied in our registered mask image in figure 1. Figure 1(a) is the result of nearest interpolation method, figure 1(b) is the result of inverse distance weighted interpolation method. As can be seen from the interpolation result, by inverse distance weighted interpolation method, some invalid points had been removed effectively and smoothness image can be achieved.

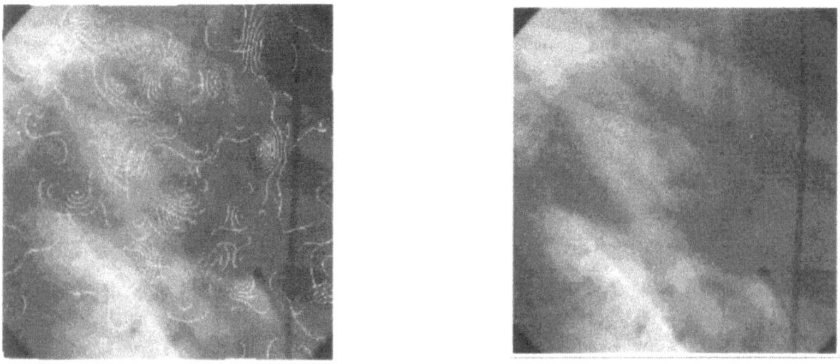

(a) Nearest interpolation method.

(b) Inverse distance weighted interpolation method

Fig. 1. Interpolation Results

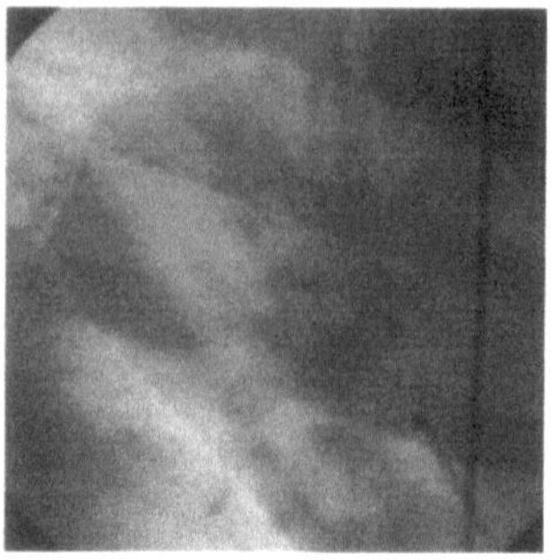

(a) Mask image

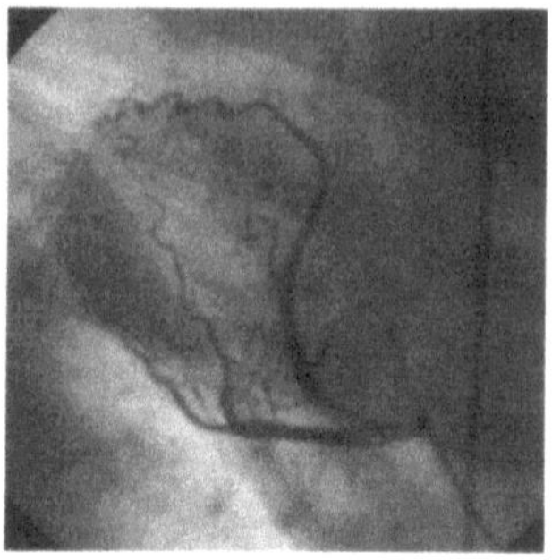

(b) Live image

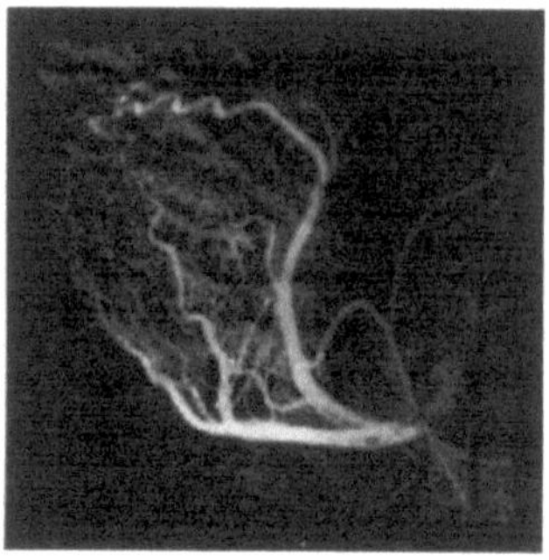

(c) Registration result

Fig. 2. Registration result of DSA images

5. Experiment Result and Discussion

The algorithm as presented in the previous sections was implemented in the C++ programming language and all experiments were carried out on a relatively low cost PC with a PIV 2.4MHz processor, 512MB memory.

We have applied our rigid registration method based on mutual information to several images which only have rigid transformations (transformation and rotation); these images are of 320×240 resolution. Experiment results (can be seen from table 1) indicate that our method is valid for image registration and can reach sub-pixel precision.

Table 1. Rigid registration results based on mutual information. The data sets in the left three columns represent true transformation between origin images and transformed images. The data sets in the right three columns are computational transformation of our algorithm.

Real Transformation			Registration Result		
dx (pixel)	dy (pixel)	θ (degree)	dx (pixel)	dy (pixel)	θ (degree)
2.0	2.0	0.0	1.603	1.720	0.000015
-6.0	-3.0	0.0	-5.969	-3.232	0.0002
1.0	-2.0	1.0	0.576	-1.534	0.913

Table 2. Registration results (mutual information and calculation time) decided by the size of sub-image blocks. Mutual information represents the criterion of registration precision.

Block Size (Pixel)	Mutual Information (Mask and Live Images)	Calculation Time (min)
32×32	0.680939	6.3431
64×64	0.726838	2.9772
128×128	0.639620	1.3697
256×256	0.613610	0.8612

Then the proposed approach is applied to an image of coronary angiography (512×512 pixels and coded in 8 bits). The mask image is shown in Figure 2(a), and the live image in Figure 2(b), and subtraction result after registration is given in Figure 2(c). The mutual information (MI) after registration is 0.726838 (sub-image block size: 64×64), but before registration MI is 0.6456229, It indicate that the correlation of the mask and live image have been greatly improved. Also registration results (mutual information and calculation time) decided by the size of sub-image blocks are compared in table 2. The information given in the table 2 indicates that the smaller the size of sub-image blocks the more calculating time is needed. The calculated mutual information is augmenting with the fine partition of the sub-image blocks, but it doesn't mean that the best registration result will be achieved in the smallest block size. There exist more than one values of the block size between 32×32～128×128 which can get best registration result.

We also implemented the proposed registration algorithm to several DSA images with different warping scale. Experimental results showed that the proposed method is robust and effective. The proposed algorithm is highly data independent and fully automatic, which is therefore well suitable to process the clinical practice. Future research will focus on better understand the influence of implementation issues, such as leaking of angiography substances, low luminance contrast of sub-level vessels and so on. Also our performance results need to be compared to other registration methods.

Acknowledgments

This work is supported by National 973 Program of China (2003CB716105), the National Natural Science Foundation of China (Project No. 60025513) and the Research Foundation of Beijing Institute of Technology (Project No.BIT-UBF-200306C04). We would like to thank Dr. WeiDong Wang (Chinese PLA General Hospital, Beijing) for providing us with the data sets used in the experiments.

References

1. Brody W.R (1982) Digital Subtraction Angiography. IEEE Transactions on Nuclear Science 29(3):1176-1180
2. Meijering E.H.W, Niessen W.J, Viergever M.A (1999) Retrospective Motion Correction in Digital Subtraction Angiography: A Review. IEEE Transactions on Medical Imaging, 18(1):2-21
3. Ashoke. S.T, David. L.W. (1997) Image Quality Optimization for Automatic Warping Registration in X - Ray DSA. Proceedings of the 19th Annual International Conference of the IEEE. 2:549-552
4. Frederik. M, Andre. C. (1997) Multimodality image registration by maximization of mutual information. IEEE Transactions on Medical Imaging. 16(2):187-198
5. Zitova B, Flusser J. (2003) Image Registration Methods: a Survey Image and Vision Computing. 24:977-1000
6. Cover T.M, Thomas J.A. (1991) Elements of Information Theory. Wiley and Sons. New York
7. Powell. M.J.D. (1964) An efficient method for finding the minimum of a function of several variables without calculating derivatives. The Computer J. 7:155-162
8. Chui H, Rangarajan A. (2003) A new point matching algorithm for non-rigid registration. Computer Vision and Image Understanding, 89:114-141
9. Bookstein F.L, Principal W (1989) Thin-plate splines and the decomposition of deformations. IEEE Transaction on Pattern Analysis and Machine Intelligence. 11(6):567-585
10. Hoope H, DeRose T, Duchamp T. (1992) Surfacere construction from unorganized points. Proceedings of the SIGGRAPH'92. Danvers. 71-78

Development of a Simulator of Cardiac Function Estimation for before and after Left Ventricular Plasty Surgery

Tatsushi Tokuyasu[1], Akito Ichiya[1], Tadashi Kitamura[1],Genichi Sakaguchi[2], and Masashi Komeda[3]

[1]Kyushu Institute of Technology, Fukuoka, Japan
[2]Cardiovascular Surgery, Kurashiki Central Hospital, Okayama, Japan
[3]Cardiovascular Surgery, Kyoto University, Kyoto, Japan

Chapter Overview. Cardiac function is evaluated with pressure-volume diagram, wall thickening function, and cardiac output. These are approximately estimated by using echocardiography and cardiac catheterizations. In order to make a surgical plan for left ventricular plasty, surgeons palpate the ventricular wall i.e. cardiac muscle of a patient's heart, where the ventricular thickness, elasticity, and contractile stiffness of the left heart are diagnosed. Cardiovascular surgeons are requested to have the skill to estimate postoperative cardiac function according to the surgical plan determined in the surgery.

We present a simulation system for cardiac function evaluation, where left ventricle models made of finite elements are built and estimate surgical efficiency for left ventricular plasty. It is assumed that the ventricular wall of the models is composed of one-layer, elastic and isotropic material. We employed a simple systemic circulation model, Windkessel, in order to compute the PV diagram of the left heart model. This paper presents a heart modeling method based on finite elements, simulation algorism, and its online use of the FEM computation results. Realistic trends of PV diagram are computed in the simulation.

Key Words. Cardiac function, finite element method, and heart modeling.

1. Introduction

Recently, left ventricular plasty is conducted on patients' heart suffering from myocardial infarction and dilated cardiomyopathy. These diseases significantly impair a pump function of heart, because the ventricular wall loses its thickness according to the progression of disease. Laplace theory shows the stress of a ventricular wall increases depending on the radius in the minor axis of the ventricle in the same ventricular pressure, so that the surgeons remove the cardiac muscle around diseased area and sew up the removed area in the minor axis direction of the left ventricle.

We consider that an accurate estimate of postoperative cardiac function before surgery would yield surgical efficiency and enhance safety, so that we have developed an evaluation simulator for cardiac function [2]. It is well known that finite element method (FEM) is one on the most effective method for heart analysis. Therefore, we also employed FEM to simulate surgical efficiency evaluated from postoperative cardiac function. Left ventricular models for before and after surgery are simply constructed, where cardiac muscular characteristics, e.g. muscular fiber direction and cardiac muscular layer, are ignored. Watanabe et al. showed that one-layer and isotropic material of a finite element heart model do not effect for PV diagrams.

In this paper, we present a heart modeling based on finite elements, a simulation method, and results.

2. Method

2.1 Finite element heart model

Geometric data of a left ventricle is measured from the CT image of a patient suffering from a slight dilated cardiomyopathy. The coauthor of cardiac surgeon judged that image does not impair to make a normal left heart model. Firstly, we measured 800 three dimensional coordinates of the inner and outer wall of the left ventricle, and arranged them as the nodes of a finite element heart model. Finally, 400 hexahedral elements compose a normal left heart model as shown in Fig. 1. The material property of the model is isotropic, elastic, and incompressible, and the Young's modulus of the elements provides contractile stiffness [1] of heart. A homogenous Poisson's ratio of 0.4 is assigned on all elements. We used a pre-post processor MENTAT (MSC., Inc.)[3] to process mesh generation of the left heart model, and give it material characteristics.

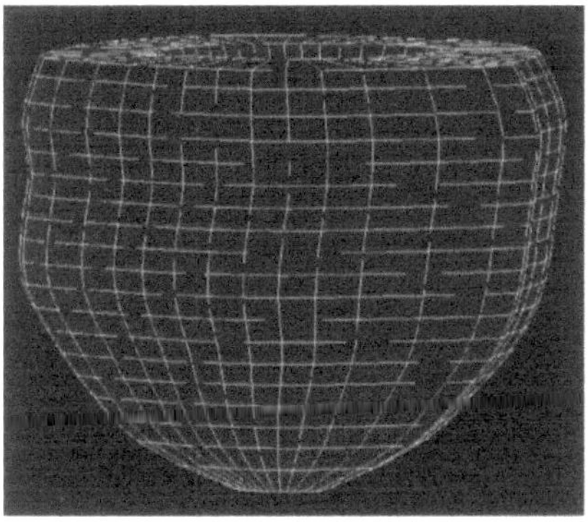

Fig. 1. Left ventricular model based on finite elements

2.2 Pre-Post operative model

Refer to the medical literature [1] and surgeon's experience, a preoperative model for myocardial infarction was built as shown in Fig. 2(a). The wall thickness around infarcted area decreases, because cardiac muscle does not have any regenerative power, so that the ventricular wall of infarcted area is thinner than the other normal part. The infarcted cardiac muscle loses its expansive and contractile properties. In addition, most myocardial infarction patients have a tendency to suffer from high blood pressure and arterial sclerosis.

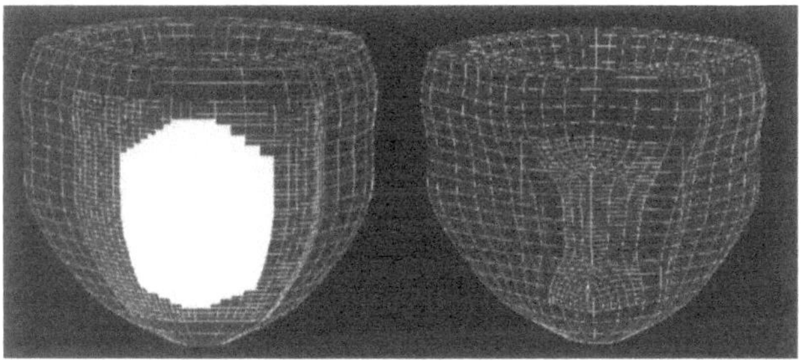

(a) Preoperative model (b) Post operative model

Fig. 2. Finite element models before and after left ventricular plasty for myocardial infarction

In Fig. 2(a), the colored part indicates the thinned and stiff area due to myocardial infarction, where the wall thickness is 30 % thinner than the other normal wall and the constant Young's modulus of 150000 Pa is given through one cardiac cycle.

The postoperative model is shown in Fig. 2(b). It is built from the preoperative model, where the infarcted part is removed firstly and all nodes around the opened area are connected forcedly with the opposite nodes for closing the cut edge in the minor axis direction.

2.3 Cardiac function evaluation

Generally, a pressure-volume (PV) diagram, Emax, and stroke volume (SV) of left heart are used for evaluating cardiac contractile function. This simulator also applies them to evaluate the cardiac function of the finite element models. Fig. 3 shows a PV diagram, where the Emax shows the maximum contractile force of a heart and is derived by the gradient of the line which passes V_0 and the end-systole point.

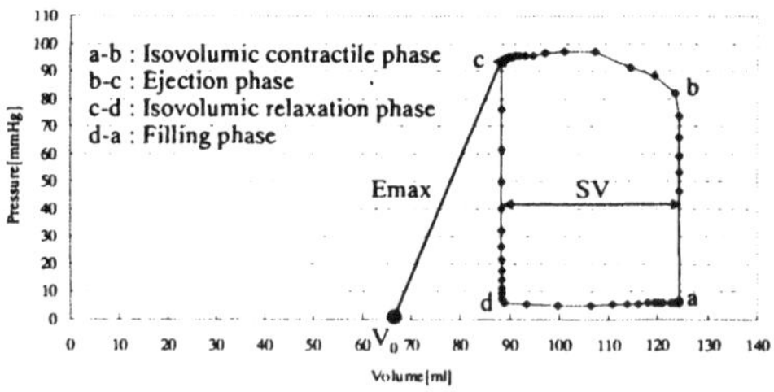

Fig. 3. Pressure and volume diagram for one cardiac cycle and its explanation

2.4 Young's modulus

We assumed that cardiac muscular property tends to nonlinear, so that updated Lagrange method is employed to analyze the heart models.

In this study, the Young's modulus of the model shows contractile force of cardiac muscle. As mentioned above, a patient suffering from myocardial infarction tends to high blood pressure and arterial sclerosis, consequently we built a preoperative model as shown in Fig. 2. The preoperative model is composed of not only normal wall but also infarcted wall. In order to identify the Young's modulus of the normal muscle for one cardiac muscle, a simple systemic circulation model, Windkessel, computed a PV diagram, as shown in Fig. 4, corresponding with a patient affected with high blood pressure and arterial sclerosis, and Fig. 5 shows a thin wall model whose ventricular thickness is the same as the normal part of the preoperative model. By using both of Fig. 4 and Fig. 5, we determined the Young's modulus of the normal muscle for one cardiac cycle, where a relationship between the ventricular pressure and volume of the thin wall model traces the PV diagram of Fig. 4. The determined Young's modulus is shown in Fig. 6.

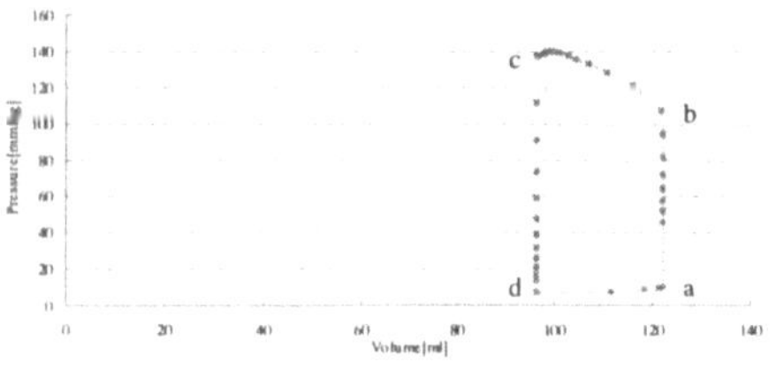

Fig. 4. PV diagram of a patient affected with high blood pressure and arteriosclerosis

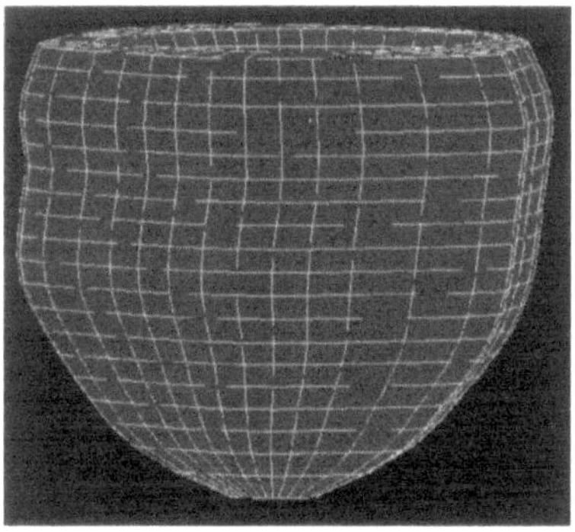

Fig. 5. Thin wall model of left ventricle

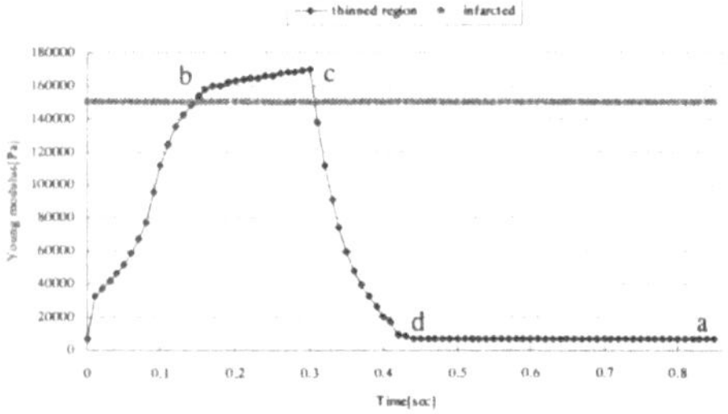

Fig. 6. Young's modules for cardiac cycle

2.5 Simulation Method

In our simulator, MSC.Marc (MSC., INC.)[4] computes a behavior of the heart models, and a Windkessel model as shown in Fig. 7 is linked to FEM in order to simulate the ventricular pressure and volume for one cardiac cycle of the heart models. Fig. 7 shows an electrical circuit of Windkessel linked to the left heart model, where the differential equations for pressure, blood flow volume, and ventricular volume are derived and processed with Eular method. The normal values of the cardiovascular parameters of Windkessel for a human weighing 60[kg] are shown in Table 1.

A simulation starts from the iso-volumic phase as shown in Fig. 8.

Table 1. Cardiovascular valve resistance and aorta compliance

Rli	:	0.001[mmHg/ml/sec]
Rlo	:	0.01[mmHg/ml/sec]
Rs	:	1.0[mmHg/ml/sec]
Rco	:	35.2[mmHg/ml/sec]
Rb	:	15.0[mmHg/ml/sec]
Csa	:	2.0[ml/mmHg]
Rla	:	6.3[mmHg]
Rra	:	3.0[mmHg]

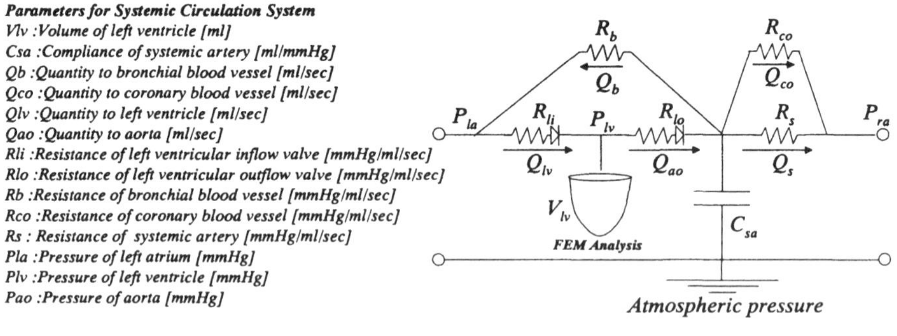

Fig. 7. Electrical circuit of Windkessel model

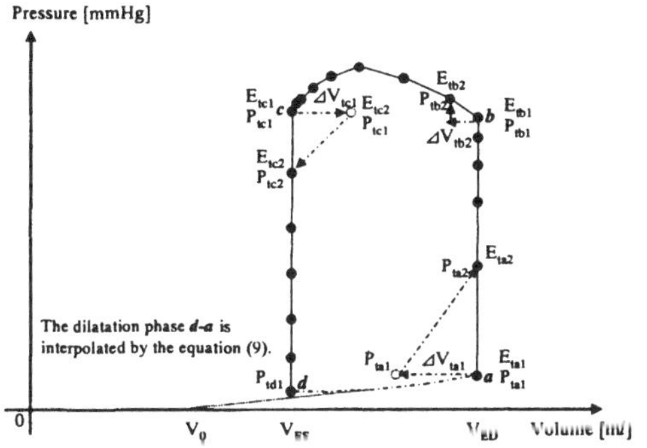

Fig. 8. Simulation algorism for one cardiac cycle

2.5.1 Isovolumic systole

Isovolumic systole starts at point ***a*** shown in Fig. 8, where E_{ta1} shows the Young's modulus, P_{ta1} is the left ventricular pressure, and V_{ED} is the end-diastolic volume. At the next time increment, $t_{a2}=t_{a1}+dt(=0.01[\text{sec}])$, Young modulus is increased to E_{ta2} in accordance with the cardiac cycle as shown in Fig. 6, the pressure P_{ta1} brings the difference of volume ΔV_{ta1}. Pressure P_{ta2} which zeros the difference of volume ΔV_{ta1} is obtained and inputted to the Windkessel model in order to calculate Pao_{ta2}. This process is repeated until *Plv* exceeds *Pao*.

2.5.2 Ejection phase

The difference of the volume ΔV_{tb1} is calculated by the Windkessel model. As Young modulus increases from E_{tb1} to E_{tb2}, the pressure P_{tb2} which corresponds to ΔV between the two volumes from Windkessel and the FEM model analysis is identified and inputted to Windkessel for simulating Pao_{tb2}. These processes are repeated until *Pao* exceeds *Plv*.

2.5.3 Isovolumic relaxation and diastole filling phase

The same analysis as the one for isovolumic systole is applied to isovolumic relaxation phase until *Plv* lowers than *Pla*. Equation (1) interpolates diastole filling phase of the PV diagram, where *Vsvm* and *Px* are needed to pass the two points ***d*** (P_{td1} and V_{ES}) and ***a*** (P_{ta1} and V_{ED}) in Fig.2.

$$VLV = V0 + V_{SVM} \exp\left(1 - \left(PLV \big/ Px\right)\right) \tag{1}$$

$$Elastance[mmHg/ml] = \frac{P_{LV}(t)}{V_{LV}(t) - V_0} \tag{2}$$

Since we focused on the contractile function evaluation of the left ventricle, the dilatation phase is interpolated with equation (1) [5], where V_{SVM} is a parameter depending on the patient weight and Px is estimated by the ventricular shape and disease progress. We assumed V_{SVM}=150 and Px=12.9 for the normal heart.

3 Results

Fig. 9 shows the preoperative model at the end-systole, where the left ventricle is expanded partially due to the blood pressure, because the infarcted part does not contractile function. This phenomenon is also found in clinical cases.

The results of the postoperative model shown in Table 2 show that both Emax and SV were improved. The end diastole volume of the postoperative model approximately equals 103 [ml]. These results meet the typical effects of ventricular plasty for myocardial infarction.

Table 2. Simulation results for cardiac functional indices

Cardiac function	*V_0[ml]*	*SV[ml]*	*Emax[mmHg/ml]*
Thin wall	*68.1*	*27.0*	*4.81*
Pre operative	*80.9*	*15.5*	*3.51*
Post operative	*55.2*	*21.1*	*5.79*

In order to compute one PV diagram by using the presented method, we needed about 20 hours including the time to build a finite element model and output the postoperative PV diagram. It is necessary for training use of this simulator to shorten the computation time for the cardiac function of the postoperative models, so that we present a method for online output of the postoperative PV diagrams in the next chapter.

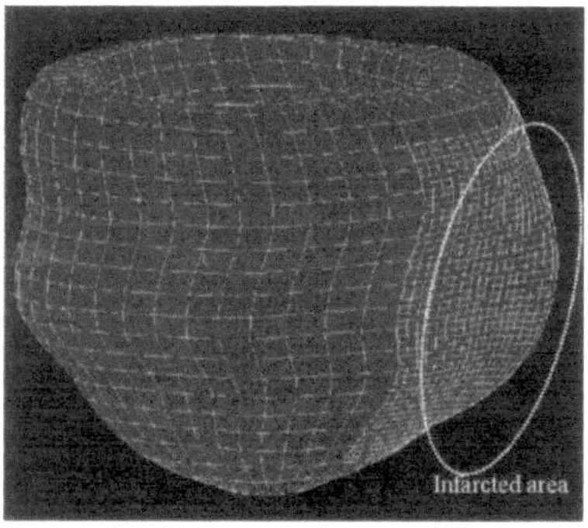

Fig. 9. Simulation result of postoperative ventricle at end systole period

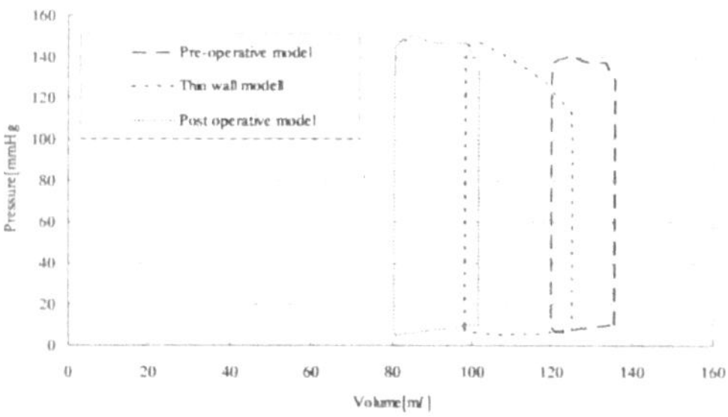

Fig. 10. Simulation result for PV diagram before and after surgery

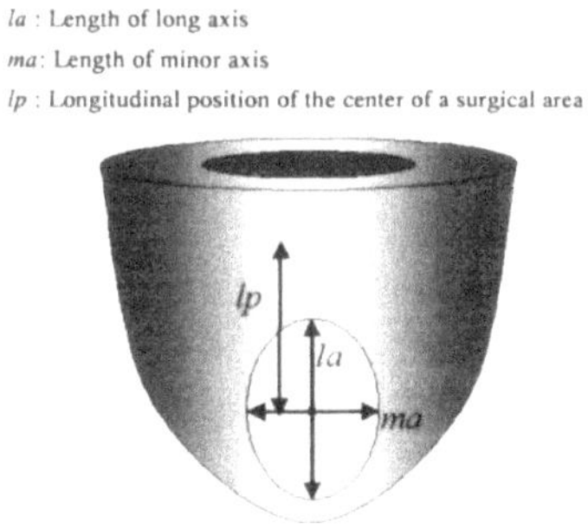

Fig. 11. Definition of a surgical region for online use of the simulator

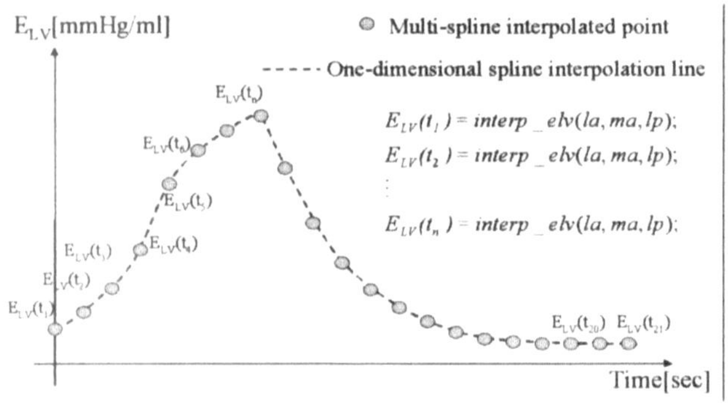

Fig. 12. Interpolation method for postoperative elastance curve based on a multi-dimensional interpolation technique

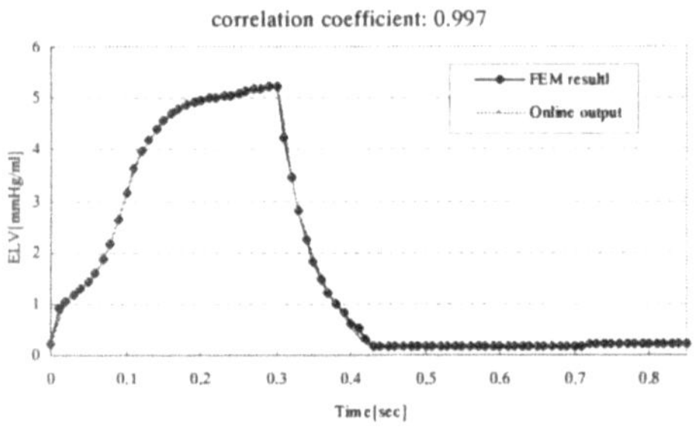

Fig. 13. Simulation results for interpolating post operative elastance curve

4. Online Use

For a quick output of the postoperative cardiac function, we employed a multi-dimensional spline function. Firstly, a surgical area is defined by an ellipse which is derived with 2 variables (lengths of long axis and minor axis) as shown in Fig. 11. The parameter ***lp*** shown in Fig. 11 shows a center position of the ellipse on the left ventricle.

A PV diagram and equation (2) derive an elastance curve for one cardiac cycle. We defined a function, *interp_elv{la, ma, lp}*, based on the three dimensional interpolation function, where the elastance values $E_{LV}(t)$ at every 0.04sec are interpolated and outputted. Finally, one dimensional interpolation makes their elastance data into an elastance curve for one cardiac cycle as shown in Fig. 12. In this simulation, the calculation for interpolation function is computed by MATLAB Spline Tool Box.

Previous to the interpolation for the elastance curve, 24 kinds of the postoperative models where the combinations of the range of *la*: 0~35mm, ma: 0~47mm, and lp: 0~35mm were spesified were simulated to make the tensile product for the *interp_elv{la, ma, lp}*.

This online output method enabled us to compute the postoperative cardiac function within about 3 sec. Fig. 13 shows the online output result for elastance curve, where the surgical region was specified with la: 22mm, ma: 35mm, and lp: 10mm. The coefficient of correlation was 0.997.

5. Discussion & Conclusion

This study simplified the characteristics of the cardiac wall, so that twist behavior for heart contraction, fibrous direction, and excitation conducting system of left heart were ignored. However, the simulation results could get good evaluation from the cardiac surgeon coauthors. In Fig. 7, the preoperative model shows typical cardiac function for myocardial infarction, where both the muscular contraction and the stroke volume decreased.

For building a preoperative model, we simplified the area of myocardial infarction into a set of one-layer elements with the constant Young's modulus. However, in order to simulate a more realistic shape of the infarcted area with smooth change of thickness along the circumference, the diseased area of the preoperative model needs to be meshed in more detailed.

In order to save the calculation time of the simulator, we adopted 3 dimensional interpolation method, where a surgical area was determined with 3 variables. This method enabled us to compute the postoperative cardiac function within about 3 seconds. We have to install a parameter that consider the models' dissymmetry, for example, a position of surgical area in the minor axis direction of the heart model, though we need huge computation time to make the four dimensional interpolation tensile.

The cardiovascular surgeons desire to research curative effects of the left ventricular plasty for dilated cardiomyopathy, because the pathogenesis of dilated cardiomyopathy is still unknown. We will build a dilated cardiomyopathy model based on a patient's MRI data, and evaluate not only contractile function but also dilated function of left heart.

Acknowledgment

This work partly was supported by Research Fellowships of the Japan Society for the Promotion of Science for Young Scientists.

References

1. Yutani Chikao (2002) Atlas of Cardiovascular Pathology.
2. T. Tokuyasu, A. Ichiya, T. Kitamura, G. Sakaguchi, and M. Komeda (2004) A Simulation for Cardiac Function Evaluation before and after Left Ventricular Plasty, Proc. of Mechanical Engineering Congress 69-70 (In Japanese)
3. http://www.mscsoftware.co.jp/solutions/software/p_marc_m.htm
4. http://ecust.isid.co.jp/public/product/marc/outline/index.html
5. The Japan Society of Mechanical Engineers (1991) Bioengineering Division

Key Word Index

MIX
Papier aus verantwortungsvollen Quellen
Paper from responsible sources
FSC® C105338

If you have any concerns about our products,
you can contact us on
ProductSafety@springernature.com

In case Publisher is established outside the EU,
the EU authorized representative is:
Springer Nature Customer Service Center GmbH
Europaplatz 3, 69115 Heidelberg, Germany

Printed by Libri Plureos GmbH
in Hamburg, Germany